Corrosion, Environmental Degradation, and Surface Engineering

Corrosion, Environmental Degradation, and Surface Engineering: Understanding, Mitigation, and Prevention offers an insightful exploration into the causes, mechanisms, and mitigation strategies of material degradation. This comprehensive work unravels the complexities of corrosion, wear, and fatigue, while showcasing cutting-edge techniques like surface engineering and predictive analytics. Packed with real-world case studies and innovative solutions, the book bridges the gap between theoretical concepts and practical applications.

This book:

- Explores surface degradation mechanisms and commercial perspectives on surface degradation.
- Discusses dimensionless parameters in wear phenomena, fracture surface degradation, and the influence of heat and radiation on surface degradation.
- Covers environmental assisted surface deterioration, and factors affecting the atmospheric corrosion.
- Highlights the importance of digitization and data-driven approaches in engineering.
- Assesses the economic losses caused by corrosion degradation, and quantifies economic losses attributed to corrosion degradation.

Ideal for students, researchers, and industry professionals, the book provides the tools and knowledge necessary to combat surface degradation and promote sustainability across critical sectors.

Corrosion, Environmental Degradation, and Surface Engineering

Understanding, Mitigation, and Prevention

Harish Hirani

CRC Press
Taylor & Francis Group
Boca Raton London New York

CRC Press is an imprint of the
Taylor & Francis Group, an **informa** business

Designed cover image: Harish Hirani

First edition published 2026
by CRC Press
2385 NW Executive Center Drive, Suite 320, Boca Raton, FL 33431

and by CRC Press
4 Park Square, Milton Park, Abingdon, Oxon, OX14 4RN

CRC Press is an imprint of Taylor & Francis Group, LLC

ISBN: 978-1-032-89866-7 (hbk)
ISBN: 978-1-032-89865-0 (pbk)
ISBN: 978-1-003-54501-9 (ebk)

DOI: 10.1201/9781003545019

Typeset in Sabon
by Newgen Publishing UK

Access the [Instructor and Student Resources/Support Material]: www.routledge.com/9781032898667

Contents

Preface *xi*
Acknowledgments *xvii*
About the author *xix*

1 Introduction to surface degradation **1**
1.1 *Introduction 1*
1.2 *Exploring surface degradation mechanisms 4*
1.3 *Commercial perspectives on surface degradation 8*
1.4 *Conclusions 18*
1.5 *Questions and answers 19*

2 Surface degradation mechanisms **21**
2.1 *Introduction 21*
2.2 *Polishing wear 25*
2.3 *Understanding adhesive wear 28*
2.4 *Exploring abrasive wear 34*
2.4.1 *Two-body abrasion 34*
2.4.2 *Three-body abrasion 38*
2.4.3 *Wear equation for two-body abrasion 39*
2.4.4 *Effect of roundness factor on abrasive wear 42*
2.4.5 *Effect of lubrication on abrasive wear 42*
2.4.6 *Effect of hardness on abrasive wear 43*
2.4.7 *Examples of abrasive wear 43*
2.4.8 *Epilogue 44*
2.5 *Dimensionless parameters in wear phenomena 45*
2.6 *Erosive and cavitation wear 47*
2.6.1 *Evaluating erosive resistance: Experimental setups and considerations 54*
2.6.2 *Cavitation wear 56*
2.7 *Fatigue, fretting, and melting wear 59*
2.7.1 *Fatigue wear 59*

2.7.2 *Fretting wear* 65
2.7.3 *Melting wear* 70
2.8 *Fracture surface degradation* 75
2.9 *Influence of heat and radiation on surface degradation* 89
2.10 *Role of fatigue and fracture mechanisms in wear debris formation* 97
2.11 *Conclusions* 103

3 Corrosion **104**
3.1 *Introduction* 105
3.2 *Corrosion fundamentals* 108
3.3 *General corrosion* 110
3.4 *Galvanic corrosion* 116
3.5 *Localized corrosion* 117
3.5.1 *Pitting corrosion* 119
3.5.2 *Crevice corrosion* 120
3.6 *Stress corrosion cracking* 123
3.7 *Conclusions* 125

4 Corrosion and surface deterioration and fractography **127**
4.1 *Synergistically combined degradation mechanisms* 128
4.1.1 *Mechanical and chemical actions* 134
4.1.2 *Mechanical and heat/radiation actions* 137
4.1.3 *Chemical and heat/radiation synergy* 138
4.1.4 *Mechanical heat/radiation and chemical actions* 142
4.2 *Fractography* 144
4.2.1 *From erosion and wear mechanisms to fractography* 148
4.2.2 *Understanding fracture and wear debris using fractography* 150
4.3 *Case study 1* 155
4.4 *Case study 2* 158
4.5 *Conclusions* 160

5 Environmentally assisted surface deterioration **162**
5.1 *Environmental factors contributing to surface degradation* 164
5.1.1 *Water and humidity* 164
5.1.2 *Temperature variations* 164
5.1.3 *UV radiation* 164
5.1.4 *Chemical exposure* 164
5.1.5 *Abrasive particles* 164
5.2 *Signs of surface degradation* 165
5.2.1 *Erosion and material removal* 166
5.2.2 *Peeling, delamination, or cracking* 166
5.2.3 *Discoloration and corrosion* 166

5.3 *Historical perspectives on environmental degradation* 166
5.4 *Influence of environmental conditions on corrosion rates* 169
5.4.1 *Rural environment* 169
5.4.2 *Urban environment* 169
5.4.3 *Industrial environment* 170
5.4.4 *Marine environment* 170
5.4.5 *Indoor environment* 171
5.5 *Factors affecting atmospheric corrosion* 171
5.5.1 *Wetness duration* 172
5.5.2 *Precipitation* 172
5.5.3 *Dust particles* 172
5.5.4 *Temperature* 172
5.5.5 *Initial exposure conditions* 172
5.5.6 *Sheltering* 172
5.5.7 *Wind velocity* 173
5.5.8 *Pollutants* 173
5.6 *Primary atmospheric contaminants affecting corrosion* 173
5.7 *Understanding environmental variations for effective design* 174
5.8 *Case studies* 175
5.8.1 *Case study 1* 176
5.8.2 *Case study 2* 178
5.8.3 *Case study 3: What caused the Titanic to strike the iceberg?* 182
5.8.4 *Case study 4: Sinking of the Erika* 188
5.8.5 *Case study 5: Impact of corrosion and cracks on steel bridges* 190
5.8.6 *Case study 6: Atmospheric-assisted surface degradation of aircraft* 193
5.8.7 *Case study 7: Building degradation* 195
5.8.8 *Case study 8: Surface degradation due to ozone* 198
5.8.9 *Case study 9: Surface degradation due to chloride* 200
5.8.10 *Case study 10: Hydrogen embrittlement* 203
5.8.11 *Case study 11: Thermal mechanical fatigue for turbine blade superalloy: Temperature-assisted oxidation* 204
5.9 *Final remarks* 204
5.10 *Conclusions* 206

6 Characterization techniques and prevention **208**
6.1 *Introduction* 208
6.2 *Analyzing surface degradation* 209
6.2.1 *Role of testing in analyzing surface degradation* 210
6.2.2 *Corrosion testing: Assessing weight loss and surface changes* 210

6.2.3 *Corrosion mitigation techniques and testing requirements* 211
6.3 *Testing methods in engineering* 211
6.3.1 *Scaled-down laboratory models* 211
6.3.2 *Realistic performance evaluations* 213
6.3.3 *Sustainability considerations* 214
6.4 *Categories of testing methods* 216
6.4.1 *Component-level and assembly-level testing* 217
6.4.2 *DT versus NDT* 217
6.5 *Classification of NDT methods* 223
6.5.1 *Visual inspection* 226
6.5.2 *Liquid penetration testing* 228
6.5.3 *Magnetic particle testing* 229
6.5.4 *Eddy current testing* 233
6.5.5 *Ultrasonic testing* 237
6.5.6 *Radiography testing* 244
6.5.7 *Thermographic non-destructive testing* 250
6.5.8 *Acoustic emission testing* 253
6.5.9 *Stress–strain DIC testing* 258
6.6 *Selection criteria for NDT techniques* 262
6.6.1 *Specific defect characteristics* 263
6.6.2 *Nine NDT techniques: Key insights and limitations* 264
6.6.3 *WO method for selecting NDT techniques* 266
6.7 *Conclusions* 277

7 Digitization of established curves and data-driven approaches **280**
7.1 *Importance of digitization and data-driven approaches in engineering* 280
7.2 *Traditional curve representations in engineering* 283
7.2.1 *Key components of the wear map* 285
7.3 *Challenges associated with manual data handling and analysis* 286
7.4 *Digitization techniques* 290
7.4.1 *Image digitization* 290
7.4.2 *Trendline fitting using Excel 365* 291
7.4.3 *Plotting using MATLAB® 2020b* 292
7.5 *Leveraging data-driven approaches for predictive maintenance* 296
7.5.1 *Analytical techniques and predictive modeling* 297
7.5.2 *Data-driven simulations* 301
7.6 *Case studies* 303
7.6.1 *Case study 1: Thermographic analysis* 303

7.6.2 *Case study 2: Corrosion analysis 304*
7.6.3 *Case study 3: Gear debris analysis 305*
7.7 *Conclusions 307*

8 **Failure analysis and life-cycle assessment approaches** 309
8.1 *Introduction 309*
8.2 *Failure and fault analysis 311*
8.3 *Failure mode and effect analysis 322*
8.3.1 *FMEA processes key steps 323*
8.3.2 *Types of FMEA 324*
8.3.3 *Benefits of FMEA 325*
8.3.4 *Challenges in implementing FMEA 325*
8.3.5 *Future directions in FMEA 326*
8.3.6 *Case study: Journal bearing system 326*
8.3.7 *Case study 2: Permanent magnetic bearing 337*
8.4 *Life-cycle assessment 344*
8.4.1 *Understanding LCA 344*
8.4.2 *Types of LCA 345*
8.4.3 *Benefits of LCA 345*
8.4.4 *Applications of LCA in industry 346*
8.4.5 *Challenges in implementing LCA 346*
8.4.6 *Future directions in LCA 347*
8.5 *Principal component analysis 347*
8.5.1 *Understanding PCA 348*
8.5.2 *Mathematical foundations of PCA 350*
8.5.3 *Benefits of PCA 350*
8.5.4 *Challenges in implementing PCA 351*
8.5.5 *Future directions for PCA 351*
8.5.6 *Case study: Fault monitoring using PCA 351*
8.6 *Conclusions 361*

9 **Mitigation strategies for surface degradation** 363
9.1 *Introduction to maintenance strategies 363*
9.1.1 *Maintenance strategies along P–F curve 365*
9.1.2 *"Time-dependent" versus "time-independent" maintenance strategies 370*
9.1.3 *Overview of condition-based maintenance 371*
9.2 *Condition monitoring and informed decision-making 372*
9.2.1 *Case study: Condition monitoring of spur gears 373*
9.2.2 *Case study: Data-driven failure mode and effect analysis to enhance maintenance planning 381*
9.3 *Life-cycle impact assessment 384*
9.3.1 *Life-cycle assessment: 385*
9.3.2 *Case study on life-cycle assessment 392*

9.4 *Surface engineering techniques 395*
- 9.4.1 *Understanding the surface 395*
- 9.4.2 *Understanding surface engineering 395*
- 9.4.3 *Understanding surface roughness 396*
- 9.4.4 *Importance of surface texture modifications 400*
- 9.4.5 *Discussion on surface texturing techniques 402*
- 9.4.6 *Case study: Tribological characteristics of textured surfaces 403*
- 9.4.7 *Discussion on surface texturing 405*

9.5 *Surface coatings 406*
- 9.5.1 *Surface coating techniques 408*
- 9.5.2 *Customizing coating performance 408*
- 9.5.3 *Classification of coating methods by state 411*
- 9.5.4 *Key coating techniques 412*
- 9.5.5 *Considerations for coating selection 417*
- 9.5.6 *Examples of coating applications and innovations 420*

9.6 *Assessing economic losses caused by corrosion degradation 421*
- 9.6.1 *Material origins and corrosion tendency 421*
- 9.6.2 *Managing corrosion to extend the material lifespan 421*
- 9.6.3 *Role of natural oxide layers in corrosion management 422*
- 9.6.4 *Corrosion mechanisms and predictive strategies 422*
- 9.6.5 *Real-time data and condition monitoring in corrosion management 423*
- 9.6.6 *Challenges and economic implications of corrosion 423*
- 9.6.7 *Financial analysis of life-cycle costs and materials degradation 435*
- 9.6.8 *Comprehensive economic analysis of corrosion 437*
- 9.6.9 *Corrosion analysis and mitigation strategies 441*

9.7 *Quantifying economic losses attributed to corrosion degradation 443*
- 9.7.1 *Framework for evaluating corrosion costs 444*
- 9.7.2 *Case study 1 448*
- 9.7.3 *Case study 2 453*
- 9.7.4 *Case study 3 454*
- 9.7.5 *Case study 4: Sector-wise corrosion costs in India 455*
- 9.7.6 *Case study 5: Estimation of corrosion costs in India using I/O analysis 457*
- 9.7.7 *Comprehensive corrosion management system framework 459*

9.8 *Conclusions 463*

Index 465

Preface

The book *Corrosion, Environmental Degradation, and Surface Engineering: Understanding, Mitigation, and Prevention* serves as an essential resource for comprehending the multifaceted challenges posed by material degradation and for implementing effective strategies to mitigate its effects. Each chapter systematically addresses core aspects of degradation mechanisms, providing a comprehensive framework for understanding and managing the impact of these processes across various industries.

Chapter 1, "Introduction to surface degradation," establishes a foundation by exploring the primary mechanisms of material degradation, including corrosion, wear, and fatigue. It emphasizes their significant impact on industries such as automotive, aerospace, infrastructure, etc. The chapter examines how environmental and mechanical factors interact to accelerate surface deterioration and introduces essential concepts like wear debris analysis and failure modes. These discussions set the stage for understanding the broader implications of surface degradation in real-world applications.

In addition to technical insights, the chapter explores the commercial challenges posed by material degradation, highlighting its economic consequences and the growing demand for innovative solutions. By addressing both the scientific and practical aspects of surface degradation, this chapter provides a comprehensive introduction that prepares readers for the detailed analysis and strategies covered in subsequent chapters.

Chapter 2, "Surface degradation mechanisms," provides a comprehensive exploration of wear processes and their relevance to industrial applications. The chapter emphasizes the importance of understanding these degradation modes to optimize material performance and system reliability. Through detailed discussions, it breaks down abrasive wear into two-body and three-body abrasion, examining micro-level phenomena such as cutting, fatigue, fracture, and grain removal. Analytical tools like the "bathtub curve" are introduced to predict failure patterns and inform maintenance strategies, making the theoretical concepts applicable to real-world challenges.

Additionally, the chapter delves into specialized wear mechanisms, including polishing wear, cavitation wear, and fatigue-related deterioration, highlighting the impact of factors such as lubrication, hardness, and surface characteristics on degradation rates. Real-world examples and experimental setups illustrate how tailored strategies can mitigate these effects. By integrating case studies with predictive models and maintenance frameworks, this chapter equips engineers and researchers with the tools to address surface degradation effectively and adapt to the evolving demands of industrial systems.

Chapter 3, "Corrosion," provides a detailed exploration of corrosion, a widespread degradation phenomenon affecting materials across various industries. It begins by explaining the fundamentals of corrosion processes, including electrochemical reactions and oxidation mechanisms, while emphasizing the role of electrochemical potentials in determining corrosion rates. The chapter also addresses general corrosion forms, such as aqueous corrosion and rusting, and discusses protective strategies like coatings and material selection to mitigate damage. These discussions are supported by practical insights.

The chapter delves into specific types of corrosion, such as galvanic corrosion, highlighting the importance of understanding the galvanic series and the influence of surface area ratios. Localized corrosion forms, including pitting and crevice corrosion, are examined, alongside stress corrosion cracking (SCC), with a focus on its mechanisms, the role of hydrogen embrittlement, and approaches for material resistance. By integrating theoretical knowledge with case studies and preventative strategies, the chapter equips engineers and researchers with practical tools to address corrosion challenges effectively, ensuring material durability and long-term system reliability.

Chapter 4, "Corrosion and surface deterioration and fractography," examines how the combined effects of mechanical, thermal, and chemical factors accelerate material degradation. The chapter provides a comprehensive exploration of degradation mechanisms, including the impact of radiation on surface integrity and the role of hydrogen embrittlement in industrial applications. It addresses structural failures caused by thermal cracking and emphasizes the importance of understanding material properties to predict and prevent surface deterioration effectively. Fractographic techniques are introduced as a tool for identifying surface degradation patterns and enabling predictive analysis to mitigate failure risks.

The chapter further bridges theoretical concepts with practical applications, offering strategies like the use of digital twins, specialized coatings, and environmental control measures to enhance durability. Case studies highlight real-world challenges in sectors such as aerospace, oil and gas, and nuclear energy, showcasing how integrated approaches improve outcomes in high-risk industries. Surface engineering techniques and innovative mitigation strategies are discussed to ensure reliability and extend the lifespan of critical systems. By combining technical insights with actionable solutions, this chapter equips engineers and researchers to address complex surface degradation issues across diverse applications.

Chapter 5, "Environmentally assisted surface deterioration," explores the significant impact of environmental factors such as UV radiation, pollutants, and moisture on material degradation. It systematically categorizes environmental conditions into rural, urban, industrial, marine, and indoor settings, each presenting unique challenges to material performance. The chapter highlights innovative mitigation techniques, including nanocoatings, designed to combat the adverse effects of these conditions. By analyzing the influence of atmospheric factors such as wetness duration, precipitation, dust particles, temperature, and wind velocity, it provides a comprehensive framework for understanding how materials deteriorate in diverse environments.

The chapter also delves into specific contaminants that drive atmospheric corrosion and underscores the importance of accounting for environmental variations in effective design. Case studies offer practical insights into real-world failures, such as lead corrosion in humid environments, satellite degradation in low Earth orbit, and the structural collapse of steel bridges due to corrosion and cracks. Notable historical events, including the sinking of the *Titanic* and *Erika*, further illustrate the catastrophic consequences of surface degradation. Through this analysis, Chapter 5 emphasizes the need for informed engineering strategies to enhance material resilience and ensure long-term sustainability in challenging environments.

Chapter 6, "Characterization techniques and prevention," focuses on advanced methodologies for evaluating surface integrity. It emphasizes the critical role of both destructive and non-destructive testing techniques in identifying material weaknesses and assessing performance. Techniques like ultrasonic testing and sensor-based diagnostics are highlighted for their ability to support predictive maintenance, reducing the risk of system failures and extending operational lifespans. The chapter also explores corrosion testing methods, which assess weight loss and surface changes, along with corrosion mitigation techniques to enhance material durability in adverse environments.

The chapter further categorizes testing approaches into scaled-down laboratory models, realistic performance evaluations, and sustainability considerations. It details various methods, including fatigue and creep testing, fracture toughness evaluation, and aerodynamic behavior testing using wind tunnels. It distinguishes between destructive testing, such as the Charpy Impact Test, and non-destructive testing (NDT) methods like visual inspection, liquid penetrant testing, magnetic particle testing, and eddy current testing. The importance of real-world testing is underscored, particularly in assessing material performance under corrosive conditions and thermal stress, ensuring practical insights for engineers and researchers aiming to develop sustainable and resilient systems.

Chapter 7, "Digitization of established curves and data-driven approaches," highlights the transformative impact of digital technologies in modern engineering analysis. It underscores how tools like digital twins and predictive analytics enhance traditional methods by providing real-time insights for monitoring and maintenance. The chapter begins by exploring traditional curve representations, such as wear maps, and delves into key components like pressure-velocity fields, wear rate contours, and oxide layer dynamics. It also addresses challenges associated with manual data handling, emphasizing the value of digitization in overcoming inefficiencies and inaccuracies. Digital twin models are introduced as powerful tools for asset management and operational optimization.

The chapter further details advanced digitization techniques, including trendline fitting with Excel, and MATLAB-based plotting, which streamline data visualization and analysis. It also explores predictive maintenance through data-driven approaches like regression, clustering, and dimensionality reduction, offering robust frameworks for fault prediction and prevention. Case studies, including thermographic, corrosion, and gear debris analyses, illustrate the practical application of these methodologies in various engineering scenarios. By combining traditional engineering knowledge with innovative data-driven techniques, the chapter provides a comprehensive roadmap for leveraging digital technologies to improve efficiency, reliability, and sustainability in engineering systems.

Chapter 8, "Failure analysis and life cycle assessment approaches," serves as a crucial intersection of engineering performance and sustainability. By incorporating frameworks such as failure mode and effect analysis (FMEA) and life-cycle assessment (LCA), the chapter provides engineers with tools to enhance system resilience while minimizing environmental impact. It begins with an introduction to failure and fault analysis, focusing on methodologies for fault detection and diagnosis, supported by real-world case studies on mechanical failures. Through detailed discussions on the process, types, and benefits of FMEA, the chapter highlights its role in identifying potential system vulnerabilities and prioritizing corrective actions. Challenges and future directions for FMEA are explored, alongside illustrative case studies, such as applications in journal bearing systems and permanent magnetic bearings.

The chapter extends its focus to LCA, explaining its types, benefits, and industrial applications. It emphasizes the integration of FMEA and LCA to form a comprehensive approach, allowing engineers to address both system reliability and environmental sustainability simultaneously. Practical applications of combined approaches are discussed, showcasing their advantages in real-world engineering scenarios. The chapter concludes by examining challenges and future directions, including advancements in technology, data analytics, and the alignment of engineering practices with sustainability goals.

Chapter 9, "Mitigation strategies for surface degradation," serves as a comprehensive synthesis of insights from earlier chapters, translating theoretical understanding into actionable solutions to address and prevent material degradation. The chapter begins with a foundational discussion on maintenance strategies, highlighting their role in preserving machinery and preventing failure. It underscores the importance of early issue detection along the P–F curve and explores both time-dependent strategies, such as scheduled inspections, and time-independent approaches like real-time monitoring. Condition-based maintenance is presented as a proactive framework, supported by tools such as sensors, historical data analysis, and advanced analytics. Case studies, such as condition monitoring of spur gears and data-driven FMEA, provide practical applications of these methodologies in enhancing maintenance planning and reducing risks.

The chapter also examines LCA, focusing on environmental and economic evaluation of materials over their life cycle. The importance of surface engineering techniques is addressed, emphasizing surface treatments, coatings, and modifications. The discussion extends to understanding surface parameters, engineering processes, and the impact of surface roughness on material performance, reinforcing the importance of these strategies in mitigating degradation and ensuring durability. This section delves into cutting-edge methods for combating surface degradation by integrating novel technologies and research-backed methodologies. It highlights advancements in nanotechnology, smart coatings, and adaptive surface treatments, focusing on their ability to enhance material performance. For instance, nanocoatings offer superior protection with minimal material usage, providing increased resistance to wear, corrosion, and environmental stresses. The integration of machine learning and artificial intelligence for real-time monitoring and analysis further augments the effectiveness of these strategies, enabling predictive adjustments in maintenance and material selection. Additionally, the section emphasizes the importance of interdisciplinary approaches, combining insights from materials science, mechanical engineering, and environmental studies. By leveraging these advancements, industries can significantly extend the life cycle of materials, reduce operational costs, and ensure sustainable practices. Key focus areas include the development of self-healing materials, smart lubricants, and environmentally friendly surface treatments. It envisions a future where digital twins and Internet of Things (IoT)-enabled monitoring systems play a central role in proactive maintenance, allowing for precise interventions before failures occur.

Acknowledgments

The completion of this book, *Corrosion, Environmental Degradation, and Surface Engineering: Understanding, Mitigation, and Prevention*, would not have been possible without the unwavering support and encouragement of several individuals, my "Mechanical Engineering Department" and my organization "Indian Institute of Technology Delhi, India".

First and foremost, I express my heartfelt gratitude to my daughters, Daksha and Netra, for their "emotional and moral support" and "create inputs". Their encouragement provided me with the motivation to persist through challenges and stay focused on the work.

I would also like to extend my deep appreciation to my students, Mr. Kishan Nath Sidh and Dr. Dharmender, for their invaluable assistance in arranging the necessary references and their meticulous efforts in organizing and preparing the figures that have enriched this book. Their dedication and contributions were instrumental in shaping the quality of this work.

A special acknowledgment is due to the "National Programme on Technology Enhanced Learning (NPTEL)", an exceptional Indian e-learning platform that allowed me to teach courses on "Corrosion, Environmental Degradation Surface Engineering." The resources and opportunities provided by NPTEL were pivotal in helping me gather and develop the material that formed the foundation of this book.

Lastly, I would like to express my gratitude to ChatGPT for assisting with language refinement, ensuring clarity, and enhancing the readability of the manuscript. The support in correcting sentences and improving the overall presentation was invaluable.

To everyone who has contributed in any way to the realization of this book, I extend my sincere thanks and appreciation.

About the author

Harish Hirani is a distinguished mechanical engineer with extensive experience in design and manufacturing. Currently a Senior Professor at IIT Delhi, he has served as the Director of CSIR-CMERI, where he led groundbreaking projects in sustainable engineering and technology innovation. A prolific author, Professor Hirani has contributed significantly to the field through his influential books, numerous research papers, and intellectual property rights (IPR), earning him international acclaim as a leading scientist. His teaching roles at premier institutions such as IIT Bombay and IIT Delhi, coupled with his impactful contributions to global engineering practices, underscore his unwavering commitment to advancing education, industry, and environmental sustainability.

About the author

Chapter 1

Introduction to surface degradation

1.1 INTRODUCTION

System failures are a frequent occurrence across various infrastructures, ranging from bridges and gear systems to steel structures, oil pipelines, and automobiles (as illustrated in Figure 1.1). These failures can stem from a multitude of issues, including structural inadequacies, excessive loads, fatigue, corrosion, or design flaws. Often, these components fail due to a combination of factors, presenting significant challenges to infrastructure integrity.

Although detailed fracture analysis and fractography will be discussed later, a pragmatic yet effective approach involves examining wear debris. Mechanical systems typically produce debris that can provide valuable insights into the causes of failure, manifesting as chips, flakes, particles, or even nano-sized dust. Although not always visible to the naked eye, specialized setups can analyze and diagnose these particles, facilitating early identification and resolution of potential failures.

The reliability of a system and its processes depends on detecting early signs of failure, such as cracks, fatigue, or corrosion. Failure analysis and life cycle assessment provide crucial insights into the causes of failures, allowing for the timely implementation of corrective measures. While fishbone diagrams can be valuable in this process, other methods for mitigating failures and exploring their causes will be discussed in subsequent chapters.

A fishbone diagram is a visual tool used to systematically identify and examine potential causes of a specific failure. It categorizes causes into groups such as people, machinery, materials, measurements, and environment to facilitate brainstorming. For instance, excessive wear in a bearing system might result from improper handling and maintenance by untrained workers or errors during maintenance. Incorrect installation methods can lead to misalignment and unequal load distribution, whereas irregular maintenance can cause wear and contamination buildup. Abrasive wear may occur due to impurities like dirt or metal particles in the lubricant or

DOI: 10.1201/9781003545019-1

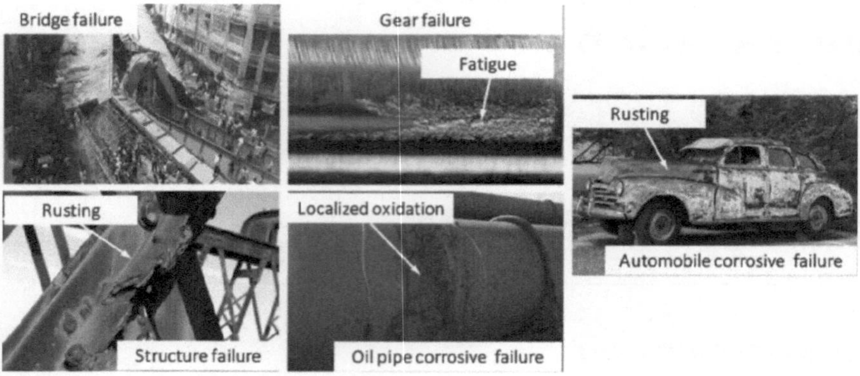

Figure 1.1 Surface degradation in automobiles, gear systems, oil pipelines, bridge failure,[1] and steel structures.

bearing material, and low-quality lubricants may not provide adequate protection against wear. Environmental factors can accelerate the degradation of materials and lubricants, increasing wear, or introduce particulate matter into the bearing system, causing abrasive wear. Each of these causes can be systematically examined to identify the primary reason for failure.

In my experience, when it comes to these failures, surface degradation tends to be a bigger player than the breakdown of the material as a whole. Examples of surface degradation include rust formation on metals exposed to moisture and oxygen, the development of cracks or scratches on plastics subjected to repeated use or high temperatures, and erosion of composite materials due to ultraviolet (UV) radiation.

Surface damage significantly impacts a system's performance and lifespan, rendering it weaker, less robust, and more susceptible to external forces, ultimately leading to failure. Materials with exterior layers, such as metals, polymers, or composites, can undergo surface degradation when these layers wear down or sustain damage. This degradation can manifest as cracks, imperfections, acid corrosion, water erosion, or wind damage. Factors such as environmental conditions, mechanical wear, chemical reactions, and biological influences can accelerate this process. For instance, materials used in manufacturing may corrode if exposed to extreme heat or chemicals during production. Careless handling of corrosive substances can further exacerbate surface deterioration, weaken materials, and ultimately lead to product failure.

Upon completion of production, surfaces are particularly susceptible to corrosion when exposed to environmental factors such as moisture, humidity, and salt spray. Corrosion can adversely affect material performance, leading to flaws such as pits and cracks. To combat this degradation, materials are often treated with coatings or protective measures to shield against environmental conditions and mechanical stresses. Throughout service, environmental conditions continue to influence surface degradation rates. Products deployed in marine environments, for instance, are highly vulnerable to saltwater exposure, while those in chemical processing plants face risks from corrosive chemicals. Effective time management is crucial to mitigate such failures, especially those of a similar nature.

Despite routine maintenance efforts, unforeseen failures can still occur, introducing unexpected complications. Moreover, environmental factors like corrosion, which manifests as crack formation and electrochemical cell generation, exert a significant structural influence, especially in adverse weather conditions such as strong winds and heavy precipitation. Instances of corrosion are frequent in various oil pipeline contexts, progressively rendering structures porous and susceptible to layer-by-layer material erosion. Rusting exacerbates surface deterioration, diminishing component thickness despite attempts at remediation through polishing or repainting. As thickness diminishes, material erosion increases vulnerability to failure. This gradual degradation not only results in fluid leakage but also carries substantial global environmental repercussions. Hence, instituting effective surface engineering methodologies becomes imperative to prevent bridge collapse, structural deterioration, corrosion-induced failure, or fatigue-related issues. The judicious implementation of surface engineering techniques is essential for ensuring structural integrity and longevity.

Breakdowns in gears, bearings, and various components have been observed, caused by fatigue. While fatigue failure is commonly associated with material deterioration and degradation, instances of fatigue failure emerging from the surface have also been noted. Many fissures followed by depressions on the surface have been observed, ultimately resulting in the collapse of the material. Therefore, this failure is primarily attributed to the interaction between surfaces, leading to surface fatigue and subsequent material breakdown.

In many cases, the root cause of failure can be traced back to a single factor, such as a surface crack. Initially, this crack may propagate gradually toward the worn side, interacting with other wear mechanisms along the way. However, the ultimate failure typically results from the convergence of multiple contributing factors. Therefore, it is essential to develop characterization techniques aimed at quantifying surface degradation and systematically exploring methods to optimize the identification and measurement of surface deterioration. Surface visualization serves as the initial and most straightforward approach to detect visible degradation, such as abrasion marks caused by physical contact, corrosion, or fatigue on the surface. However, to identify cracks resulting from the fatigue process in the underlying layer, particularly in the intermediary layer used in coatings to bond to the substrate, alternative methods are necessary. Fortunately, numerous destructive and non-destructive methods are available in the literature, and a thorough understanding of these characterization procedures is imperative. It is also crucial to evaluate the effectiveness of different solutions across diverse contexts and consider their associated costs. With a comprehensive understanding of the problem's scope and severity, a range of remedies can be proposed. Sometimes, using multiple characterization methodologies can achieve optimal efficacy and offer improved solutions, resulting in reduced pollution and a diminished environmental impact.

1.2 EXPLORING SURFACE DEGRADATION MECHANISMS

Failure mechanisms can be classified into three principal domains: the physical domain, encompassing both mechanical and thermal properties; the chemical domain, involving molecular-level interactions of substances; and the biological domain, which pertains to biological phenomena.

For the scope of this book, the focus will be on the examination of physical and chemical phenomena, as the biological domain falls beyond the current purview of this text and will therefore be omitted. The exploration will delve into various aspects related to physical phenomena, including wear phenomena arising from tribological conditions, surface fatigue conditions, and the effects of temperature.

There are various wear mechanisms, each distinct in nature, and addressing just one or two mechanisms may not be sufficient. Therefore, it is essential to consider the multitude of wear mechanisms that may occur. Let us consider "adhesive wear," which occurs in the form of "cold joining" or "cold welding," characterized by high contact stresses and the possibility of plastic deformation at the interface of contacting surface. When subjected to relative velocities, surfaces undergo significant increases in friction and surface degradation. Another prominent wear mechanism is abrasive wear, which can be triggered by debris generated by adhesive wear or by external particles in the environment. In addition, fatigue wear, erosive wear, fretting wear, and high-temperature surface degradation are included in the physical domain of surface failure mechanisms. Fatigue wear is often attributed to surface defects. Erosive wear occurs when particles collide with high velocity and at a precise angle. Fretting wear can occur when two interconnected systems experience micron-level amplitude or vibration, which is often invisible. Consequently, this type of wear does not provide advanced warning before failure. A significant increase in temperature softens the material surface, causing particles to spread out and resulting in the melting and erosion of the material. An important problem that frequently occurs in various electricity-related devices is the occurrence of high-temperature corrosion. These failures happen because of high temperatures and the presence of fly ash and other corrosive compounds, which cause severe damage in these conditions. Concrete structures are prone to corrosion and environmental degradation. To properly understand, it is crucial to have a deep understanding of the corrosion mechanisms.

Although wear is generally perceived as a gradual process, the presence of acidic or corrosive media can significantly accelerate wear rates. This underscores the importance of considering factors such as salt formation, various pollutants—especially gaseous emissions with acidic properties—and the direct impact of chlorine on surfaces. When these pollutants interact with water, they trigger chemical reactions that produce acids, accelerating corrosion. Therefore, it is crucial to recognize and address these interrelated factors. Identifying the primary sources and understanding their correlation with subsequent phenomena is essential. It is important to comprehend the involved complex interactions, which follow a sequential theme as elaborated in Figure 1.2.

Traditionally, automobiles were predominantly crafted from metallic materials. However, modern advancements in the automotive industry have introduced plastic and composite materials, resulting in reduced yet still notable damage from environmental factors. Dust particles significantly contribute to surface degradation, as do wind gusts and environmental humidity reacting with atmospheric gases. The reactivity of materials with environmental factors also determines the extent of surface deterioration, with different materials exhibiting different effects. For instance, plastic demonstrates reduced reactivity compared to steel. Steel experiences accelerated deterioration due to corrosion, whereas plastic suffers greater surface damage from physical factors. Composites, while useful, are prone to defragmentation or delamination and are susceptible to UV radiation, which can break the lamellar structure into smaller fragments, compromising structural integrity. If composites are manufactured layer by layer, delamination can exacerbate existing issues. These phenomena can be interrelated, with radiation-thermal, UV, or other types—often playing a role.

In many cases, heat or thermal damage, as well as radiation damage, manifest in the presence of corrosive water content, potentially leading to significant photochemical deterioration. A comprehensive assessment of the entire structure reveals an inevitable exposure to some form of water

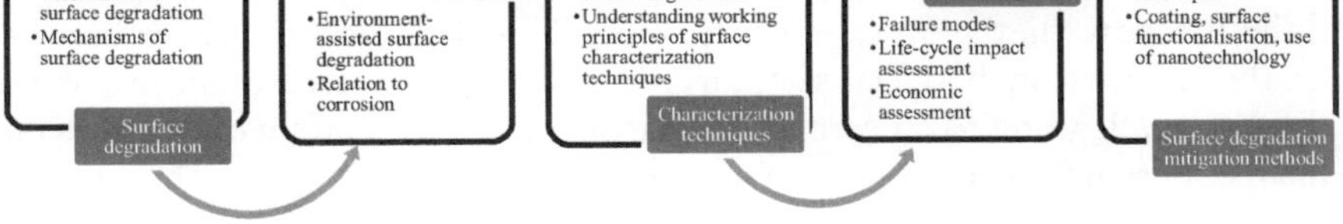

Figure 1.2 Overall framework of book.

corrosion or surface degradation, presenting a substantial challenge. Furthermore, localized faults can initiate corrosion due to elevated stresses or acidic reactions in specific areas. Flow-assisted corrosion poses a notable concern, particularly when water movement exacerbates the corrosion process. Notably, there exists a correlation between temperature and oxidation, with higher temperatures markedly accelerating oxidation. This phenomenon, commonly referred to as "fatal oxidation." In scholarly discourse, the term "oxidation" often supplants "corrosion." Oxidation delineates a chemical process characterized by electron exchange or the modification of atoms or molecules through either loss or gain. Corrosion, on the other hand, represents a broader field encompassing various processes, within which oxidation serves as a specific manifestation. To enhance the conceptual clarity surrounding corrosion, it proves valuable to scrutinize oxidation as a discrete phenomenon within the broader corrosion framework. For instance, preventing corrosion on automobiles often involves applying protective coatings. This underscores the principle of surface engineering, wherein techniques are used to shield surfaces from environmental factors by establishing protective barriers. Such measures mitigate environmental harm to surfaces and prevent surface deterioration induced by environmental factors.

Moreover, it is acknowledged that the interplay of mechanical, thermal, and chemical processes may precipitate distinct forms of severe failure. Consequently, four domains of action are categorized: mechanical, thermal, chemical, and combined degradation mechanisms, as shown in Figure 1.3. Each domain encompasses viable reactions, necessitating meticulous examination. Evaluating various failure modes is imperative for assessing their impact on the system's longevity. An integral aspect in evaluating overall success or failure lies in discerning the relationship between longevity and economic considerations.

Establishing a robust correlation between wear, corrosion, and surface degradation is essential. Although wear phenomena and physical mechanisms are significant, the presence of corrosion dramatically accelerates wear. Temperature plays a crucial role in almost every corrosion event, with higher temperatures increasing the severity of corrosion. Therefore, it is vital to determine the correct methodology or appropriate chemical for each specific situation. A comprehensive understanding of these types of failures is crucial.

Let us consider an example of machine tools, which by and large possess a finite lifespan, they require re-sharpening. This degradation occurs because of the interaction between the surface and the corrosive qualities of the fluid/water, leading to a decrease in sharpness and ultimately dullness. Tools may experience wear due to excessive pressure, resulting in the ejection of minute particles. Consequently, the degree of sharpness progressively diminishes. Similarly, automobiles are fitted with tires that release various

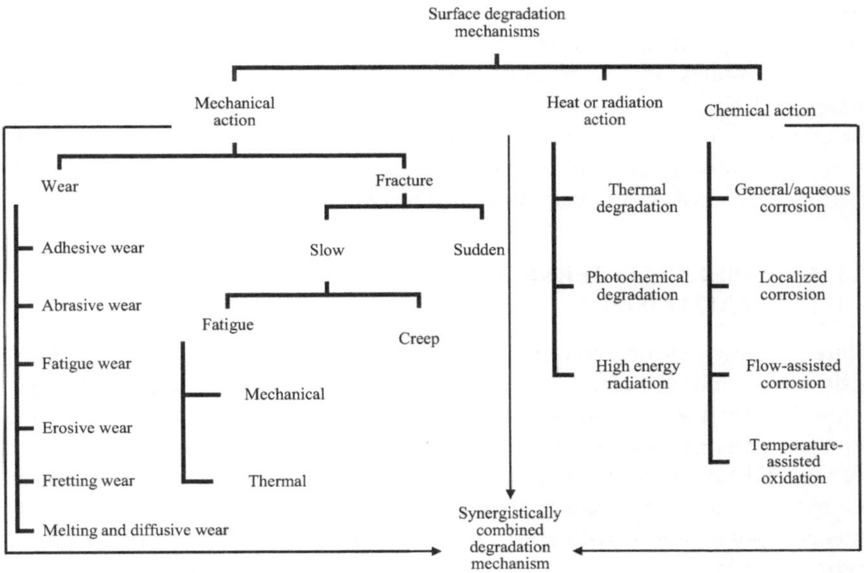

Figure 1.3 Mechanisms of surface degradation.

particles upon contact with the environment. This interaction causes the tires to gradually deteriorate, leading to the emission of particles that contribute to environmental pollution.

A comprehensive understanding of corrosion processes, especially their environmental implications and surface interactions, as well as the mechanisms behind surface corrosion and damage, is crucial to our goals. Improving the quality of materials used in automobiles—including motorcycles, cars, and trucks—by following the framework outlined in Figure 1.2 can significantly reduce environmental harm and lead to cost savings. This strategic approach provides a clear differentiation and serves as a critical interface. Using this methodology allows us to explore the most effective ways to achieve this distinction and extend the functional lifespan of products or systems, all while ensuring environmental sustainability.

A solid grasp of surface engineering enables the development of refined treatments and holistic solutions to address systemic issues. With an understanding of surface engineering, knowledge about specific raw material requirements, precise energy quantities, and optimal techniques can be acquired to effectively prolong product lifespan while minimizing environmental impact. By considering the perspective of environmental contamination, the aim is to improve product longevity and preserve the environment.

The objective is to develop a thorough scientific understanding and adopt a systematic approach to understand the occurrence of failures, mitigate their impact, and significantly extend the lifespan of systems, components, or subassemblies. This book will explore various methods for characterizing material surfaces, as detailed in existing literature, including diverse characterization techniques, coating procedures, and surface modification methodologies. In addition, it will examine different damage processes, mitigation methods, preventive tactics, and cost-saving measures.

1.3 COMMERCIAL PERSPECTIVES ON SURFACE DEGRADATION

Material degradation, resulting from environmental conditions, mechanical strain, and chemical reactions, poses significant financial and operational challenges. Financially, it leads to expenses for maintenance and replacements, as well as indirect costs from reduced productivity, product recalls, and customer dissatisfaction. Moreover, there are insurance expenses related to liability and safety concerns, given that compromised structural integrity poses safety risks and potential legal consequences. In addition, some degradation processes contribute to environmental pollution, underscoring the importance of minimizing the ecological impact through effective control strategies. Addressing material degradation is not only essential for economic efficiency but also for maintaining high standards of performance and environmental responsibility. Therefore, implementing robust control measures to mitigate degradation is crucial to prolong the service life and resilience of industrial assets, ultimately reducing the frequency of replacements and long-term costs.

To manage material degradation, various strategies can be implemented, including regular inspections, maintenance, cleaning, and selecting materials with enhanced resistance to specific forms of degradation. Protective measures such as coatings, corrosion inhibitors, and effective lubrication can also be used. Investing in environmental controls, such as temperature and humidity regulation, as well as advanced monitoring systems like sensors and predictive analytics, allows for real-time tracking of material conditions. This investment can drive the development of innovative materials, coatings, and technologies, helping industries stay ahead in mitigating degradation effects. However, ensuring regular maintenance and inspection requires investments in monitoring systems, specialized equipment, skilled personnel, and preventive measures.

Although some may argue against striving for a perfect or ideal surface, there are instances where creating surface irregularities can enhance performance. However, we will address these points in a separate chapter focused on surface design. For now, it is important to acknowledge that surface degradation is undesirable, as deviations from the ideal surface can lead to economic implications and associated penalties. In specific scenarios, a sophisticated technique known as a three-dimensional (3D) optical profilometer may be used. This technique is essential when surface roughness is critically important and requires meticulous examination. Is it feasible to conduct 2D surface roughness measurements at 40 to 50 designated spots and subsequently utilize these data to generate a comprehensive 3D profile? To optimize resource utilization, a profound understanding of scientific principles and practical expertise is indispensable. In certain contexts, transitioning from micro to nano levels may prove more suitable. The initial steps involve establishing a framework and then formulating a cost-effective methodology for detecting and diagnosing the root cause of failure. Once identified, a comprehensive analysis of the failure can ensue. It is wise to reserve the use of expensive machinery for advanced stages, understanding that intricate nano-level details may not be crucial during the initial phases.

Gaining comprehensive knowledge is paramount for us. Therefore, books like this play a pivotal role, and it is crucial for us to diligently engage in them to optimize costs and ensure overall economic efficiency. Whether a suggestion is economically viable or not, informed decisions must be made based on it. The aim is to equip individuals with ample knowledge so that they can confidently navigate such decisions. Situations may vary, with some being favorable and others not. However, armed with knowledge, we can discern and make these pivotal decisions effectively.

Material degradation bears substantial financial and industrial consequences, underscoring the critical necessity of its management to ensure the enduring resilience of materials and structures. Corrosion, as one of the key degradation mechanisms, accounts for approximately 5.95% of the Indian GDP,[2] equivalent to 1.667 lakh crore INR. The price is significant, particularly when compared to the 2023 education budget of 1.16 lakh crore INR,[3] which is lower than the cost of the corrosion.

The costs can be divided into two main components (Figure 1.4): direct losses and indirect losses. The direct losses (replacement or repair, procurement of new materials, expenditure on monitoring systems, specialized equipment, skilled personnel, preventive measures aimed at mitigating degradation effects) contribute around 2.3% to the GDP. The indirect losses, such as a corporate shutdown or the entire closure of a factory (the labor costs, the downtime endured during the replacement process, environmental repercussions, liability, and safety concerns leading to safety hazards and legal repercussions), incur significant costs.

In addition, the continuous employment of the system leads to a gradual drop in efficiency over time. Initially, with an efficiency of 99%, the efficiency may decrease to 98% after 1 year and further decline to 95% after 2 years. In fact, as efficiency decreases over time,[4] as illustrated in Figure 1.5, there is a corresponding waste of energy, which is not economically justifiable due to its high cost. Occasionally, it becomes necessary to substitute an entire

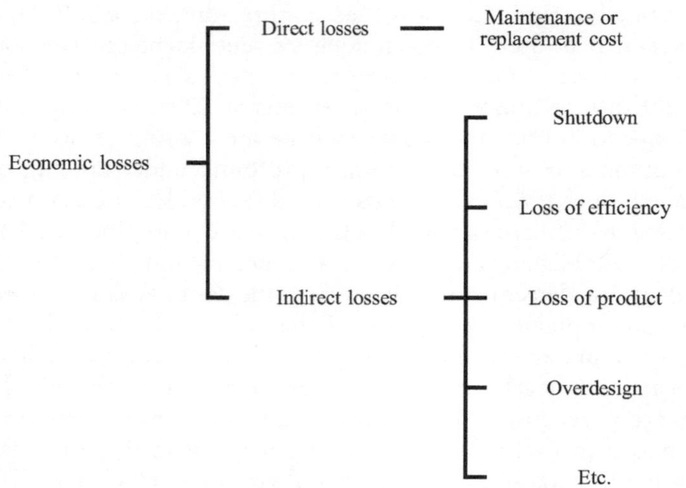

Figure 1.4 Economic impact of product failures.

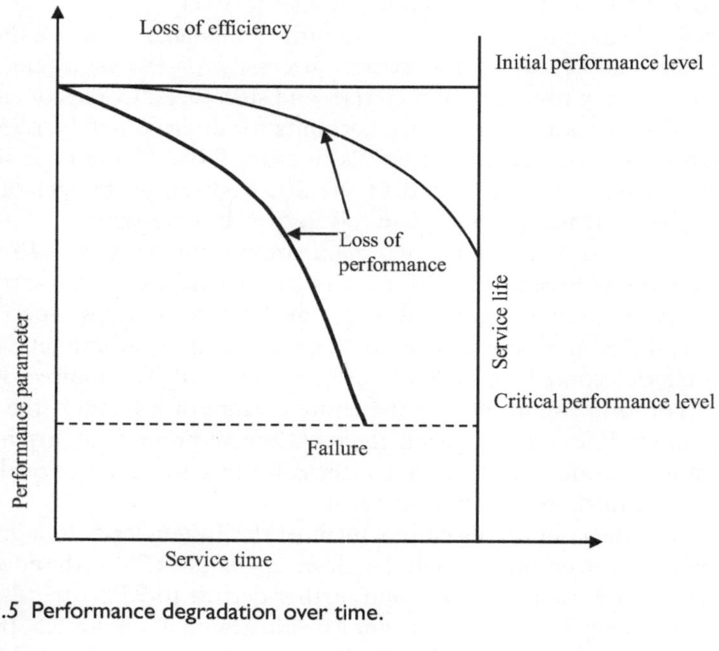

Figure 1.5 Performance degradation over time.

Figure 1.6 Images depicting few tragedies in India (Bhopal gas tragedy[6]; distress of Thane creek bridge[7]; derailment of Charminar express[8]).

product or to excessively engineer the systems, which results in additional costs. It is anticipated that such circumstances will occur. Therefore, by improving and developing a more advanced scientific approach and gaining a deeper comprehension of the process, it is possible to greatly diminish these losses. By gaining a comprehensive understanding of science, we may devise more efficient strategies to address these problems. To get an economical or cost-efficient resolution, it is important to have a comprehensive understanding of the contents within this book.

Let us consider agriculture as an example. Environmental dust and silica particles, when combined with moisture, can be extremely detrimental unless we utilize our expertise in surface engineering. Agriculture alone contributes to over 6.1% of the overall corrosion expenses.[5] Similarly, in the field of mining, the presence of particles and reactive gases, along with environmental moisture or water content, can induce corrosive-wear, resulting in significant financial costs. In the field of manufacturing, we have observed multiple instances where the manufacturing tool undergoes a process of deterioration over time, resulting in a loss of tool sharpness and a decrease in productivity. This one factor contributes to around 5.7% of the whole expenditure on corrosion. The tribo-corrosion failures in the oil and gas sector account for an additional 1.5%.

Figure 1.6 presents visuals capturing a few tragedies. The Bhopal Gas Tragedy, occurring in December 1984, stands out as one of the most catastrophic industrial disasters in history, stemming from the release of methyl isocyanate gas due to safety system failures, notably the refrigeration system and gas scrubber. Similarly, impactful events include the distress witnessed in the Thane Creek Bridge and the derailment of the Charminar Express, underscoring the critical need to promptly address issues like corrosion in engineering infrastructure to uphold public safety. As the complexity of these issues intensifies, understanding them becomes increasingly paramount, particularly in averting potential catastrophes that could lead to significant national losses and endanger lives.

Exploring risk analysis becomes crucial, necessitating a comprehensive study of surface damage from a risk perspective. Risk management evaluation involves assessing two key factors: the likelihood of failure and the resulting consequences, represented as $R = P \times C$, where R signifies Risk, P denotes Probability, and C represents Consequence.

Risk assessment matrices simplify the evaluation of failure probability and consequence levels, typically denoted as A to D for probability and 1 to 4 for consequence severity. In the event of a failure, what will be the financial consequences of such failure? A matrix (as depicted in Figure 1.7), such as a probability and consequence matrix, can be utilized to assess the likelihood of an event occurring on a scale ranging from A to D. In addition, the repercussions resulting from the event can be rated.

For example, when the likelihood is extremely small and the repercussions are nearly inconsequential, there is no cause for concern. Nevertheless, in the top-right quadrant, excessive hazards are there, which will result in highly significant repercussions. As the likelihood increases, more red color becomes visible, indicating the need for a thorough investigation of the variable R. On the other hand, if the repercussions are not severe, as observed in this scenario, but there is a growing likelihood of failure, then we are transitioning into a state of moderate risk. It is necessary to evaluate the surface treatment options and identify the one that will provide a cost-effective benefit analysis for us. Therefore, it is imperative to approach this situation from a risk management perspective.

Let us reconsider the topic of Bhopal gas tragedy. The release of a substantial amount of isocyanide was because of three main failures. To minimize operational expenses, which has been the precedence over other criteria in most of the industries in India, operators deactivated the refrigeration equipment. Another concern arose over the gas scrubber, which was specifically engineered to eliminate the noxious gases. The scrubber effectively hinders the passage of cyanide in the event of a leak, hence minimizing its detrimental impact. Despite the failure of the scrubber unit, the operators were aware of the issue but believed it would not be problematic as there

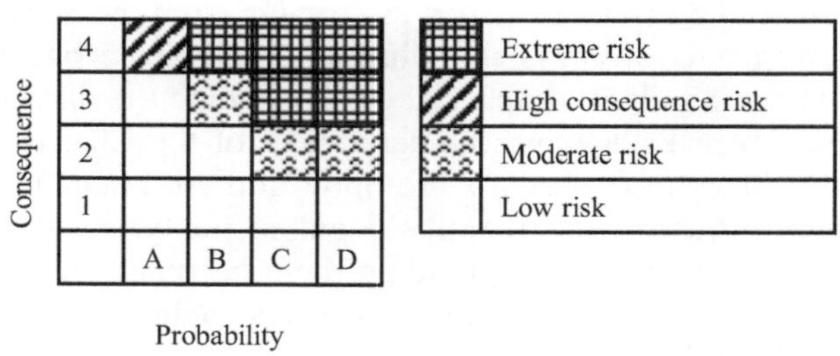

Figure 1.7 Risk management matrix.

was no gas leakage at that moment. Finally, an additional problem was a pipe that had become rusted.

The "Bhopal gas tragedy" indicates that failures in three subsystems led to a significant catastrophe, resulting in the loss of many lives and causing both long-term and short-term injuries. Many individuals continue to suffer long after the immediate threat has passed. Therefore, it is crucial to approach such situations from a risk management perspective, considering the potential repercussions and devising appropriate solutions while aligning strategies properly.

The Charminar express derailment is a well-known example. Insufficient maintenance and frequent upkeep of the tracks will inevitably lead to issues. Even the railway track experiences the phenomenon of wear, which ultimately results in failure. Bridge failure occurs due to structural weaknesses, such as inadequate bolt connections or the existence of gaps or cracks. Early identification of these issues might mitigate the occurrence of such damage.

Let us reconsider our roles from a slightly different angle: What should be the industry's function or purpose? What role should the government play? And what responsibilities fall upon the user? The advocate proposes the implementation of life cycle assessments (as shown in Figure 1.8) by both industry and government, as well as individual users. Why? To effectively introduce any product or procedure, a thorough comprehension of its objectives and implications is crucial. How much raw material will be required? What energy quantities are needed? Naturally, improving the lifespan of a product reduces the demand for raw materials and long-term energy usage.

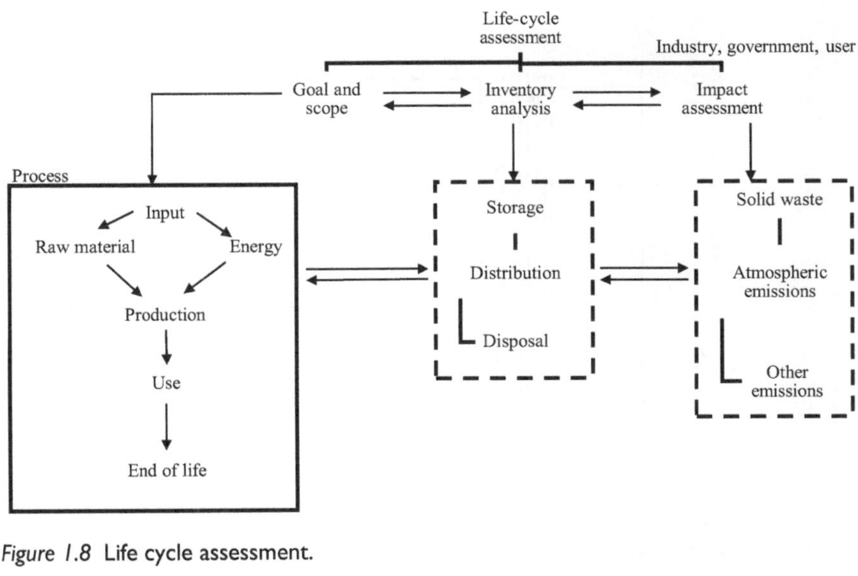

Figure 1.8 Life cycle assessment.

This signals an intention to extend product lifespans and transition toward environmentally friendly options. Manufacturing processes must include inventory analysis to determine storage duration, choose distribution systems, and plan for end-of-life disposal methods. This aspect carries significant weight from a governmental standpoint. Evaluation of manufacturing processes and surface engineering techniques is essential, given the escalating solid waste accumulation globally, particularly in regions like India. Efficient waste disposal techniques are imperative, allowing us to assess available unprocessed material for utilization and ensure its uninterrupted use. Reducing solid waste is vital to mitigate environmental pollutants from processing. The aim is to minimize the necessity for extensive manufacturing processes to refine and enhance the functionality of the product. Undoubtedly, the production process also exerts a significant impact.

The upcoming presentation will feature several concrete illustrations derived from research study,[9] emphasizing the surface's roughness. The initial case had a surface roughness of around 0.362 μm. By adjusting the surface roughness to 0.051 μm, we observed a 73% decrease in failure rate (as shown in Figure 1.9), resulting in a lifespan extension of around three to four times. This means that by simply modifying one manufacturing technique, the lifespan is increased by fourfold.

Let us consider one more illustration from reference.[10] The researchers applied a surface coating, observing a wear reduction of over 50%. This method proved highly effective, doubling the material's lifespan, and reducing both raw material usage and processing energy. Yet, the coating's cost

Figure 1.9 Effect of roughness on wear.

is a crucial factor. Therefore, a thorough economic assessment is imperative. Such analysis guides decision-making on whether to proceed. It necessitates a deep scientific understanding, which can be gained from the contents of this book.

Now, let us discuss the economic analysis of controlling surface degradation. Although the goal is to minimize surface degradation effectively, it is essential to evaluate the advisability and potential outcomes of such actions. Continuously improving or minimizing surface degradation remains crucial, and mitigation measures should be grounded in informed decision-making. Protective measures entail costs, necessitating an assessment of their cost-effectiveness and overall benefits. If the cost savings from reduced material degradation exceed the cost of protective measures, the outcome is favorable. However, if the costs outweigh the benefits, it may not be prudent to implement the mitigation measures. This simple rule of thumb guides decision-making, although the full impact of protective measures may not be immediately apparent. It may take months or even years to fully evaluate the effectiveness of these measures.

According to financial theory, the value of an expenditure is influenced by the passage of time. Hence, when evaluating the allocation of funds, it is essential to consider the current worth. Present value refers to the practice of decreasing the worth of future cash flows by utilizing a pre-established rate of return. Immediate cash flows are prioritized over cash flows of equal nominal value received later. The value of preventative measures can be quantified using the following equation:

$$\text{Value} = -\text{Cost} + \text{Savings} \times A(N, R)$$

$$A(N,\ R) = \frac{\left(1-(1+R)^{-N}\right)}{R}$$

Here, A stands for annuity (sequence of uniform payments given at consistent intervals), N reflects the duration in years. The variable R represents the necessary rate of return, which is typically stated as a decimal or percentage per year. For instance, it could be written as 0.1 or 10% annually. Value denotes the present monetary value of preventive measures, Cost relates to the initial expenditure incurred for establishing preventive measures, while savings indicates the annual monetary amount saved as a result of the decreased degradation of materials.

The annual net return (A(N, R)) is the proportion of the total value of yearly savings compared to the value of a single yearly saving. To provide an example, let us consider a scenario where the value of N is 10 and the value of R is 0.05. In this specific situation, the value of A(N, R) can be calculated by applying the formula $(1 - (1.05)^{-10})/0.05$, resulting in a precise value of 7.72. The current value of this annuity, which is 7.72, is less than the value of 10 that would be obtained without applying any capital discounting. This implies that the total value of the savings after 10 years is 7.72 times the annual savings, rather than 10 times the annual savings. It is imperative to recognize that increasing values of R constantly decrease the value of an annuity for a given N. As the rate of return increases, the annuity decreases, which in turn affects the savings. If the multiplication factor is lower, the cost may exceed the savings, resulting in a negative value, indicating an unprofitable situation where measures to control material or surface degradation may not be feasible.

Now, let us discuss the question of the rate of return. In some cases, the assumed rate of return may be 0.05, 0.1, or 0.15. The resulting annuity varies accordingly, with higher rates leading to higher annuities. However, if the rate of return is very high, it may not be advisable to implement measures to prevent surface degradation, as it could be unprofitable. Conversely, if the rate of return is low, implementing such measures becomes more viable.

For new businesses, especially those with inherent risks, the pursuit of fast returns is often a priority. To mitigate this, governments often provide support to high-risk projects, including startups. These initiatives aim to encourage innovation by providing financial assistance and resources to new ventures. Incubation centers and similar institutions also play a crucial role in supporting startups by offering resources and guidance, facilitating the development of new technologies and processes without undue financial strain on the entrepreneurs.

Now, the question arises: what exactly do we mean by "saving"? Saving, in this context, is a probabilistic measure. We cannot determine an exact value; instead, we express savings as probabilities. These probabilities represent the likelihood of failure in two scenarios: one where no preventive measures are taken, and another where preventive measures are implemented against surface degradation. An effective method for assessing the cost benefits of implementing protective measures on a system involves analyzing the probability of system failure caused by a certain type of material deterioration, such as wear, and the financial repercussions of such failure. The value of savings over a specific period can be calculated by multiplying the reduction in the probability of failure by the cost of the failure. The mathematical expression that represents this relationship is as follows:

$$\text{Savings} = \left(\text{Prob}(\text{U}) - \text{Prob}(\text{P})\right) \times \text{Cost}(\text{F})$$

The savings formula is derived by subtracting the likelihood of an unfavorable outcome (U) from the probability of a favorable outcome (P), and then multiplying the result by the cost of performing a specific action (F). Prob(U) denotes the probability of failure happening without any preventative measure. P represents the likelihood of a failure occurring while a protective measure is being implemented. Cost(F): The financial worth linked to the event of failure. We discuss savings in terms of probabilities because it is not based on precise or deterministic numbers. We distinguish between the probability of failure without preventive measures and the probability of failure with protective measures, each associated with its respective costs. The cost of failure, in this case, refers to the expenses incurred upon failure. To illustrate, let us consider an example where the cost of each failure is \$10,000, and the likelihood of failure without protection is 0.3, while with protection, it drops to 0.1. The difference in probabilities multiplied by the cost yields the savings. In this case, it would be (0.3 – 0.1) * \$100,000, resulting in savings of \$20,000. Therefore, the estimated cost reduction achieved by implementing the preventive measures for the stipulated period would be \$20,000.

The required rate of return is established by the financial framework of a corporation or government agency, rather than being directly influenced by criteria related to materials deterioration. The effectiveness of any strategy for controlling materials degradation is not absolute but rather dependent on the current financial conditions. Estimating rates of materials degradation is a challenging problem, as we have already described. Despite the use of advanced numerical modeling tools, determining the quantity and whereabouts of degraded material does not yield conclusive evidence on whether material deterioration should be managed or disregarded. To determine the extent of materials degradation, one must either adhere to the previously stated model or utilize a more sophisticated model that assesses the suitable course of action. Typically, organizations that have more business risks generally have higher rates of return that are necessary. As a result, they may discover that it is more advantageous to disregard the process of materials degradation. The schematic Figure 1.10 illustrates the relationship between the decrease in the value of an investment in materials degradation and the growth in the needed rate of return.

We assume that these probabilities are not time-dependent for simplicity in our analysis. This probabilistic approach allows us to better understand the potential savings associated with preventive measures against surface degradation.

Calculating the rate of material degradation is no simple task, primarily due to the limited information available regarding surface degradation types. Unless significant research has been conducted on a specific product, gathering relevant data remains a challenge. This underscores the critical importance of using a feedback loop. Continuously collecting data allows us to develop mathematical or probabilistic models for better assessment.

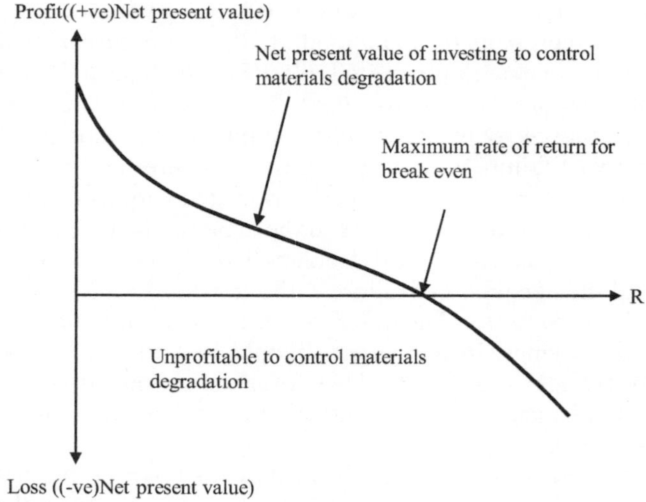

Figure 1.10 Effect of rate of returns on profitability.

In essence, a comprehensive feedback loop, along with hands-on experience and knowledge, is indispensable. The emphasis is on the necessity of continuous learning, encouraging participation in lectures and engagement in exercises to broaden understanding. Armed with knowledge, informed decisions can be made in the future with greater confidence. With that, I thank you for your attention.

1.4 CONCLUSIONS

The surface degradation plays a critical role in various mechanical components and systems across multiple industries. Effective management and mitigation of these degradation processes are essentially required to extend the service life of materials and structures, thereby reducing economic losses and environmental impact. Surface engineering methodologies emerge as vital tools for protecting materials from harsh operational and environmental conditions. Commercial considerations, including the costs associated with degradation and preventive measures, underscore the importance of informed decision-making to optimize maintenance practices and reduce long-term costs. Through a comprehensive understanding of degradation mechanisms and cost-effective control strategies, it is possible to improve product performance, ensure safety, and promote sustainable practices in material management and engineering.

1.5 QUESTIONS AND ANSWERS

Question 1: What are the implications of surface degradation on product performance, and what costs are associated with addressing it?

Answer: Surface degradation, if the surface is not ideal, can lead to noticeable changes in product performance, particularly in terms of friction and wear. These changes can have severe economic implications, resulting in substantial expenditures on replacement, maintenance, and productivity. In addition, implementing a condition monitoring system and using qualified workers to address these issues can also incur significant costs.

Question 2: Why is it crucial for industries such as automotive, aerospace, electronics, and manufacturing to address surface degradation, and how does it impact productivity, quality, and safety?

Answer: Addressing surface degradation is paramount in industries like automotive, aerospace, electronics, and manufacturing to maintain product quality standards. Surface degradation directly affects productivity, quality, and safety. By making informed decisions about the viability of surface treatment from the outset, industries can ensure cost-effectiveness and uphold product quality requirements, leading to enhanced productivity, improved quality, and greater safety measures.

Question 3: How are environmental humidity and natural rainfall linked to material degradation and corrosion, and what effects do they have on different modes of transportation?

Answer: Water management poses significant challenges due to the dynamic and corrosive nature of water. Steel or iron materials are prone to rust, leading to inevitable damage. Even if we substitute them with materials like plastic pipes, tribo-corrosion remains a serious issue. Various forms of transportation, including cars, buses, and railways, also encounter notable difficulties, necessitating an understanding of their failure modes. Aqueous corrosion, a common occurrence driven by the omnipresence of moisture in the environment, is frequently observed, contributing to material degradation and posing challenges across different sectors.

REFERENCES

1. Newspaper article published on 01.04.2016, https://punchng.com/22-die-in-india-bridge-collapse/
2. Bhaskaran, R., Palaniswamy, N., & Rengaswamy, N. S. (2009). Cost of corrosion and corrosion auditing in Indian industries. *Corrosion Reviews*, 27(Supplement), pp. 1–35. https://doi.org/10.1515/CORRREV.2009.27.S1.1
3. Expenditure of Government of India, www.indiabudget.gov.in/doc/Budget_at_Glance/bag6.pdf

4. Batchelor, A. W., Lam, L. N., & Chandrasekaran, M. (2011). *Materials Degradation and Its Control by Surface Engineering*. World Scientific.
5. International measures of prevention, Application, and Economics of Corrosion Technologies Study, http://impact.nace.org/documents/Nace-International-Report.pdf
6. Newspaper article published on 16.06.2025, https://telanganatoday.com/western-railways-156-year-old-thane-creek-bridges-being-dismantled
7. 70 milestones in Independent India's business history. Mint. livemint.com
8. Newspaper article published on 10.01.2024, https://indianexpress.com/article/cities/hyderabad/six-injured-as-few-coaches-of-charminar-express-derail-at-hyderabad-station-9102929/
9. Sidh, K. N., Jangra, D., & Hirani, H. (2023). An experimental investigation of the tribological performance and dispersibility of 2D nanoparticles as oil additives. *Lubricants*, *11*(4), 179. https://doi.org/10.3390/lubricants11040179
10. Kumar, S., Goli, R., & Ghosh, S. K. (2022). Performance analysis of SiC–Ni–P nanocomposite electroless coated brake pad. *Materials and Manufacturing Processes*, *37*(7), 764–781. https://doi.org/10.1080/10426914.2021.1981932

Chapter 2

Surface degradation mechanisms

2.1 INTRODUCTION

In the preceding chapter, we discussed the topic of degradation by emphasizing the distinction between surface and material degradation. Surface degradation predominantly contributes to failures, often leading to material degradation. Notably, corrosion plays a pivotal role, with India suffering an annual loss of nearly INR 1.67 trillion due to corrosion-related failures, surpassing the entire educational budget. In addition, we highlighted the interrelationship between environmental and surface degradation. To mitigate this interaction, integrating surface engineering technologies becomes imperative, serving to either shield the environment from surfaces or isolate surfaces from environmental factors. In this chapter, our focus is on exploring surface deterioration mechanisms.

Besides corrosion, mechanical action also plays a significant role in surface degradation. This encompasses the effects of mechanical forces—both static and cyclic or oscillating—that exert pressure and induce fatigue, resulting in localized stresses and internal accumulated stresses within the uppermost layers of the surface. Such pressure initiates wear and deterioration, leading to surface degradation. Wear, alongside fracture, is one notable mechanism of surface degradation that can occur gradually or suddenly, causing the part or system to break into fragments.

In this chapter, we delve into various forms of surface degradation mechanisms, including adhesive wear, abrasive wear, fatigue wear, erosive wear, melting and diffusive wear, as well as fracture-induced surface degradation. Furthermore, we investigate the influence of heat and radiation on surface degradation, as well as the role of fatigue and fracture mechanisms in the formation of wear debris. The severity of such damage is determined by a variety of factors, including material properties, the intensity and frequency of mechanical forces, and the material's surface characteristics.

Mechanical wear refers to the gradual removal of surface materials. Figure 2.1 illustrates four simple day-to-day examples that we have all likely observed. The first example of this phenomenon is the wear of brake pads in

DOI: 10.1201/9781003545019-2

automobiles (Figure 2.1(a)), where high friction and low wear are desired. The brake system relies on these pads, which wear down over time, affecting braking efficiency. Factors such as driving technique and road conditions influence the rate of wear. Once the wear depth reaches 3 to 4 mm, friction properties diminish, leading to less effective braking. Replacements become necessary to maintain safety. Signs indicating a need for replacement include abnormal noises and decreased efficiency.

Vehicle tire wear, as shown in Figure 2.1(b), is another example of surface degradation. Tire wear refers to the gradual loss of material from the tire surface due to regular use. Furthermore, heat-related wear issues may arise, potentially causing tire expansion and associated localized irregularities on the surface. These irregularities can increase sliding, and the tire may become increasingly worn out and ultimately unusable, necessitating replacement. The decision to replace a tire and its urgency depend on a variety of factors, including driving habits, road conditions, tire pressure, and wheel alignment, all of which influence tire wear. Monitoring tire wear typically involves assessing tread depth using specialized tools. Wear bars, located within the tread grooves, indicate the need for tire replacement once the tread wears down to their level. Implementing maintenance practices such as proper inflation, rotation, and alignment checks can extend the lifespan of tires.

Figure 2.1(c) depicts shoe wear as another illustration of surface degradation. This process involves the gradual deterioration of a shoe's sole over time due to regular usage. The wear typically occurs evenly, but certain factors, such as the uneven distribution of mechanical force, may lead to accelerated degradation on one side, particularly evident in cases where the right-hand shoe experiences higher force. Variations in shoe type, materials, walking or running techniques, and the types of surfaces traversed influence the extent and pattern of wear. As a shoe's sole deteriorates, traction may diminish, and the development of holes or cracks increases the risk of slips, falls, and foot injuries. Selecting appropriate footwear tailored to the activity and terrain can help mitigate wear and maintain comfort, support, and stability throughout its lifespan.

The scratches on spectacle lenses (Figure 2.1(d)) exemplify yet another form of surface degradation. Regular cleaning with a microfiber cloth or lens cleaning solution helps remove debris and oil buildup, which can cause scratches. It is important to handle lenses with care and store them in a protective case when not in use to prevent scratches and other damage. Proper care not only ensures clear vision but also extends the longevity of eyewear. In addition, selecting high-quality materials can significantly reduce wear and enhance the lifespan of eyewear. From this perspective, wear emerges as a significant contributor to surface degradation. There is a need to understand the wear mechanisms and establish a precise definition of wear.

(a) Wear of brake pads (b) Tyre wear
(c) Shoe wear (d) Scratches on spectacle lenses

Figure 2.1 Instances of wear patterns.

In operations, wear is defined as the unwanted removal of material from a solid. We use the term "undesirable" to recognize two distinct definitions: zero wear and measurable wear. Zero wear denotes a scenario where wear processes occur within the tribo-pairing, yet they do not compromise performance or service life. In fact, such occurrences may enhance performance, rendering them desirable. Conversely, measurable wear refers to quantifiable material loss that can be precisely assessed, but it causes a reduction in the required performance. To understand "zero wear," let us consider a hypothetical scenario involving physical contact between the surfaces of two materials. In this scenario, one material ("material 1") moves against the second material ("material 2") under the applied force. After several cycles, some asperities detach from both materials, constituting zero wear. This asperity detachment helps to smooth surfaces and fill gaps or valleys, ultimately reducing overall roughness to negligible levels.

The "bathtub" curve (Figure 2.2) can help us understand the idea of "measurable wear." This curve shows how the failure rate might be very high at first, but then it slowly goes down until it reaches a steady state. This nomenclature originates from the curve's resemblance to the shape

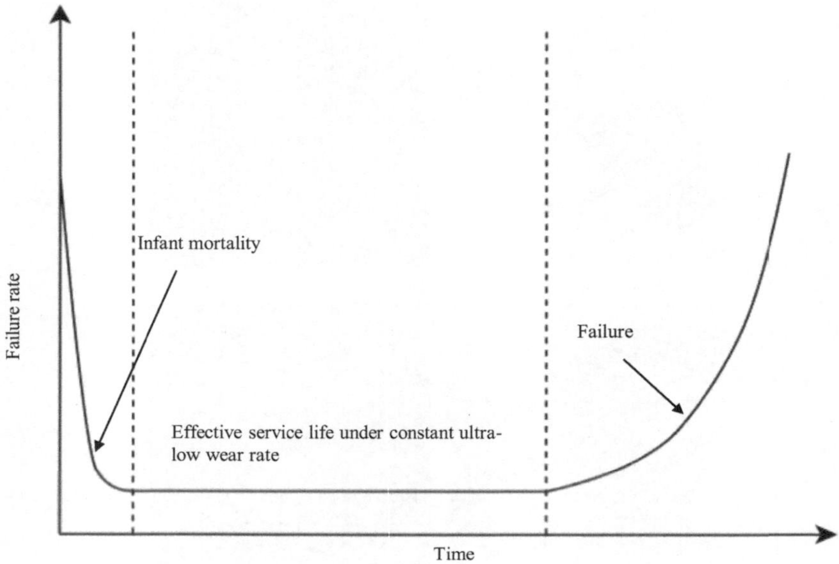

Figure 2.2 Bathtub curve.

of a bathtub when plotted with time on the x-axis and failure rate on the y-axis.

Under these circumstances, the initial failure rate can be characterized as "infant mortality" rather than "zero wear." This term refers to a higher failure rate in which defective or weak components succumb prematurely due to manufacturing flaws, design inadequacies, or mishandling during installation or operation. In this phase, manufacturers or sellers typically issue product warranties, committing to rectify any issues within a specified timeframe following purchase.

In other words, "infant mortality" pertains to the premature failure of a product or system shortly after its commencement of use, attributed to latent defects or weaknesses present from inception but not immediately discernible. Conversely, "zero wear" refers to the absence of measurable deterioration or erosion on a surface or material, indicating that it retains its pristine condition without any discernible signs of wear or degradation over a designated period.

The material enters its normal operational life after the "infant mortality" phase, during which failure rates stabilize in terms of "wear rate," allowing for the measurement or accounting of wear to estimate the product's useful life. Eventually, wear rates escalate as the component nears the end of its serviceable life. Therefore, we focus our efforts on estimating the life expectancy of the component by analyzing the wear behavior of the material pair under the specified operating conditions. It is important to note that while this represents a theoretical scenario, not all wear curves adhere to this pattern, as some may exhibit continuous wear rate escalation, never reaching a steady-state condition.

Reliability engineers frequently use the bathtub curve as a graphical representation to illustrate the trajectory of product failure rates over their lifecycle. By comprehensively understanding the underlying reasons for premature and prolonged breakdowns, engineers can strategize to enhance product reliability and longevity. To gauge the reliability of a product, it is crucial to assess the "wear rate," which correlates with the "mean time between failures," a pivotal metric in determining product reliability. To understand the wear rate, one can assume that two surfaces are in mechanical contact. When two solid bodies come into contact, the actual contact occurs primarily at the peaks of surface roughness, termed asperities. The contact region experiences markedly elevated pressure and stress compared to smoother surfaces. As relative motion between the bodies ensues, these asperities can fracture, plough through softer surfaces, or form cold junctions, resulting in material loss and surface deterioration that ultimately affects their performance and longevity.

We use a polishing process to minimize the negative effects of a rough surface. Polishing increases the real area of contact by gradual abrasion of surfaces through continuous rubbing, commonly observed in metalworking, machining, and optical lens polishing. However, there are examples to illustrate the potential drawbacks, complexities, and ramifications of surface alteration.

2.2 POLISHING WEAR

Polishing wear refers to the gradual smoothing of material from a surface due to the mechanical action of polishing. Polishing wear can result in the reduction or elimination of surface texture, which may have been intentionally designed to fulfill specific functional requirements such as retaining lubricant or enhancing frictional properties. While polishing wear can improve aesthetics and surface smoothness, it may also inadvertently lead to performance degradation or functional issues, depending on the intended application and the importance of surface characteristics.

Let us examine the illustration of a cylinder liner featuring a textured surface, as portrayed in Figure 2.3. When this type of cylinder liner undergoes polishing, the intentional removal of surface texture, which was originally designed to retain the lubricant, results in performance deterioration. Despite the wear rates remaining within acceptable thresholds, the polishing process gradually erodes the texture, potentially causing an increase in lubricant consumption due to the loss of the desired surface texture.

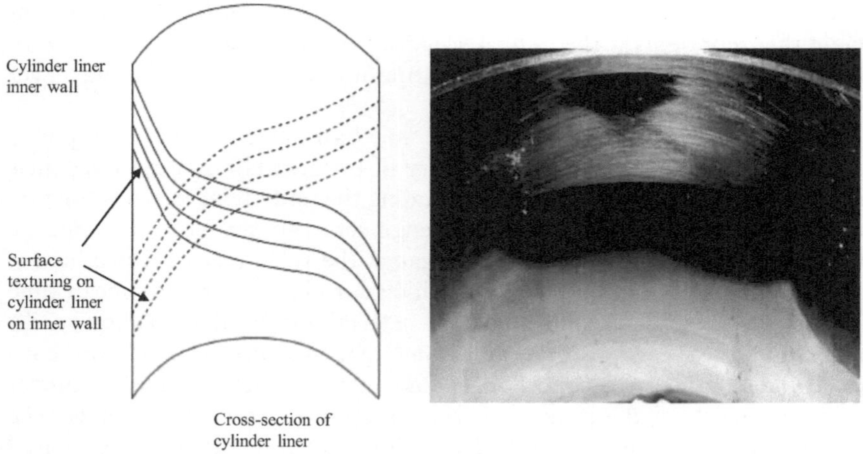

Figure 2.3 Polishing of surface texture.

Figure 2.4 Undesirable removal of carbon graphite material from operating stainless surface.

Furthermore, polishing decreases surface wetting, which in turn increases lubricant leakage from the crankcase combustion chamber. Chapter 9 will delve deeper into this topic.

In rotary joints, a similar phenomenon occurs at the interface of mechanical seals and stainless-steel shafts, where the transfer of lubricant layers is critical for maintaining sealing integrity. The interaction between carbon graphite and stainless-steel surfaces (shown in Figure 2.4), wherein deliberate transfer of carbon graphite layers onto stainless steel surfaces is used to reduce friction. Initially, wear rates may spike due to the transformation from carbon graphite to stainless steel; however, subsequent friction reduction results in lower wear rates. This deliberate deposition of carbon graphite layers is imperative for achieving the desired frictional properties. Attempting to remove these deposited layers under the mistaken assumption of surface contamination would precipitate unnecessary component failure, surface wear, and increased power consumption.

Numerous factors, including surface hardness, roughness, rubbing speed, and lubrication, influence the polishing mechanism's extent and outcomes. Excessive polishing leads to scratches, surface deformations, or dimensional losses ranging from 15 to 50 µm, each of which carries significant implications. At the microscale, those scratches appear as clusters of pits. These pits, typically smaller than 50 µm, remain imperceptible to the naked eye. This specificity arises because anything below 50 µm eludes visual detection. Figure 2.5 illustrates worn-out rolling bearing rollers. Figure 2.6 illustrates one possible mechanism of pit formation. According to the sketches shown in Figure 2.6, pit formation typically progresses through four stages. Initially, cyclic loading causes subsurface cracking, as shown in Figure 2.5(b), where material cracks begin to manifest. Subsequent sketches, Figure 2.5(c) and 2.5(d), depict the fusion of cracks and the detachment of wear particles, demonstrating how wear transforms surface profiles. Pitting is deemed unfavorable due to the marked escalation in the sliding-to-rolling ratio, heightened noise levels, motion irregularities, and increased frictional resistance. A thorough understanding of wear mechanisms becomes

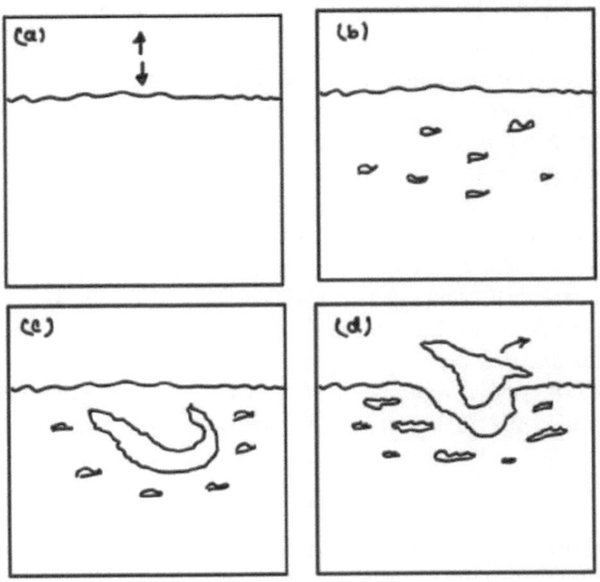

Figure 2.5 Worn rollers.

Figure 2.6 Mechanism of pit formation (15 to 50 μm).

indispensable. The timely detection of pit formation enables proactive intervention, considerably prolonging the bearing's operational lifespan from 4000 hours to as much as 40,000 hours. Conversely, failure to detect faults in a timely manner may result in significant bearing damage, drastically reducing its lifespan to a mere 400 hours.

2.3 UNDERSTANDING ADHESIVE WEAR

Adhesive wear is a common occurrence in metals. For instance, when you force a steel pin into an indium block under load and then withdraw it, it transfers a thin layer of indium onto the steel pin. We also observe this phenomenon when we push brass against indium. Such behavior illustrates the loss of indium material, driven by the adhesive force between steel and indium. However, by introducing a thin layer of lubricant at the interface between the indium and metal, the amount of transferred indium decreases substantially. A possible reason for this change could be the lubricant layer's significantly lower shear strength compared to the indium metal's shear strength, which results in a weaker interface between the steel and indium, making it more susceptible to shearing and lowering the indium wear rate. To predict the wear rate, the area of contact must be accurately determined. An approximation of the real area of contact (A) can be achieved by using the following equation:

$$A = W/H \tag{2.1}$$

where W represents the normal load (load normal to the surface area) exerted to press one surface against another, and H signifies the hardness of the softer material.

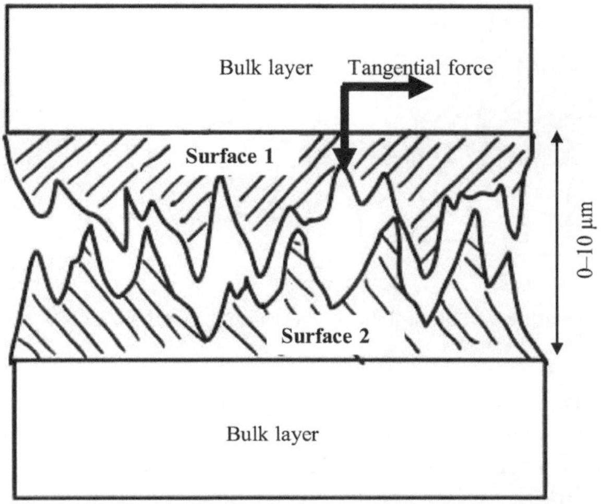

Figure 2.7 Observation of practical surfaces on micrometer scale.

It is well known that apparently smooth surfaces at the macroscopic level possess microscale roughness (as shown in Figure 2.7). When two surfaces make contact, they do so at relatively few isolated points called asperities. This leads to concentrated stress, as the actual contact area typically ranges from 1% to 15% compared to the apparent area. When a normal load is applied, the pressure of these asperities increases significantly. Consequently, higher asperities may undergo plastic bending or fracture, while those with lower contact experience elastic deformation. Moreover, in cases where metals undergo plastic deformation and come extremely close to each other, with a separation of less than 1 atomic diameter, electron transmission takes place. This electron transmission has the potential to lead to microwelding. In such a situation, applying tangential force to move one metal surface relative to another increases the likelihood of material transfer between them, initiating adhesive wear, which could eventually progress to severe wear. In addition to the phenomenon known as "electron transfer adhesion between metal surfaces," another mechanism exists: diffusion. Adhesion in electron transfer happens spontaneously and instantaneously, unrestricted by high temperatures. Conversely, diffusion-based adhesion results from the diffusion-driven mingling of the contacting bodies. Typically, this form of adhesion only occurs at elevated temperatures and progresses much slower.

Why does not adhesive wear occur in every situation when two metals come into contact? The types of metals involved, and their cleanliness often influence adhesive wear. Adhesive wear is more likely to occur on

Figure 2.8 Formation of contamination layers on a metal surface.

clean metals, particularly when they undergo manufacturing and immediate use. However, with time, a small amount of contamination builds up (shown in Figure 2.8) on metal surfaces and forms surface films. With a surface film, there is a reduced possibility of electron transmission and, consequently, less metal transfer and wear. This happens when a finished or ready-to-use metal comes into contact with environmental pollutants or agents such as oxygen, fluoride, or water on its surface. This leads to the formation of a contaminated layer, which is typically porous and weaker at the interface, resulting in lower shear strength. Therefore, it is a beneficial situation, as the environment helps reduce interfacial strain when metals meet each other. Nevertheless, relative tangential motion at the interface can disperse contaminant films at the points of contact, potentially leading to junction welding. As the sliding process persists, these junctions experience shearing, resulting in measurable wear and the formation of a few new junctions. The extent of wear depends on the location where the junction is sheared. If the interface is weaker than its surrounding sides, interface shearing occurs (as shown in Figure 2.9(a)), resulting in minimal surface damage and wear. Conversely, if shearing takes place away from the interface, metal is transferred from one surface to the other (as shown in Figure 2.9(b)). The continuous increase in particle transfer and accumulation eventually results in the formation of large lumps (as shown in Figure 2.9(c)). When these lumps exceed the size of asperities, they come under stress and may fracture from the surface, as shown in the fractured surface or removed particle (as shown in Figure 2.9(d)). If the loose particles are soft in nature, they may act as solid lubricants. However, if they are hard, they could lead to abrasive wear, a topic we will discuss in the next subheading.

Archard postulated that the contact between the surfaces of the tribological pair involves junction formation and breakage. Figure 2.10 illustrates the Archard model, assuming a circular cross-section of asperities after plastic

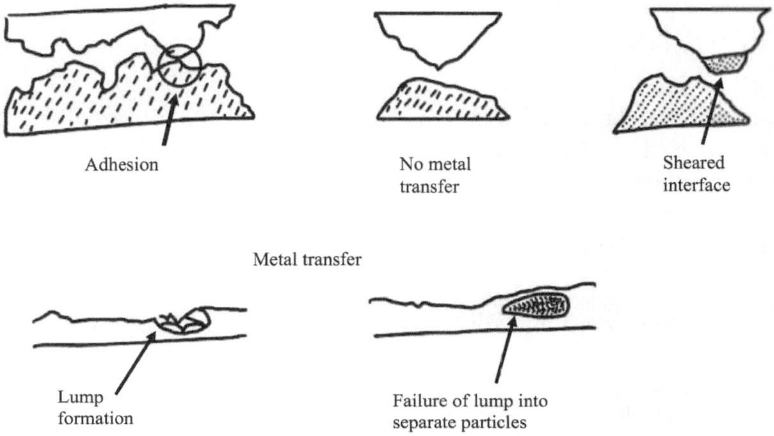

Figure 2.9 Illustration of adhesive wear mechanism.

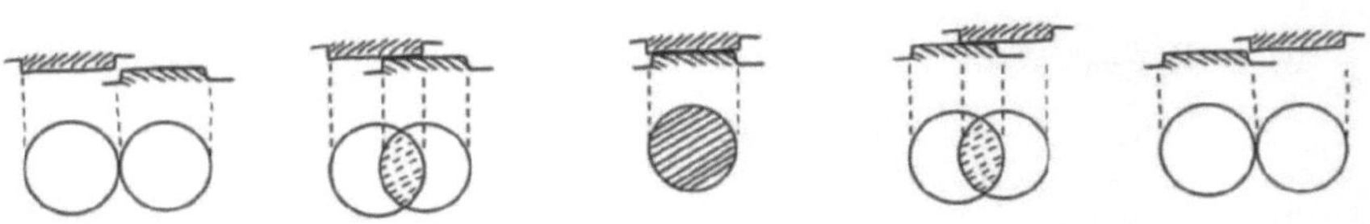

Figure 2.10 Archard's wear model.

deformation. The first sketch illustrates how asperities form junctions. As the sliding distance increases, the area of contact decreases. However, this process is continuous and occurs among several asperities. We assume an average of "n" asperities in contact at any given moment. The number of asperities can vary greatly, whether it is 100, 1000, or 10,000, acknowledging that this number may continuously vary due to the presence of both high and low peaks, with the possibility of continuous variation. Now, considering the real area of contact denoted by A, it is the summation of the contact areas at each asperity, given the total number of asperities, n. This means we need to consider the summation of the contact area at each individual asperity.

As per Figure 2.10, the radius of asperity is "a;" therefore the maximum contact area for each asperity will be πa^2. If we multiply this by the hardness, it will give us the amount of load carried δW (πa^2.H) by each asperity. Similarly, wear will cause loss of volume from the surfaces. If 100% of the asperity has been damaged or transferred from one metal to the other, the volume can be given by $2\pi a^3/3$. However, 100% damage to asperity is questionable. Therefore, one can consider a probability or multiplication constant (K_1) and wear volume can be expressed as:

$$\delta V = K_1 \left(2\pi a^3/3\right) \tag{2.2}$$

To establish a relation of wear volume with the applied normal load, we can rearrange Eq. (2.2) in wear volume per unit sliding distance for each asperity contact:

$$\delta v = \delta V/(2a) = K_1 \left(\pi a^2/3\right) \tag{2.3}$$

The wear volume per unit sliding distance for all the “n” asperities:

$$v = K_1 \sum_{i=1}^{n} \left(\pi a_i^2/3\right) \tag{2.4}$$

Total load carried by “n” asperities can be expressed as:

$$W = \sum_{i=1}^{n} H.\pi a^2 \tag{2.5}$$

By substituting Eq. (2.5) in Eq. (2.4) and rearranging the wear volume per unit sliding distance, it can be expressed as:

$$v = K_1 \frac{W}{3H} \tag{2.6}$$

These are the important factors: the applied load, the hardness, and the wear constant (K_1). If the total sliding distance is L, one can determine the overall volume worn out from a surface:

$$V = K_1 \frac{WL}{3H} \tag{2.7}$$

This way of expressing wear volume is the simplest because it does not require guessing the number of sharp edges touching, the shear strength of each elastic and plastic junction, the effect of lubricant thickness, surface roughness, and other factors. According to Archard, wear volume is directly proportional to sliding distance (L), directly proportional to applied load (W), and inversely proportional to the softer material's hardness. The constant K_1 is dimensionless and represents the probability of removing a wear particle, which may be considered an index of severity. When K_1 equals 1, every junction involved in a mechanical interaction will result in a wear fragment. This implies that 100% of the junctions coming into contact will contribute to wear. However, if K_1 equals 0.1, only 10%, or one-tenth, of the friction junctions produce wear fragments. Similarly, if K_1 equals 10^{-7}, it indicates that only one failure occurs out of every 10 million junctions. So, the goal is to achieve the lowest possible value for K1. Engineering weaker junctions to minimize K_1 would significantly benefit overall product design and surface engineering.

Localized stress and relative motion between two surfaces initiate adhesive wear, raising the question: what specific material is abrading from these surfaces? To answer this question, let us consider freshly manufactured surfaces, which often exhibit high surface energy. The environment's immediate interaction with surface energy is an advantage. Exposure to oxygen in the atmosphere results in the formation of a thin oxide layer on the surface of almost all the metal components. It is crucial to note that the surface is observed through the contamination (oxide) layer. For instance, iron, a widely used metal, has three oxides. To evaluate the wear volume of iron, it is important to know the type of iron oxide and its stoichiometry, which refers to the relative amounts of iron and oxygen in the compound. Iron oxide exists in various forms, such as FeO (iron(II) oxide), Fe_2O_3 (iron(III) oxide), and Fe_3O_4 (iron(II,III) oxide), each with different stoichiometries. We can calculate the volume of iron in the oxide based on its stoichiometry using the molar volume of each compound.

To reduce the wear constant, it is recommended to change the phases from iron to steel. Iron has fewer phases than steel due to compositional differences; steel is less prone to adhesive wear compared to iron. Therefore, using less pure metals is preferable, or else using a lubricant layer becomes necessary. The presence of a lubricant almost eliminates adhesive wear. If a lubricant is not feasible, creating an oxide layer using oxidizing agents is another option. The effectiveness of these measures varies depending on the specific situation. In some cases, oxidation is crucial, whereas in others, lubrication or hardness plays a key role.

To summarize adhesive wear, we often recommend reducing microstructure complexity to reduce adhesion. This emphasizes the importance of avoiding identical phases, as demonstrated in the comparison between steel and iron. The wear coefficient is commonly used as a metric to quantify the severity of adhesive wear. Materials with high hardness exhibit lower susceptibility to adhesive deterioration, as their hardness impedes microscopic contact between surface atoms and restricts the transfer of electrons between surfaces. We also discussed the importance of increasing the complexity of the structure to minimize adhesive wear. I provided an example comparing steel to iron, highlighting steel's superior performance in certain contexts. We also introduced the concept of the wear coefficient, a measure of the probability and severity of adhesive wear. A higher value indicates greater chances of adhesive wear, whereas a lower value is preferable for product durability. Therefore, selecting appropriate materials, increasing hardness, and aligning operating parameters with requirements are essential for minimizing wear.

2.4 EXPLORING ABRASIVE WEAR

In the preceding section, we examined adhesive wear, a phenomenon arising from the exchange of electrons between metal surfaces, leading to the formation of strong bonds. However, achieving the precise distance necessary for electronic transmission, typically spanning one to two atomic distances, poses a challenge without some degree of plastic deformation. Consequently, to mitigate such adhesion, maintaining high surface hardness and relatively low loads is imperative. This approach impedes deflection and deformation, thus averting close contact between surfaces and adhesive wear.

In the present heading, we are going to discuss abrasive wear, which is an extensively observed phenomenon in material degradation processes and is emerging as one of the main reasons for surface degradation. Sometimes, adhesive wear initiates the process by generating debris, which paves the way for abrasive wear to take root. Contributing factors, such as dust and environmental elements, further compound its occurrence. We classify the abrasive wear phenomenon as either "two-body abrasion" or "three-body abrasion" based on the number of surfaces involved.

2.4.1 Two-body abrasion

This type of wear entails the interaction between two surfaces, each characterized by their respective asperities. The application of a normal load causes harder asperities to penetrate softer surfaces, causing plastic deformation. In addition, the introduction of relative motion facilitates the displacement of embedded asperities, potentially leading to the formation of grooves and the accumulation of edges. Similar to cutting tools, the asperities of harder surfaces interact with softer ones, leading to the removal of material from the softer material surface and consequently an elevated wear rate. The overall wear rate is contingent upon factors such as the material's nature (ductile or brittle) and the pattern of asperities (conical, spherical, etc.). Tribological conditions (normal load and relative velocity) cause two main effects in ductile materials: micro-fatigue, which is marked by frontal buildup, and/or micro-cutting, which mostly causes grooves to form. When it comes to brittle materials, the basic processes are similar to those seen in ductile materials; however, the results are usually micro-fracture or the removal of grains from softer surfaces. This leads to four different types of abrasive wear: "micro-cutting," "micro-fatigue," "micro-fracture," and "removal of grains."

2.4.1.1 Micro-cutting

Abrasive wear in ductile materials stems from the sharpness of asperities, whereas sharper ones have the ability to penetrate softer materials, similar to the microscale action of metal chips. This process results in the chipping off of soft material (as shown in Figure 2.11), consequently contributing to the formation of wear debris. The extent of abrasion is contingent upon several factors, including the relative hardness of the surfaces involved, the relative motion between them, and the applied normal load. These factors together influence the degree to which abrasive wear manifests on the surfaces under consideration.

2.4.1.2 Micro-fatigue

The bluntness of the asperities primarily causes this type of abrasive wear in ductile materials. A blunt asperity hinders the cutting action of a ductile material, leading to the formation of built-up edges. However, over time, continuous relative motion causes cracks to form within or beneath the asperities. These cracks eventually propagate into fractures after repeated cycles (as shown in Figure 2.12), often resulting in the detachment of built-up edges. Consequently, the generation of loose particles or debris influences the overall wear rate, which is dependent on various operational parameters.

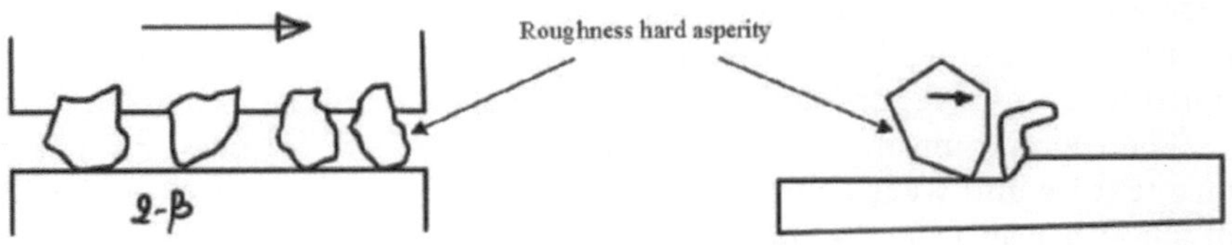

Figure 2.11 Illustration of micro-cutting wear phenomenon.

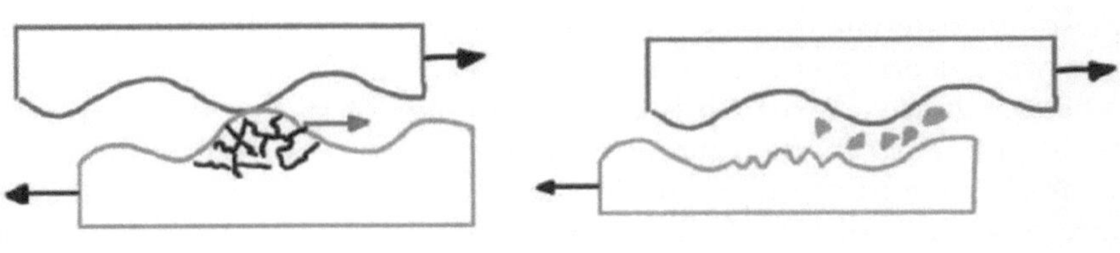

Figure 2.12 Illustration of micro-fatigue wear phenomenon.

2.4.1.3 Micro-fracture

Brittle materials exhibit this phenomenon, where fracture creates a fresh surface with varying degrees of smoothness or roughness, depending on the applied loading rate and other parameters. Initially, minute cracks develop beneath the surface, with higher loading rates increasing the likelihood of surface microfracture (pitting), ultimately resulting in the formation of micro-sized pits on the surface due to the fusion of small cracks. This mechanism also entails the relative sliding of brittle materials, generating micro-cracks beneath the surface. Subsequent sliding actions cause these micro-cracks to propagate and interconnect, eventually resulting in fragmented surface fractures.

2.4.1.4 Direct grain removal

In brittle materials like ceramics, grain removal happens because the boundaries between grains are not strong enough. This means that particles do not stick together well, which causes whole grains to come off the parent surface and be worn away as wear debris. Under operational wear conditions, this process can potentially lead to the complete removal of entire grains, hence the term "grain removal." It is worth noting that "wear rate" in this kind of mechanism depends on the "homogenous" or "heterogenous" phase (Figure 2.13). We observe a wear rate when working with a heterogeneous surface that contains grains of varying sizes, such as grain 1 > grain 2 > grain 3. Therefore, it is advisable to avoid a heterogeneous phase structure when subjected to two-body abrasion of ceramic-like materials. Hence, it is crucial to opt for a homogeneous phase. Furthermore, we prefer a small grain size. For example, when the grain size is less than 1 μm, there is a notable decrease in the wear rate. Conversely, an increase in grain size, for example, by 12 times, can lead to a surge in failures by 7 to 8 times. Therefore, it is imperative to strive for a low grain size, ideally as minimal as feasible.

When comparing abrasive wear and adhesive wear, differences arise in surface texture and wear debris characteristics. Adhesive wear usually results in smoother surfaces, producing flakes or plate-like shapes with minimal thickness. Abrasive wear, on the other hand, creates more irregular surfaces with particles of varying sizes because of cutting actions. Soft surfaces, particularly those susceptible to abrasion, exhibit pronounced irregularities as softer materials dislodge particles of varying sizes. In addition, adhesive wear debris tends to be flatter and more uniform with shorter cuts, whereas abrasive wear debris displays a range of shapes and sizes, often with greater hardness. Figure 2.14 consists of two subfigures illustrating wear characteristics on different materials. Figure 2.14(a) shows the surface appearance of cam after the removal of abrasive (due to fatigue action) wear debris. The highlighted region demonstrates surface roughness and the

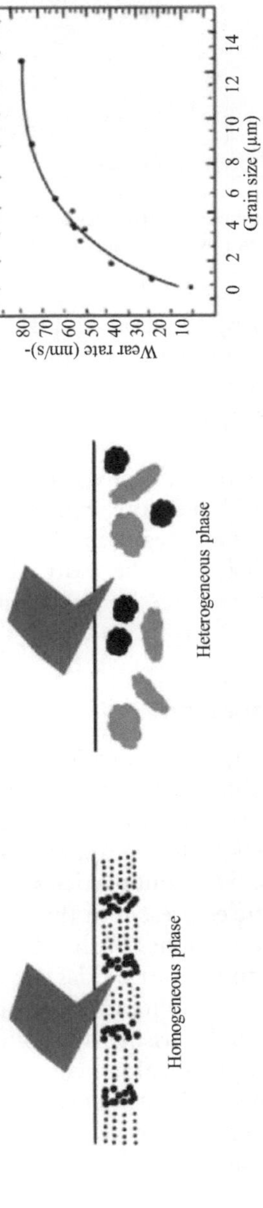

Figure 2.13 Depiction of homogeneous and heterogeneous phases, and the influence of grain size.

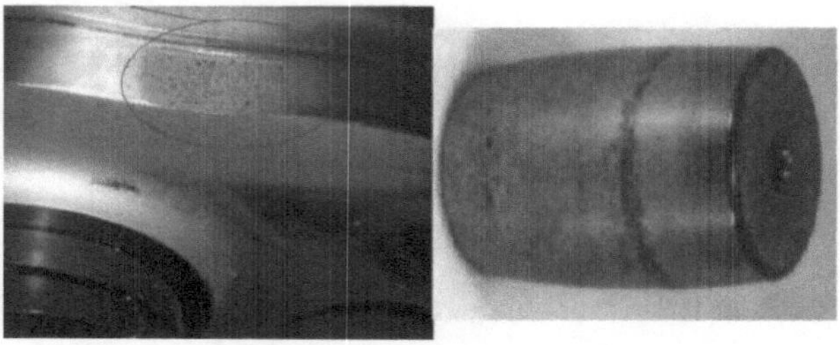

Figure 2.14 (a) Surface appearance post-removal of abrasive wear debris. (b) Surface appearance post-removal of adhesive wear debris.

formation of pits caused by material removal through mechanical action. Figure 2.14(b) depicts the surface appearance of roller (of the roller bearing) after the removal of adhesive wear debris. The cylindrical surface exhibits marks and smears caused by material transfer and localized bonding during sliding contact, typical of adhesive wear.

Real-world scenarios frequently involve a combination of these debris types, leading to a hybrid of wear mechanisms. Furthermore, particles dislodged from surfaces may reattach through welding, illustrating the adhesive wear characteristics. Furthermore, dislodged particles can transition from two-body wear to three-body abrasion, further complicating the wear process.

2.4.2 Three-body abrasion

Unlike two-body abrasion, which involves only two surfaces, three-body abrasion introduces the additional particles' surfacing, altering the dynamics of wear. Several factors contribute to the complexity of three-body abrasion. First, the third body refers to foreign particles introduced during the process. These can be oxides, worn-out particles generated by adhesive or abrasive actions, or hard particles present in the environment. The composition of hard or soft materials determines the hardness of this oxide. For instance, in fretting fatigue wear, some iron oxides act as solid lubricants, whereas others become rigid and contribute to three-body abrasion. We will delve deeper into this topic when discussing fretting fatigue wear.

These particles' shape and size, along with operating parameters like load and sliding, intricately link to the damage they cause. The characteristics of abrasive particles generated during wear, particularly their shape (whether round or sharp), have a significant impact on their abrasive nature. Even particles of the same size, such as 50 μm, can vary in abrasiveness based on their shape. Initially, sharp-edged abrasive particles tend to round off with use, gradually smoothing their edges and increasing their roundness factor, leading to less severe abrasion compared to sharp-edged particles, which inflict more damage. These particles adapt to situations with less resistance or scratching, sometimes proving less harmful than foreign particles. For instance, silica dust particles, with a hardness of 1100 Vickers hardness (VHN), are harder than steel surfaces and cause more severe wear.

Research reveals that three-body abrasion is actually less harmful than two-body abrasion, despite initial concerns about its severity. This is because the particles are not confined to either surface 1 or surface 2; they can roll and slide, with approximately 80% of the kinetic energy being consumed in rolling motion. Consequently, the third body adopts a rolling action, which lowers the sliding-to-rolling ratio to less than 0.2, thereby reducing the severity of three-body abrasion. For instance, while the wear coefficient for two-body abrasion ranges from 5×10^{-3} to 50×10^{-3}, under similar conditions, this coefficient might reduce to approximately 10% to 20% for three-body abrasion. Examples of three-body abrasion include: (1) earth-moving equipment; (2) slurry pumps; (3) rock drilling; (4) crushers; (5) pneumatic transportation systems; (6) dies used for powder metallurgy purposes, etc. Given the significant costs associated with maintaining such equipment, addressing three-body abrasion issues is crucial. In addition, failures due to two-body abrasion can eventually escalate into three-body abrasion scenarios. Therefore, a comprehensive understanding of three-body abrasion and conducting practical experiments is essential for mitigating such failures.

To mitigate abrasive wear, it is necessary to develop and implement strategies that adjust and enhance resistance. An effective approach involves formulating an abrasive wear equation, which is essential for quantitative analysis, prediction, and optimization of wear behavior.

2.4.3 Wear equation for two-body abrasion

Let us attempt to quantify two-body abrasive wear. Here, we will delve into a detailed discussion of an equation that Rabinowicz introduced. Rabinowicz assumed a conical asperity, as shown in Figure 2.15, with an angle α, making the total asperity angle 2α. Under load, the hard asperity penetrates the soft surface to a depth of x, as illustrated in Figure 2.15. Rabinowicz's assumption is that, under load, this conical asperity indents the soft surface, and, during sliding motion, the asperity can move either to the right or left, depending on the circumstances.

If we assume there are "n" asperities penetrating into soft materials and each asperity has the same cone radius, the same depth of penetration, and the same cone angle. Furthermore, each asperity experiences a load W_i (where i ranges from 1 to n); this load is equivalent to the hardness of the soft material (H) multiplied by the contact area. We assume that the soft material has a specific area of penetration, represented by the cone's diameter

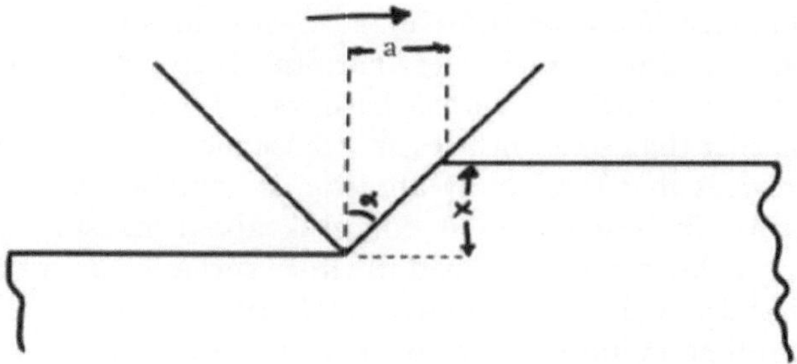

Figure 2.15 Penetration of a conical asperity into a soft surface.

of 2a. Denoting the radius as "a," the area is πa^2. Therefore, the average contact area turns out to be $0.5 \times \pi a^2$. The ith asperity sustains the load when we multiply this by the hardness, as expressed in following equation.

$$W_i = H^*(0.5\pi a^2) \tag{2.8}$$

$$\text{Rearranging,}\quad a^2 = W_i/(H^*0.5^*\pi) \tag{2.9}$$

To estimate the wear volume, there is a need to find the volume swept by the asperity penetrating the soft surface:

$$\Delta v = a.x.L \tag{2.10}$$

This volume is calculated by multiplying the radius by the depth of penetration and the sliding distance. As illustrated in Figure 2.15, we can express "x" in terms of "a" and α:

$$X = a/\tan\alpha.$$

On substituting the expression of x in Eq. (2.10):

$$\nabla v = a.a/\tan\alpha.L \tag{2.11}$$

On substituting the expression of a^2 in Eq. (2.11):

$$\nabla v = W_i/(H^*0.5^*\pi^*\tan\alpha).L \tag{2.12}$$

Total wear is the sum of the wear caused by individual asperity:

$$V = \frac{L\sum_{i=1}^{n} W_i}{0.5H\pi\tan\alpha}$$

$$V = \frac{LW}{0.5H\pi\tan\alpha}$$

$$V = \frac{2}{\pi\tan\alpha}\frac{WL}{H}$$

$$V = K\frac{WL}{H} \quad (2.13)$$

It is essential to note K $(= \frac{2}{\pi\tan\alpha})$ can be regarded as a wear coefficient, analogous to the wear coefficient in adhesive wear. The principles applied to adhesive wear are also applicable to abrasive wear. In both cases, the total wear from a surface is proportional to the normal load and inversely proportional to hardness, while being proportional to the sliding distance. In addition, the microstructure, specifically the asperity angle (α), influences abrasive wear.

If we calculate the value of K in terms of α, we can consider various angles: 30 to 80 degrees. This range represents a progression from sharp to flat, eventually approaching a round shape of asperity. A round surface typically requires minimal cutting and is more resistant to wear over many cycles. For instance, with an angle of 80 degrees and an overall angle of 160 degrees, the wear coefficient is just 0.11. Conversely, in a case with an angle of 32.5 degrees (total asperity cone angle of 65 degrees), the wear coefficient increases to 1. The Rabinowicz equation might give more accurate predictions of wear volume than expected because initially sharp particles become less sharp over time, and micro-cutting changes into micro-fatigue, which lets materials last through more than one cycle.

This example illustrates how the wear coefficient changes dramatically as the angle α increases. Therefore, it is crucial to design surfaces that avoid sharp asperities to minimize abrasive wear. In addition, wear debris generated during abrasion action can increase in size due to adhesion, and controlling this particle size growth is challenging. Initially, debris might be 1 μm, but continuous sliding can enlarge to 3 μm, 5 μm, 10 μm, or even larger if particles adhere due to high surface energy. To minimize damage from three-body abrasion, it is essential to consider particle size and clearance. When the clearance surpasses the particle size, damage decreases. Filtration plays a crucial role in minimizing wear by removing particles from the interface as they are generated. One approach is to texture the surface to create spaces where particles can be trapped as they roll.

Abrasive wear is influenced by several factors, including the roundness of the particle shape, the presence of lubrication or an oxide layer, the effect of the particle on the surface hardness, the ductility or brittleness of the metal, etc.

2.4.4 Effect of roundness factor on abrasive wear

Typically, wear debris in abrasive wear comprises angular and uneven fragments. As abrasive particles scour the surface, they dislodge small, angular fragments, constituting wear debris. Angular particles may excel in material ablation, whereas circular particles often yield a refined surface texture. The wear rate of abrasive particles and the abraded surface can be influenced by the roundness factor, which is expressed as:

$$\text{Roundness factor } F = \frac{4\pi A}{P^2}; \text{ where } P = \text{Perimeter}$$

The "roundness factor" stands out as a crucial measure for assessing the geometrical characteristics of abrasive particles, encompassing variations from circular to angular to irregular shapes. Abrasive particles with a higher roundness factor may present advantages in specific scenarios, such as facilitating a smoother surface finish or reducing vulnerability to surface impairment. Sometimes spherical particles stick to a surface and cause abrasion. Abrasive wear by spherical particles can manifest under sliding conditions, such as during braking, as evidenced by the incorporation of spherical particles into the brake disc after undergoing plastic deformation, thereby transitioning abrasion from three-body to two-body interactions.

2.4.5 Effect of lubrication on abrasive wear

Lubrication is critical to mitigating abrasive wear because it reduces direct contact between surfaces and acts as a barrier that minimizes the interaction of rough surface imperfections. The flow of lubricant helps to remove abrasive particles, preventing them from embedding and causing further damage. For instance, cylinder and liner surfaces are susceptible to abrasive wear when hard particles, such as dust or debris, infiltrate the lubrication system, resulting in scrapes and dents on the metal surfaces. To address this issue, high-quality lubricants that can withstand the operational demands of the engine are essential for slowing the wear rate. In addition, regular maintenance practices, such as oil changes and filter replacements, are crucial in postponing wear.

2.4.6 Effect of hardness on abrasive wear

When hard abrasive particles engage with a softer material, they can plough, cut, or tear the surface, resulting in the formation of irregularly shaped debris with sharp edges. Silica, being one of the most common abrasives, boasts an approximate Vickers hardness (VHN) of 1100, posing a considerable challenge for mitigating abrasive wear in most practical engineering materials. Conversely, when abrasive particles are softer than the material being abraded, they tend to deform and round during the wear process, producing smoother, more rounded wear debris. Abrasive wear halts entirely when the material's hardness surpasses the particle's hardness by a factor of 1.2; however, some scratching may occur when the hardness levels are comparable. Implementing hard coatings with a hardness exceeding 1.2 times that of the particles presents an effective solution for combating abrasive wear. It is important to note that an excessively high hardness level may not significantly decrease the rate of wear. For instance, the literature, such as the ASM handbook, provides an example of how to treat an automobile engine camshaft. Here, they deliberately avoid full-hardening processes. Why? Such procedures may cause cracking or reduce the material's toughness, thereby increasing the wear rate. Although wear equations may imply that increased hardness correlates with better outcomes, exceeding a certain threshold of hardness jeopardizes the material's toughness, transitioning it from ductile to brittle.

In the context of two-body abrasion, maintaining a hardness ratio close to one between two surfaces minimizes the wear rate, typically exhibiting a ±20% variation in experimental data within this range. However, a significantly high hardness ratio between surfaces, where one surface's hardness predominates over the other, can lead to a rapid escalation in the wear rate. Sometimes, introducing a sacrificial element allows for the use of an economically feasible material with lower hardness, making it more affordable and easier to replace as needed. This method is frequently utilized in bearings, especially those with extended stainless-steel shafts that are expensive and labor-intensive to replace or reassemble. Nonetheless, when using sacrificial bearing materials, lubrication becomes critical to minimize wear and maintain peak performance. This underscores the critical importance of maintaining a balanced hardness ratio in two-body abrasion scenarios.

2.4.7 Examples of abrasive wear

Now, let us consider a few examples of abrasive wear. We often observe two-body abrasion in manufacturing processes, industrial equipment, and automotive parts. This mechanism is prevalent across various machines and is a dominant failure mechanism among wear mechanisms.

Figure 2.16 Illustration of two-body abrasion.

For instance, when using emery paper (as depicted in Figure 2.16), many individuals have encountered the process of material removal from a surface. This exemplifies a form of two-body abrasion, in which the surface asperities of the emery paper engage with a soft surface, resulting in material removal and deposition onto the paper's surface. The extent of abrasion is contingent upon factors such as velocity, applied pressure, hardness, surface texture, the presence of lubricants, and contaminants.

Dust and dirt particles infiltrate both the brake pad and the disc during braking, fueling abrasive wear that results in high-stress conditions and irregular control, underscoring the need for preventive measures. Similarly, three-body abrasion in cars, especially internal combustion engines, often results in lost efficiency due to scratches on main bearings and piston liners caused by debris in the lubricating oil. This shows how important it is to keep the surface texture and oil quality in good shape for the best performance.

2.4.8 Epilogue

To summarize, in ductile materials, wear mechanisms include micro-fatigue and micro-cutting, whereas in brittle materials, micro-fracture or grain removal is considered. Regardless of the material type, wear debris generation occurs, eventually leading to a three-body abrasion scenario where the wear particles act as a third body after initially discussing two-body abrasion. Another noteworthy aspect of understanding wear mechanisms is the similarity between abrasive wear and two related mechanisms: erosion wear and cavitation wear. The resemblance between abrasive wear and erosion wear stems from the particles' interaction with a surface and velocity. In erosion wear, fast-moving particles hit at different angles, called the angle of attack. This causes erosive wear, and the particles that are thrown out look like those that are thrown out in abrasive wear, showing that the two types of wear are very similar in many ways. The main difference is the kinetic energy of the particles involved. In erosive wear, particles hit materials at different speeds, which could lead to different wear mechanisms for materials that are flexible and those that are hard. Similarly, cavitation wear involves liquid particles hitting soft surfaces due to negative pressure, with the collapse of liquid bubbles generating high-pressure impacts that remove material from the surface. It entails the formation and bursting of bubbles due to negative pressure, resulting in high-pressure fluid impacting and damaging the surface, akin to abrasive wear mechanisms. However, we will discuss these topics—erosive wear and cavitation wear—in detail in subsequent subheadings, highlighting their similarities with abrasive wear.

2.5 DIMENSIONLESS PARAMETERS IN WEAR PHENOMENA

In the previous sections, we examined two significant wear mechanisms: adhesion and abrasion. Adhesion occurs when two surfaces come into close contact, resulting in bonding forces that cause material to be removed or transferred from one surface to another. This process generates wear debris and material loss, particularly prevalent in metal-to-metal contacts. On the other hand, abrasive wear occurs when hard particles or asperities on one surface contact, penetrate, or sever a softer surface, resulting in the production of wear debris and material loss. Factors, such as the size, shape, and concentration of abrasive particles, applied stress, and the relative motion of surfaces, influence the extent of abrasive wear.

Adhesion and abrasion often occur simultaneously and interact with each other. For instance, the presence of abrasive particles can accelerate adhesive wear, while adhesive wear can produce debris that initiates abrasive wear. In adhesive wear, the bonding force between two surfaces is crucial, leading to the formation of a plastic zone or micro-welding. Applying relative motion to one of the surfaces can cause material to transfer or detach from the softer surface and attach to the harder one, leading to the formation of wear debris. This interaction underscores the complexity of wear mechanisms in practical applications. We can utilize the wear coefficient (K) in relation to specific wear phenomena to simplify this complexity and ensure proper wear comparison and analysis. The wear coefficient is inherently subject to probability; it fluctuates and must consistently remain below 1. We recognize that various factors, often challenging to accurately quantify, influence the complexity of wear processes. Consequently, wear phenomena inherently possess a statistical nature, making the wear coefficient a reflection of this statistical essence. Utilizing the wear coefficient facilitates the attainment of consistent results, albeit recognizing its nature as a comparative tool rather than an absolute measure. This allows for qualitative comparisons between different materials or lubricants, providing valuable insights through the use of non-dimensional parameters.

When comparing different types of wear, such as adhesive and abrasive wear, expressing them as dimensionless parameters allows for a quantitative evaluation of wear severity across different systems. There are several advantages to using wear coefficients:

- Comparing wear coefficients provides insights into relative severity and material loss impact.
- Knowledge of wear coefficients aids in selecting appropriate materials, coatings, lubricants, and surface treatments to mitigate wear.
- Wear coefficients help optimize maintenance intervals, enhance product reliability, and reduce downtime.
- Wear coefficients are important because they eliminate the influence of specific test settings, allowing for consistent comparisons across different testing techniques and setups.
- Product reliability can be related to the wear coefficient. A very low wear coefficient improves product reliability. This decrease in downtime leads to cost savings, rendering it economically advantageous.
- This approach offers the following benefits:
    - Providing numerical values for quantitative assessment
    - Ensuring repeatability
    - Optimizing performance
    - We are offering a standardized method for evaluating and comparing the performance of lubricants and materials.

Using dimensionless wear coefficients from a pin-on-disc tribometer and a four-ball tester effectively allows for a standardized and quantitative comparison of lubricant and material performance. This method facilitates the ranking and evaluation of various lubricants and materials based on their efficacy, with lower coefficients indicating better performance. We can assess various lubricants based on their performance under identical conditions, even when we use the same material.

Figure 2.17 shows a modified bathtub curve, with zones A to B and C to D identifying areas experiencing rapid wear. Line B to C exhibits comparatively stable wear with very minimal slope (gradual increase). To ensure consistency, it is critical that the wear constant K corresponds primarily to the BC zone. K represents a probability (necessitating multiple experiments to yield an accurate result of K). For instance, if K equals 0.1, it means that one-tenth of the junctions in contact will generate wear debris. A high wear coefficient indicates a notable wear rate, potentially prompting the need for

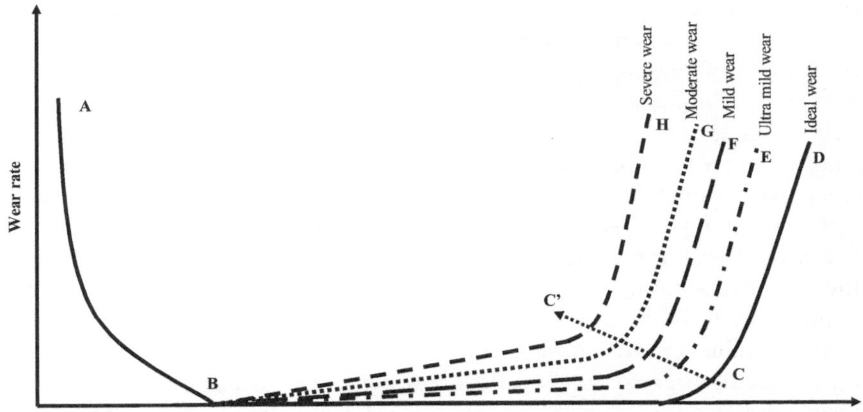

Figure 2.17 Modified bathtub curve.

design alterations, material enhancements, temperature adjustments, or the implementation of coatings or lubricants to alleviate wear-related issues.

On the other hand, if K is equal to 10^{-7}, only one wear fragment will occur in every 10 million contacts or repeated interactions. It is advisable for the wear factor to be 10^{-7} or even lower (often representing the micro-fatigue failure process), as it ensures prolonged product life. One can reduce the wear coefficient by targeting values ranging from 10^{-14} to 10^{-16}, thus entering the realm of ultra-mild wear. It is important to remember that wear is inevitable when two surfaces come into contact. Our objective is to reduce wear through a combination of engineering and scientific endeavors, recognizing that the wear coefficient (K) is influenced by various factors, including sliding speed, volume alterations, hardness, temperature, and vibration. Interestingly, a lesser or slower speed can sometimes correspond to higher wear coefficient values. For instance, certain brake materials exhibit accelerated wear at lower speeds. Despite the common understanding that higher speeds increase sliding distance and wear, this is not always the case. Unique configurations and grain structures can lead to increased wear at lower speeds, underscoring the importance of understanding various wear mechanisms.

2.6 EROSIVE AND CAVITATION WEAR

Erosive and cavitation wear involve the erosion of material from a solid surface due to the impact or impingement of solid particles or fluid jets, resulting in surface deterioration. Abrasive wear, which shares similarities with erosive and cavitation wear, occurs when solids or solid particles slide between each other. Two different types of wear are erosive wear, which is caused by particles hitting at high speeds with a lot of kinetic energy, and cavitation wear, which is caused by bubbles forming and collapsing in a liquid and applying high-pressure forces to the surface. In erosive wear, the consideration of loose impacting particles, akin to three-body abrasive wear, is vital. These particles, transported by the liquid or gas, possess velocities often exceeding 10 m/s. Moreover, a critical aspect is the 5 μm particle size; particles smaller than this typically do not cause severe erosive wear. Impingement angles also play a crucial role, influencing the type and severity of erosive wear. For instance, small impingement angles, like 5 degrees, elicit different effects compared to angles of 30 or 60 degrees. Various factors, including material properties, particle size, shape, velocity, and exposure duration, influence erosive wear. The ductile mode maximizes impact energy at an angle of approximately 30 degrees, while the brittle mode peaks significant erosive wear at a 90-degree angle. Larger particle sizes exacerbate wear due to increased impact force.

Understanding the science behind erosive and cavitation wear allows us to apply this knowledge to develop superior products for society. This raises the question: What precautions are essential to withstand erosive and cavitation wear? To address this, it is necessary to develop an erosive wear equation that incorporates impact angle, particle velocity, and particle size, with the impact angle determining the extent of moment transfer. The wear rate due to erosion (V_e) can be expressed as:

$$V_e = K * A * \alpha * i(v) * M \tag{2.14}$$

$$i(v) = (\text{particle\_vel})^n \tag{2.15}$$

where n is a constant that depends on the material:

- n = 2 to 2.5 for metals
- n = 2.5 to 3 for ceramics

$$M = (\text{particle\_size})^3 \tag{2.16}$$

These equations can be combined to express V_e as a function of the particle velocity and size (combined equation):

$$V_e = K * A * \alpha * (\text{particle\_vel})^n * (\text{particle\_size})^3 \tag{2.17}$$

where K represents the wear coefficient and signifies the probability of material removal. In the erosive wear equation, the wear rate and impact velocity have a power law relationship; v denotes the impact velocity in meters per second (m/s), and n represents the velocity exponent. The last

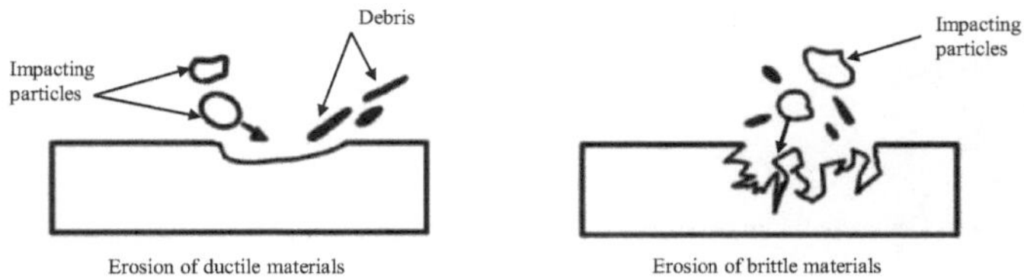

Figure 2.18 Illustration of erosive wear.

parameter, M, represents the particle size. Doubling the particle size can potentially increase the wear rate by a factor of 8. Doubling the particle size can significantly impact the wear rate.

Depending on the nature of the impact, whether it is direct, normal, or glancing, various wear mechanisms manifest. Direct impacts typically induce plastic deformation, surface fracturing, and material ejection, whereas glancing impacts generate cutting actions that lead to material removal along the particle's trajectory. The velocity of impact particles significantly influences energy dissipation, contingent upon the material's behavior. Brittle materials frequently exhibit radial crack formation, as depicted in Figure 2.18, whereas ductile materials undergo plastic deformation and surface cutting, as illustrated in Figure 2.18. The study of these phenomena is imperative due to their widespread occurrence in nearly all mechanical systems. Systems with relative movement are inevitably exposed to dust particles, with gas turbine blades being a notable example, frequently failing due to particle impingement or erosion. Slurries represent another common scenario where high-hardness particles, combined with impingement angles, can result in pump failures, impeller damage, and housing deterioration. Figure 2.19 demonstrates the maximum energy transferred to impacting surfaces.

One has probably encountered mountains sculpted by water over time. This natural process, known as water erosion, is widespread in nature. With its persistent flow, water gradually wears away rocks and various materials. Rock, being a harder material, often becomes brittle. This brittleness reduces resistance to erosive wear, potentially causing fragmentation and groove formation, ultimately increasing failure rates. Essentially, erosive wear shares similarities with abrasive wear, which, as previously discussed, involves surface abrasion processes. However, we categorize erosive wear differently because it emphasizes the effective energy transfer to the surfaces of the objects under consideration, unlike abrasive or adhesive wear, which emphasizes load. It is essential to note that in studying rock wear, hardness is not the primary factor; instead, high toughness is crucial. Adequate toughness, surpassing a minimum requirement, can effectively prevent or mitigate erosive wear.

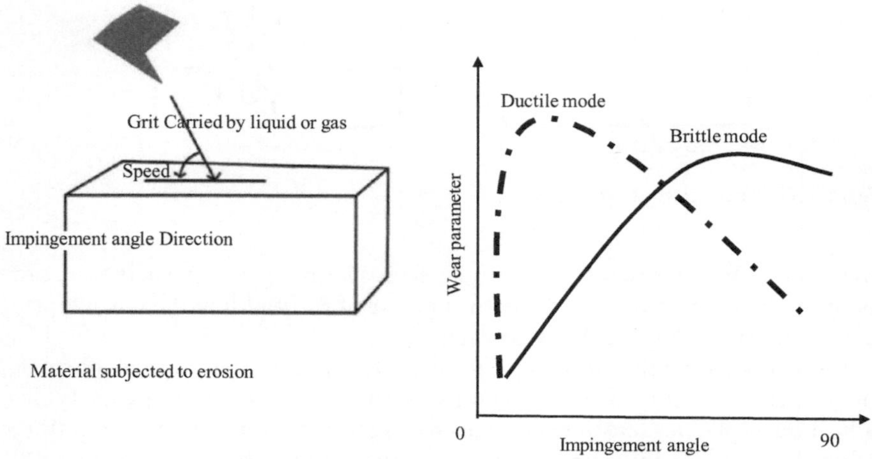

Figure 2.19 Impact angles and erosive wear in ductile and brittle modes.

In the case of liquids, the wear rate is often proportional to V^5. Consequently, fluid erosion can be more destructive compared to solid erosion. Rubber, known for its elasticity, can absorb even more energy, sometimes outperforming materials like ceramics. Thus, in certain situations, opting for rubber instead of ceramics or other sturdy materials can be advantageous. However, if the energy surpasses a certain threshold, the rubber may tear apart, accelerating wear. Hence, to effectively apply our scientific knowledge, understanding the threshold limit—how much energy rubber can truly absorb—is crucial. Each type of rubber will have its own unique threshold limits, necessitating a tailored approach to part design. Often, a simple rubber coating, even just a few millimeters thick, can significantly enhance the overall system design.

The erosive wear caused by liquid particles can lead to severe failures, particularly when these particles are propelled at high velocity. This phenomenon is evident in water jet cutting machines, renowned for their ability to cause brittle failure. However, this method also enables precise machining, facilitating the creation of high-quality products. The wear rate in cases of liquid particle erosion is highly sensitive, following a power law (V^5), making it far more sensitive compared to the velocity required for solid erosion, which follows V^2. The distinction between V^5 for fluid erosion and V^2 for solid erosion is crucial.

When discussing these phenomena, one notable example is aviation. Aircraft frequently traverse through clouds containing diverse particles, including water droplets, at high relative velocities as they speed from below to above the clouds. This occurrence, termed fuselage erosion, presents a significant concern in the aviation industry. Fuselage erosion is the gradual loss of material from the aircraft's fuselage surface due to liquid particle wear. It evolves over time, influenced by various weather conditions such as high-velocity winds, rain, sand particles, and other foreign objects. Addressing fuselage erosion often involves implementing design changes such as corrosion-resistant and erosion-resistant coatings. These measures effectively mitigate erosion damage, thereby prolonging the aircraft's lifespan.

Once again, the presence of water, which triggers corrosion processes, compounds the threat of corrosion and remains a primary concern. These occurrences can manifest sequentially or concurrently, influenced by factors such as erosive wear and environmental conditions. To mitigate such risks, regular non-destructive inspections are indispensable to identify minute cracks that could evolve into brittle failures. Consistent non-destructive testing (NDT) examinations facilitate early erosion detection, allowing proactive replacement of components before major failures ensue. Hence, timely action or consideration of coating adjustments or reconditioning becomes essential when warranted. This comprehension bears significant importance, especially within the aviation sector, where costs are substantial. Even minor failures can lead to considerable expenses, underscoring the need for robust preventive measures.

Figure 2.20 Water-induced erosion on rock.

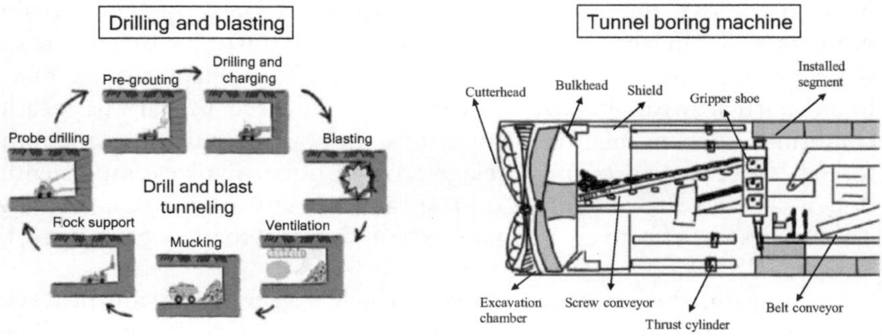

Figure 2.21 Different types of machines for tunneling purposes.

An essential inquiry emerges: are factors associated with erosion solely detrimental, or could they also yield benefits? Can they result in significant failures, or might we utilize this knowledge to improve product development? Let us consider a positive perspective and explore the water-assisted tunneling method. This method is relatively novel, currently under exploration by researchers, and holds the potential for substantial advancements. So, what exactly is tunneling? We have encountered numerous tunnels across different regions, often indicating developmental progress.

Various machines are used for tunneling (as shown in Figure 2.21), with drilling and blasting and the Tunnel Boring Machine (TBM) being two common methods. TBMs, albeit effective, are costly and primarily suited for tunnels exceeding 2 km in length. For tunnels shorter than this, such as those spanning 600 m to 1.5 km, drilling and blasting are the more prevalent approaches. This technique involves drilling holes into the rock or designated tunnel area, followed by grouting and the placement of explosive charges before initiating the blasting process.

The blasting process will generate a significant number of particles, which will impact the environment. Therefore, ventilation becomes critical to ensuring a clean work environment and personnel safety. Additionally, mucking is necessary to clear away all debris from the site, either transporting it to a distant location or to a designated landfill area. Following this, rock support becomes imperative, often involving concreting or reinforcement to stabilize the tunnel. This process continues in a cyclical manner, with an average advancement rate of about 3 m/day.

The "drilling and blasting" method, suitable for boring shorter tunnels, is environmentally taxing. In addition, blasting can induce deep-stress fractures. Even if we successfully construct and reinforce a tunnel with concrete today, these deep-seated cracks may lead to its failure within 3 to 5 years. In contrast, the TBM offers a more automated approach. The

TBM uses a large cutter head to slice through the rock, collecting debris and conveying it back for disposal. Another conveyor system transports support rings to fortify the tunnel's surface. While the TBM proves efficient, its implementation comes with higher costs, especially for shorter tunnel lengths, typically those less than 2 km.

In many developing countries, there is often little need for tunnels exceeding 2 km. So, what is a viable solution for constructing tunnels without causing environmental harm? One such method involves using water jet cutting to construct tunnels. We know that water has the ability to penetrate even sturdy rocks because of its capacity to induce brittle fractures. This occurs as the high-velocity water creates a zone of crushing force, initiating fractures. While operating at high pressure, the water jet presents a comprehensive solution. Mechanical rock-cutting techniques can be supplanted by water jet cutting due to its efficacy. To effectively cut through rock with a water jet, the pressure must slightly exceed the threshold pressure. Ideally, the pressure should be approximately 20% to 25% higher than the erosion resistance of the rock; microcracks will occur, facilitating easier grain removal. This ensures there is sufficient pressure to overcome resistance without inducing excessive cracking.

In addition to pressure, the flow rate significantly influences the cutting process, and its precise management ensures an optimal depth of cut. Higher flow rates enable deeper cuts, accommodating customized designs when necessary. Furthermore, the loose particles produced during cutting can be repurposed to enhance erosive wear. Because there are no blasting methods, there is no demolition debris, making the process more environmentally friendly, cost-effective, and efficient.

Moreover, slabs removed during tunnel cutting can be repurposed for development purposes. For example, at designated points such as the tunnel entrance and exit (marked as points 1 and 2), a water jet can carve out required rectangular shapes or other specified forms (as illustrated in Figure 2.22). These segments can then be sliced into sections ranging from 1 m to 10 mm thick, depending on specifications, simplifying slab removal. This approach reduces pollution while expediting the process, as only stones need shaping and direct transportation to desired locations, eliminating the necessity to remove muck. Ultimately, this method facilitates the conversion of rocks into tiles, streamlining their transportation to their intended destinations. Altogether, this approach represents a win–win scenario.

Wang et al.[1] conducted experiments on the rock-breaking capabilities of high-pressure water jets when coupled with TBMs. Their research revealed that integrating high-pressure water jets enhances TBM excavation efficiency while reducing disc cutter forces and wear rates. The simulation yielded optimal results with a nozzle diameter of 1.5 mm and a jet velocity of 224.5 m/s.

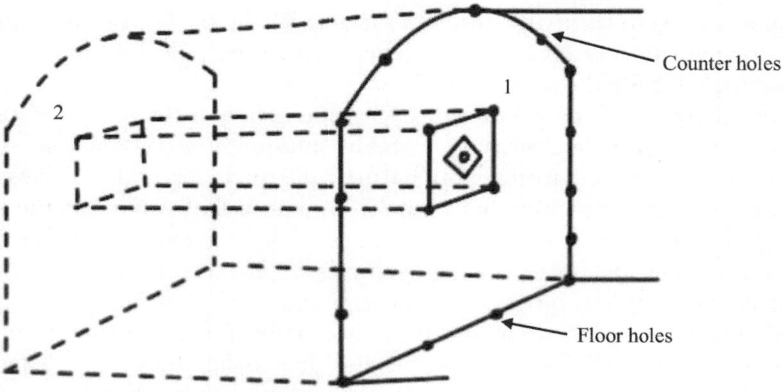

Figure 2.22 Isometric view of tunneling process.

Some researchers advocate using a diamond wire setup to cut the tunnel's back slice, as water may encounter obstacles on this side despite its smooth passage through the front. In such scenarios, combining the diamond wire cutting method with water jet cutting can offer an effective solution. Those interested in delving deeper into this subject can find additional insights by visiting relevant websites, as ongoing research continually enhances this technology. In the future, we may witness fully validated technologies for tunnel construction utilizing erosive wear mechanisms.

2.6.1 Evaluating erosive resistance: Experimental setups and considerations

To assess erosive resistance and material durability against erosion, various experimental setups are available. One such method (shown in Figure 2.23) involves using a compressor to propel gas or air, simulating erosive conditions, and evaluating material resistance. While an air compressor is typically used for these experiments, a pump can be used if a liquid carrier fluid is preferred, enabling the use of process fluid. Adjusting the impingement angle, which ranges from 10 to 50 degrees, is also possible by changing the sample's position. This versatile setup permits experimentation with different particle numbers, sizes, airflow rates, and angles.

Initially, calibrated data serves as the foundation, specifying parameters such as a pressure of 0.15 bar, a flow rate of 12 g/min, and a velocity of

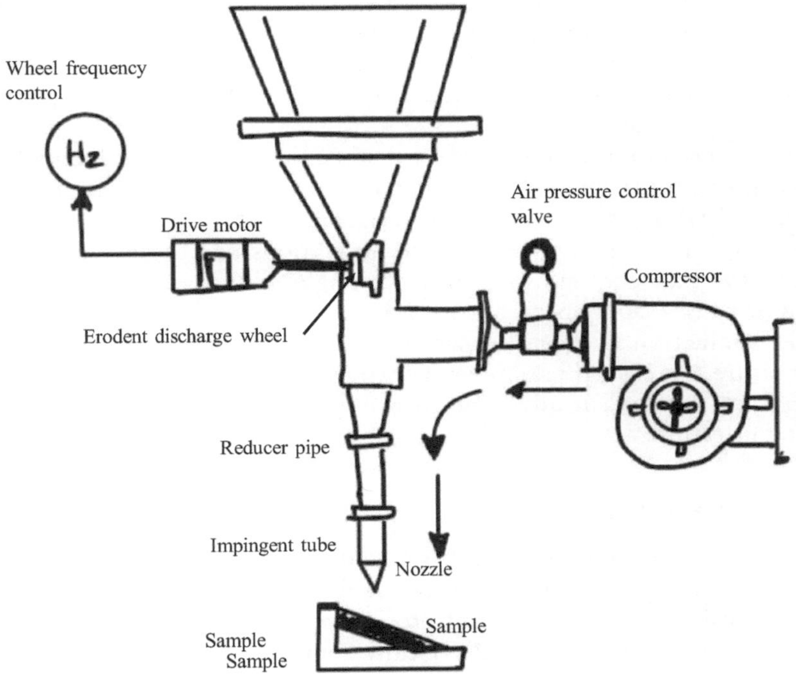

Figure 2.23 Experimental configuration for assessing material erosion resistance.

60 m/s, adhering to American Society for Testing and Materials (ASTM) standards. These standards offer guidance for establishing pressure settings and determining the necessary flow rate, which may vary depending on specific ASTM requirements.

The preparation of the test specimen involves mounting it in a holder and directing the flow onto it using a nozzle, allowing experiments to be conducted at various angles from 10 to 90 degrees. These experiments assess the material's resistance to environmental factors, particles, flow rate, or process fluid, thereby facilitating the calculation of erosion rate and wear coefficient. This methodology allows for the comparative ranking of various materials, which requires repeating each test 2–3 times to ensure accuracy. In a single set of experiments, multiple repetitions under the same conditions are imperative to ensure repeatability and minimize noise, yielding more accurate results.

Such setups allow for the study of the impact of the impingement angle on ductile materials. For instance, when alumina particles (with a mean size of 50 µm) are directed at a 10-degree angle, a notable spread in the erosion area, characterized as cutting wear, is evident. Similarly, at 75 degrees, the spread of erosion narrows, suggesting reduced severity. At 90 degrees, the erosion area notably reduces compared to the 10-degree angle. Deeper penetration results in crater formation.

Materials need to be carefully selected based on the impingement angle. At 90 degrees, brittle materials (as shown in Figure 2.20) are unsuitable, whereas at 30 degrees, even ductile materials may exhibit high wear rates. The selection process involves considering various curves (as shown in Figure 2.19) that correlate impingement angle with material behavior. In addition, this equipment allows for the testing of coatings, enabling the evaluation of material performance under various conditions. Temperature is a crucial parameter as material behavior changes at high temperatures, making it advisable to conduct experiments at the operating temperature.

2.6.2 Cavitation wear

Cavitation wear, like erosive wear, has several similarities. In this scenario, particles form in a liquid state rather than a solid state. It is important to note that liquid can break down into numerous streamers or particles, posing a concern. Additionally, every fluid typically contains some amount of gas, usually around 0.1% to 0.2% under normal conditions, which is typically not worrisome as the gas remains compressed. However, bubbles can experience intense pressure, especially in convergent–divergent domains. The schematic illustration (shown in Figure 2.24) depicts a pressurized fluid–air environment with a gas bubble. When this bubble contacts a surface, it experiences extreme pressure from all angles, leading to its eventual bursting. When the bubble bursts, it releases wave energy and disintegrates into numerous streamers and particles, effectively impacting the surface and removing debris in the form of fractures.

It is to be emphasized that the formation of fractures does not happen abruptly in a single cycle; it typically requires numerous cycles of fluid impact to manifest. Cavitation initiates as a gradual process, beginning with the formation of small pits on the surface, which then progressively expand to cover a significant area. Although these pits may not immediately result in catastrophic failure, they signify a notable weakening of material strength. A multitude of pits indicates a substantial decrease in material integrity, warranting replacement. Although cavitation does not lead to instantaneous failure, its cumulative effect over time ultimately causes material failure. This phenomenon arises from the formation of bubbles in low-pressure regions, which gradually grow and coalesce.

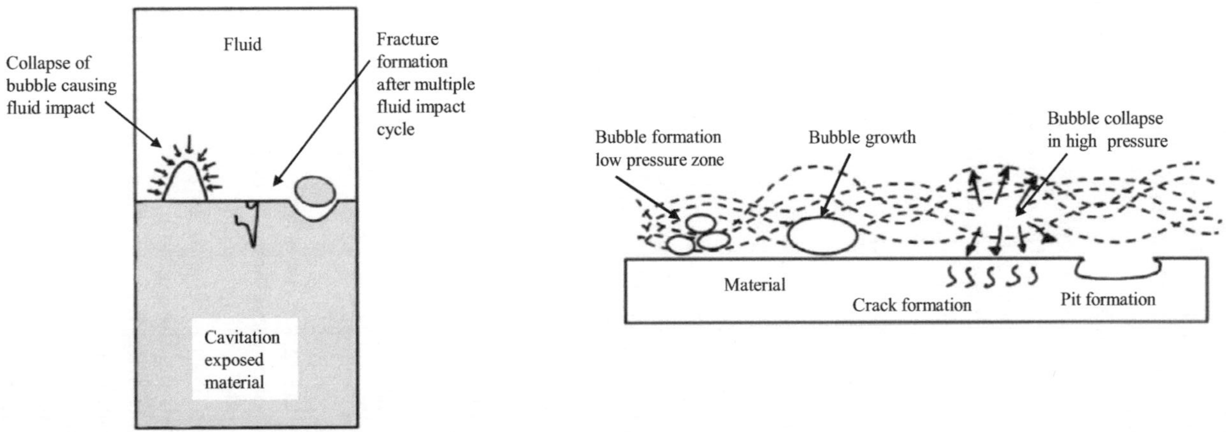

Figure 2.24 Disintegration of a compressed gas bubble and resulting surface degradation.

When individual bubbles coalesce into larger ones and collapse within a fluid medium, they generate shock waves capable of initiating crack formation. The bursting of bubbles on a surface leads to pit formation, with the potential for both pits and cracks to emerge upon collapse. This collapsing action, along with the accompanying shock waves, can directly harm surfaces by creating pits or subsurface cracks, a common occurrence in machinery operating in fluid environments like propeller water turbines and ships. The repeated collapse of bubbles causes wear, subjecting the surface to continuous impacts. Even sturdy surfaces can suffer damage from this process. While hard surfaces offer some resistance, they cannot fully absorb the energy, often resulting in the formation of surface cracks. The accumulation of these cracks may further lead to the development of larger pits. The unpredictable nature of bubble bursts causes this wear to spread continuously, not just in specific regions. Cavitation-induced wear poses a significant threat to propellers, water turbines, and various machinery components.

To review what we have already talked about, cavitation happens when negative pressure builds up on the downstream surfaces of objects that are submerged in liquid flow. This causes bubbles to form that are made up of dissolved gases and vapor. For example, in hydrodynamic bearings, vapor formation arises when pressure drops below the vapor pressure threshold, resulting in the creation of gas bubbles. When bubbles collapse in areas of negative pressure, they may either dissipate harmlessly or directly impact the surface.

So, what is our method for predicting this? Primarily, the erosion rate depends on the flow rate, which dictates how much cavitation erosion occurs on the surface. The mass flow rate, determined by density and volume, remains constant. Hence, it is essentially about how volume changes over time, expressed as dV/dt. We must devise a quantification or equation for this, factoring in the material's density (ρ), velocity (v), and more as wear rate is typically proportional to v^N, where N is generally greater than 2; it is not confined to just 2.

Figure 2.25 depicts cavitation occurring in a hydrodynamic bearing commonly used in IC (petrol or diesel) engine cases. In such bearings, how does cavitation manifest itself? Our investigation involved experiments utilizing a transparent bearing supplied with liquid. In the convergent domain, wear was absent, even on the suction side. However, we observed some liquid leakage in areas with higher volumes, such as the divergent side. The divergent zone lacked sufficient liquid to sustain negative pressure. Consequently, streamers formed within the fluid, where gas bubbles interconnected. Gas bubbles interact under high or reciprocating loads, causing pressure changes that could lead to their bursting and subsequent cavitation wear.

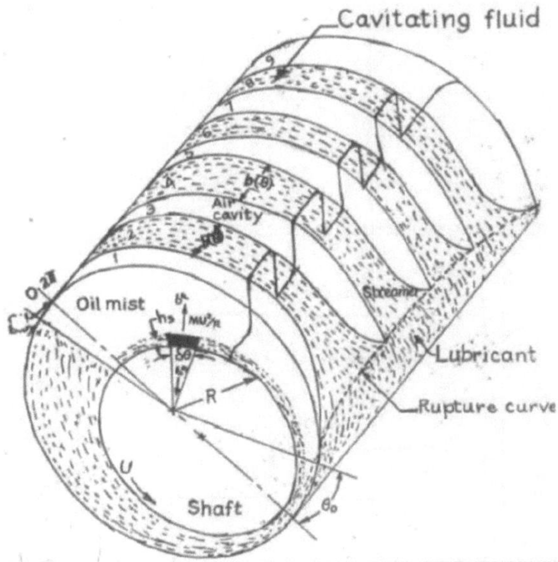

Figure 2.25 Cavitation in hydrodynamic bearing.

2.7 FATIGUE, FRETTING, AND MELTING WEAR

Dynamic stress conditions can cause fatigue, fretting, or melting, all of which lead to surface damage. Fatigue wear and fretting wear are comparable; the key distinction between the two is the magnitude of motion; fretting refers to tiny amplitudes, usually approximately 1 µm, which distinguishes it as a distinct subgroup of fatigue. Fretting involves abrasion in tight contact areas, with additional wear occurring as particles form (three-body abrasion) and accumulate in a confined space. Frictional heating also links to fatigue and fretting, and if temperatures continue to rise, it can lead to melting. When the material's temperature surpasses its melting point due to frictional heat or contact with heated surfaces, it transitions from solid to liquid. Melting wear represents an extreme form of fatigue wear. Fretting can also cause localized melting due to increased abrasion from wear particles, especially when efficient heat dissipation is lacking.

2.7.1 Fatigue wear

Fatigue wear happens due to repeated loading and unloading, with stress/strain cycling between positive and negative, σ_{max} and σ_{min}, or σ_{max} and zero,

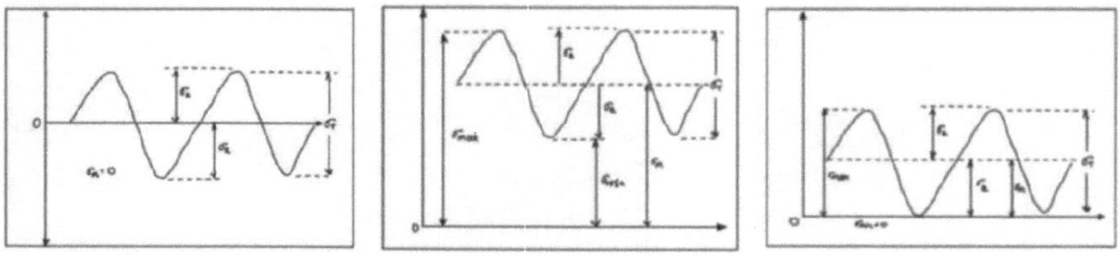

Figure 2.26 Different types of fluctuating stresses.

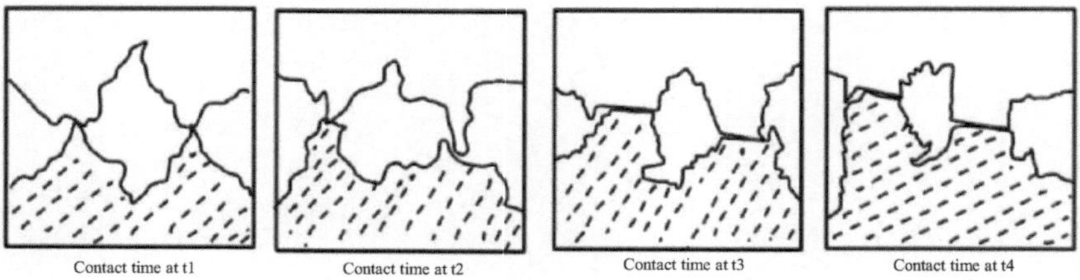

Figure 2.27 Temporal variation in contact area.

as illustrated in Figure 2.26. Even stress levels lower than the material's ultimate or endurance strength can cause fatigue. However, if the applied loading is constant (i.e., static loading), the actual stresses at the interface of asperities, due to temporal variation (Figure 2.27) in the contact area, provide fatigue load conditions. In almost every case, surfaces exhibit some form of asperity. Asperities can bend, tear, or form cracks when subjected to relative sliding, leading to localized fatigue. Fatigue wear can be divided into two categories: fatigue wear during rolling and fatigue wear during sliding, depending on the slide-to-roll ratio.

This type of failure is due to multiple reversals of contact stress, as indicated by a wear coefficient less than 10^{-7} or a failure probability of once in 10 million cycles. Over time, this leads to crack formation, which worsens as the cracks grow and multiply. Eventually, the material fails as these cracks expand.

Figure 2.28 illustrates the surface strains generated by sliding motion. As the sliding action takes place, the upper layers of the material undergo significant deformation, aligning themselves parallel to the surface. Consequently, the grain structure near the surface elongates and aligns in the same direction. This deformed top layer can be termed "Division 1," while the subsequent layer, experiencing moderate deformation, can be referred to as "Division 2." Within "Division 2," the cells are observed to be moderately strained, potentially leading to the formation of voids and nuclei, ultimately

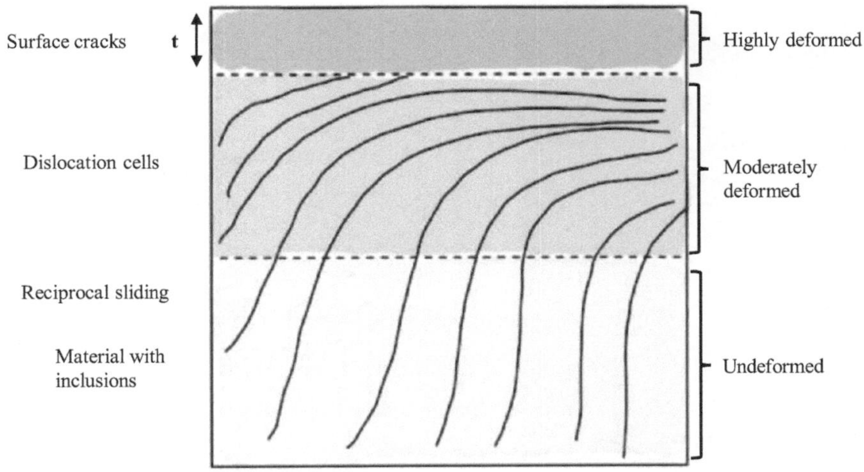

Figure 2.28 Influence of sliding-induced surface shear strain on grain structure.

disrupting the original grain structure and giving rise to dislocation cells. The development of these cells is influenced by the stacking fault energy, with higher stacking fault energy favoring their formation.

The thickness of the layer of "Division 1," denoted as "t," varies depending on the coefficient of friction. In instances of intense sliding where the coefficient of friction approaches unity, material displacement of up to 0.1 mm depth occurs. Unlubricated sliding scenarios, particularly reciprocal sliding, tend to have a higher rate of wear particle formation. However, in lubricated systems with lower friction coefficients, surface deformation is either minimal or absent. Therefore, lubrication plays a crucial role in controlling fatigue wear during sliding, reducing the coefficient of friction to levels even below 0.1.

In Division 2, where cracks may form, there is a possibility of void formation and crack manipulation. Metals such as aluminum, copper, and iron, characterized by high stacking fault energy, exhibit a propensity for dislocation cell formation. These cells' high-energy boundaries provide potential sites for void formation and secondary cracks. Research indicates that pure materials, devoid of inclusions, tend to exhibit an extended service life. For example, pure copper (99.96% purity) sliding against steel exhibits a wear rate ten times lower than other materials, despite having the highest coefficient of friction among the tested materials. Conversely, steel enriched with carbide particles has a low friction coefficient but one of the highest wear rates. Adhesion factors influence friction, while inclusion density in the material tends to escalate the wear rate. Consequently, materials with

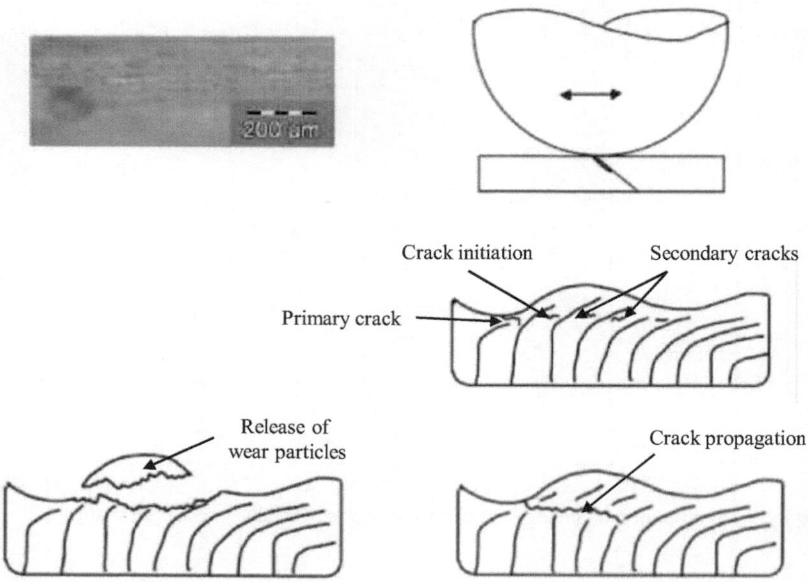

Figure 2.29 Influence of sliding-on surface crack.

complex impurities tend to exhibit lower friction coefficients but high wear rates. If we use certain alloys with additives such as high thermal conductivity or nanoparticles, we may increase the likelihood of fatigue wear or crack initiation.

It is important to note that a high coefficient of friction does not necessarily correlate with a high wear rate, nor does a low coefficient of friction guarantee a low wear rate. In different scenarios, a higher coefficient of friction may actually result in less wear, whereas a lower coefficient of friction might lead to higher wear rates.

To illustrate fatigue wear during sliding, Figure 2.29 depicts a pit on the gear surface. Primary cracks, as evidenced in Figure 2.29, typically originate from weak points at the surface and propagate downward along dislocation cell boundaries. Secondary cracks intersecting with primary cracks accelerate crack propagation rates. When these cracks reach the surface, they release wear particles. Most engineering materials contain inclusions and imperfections that serve as nucleation sites for secondary cracks. Essentially, materials without inclusions may demonstrate a longer service life.

Figure 2.30 illustrates that the contact surface is where the maximum shear stress during sliding occurs. Despite the high contact stress at the interface, the maximum resultant stress is typically located a few micrometers below

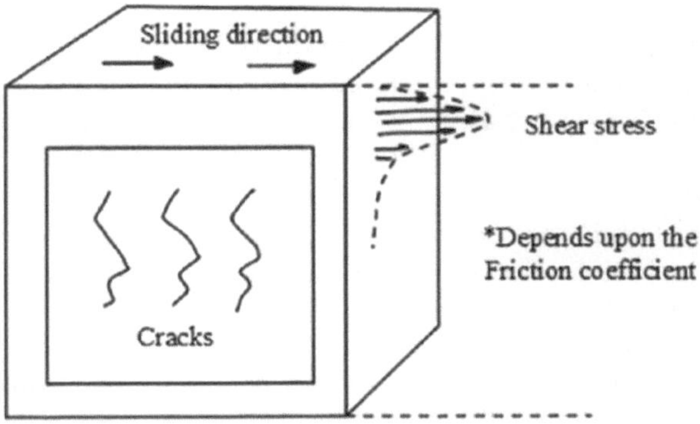

Figure 2.30 Influence of sliding-induced surface shear strain on crack formation and grain structure evolution.

the surface unless the friction coefficient surpasses a specific threshold, causing the maximum stress to manifest at the surface. This phenomenon results in delamination, which is characterized by subsurface crack propagation and the formation of lamellar wear particles.

As previously discussed, various materials exhibit differing propensities for generating dislocation cells, a tendency that varies from one material to another. Stacking fault energy commonly explains this discrepancy. For instance, metals like iron, copper, and aluminum are known to form dislocation cells due to their elevated stacking fault energies. In fatigue wear scenarios, the presence of these high-energy cells heightens the likelihood of void and crack formation, thereby accelerating wear. Furthermore, such wear may extend beyond the surface, manifesting below due to preexisting distorted cells. Fatigue-related fractures typically originate at weak points, such as grains with exceptionally high energy. From these initiation sites, fractures propagate and eventually surface, culminating in significant fatigue failure.

This underscores the critical importance of minimizing inclusions and artifacts, as their presence heightens the risk of crack initiation and propagation, ultimately leading to failure. Hence, cracks that originate from the surface tend to widen under unlubricated conditions, underscoring the necessity for effective lubrication to address fatigue-related concerns.

Lubrication can take a variety of forms, including oxides, solid lubricants, or coatings designed to act as lubricants. Adequate application of lubricant significantly reduces the coefficient of friction. This reduction effectively prevents the dislocation of the top surface, which would otherwise occur in parallel to the sliding direction. Consequently, this reduction impedes crack growth, potentially maintaining the initial crack size unchanged over numerous cycles.

By implementing appropriate lubrication techniques, surface design can be optimized to enhance service life and overall longevity.

Localized friction forces, asperity contact, and surface roughness all contribute to the development of strained layers that are prone to cracking and the subsequent discharge of wear particles. This process typically yields more irregular wear particles compared to subsurface delamination. Notably, subsurface cracks might not exhibit any surface wear symptoms before the release of a wear particle.

Fatigue wear is commonly observed in rolling bearings, gears, cams, and followers, where the rolling elements experience both rotational and slight sliding motions. In these situations, the maximum stress is notably high, suggesting that the combination of high normal contact stress and tangential shear stress is likely to lead to fatigue wear. In Figure 2.31, numerous pits are visible, indicating a high density of pit formations.

The presence of multiple pits suggests that the formation of a single pit does not lead to immediate fatigue failure. It is worth noting that pits smaller than 50 µm are invisible to the naked eye and require sophisticated methods for examination; otherwise, pits continue to form over time. Notably, pit formation rates can accelerate significantly. For instance, while it may take a certain amount of time to form one pit, ten pits might form in a much shorter period than ten times the time taken to form one pit. This accelerated progression is evident in the outer ring of a roller bearing, as shown in Figure 2.31. Despite experiencing surface fatigue, this bearing did not fail prematurely and survived over 40,000 hours of operation. Such fatigue-related issues are common after prolonged use. After 40,000 hours, the bearing can be inspected to determine whether it should be replaced or can continue to be used. Rolling bearings necessitate a smooth and undamaged contacting surface to function reliably. However, when the bearing

Figure 2.31 Development of multiple pits on inner surface of outer ring of a roller bearing.

surface contains numerous pits, it becomes unreliable. Pitted rolling surfaces cause severe vibrations due to their irregularity, preventing the bearing from effectively sustaining loads. If we reuse such pitted bearings, the vibrations can escalate, potentially leading to fracture.

In laboratory observations of real-world gear wear, fatigue (pitting) wear is a common occurrence. This is characterized by the presence of numerous pits on the surface, sometimes containing embedded wear particles. These wear particles, generated under high contact stress, embed themselves within the material. Once embedded, they indirectly contribute to increased surface roughness, leading to abrasive wear. Consequently, situations like these can result in fatigue-induced abrasive wear or severe wear.

Various factors come into play in such scenarios, including the applied stress, number of cycles, material characteristics, and environmental conditions. Moisture, for example, can accelerate surface corrosion. When the oxide layer is stripped from the surface due to high friction, the exposed surface reacts with water, forming another oxide layer and leading to corrosion. This corrosion may be more pronounced compared to normal oxide wear. Thus, environmental conditions are critical, and all these parameters collectively influence surface failure or fatigue. Therefore, stress level, number of cycles, material properties (whether brittle or ductile), environmental elements, and their chemical activity, leading to corrosion-assisted wear, are all crucial factors to consider.

2.7.2 Fretting wear

Controlling the relative sliding at the contact surfaces can reduce fatigue wear. While preventing major movements is relatively straightforward, managing small, reciprocating sliding motions with micrometer-scale amplitudes at the contact interface introduces another type of wear known as "fretting wear." Around 1927, fretting wear gained its reputation as a silent killer. Unlike pitting wear, which is visible during regular maintenance, fretting wear typically occurs in joints not typically accessed during routine inspections, posing a significant challenge. Fretting wear involves motions within a range of 1 to –300 μm, characterized by high-frequency oscillations and exposure to vibrations. Studying fretting wear is crucial, as many bridge failures initially stem from fretting wear before other wear mechanisms take over. This type of wear can cause significant damage, often without much warning.

An interference fit pair, after travelling 1000 km in a truck, may have undergone many cycles of low-amplitude, high-frequency vibrations, potentially leading to "fretting wear" and other failures. It is challenging to detect this wear because we rarely check interference fits. Fretting wear gradually removes surface material; however, particles or debris remain trapped between surfaces, preventing them from fully contacting each

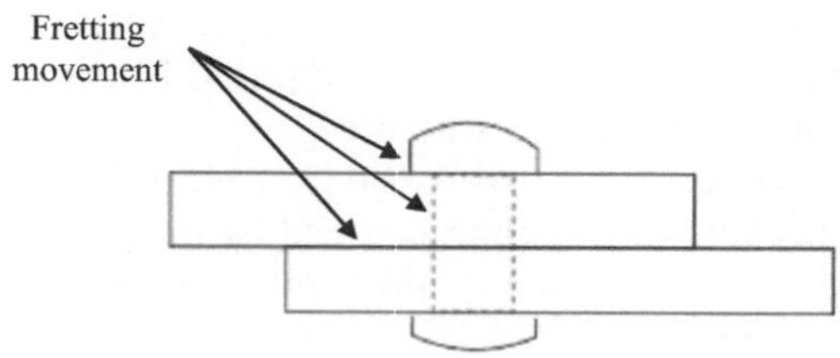

Figure 2.32 Depiction of fretting motion in a bolted joint.

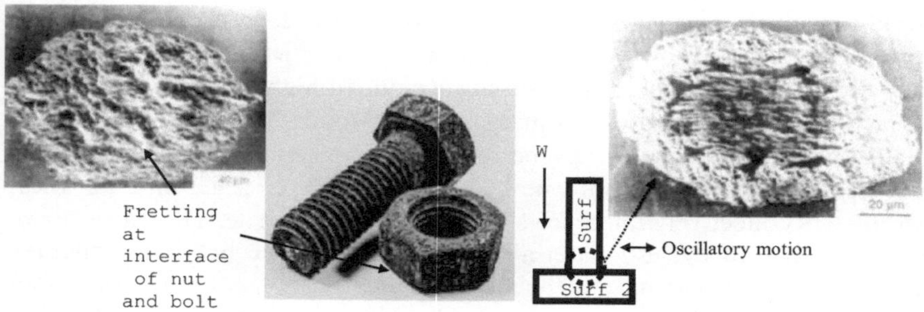

Figure 2.33 Potential areas of damage in bolted/joined components exposed to fretting conditions.

other, making it difficult for us, or the operator, to expose or observe them. This obstruction occurs because the particles become lodged between the surfaces, hindering their separation and visibility. Consequently, unless there is a joint failure, we may not be able to detect these particles unless we use NDT methods. This phenomenon is common in mechanical assemblies such as press fit, rivet (Figure 2.32), and bolt joints (Figure 2.33). In a rivet joint, multiple contact interfaces exist, providing numerous locations susceptible to fretting wear. It is nearly impossible to entirely prevent microscopic motion, as vibrations from the external environment inevitably induce such motion in assemblies like riveted ones. This phenomenon is similarly observed in nut and bolt connections; upon disassembly, debris, often varying in color, is commonly found between the mating surfaces.

Focusing on fretting (Figure 2.33), the pressure distribution between two contacting surfaces shows maximum pressure at the center. Vibration and incidental motion cause tangential movement, leading to minimal stress at the center and maximal stress at the joint's periphery. This pattern occurs when compressive stress peaks coincide with minimal tangential stress, resulting in high tangential motion in areas with reduced or nearly zero compressive stress. Normal stress is highest at the contact center and zero at the periphery, whereas shear stress is maximal at the periphery and minimal at the center. Such fatigue loading can remove surface material in tiny particles, creating debris found in bolted, riveted, or press-fitted joints. Fretting wear is common in various engineering contexts, such as riveted joints (Figure 2.31), nut–bolt connections (Figure 2.32), and light interference fits.

In the initial stage of fretting wear, the wear coefficient is usually less than 10^{-7}, indicating that fatigue phenomena are likely to occur within fretting wear mechanisms. As we are quoting, "the wear coefficient is usually less than 10^{-7}," important point to consider is whether fatigue/fretting wear is related to the Archard equation (as discussed in adhesive wear). We assert that fatigue wear causes cracks to form and grow, eventually creating wear debris. The Archard equation can be used to describe this generation of wear debris. Although many authors have suggested other mechanisms or equations, the Archard equation is notable for its simplicity and adaptability to various situations. However, the wear coefficient is a crucial parameter to be determined by performing extensive experiments. This coefficient can be used in conjunction with experimental observation and mathematical models to predict fatigue/fretting wear in advance.

In the context of fretting wear, it is important to discuss the local coefficient of friction. When compressive stress is at its maximum at the center and tensile stress is at its minimum, the coefficient of friction naturally approaches zero at the center. Conversely, toward the outer region where tensile stress peaks and compressive stress is minimal, the coefficient of friction becomes exceptionally high, potentially exceeding 1. Some sources even suggest that the coefficient of friction could theoretically approach infinity, as there is no possibility for wear debris to escape from that area.

However, rather than suggesting any value of coefficient of friction, we anticipate a scenario where slip occurs. Slip arises when the tangential stress surpasses the frictional stress. In essence, slip formation is contingent upon the relationship between the applied load, frictional stress, and tangential stress. If the applied load escalates to a point where tangential stress exceeds frictional stress, slip will ensue; otherwise, slip will not occur. In summary, within this contact region, there exists a combination of tensile stress, frictional energy, and compressive stress, all of which interact predominantly within this specific area. Generally, wear is less prevalent or even absent at the center of this region. Instead, it predominantly manifests in the peripheral domain, where wear particles attempt to escape but often remain intact due to the high tangential stress compared to the center.

In the fretting wear process, a layer of worn debris gradually accumulates on the damaged surface because the particles cannot escape, leading to continuous buildup. Initially, there may be only a few particles of 1 μm size, but over time, they can act as abrasive materials, worsening the failure by increasing the size of wear debris to between 5 and 50 μm. The wear rate depends on the formation of new debris and its effect on wear. If the new debris acts as a lubricant, the wear rate might significantly decrease. When debris is removed from the interface, the clearance increases, leading to a greater amplitude of relative motion. Upon opening the joint, a significant number of debris will be released. It is worth noting that fretting wear accelerates mechanical fatigue by inducing surface cracks, primarily manifesting as surface cracks rather than subsurface ones. These cracks tend to develop around wear scars, propagating gradually until eventual failure occurs.

To illustrate the progression of fretting wear (as illustrated in Figure 2.34), let us start with the first phase of "fretting wear." Initially, there is a "stick" phase in which the riveted, bolted, interference-fit joint maintains its connection, but wear still occurs, as represented by a coefficient of wear in the range of 10^{-16} to 10^{-14} .

As depicted in Figure 2.34, fretting wear can be classified into four distinct phases: phase 1, phase 2, phase 3, and phase 4. Despite robust construction and design, load variations and thermal expansion cannot be entirely eradicated—they persist throughout the process. Fretting wear advances through these phases, beginning with the stick phase. In this initial phase, joint surfaces remain in close contact with minimal relative movement, fulfilling the joint's intended function. However, the forceful connection of the surfaces induces high compressive loads, leading to surface deformation and the potential for micro-crack formation due to

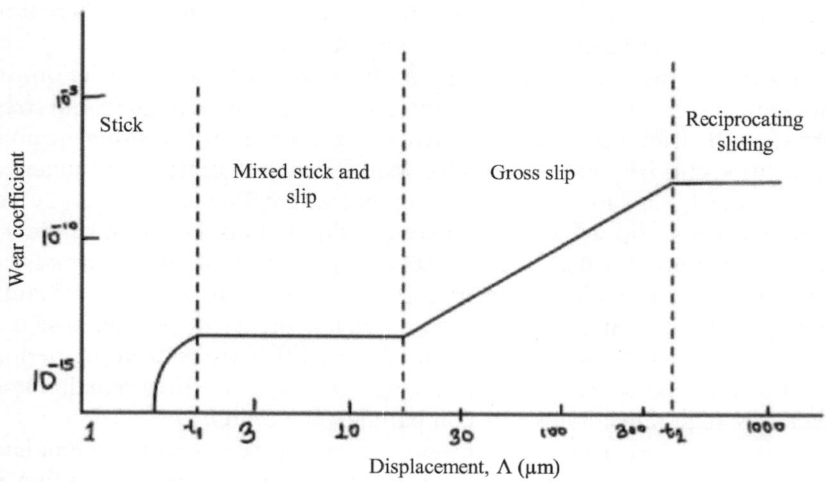

Figure 2.34 Different stages of fretting wear.

extreme contact stresses. Although there is slight movement, the contact pressure or stress remains exceedingly high, keeping the surfaces nearly stationary with no visible shifting. For example, the movement might fall within the range of 1 to t1 μm. It is possible that debris and worn-out material might have formed in this area, but it is hard to see any visible wear now. There is a chance that wear is emerging from the joint itself. Another factor could be an oxide layer that may have provided lubrication before, but now it is likely ruptured. This debris and oxide layer might act as abrasives, possibly causing further damage to the joint surface. This explains why the wear rate has increased from almost negligible (wear coefficient ~10^{-15} to 10^{-14}).

In the second stage, a combination of sticking and slipping occurs. During this phase, the condition worsens due to increased sliding motion and rubbing. Subsequent vibrations or changes in load conditions cause slight movement of the joint surfaces, characteristic of the mixed stick-slip phase. Initially minimal, this movement can progress to approximately 10 to 12 μm, which remains imperceptible to the naked eye. This defines stage 2, or phase 2.

Upon entering the gross slip phase, previously imperceptible micron-sized movements become noticeable. The motion adopts a repetitive, dynamic, and cyclic nature. This observable movement is termed gross sliding motion, as the once tightly bound or intact condition has considerably loosened, permitting relative motion to begin. With a continued increase in external stresses or vibrations, the surfaces will slide against each other, resulting in the generation of large sized wear debris. Well, the wear debris generated is very tiny, maybe initially around 0.01 to 0.1 μm, which is at the nanometer level. This could potentially cause the joint to loosen, increasing vibrations and accelerating wear.

Following this, the reciprocating sliding motion exacerbates fatigue, leading to damage. Thus, in the fourth phase, significant movement of the joint surface occurs, culminating in severe damage or material failure. Illustrated in Figure 2.33 are two contact patches: "Patch A" measures 20 μm, whereas "Patch B" measures 40 μm. These patches may contain several small pits, indicating an escalation in surface roughness that amplifies effective stresses and diminishes fatigue strength, ultimately culminating in fretting failure.

For a deeper comprehension of fretting wear, examining the wear debris is essential. Aluminum oxide debris typically presents as black, whereas iron oxide manifests in three distinct forms. Although ferrous oxide is rare, a soft black magnetite oxide acts as a solid lubricant and is the preferred option. However, under dry conditions, Fe_3O_4 is typically observed. The presence of moisture induces rust and corrosion, resulting in wear debris that are harder and more abrasive than the previous black magnetite oxide, thereby causing more severe damage.

Materials like titanium or certain solids are highly susceptible to fretting damage, so it is best not to use them in these joint connections.

2.7.3 Melting wear

Melting wear, a severe form of wear, occurs due to an increase in temperature arising from either friction or environmental conditions. When the working temperature or the temperature at the contact point exceeds the melting point of the material, localized melting will naturally occur. This elevated temperature can be caused by environmental conditions or high ambient temperatures. Frictional heat is another significant factor. When the coefficient of friction is high and the material is unable to dissipate heat effectively, inadequate cooling leads to temperature accumulation. Over time, this accumulated heat can reach the material's melting point. Once melting occurs, the hot surfaces are unable to dissipate the heat efficiently, resulting in volume loss and microstructural changes (as depicted in Figure 2.35). This temperature rise is often due to insufficient heat dissipation.

The melting wear is quite rare because it requires extremely high temperatures. However, we have observed cases, such as during four-ball tests (as illustrated in Figure 2.35), where melting wear occurs on the surface due to high temperatures. This results in cracks forming, which, when subjected to sliding conditions, can generate wear debris or even cause fragments to break off. So, what we are seeing here is the result of a ball failure in a four-ball tester. This tribology tester is used to apply extreme stress and determine if a lubricant is effective. In this example, the extreme pressure leads to frictional heat buildup. Without proper lubrication, this can cause metals to melt or soften, resulting in material flow. As a result, worn debris accumulates in ruptured forms, as seen here with numerous

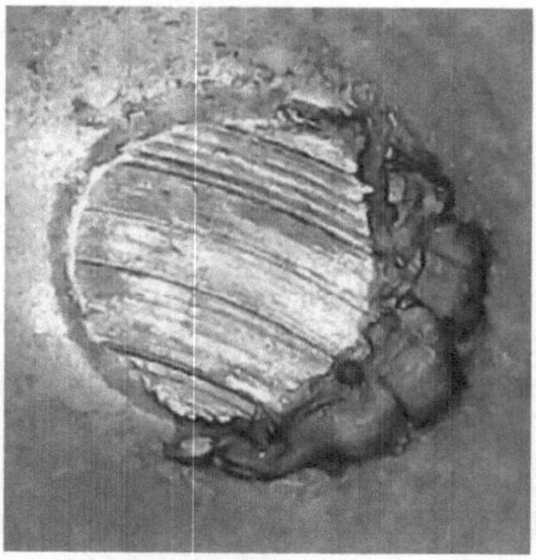

Figure 2.35 Resolidified molten layer with cracks.

cracks. Testing under sliding conditions or at high speeds can release even more wear debris from the surface. Heavy loads and high-speed sliding, along with this localized heat generation, significantly damage the surface. If this heat is not dissipated quickly, it can lead to temperatures that exceed the material's melting point, forming a molten layer known as melting wear. If you do not remove this layer immediately, it solidifies and creates an irregular surface, thereby increasing the wear rate.

Eventually, it will emerge in the form of droplets, which may also have an irregular mixed form. However, another crucial point to note is that whenever melting occurs, atmospheric oxygen is typically present. The interaction between atmospheric oxygen and the molten layer may result in the formation of an oxide layer, significantly altering the material's characteristics and behavior. Whether this oxide formation is advantageous or detrimental depends on the material and the type of oxide formed.

When two solid surfaces rub against each other, they lose energy due to friction. If this energy is not managed, it can cause increased temperatures, potentially leading to melting wear and seizure. The velocity of material movement directly correlates with the rise in frictional temperature; higher relative velocities lead to higher frictional temperatures, while lower velocities result in lesser temperature increases. At very high sliding speeds, the temperature rise is significant, whereas it is less at lower speeds. Melting wear is fundamentally about localized heating. Instead of distributing the energy across the entire surface, it causes localized heating that forms and immediately solidifies a molten layer, ultimately leading to material loss and wear. Interaction with atmospheric oxygen forms an oxide layer, which can influence subsequent wear and friction characteristics, either aiding or hindering the process.

We prefer materials with high thermal conductivity that can dissipate heat rapidly to prevent these issues. We might use lubricants or coatings to reduce friction and maintain lower temperatures if heat dissipation is insufficient. Regular maintenance can also help, but if problems like melting wear or seizure persist, alignment issues should be checked. Even a minor misalignment, such as 0.1 degrees, can increase temperatures by 5 degrees, causing significant problems.

Now, let me introduce a concept that has been prominent in literature for nearly three decades: the wear map. What precisely is a wear map? Authors typically develop wear maps for commonly used materials, and they have created such maps for various materials. In Figure 2.36, we present a wear map that is specifically focused on steel material, delineating the range from ultra-mild wear to complete seizure. The map's vertical axis denotes normalized pressure, while the horizontal axis represents normalized sliding velocity. Both normalized pressure and sliding velocity clearly influence the wear rate. Various wear mechanisms, such as "ultra-mild wear (dimensionless wear coefficient $K < 10^{-9}$)," "mild oxidational wear ($K \sim 10^{-8}$ to 10^{-6}),"

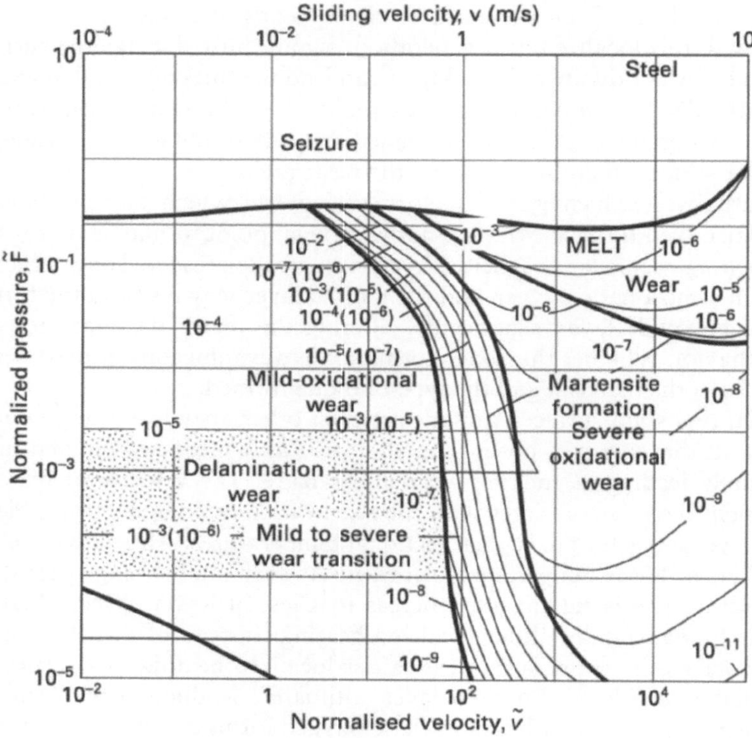

Figure 2.36 Wear mechanisms associated with steel material.[2]

"severe oxidational wear (K ~ 10^{-6})" and "melt wear (K ~ 10^{-3})", exhibit diverse behaviors without following a uniform trend. Melt wear, a form of severe wear occurring after several other wear mechanisms and heat, can lead to a complete seizure if left unchecked. Relative motion halts during a seizure, crucial for system functionality as the system absorbs all the energy input for heating. To mitigate such occurrences, careful material selection and control of normalized relative velocity and pressure are essential to reaching the ultra-mild wear zone.

Another aspect worth noting is the contours depicting nondimensional wear coefficients, such as 10^{-9}, 10^{-8}, 10^{-5}, 10^{-3}, and so on. It is crucial to understand that the value of the nondimensional wear coefficient encompasses all types of wear mechanisms—abrasive wear, adhesive wear, fatigue wear, fretting wear—culminating in the ultimate wear constant. Through wear maps or analogous graphical depictions, we have gained the capacity to discern the wear characteristics of materials under defined conditions. This suggests that we can assess the performance of a particular material across various scenarios and anticipate the magnitude of the wear coefficient. Leveraging these maps enables us to precisely determine the optimal operational conditions to alleviate wear. To steer clear of severe wear, it is essential to identify the critical factors that should never occur, even in the most adverse situations. The analysis of wear maps allows us to discern these critical factors.

Now, let us delve into another aspect concerning oxidational wear. In some cases, we discuss the formation of oxides during sliding conditions, and subsequently, this oxidation layer undergoes wear, revealing the fresh or virgin surface beneath. This exposed surface then has the potential to oxidize once more. Thus, the processes of oxidation formation and removal are ongoing. This continuous cycle underscores the importance of three elements in governing the thickness of the oxide layer: rupture time, indicating how long it takes to breach the oxide layer; reoxidation time, signifying the duration required for the layer to reoxidize; and lastly, the formation rate. When these factors are in equilibrium, the oxide layer is likely to remain intact. However, if this balance is disrupted, the oxide layer may either expand or be entirely eradicated.

Now, let us delve into quantifying melting wear. Throughout our discussions, temperature has been a recurring theme, whether it is flash temperature or bulk temperature. Melting occurs when the bulk temperature hits the melting point. To estimate these temperatures, we equate the generated friction energy with the dissipated heat flux. By comparing friction energy with heat flux and rearranging the equation, we can determine the bulk temperature (T_b). It is important to note that only a fraction of the energy—not necessarily 100% of the friction energy—contributes to the temperature increase.

$$\text{Frictional heat: } H = \frac{\mu F v}{A} \tag{2.18}$$

$$\text{Heat flux: } J = K_m \frac{T_b - T_o}{l} \tag{2.19}$$

$$\text{Bulk Temp.: } T_b = T_o + \frac{\alpha \mu F v l}{A K_m} \tag{2.20}$$

$$\text{Flash Temp.: } T_f = T_o + \alpha \mu T^* \beta \tilde{v} \frac{r_a}{r_o} \tag{2.21}$$

where μ is the coefficient of friction; ν is velocity at interface; A is cross-sectional area; v is velocity; K_m is thermal conductivity of the material; T_o is sink Temp.; l is diffusion distance; α is fraction of heat flow; r_a, r_o are asperity size; β is constant.

The fraction of heat transferred is denoted by α, which can be 0.5 to 1.0. The figure also represents the conductivity coefficient, which is associated with the surface's thermal conductivity. The flash temperature is generally higher compared to the bulk temperature. This temperature generally requires considering the level of surface roughness (asperities) and the corresponding temperature rise. If the temperature rise at the asperity level is significant, it can lead to problems like seizure or melting, indicating that most issues start at the asperity level. Now, just to emphasize this, I will give an example of a gear pair. The Hertzian formulas typically determine contact stresses in gear pairs. To estimate the flash temperature (T_f), which depends on the coefficient of friction, relative velocity, normal load, effective Young's modulus, effective radius of curvature, and face width (depth of the gear), we can use an empirical formula:

$$T_f = 0.62\mu \frac{|V_1 - V_2|}{\sqrt{\beta_1 V_1} + \sqrt{\beta_2 V_2}} \left[\frac{W_N}{F}\right]^{\frac{3}{4}} \left[\frac{E'}{R'}\right]^{\frac{1}{4}} \tag{2.22}$$

$$\frac{1}{E'} = \frac{1}{2}\left[\frac{1-v_1^2}{E_1} + \frac{1-v_2^2}{E_2}\right]; \frac{1}{R'} = \frac{1}{R_1} + \frac{1}{R_2} \tag{2.23}$$

In this formula, β_1 and β_2 represent the coefficients of thermal conductivity. Different authors have various terminologies, and I am using Bartz's terms.[3] The coefficient of thermal contact (β) can be given as thermal conductivity (λ), specific heat (C), and density (ρ).

$$\beta = \lambda C \rho \tag{2.24}$$

If the beta value for any material is lower, the temperature will be higher; if the beta value is higher, the temperature will be lower. Therefore, during material selection, we should choose a material with a high beta value. This relation also provides the effective Young's modulus and effective radius of curvature.

Here, we are using W_n. We know that we can calculate W_n for a gear from the applied torque (T). If we know the pitch radius, we can find W_t

$$W_t = \frac{T}{r_p} \tag{2.25}$$

$$W_N = \frac{T}{\cos\varphi r_p} \tag{2.26}$$

W_t can be expressed in terms of the normal load on the gear pair: $W_t = W_n * \cos(\varphi)$, where φ is the pressure angle. The radial force (W_r) can be given as $W_n * \sin(\varphi)$. Once these values are calculated, they can be substituted into the formula to determine the flash temperature. Some additional data is required, and Bartz has provided a slightly modified formula for that purpose:

$$T_f = 0.62\mu\sqrt{\frac{\omega}{\beta}}\left[\frac{W}{F}\right]^{3/4}[mZE']^{1/4} f_z \tag{2.27}$$

This formula considers the relative speed of the pinion and gear, the β conductance, the applied load (W_n), the gear's face width (F), the module (m), the number of teeth (Z), the effective Young's modulus (E′), and the factor f_z provided by the author. As the number of teeth increases, the stress level decreases, leading to a reduction in the coefficient. For instance, if the coefficient is 0.813 with 17 teeth, increasing the number of teeth to 100 reduces it to 0.124, a sixfold decrease. Consequently, if the temperature is around 100°C with 17 teeth, increasing the number of teeth to 100 could reduce the flash temperature to only 16°C or 17°C. Therefore, design changes are both possible and advisable.

In this context, the author compared two gear material pairs: nylon-nylon and nylon-steel. They selected gears with 17 teeth, a module of 3, a face width of 30 mm, an angular speed of 150 rad/s, and a power transmission of around 850 watts. The beta value for nylon-nylon is 0.417×10^6, while for nylon-steel, it is significantly higher—almost 400 times greater. However, the coefficient of friction remains the same in both cases.

In addition, the effective Young's modulus for nylon versus steel is almost double that of nylon versus nylon. For the nylon versus nylon case, the beta value was 0.417×10^6 (417,000), while for the steel versus nylon case, the beta value is 180×10^6. After altering these two parameters, we discovered that the flash temperature difference (T_f) between nylon and steel was 160.75°C, nearly 161°C, and 9.2°C for steel and nylon, respectively. This significant difference shows how changing the material pair can lead to better results.

2.8 FRACTURE SURFACE DEGRADATION

Fracture surface degradation refers to the gradual transformation that takes place on a material's surface after it fractures. This transformation can manifest in various ways: the surface may shatter into multiple fragments, or the entire solid body may fracture. Either scenario could see the surface stripped away, worn down, or reshaped, leading to the emergence of a new surface. Alternatively, the solid body itself may break apart. Occasionally, people refer to fracture as a type of segmentation, separation, or fragmentation.

If the reduction in area is less than 5%, we refer to a fracture in a solid body as brittle. Conversely, we label it as ductile failure if necking occurs and the reduction in area exceeds 5%. On occasion, we observe failures that begin as ductile but transition to brittle over time. For example, in ductile materials, if the load suddenly increases, the material may not have enough time to undergo the typical percentage reduction in area, resulting in spontaneous failure that may appear brittle. In addition, many brittle materials exhibit ductile behavior at higher temperatures. Conversely, materials that are ductile at ambient temperatures can become brittle if the temperature decreases.

In brittle failure, the fracture stress aligns with the yield stress, as shown in Figure 2.37. However, in other instances, the fracture stress might be equal to the ultimate tensile stress. In summary, brittle material failure happens without significant deformation, as evidenced by minimal percentage elongation or area reduction. As there is little to no substantial deformation in brittle failure, early indications of such failure are unlikely.

On the contrary, ductile materials, including most metals and some polymers, undergo plastic deformation prior to failure, which can be seen in changes in shape or form. These materials exhibit bending and significant plastic deformation before failure, typically when stress surpasses the yield strength. Ductile fracture occurs in materials capable of undergoing substantial plastic deformation before failure, presenting visible warning signs such as necking or observable deformation prior to final failure. This fracture type is typical of materials that are able to deform plastically and absorb significant energy before breaking. Several mechanisms contribute to ductile fracture: plastic deformation and necking, where localized reduction in cross-sectional area allows stress redistribution and further deformation; void nucleation and growth, involving microscopic void formation at imperfections like inclusions or grain boundaries, followed by void coalescence; and shear band formation, where localized intense shear deformation zones act as preferred paths for further deformation, accommodating large plastic strains before eventual fracture.

The manufacturing process often dictates the material's brittle or ductile behavior. It holds significant sway in determining the outcome. Are there voids present? Their presence can significantly influence fracture behavior, potentially causing a shift from ductile to brittle failure. In addition, the quality of sintering post-powder metallurgy is crucial. Inadequate sintering can trigger a transition from ductile to brittle failure.

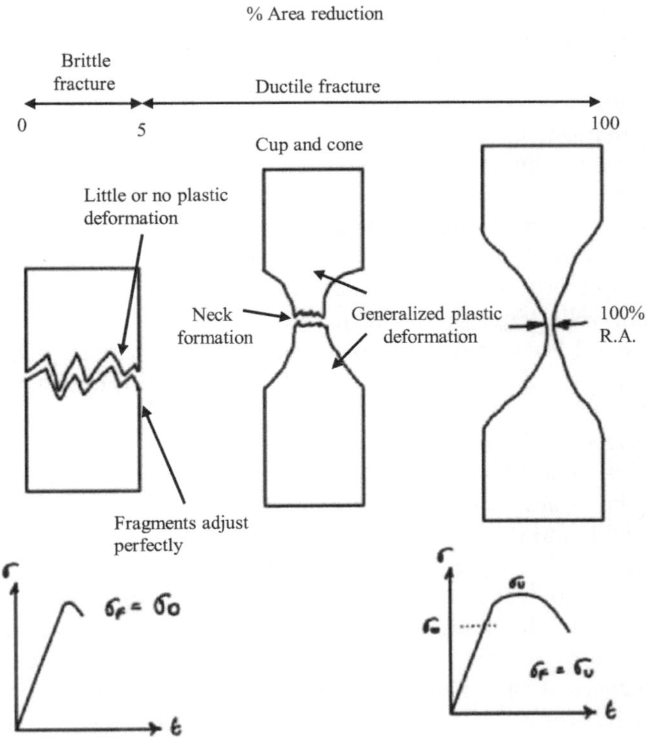

Figure 2.37 Depiction of brittle and ductile fractures.

Consider cast iron as an example. It can undergo two distinct processes: casting or powder metallurgy, resulting in shafts from both methods. When subjected to tensile or torsional loading, one may exhibit ductile failure, while the other displays brittle failure.

The surface geometry significantly influences material behavior. For instance, transforming a rectangular shape into a more intricate form with delicate sections can shift from typical ductile failure to brittle failure.

Moreover, the microstructure of the material plays a pivotal role. Manipulating the microstructure can alter a ductile fracture into a brittle one and vice versa.

In addition, the interaction between the material and its environment is crucial. Under corrosive circumstances, a ductile material may transition to brittle failure. Countless instances in the literature demonstrate how moisture, corrosion, or rust can convert ductile failure into brittle failure.

Now, let us delve into the disparity between surfaces created by brittle and ductile fracture. A brittle fracture is typified by a smooth, flat surface with minimal or no plastic deformation and negligible reduction in dimensions. This type of fracture occurs spontaneously, resulting in a flat, smooth surface devoid of characteristics like concentric circles or signs of plastic deformation, such as necking.

After fracturing, the newly exposed surface undergoes changes influenced by environmental factors, stressors, or external forces, significantly altering its properties. Environmental elements such as ultraviolet (UV) radiation, temperature fluctuations, or exposure to moisture play a crucial role in this degradation process. For instance, UV radiation can lead to surface discoloration, yellowing, or the formation of cracks. Pre-existing cracks may also expand because of UV exposure, while moisture exposure might cause swelling or foster microbial growth, thereby altering the fractured surface.

As we delve into the discussion on fracture surface degradation, a pertinent question arises: Is it truly necessary to understand fracture when the focus of this book lies on surface degradation? Fracture mechanics is a well-established field, and there is a plethora of literature, mathematical models, and experimental techniques dedicated to fracture analysis.

What is the significance of surface degradation in relation to fracture? What does this entail? To explain it, it can be said that upon fracturing, a material exposes at least two new surfaces, both susceptible to environmental influences. Consequently, there is a possibility of environmental factors affecting these surfaces. Moreover, residual stresses or heightened external temperatures may be present. Previously shielded by thermal insulation, these surfaces might now be exposed to temperature fluctuations. Hence, we assert that the newly exposed surfaces of a fractured material may undergo various transformations. We need to identify these changes and understand how they affect failure analysis. If the fractured material surface is degraded by the environment, stresses, or temperature, we might not reach the root cause of the failure. Instead, we could be misled by secondary factors unrelated to the original cause of failure. For example, the presence of cracks is not solely attributed to the fracture itself but rather to UV exposure. In those cases, without considering UV radiation exposure, accurately pinpointing the root cause of failure becomes challenging. It is imperative to either prevent UV exposure or establish protective measures to shield the surface from UV rays.

Moisture is also an important environmental factor to consider. High humidity can prompt freshly exposed material to absorb moisture, potentially leading to swelling and exacerbating issues. This can obscure the true root cause of failure, complicating the analysis process.

Prolonged observation of a fractured surface over several hours can lead to further surface deterioration, known as slow fracture surface degradation.

In such conditions, existing cracks may branch out in various directions, giving rise to new cracks. It is essential to grasp these processes, emphasizing the significance of comprehending and characterizing fracture surface degradation for precise failure analysis and preventing misinterpretation of the root cause of material failure. The term "fractography" is often used to describe the process of examining a fractured surface and identifying the ensuing changes. Fractography is the study of broken or fractured surfaces.

Fractography utilizes two main approaches: macroscopic and microscopic. Macroscopic fractography entails visual inspection, following specific rules and guidelines to discern characteristics like brittleness, ductility, high stress, temperature influence, or creep. This method relies on direct observation of the fracture surface's physical attributes. Macroscopic analysis involves identifying fracture lines on newly exposed surfaces resulting from the fracture. By scrutinizing these fracture lines, one can predict the type of failure that occurred. However, microscopic examination may be required in certain cases, using magnification tools such as microscopes capable of up to 1000× magnification or scanning electron microscopes (SEMs) with magnification levels reaching up to 10 million times.

Figure 2.38 consists of two main sections. The top section shows clear beach marks on the fractured surface of the outer ring of a four-row roller bearing. These concentric circular patterns indicate fatigue progression, highlighting areas affected by cyclic loading over time. Arrows point to the beach marks, emphasizing their location and orientation on the surface. The bottom section contains three images:

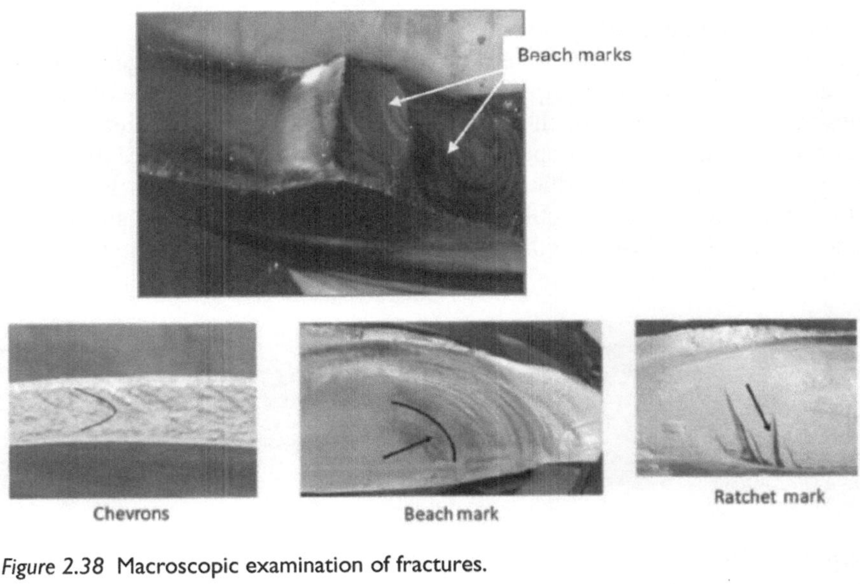

Figure 2.38 Macroscopic examination of fractures.

1. *Chevrons:* This image shows V-shaped grooves or lines, suggesting high-impact failure caused by sudden stress release or rapid velocity. These markings typically indicate brittle failure, even in ductile materials, under extreme conditions.
2. *Beach marks:* Similar to the top section, this image displays periodic curved patterns representing normal fatigue growth due to cyclic loading. These marks resemble clamshell patterns or concentric rings, though they may not be perfectly concentric. Sometimes referred to as "circular ridges," these marks generally indicate ductile failure. Such lines provide valuable insights into the type of failure: brittle or ductile. For instance, impact loading often results in brittle failure (only partial arcs), while repetitive cycle loading tends to cause ductile failure. By examining these lines, we can infer the nature of the surface failure.
3. *Ratchet marks:* This image shows stair-step-like features, indicating the merging of multiple cracks or fracture planes. Ratchet marks are commonly observed in materials subjected to torsional or cyclic stresses, where cracks initiate at different locations before merging.

These visual features are critical for understanding material failure mechanisms and are often used in failure analysis and forensic engineering.

It is worth noting that Figures 2.39, 2.41, and 2.42 depict common types of fractures. Description of these figures is commonly found in educational materials and technical articles related to material science and failure analysis. For the present chapter, various images of these figures have been taken from the public domain.[4]

In Figure 2.39, we observe vertical ridges. These elevated ridges indicate a fracture and can signify impurities in the material. The presence of impurities can initiate cracks, which are not necessarily caused by repeated loading but rather by inclusions or specific microstructural characteristics. Initially, the fracture may start as ductile but can eventually lead to brittle failure. This contrasts with the circular ridges seen in fatigue loading, suggesting that impurities, inclusions, voids, or gas bubbles contribute to the failure.

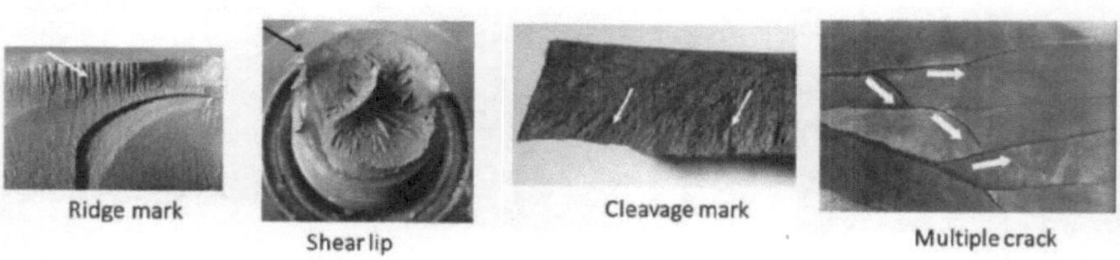

Figure 2.39 Different types of macroscopic fracture patterns.[4]

Another well-known example is the shear lip (Figure 2.39), which is indicative of ductile failure. Shear lips are curved or wavy lines that reveal shear deformation in a material, resulting from ductile fracture or plastic deformation. Another type of failure is characterized by cleavage marks (Figure 2.39). Cleavage lines are straight or slightly curved fracture lines that often suggest the presence of impurities and indicate brittle failure, typically aligning with crystallographic planes in brittle materials. Finally, the presence of multiple cracks (Figure 2.39) can also indicate brittle failure. When multiple cracks are present, the failure likely began with some form of ductile failure and then progressed to brittle failure, potentially due to variations in loading, misalignment, or changes in temperature. This indicates severe problems in the system, not just in one component, but across multiple components at the same time.

Furthermore, there is a necessity to delve into micro-level fractography. At this scale, we must consider composition, crystal structure, and process history. A thorough understanding and extensive knowledge are essential for accurate failure analysis. For example, even ductile materials may experience brittle failures under low-temperature conditions due to microstructural changes that render the material brittle.

Creep deformation is a time-dependent process characterized by the gradual stretching or extension of a material under a constant load or stress. Creep is typically associated with high temperatures, but it can occur in certain materials even at lower temperatures due to the diffusion of atoms or molecules within the material. This progressive deformation continues until eventual fracture, with failure progression persisting under unchanged loading conditions, leading to an escalating strain rate or plastic deformation over time.

One aspect to consider is that fatigue, a slow process of crack formation and development under repeated loading and unloading cycles, can be viewed as a type of creep failure. This process primarily generates microcracks, some of which may merge with inclusions, discontinuities, or voids, creating areas unable to sustain the load and resulting in a rapid increase in stress and complete failure.

Creep often occurs at higher temperatures, typically between 0.3 and 0.7 times the melting temperature, driven by atomic and molecular diffusion requiring thermal energy, often provided at elevated temperatures due to factors like high friction, inadequate lubrication, or insufficient cooling. Creep fractography involves examining fracture surfaces in materials undergoing creep deformation. Analyzing creep fractography helps determine dominant deformation mechanisms like dislocation creep, diffusional creep, and grain boundary sliding, aiding in identifying contributing factors to creep failure such as impurities, particle boundaries, and material defects. Various techniques, such as optical microscopy (OM), SEM, and transmission electron microscopy (TEM), enable the examination and analysis of features of creep fractures, including elongated voids or cavities, grain boundary sliding, and striations indicating a cyclic process of fissure growth and arrest.

The nature of creep fractures, whether transgranular (TG) or intergranular (IG), depends on the material and environmental conditions.

In Figure 2.40, three lines with the text "crack" are depicted, one showing a crack that initiates and follows the grain boundaries, indicating IG failure. In this type of failure, the crack begins at a grain boundary and spreads across grains or boundaries, leading to material failure. If the failure occurs in this manner, it is termed IG. The next type is granular failure, which can occur across the grains and may be either crystalline or non-crystalline. In non-crystalline materials, the crack does not follow a specific path and traverses different grains without adhering to crystalline lines. However, in crystalline materials, the crack follows the crystalline planes, adhering to a systematic pattern. This type of failure is more common in brittle materials because the grain boundaries are weaker, leading to IG failure.

Microstructure observation classifies fractures into two forms: IG and TG. In IG fractures, the crack moves along the grain boundaries, typically occurring in materials with weak grain boundaries. This type of failure is common in brittle materials but can also occur in ductile materials subjected to corrosion or embrittlement.

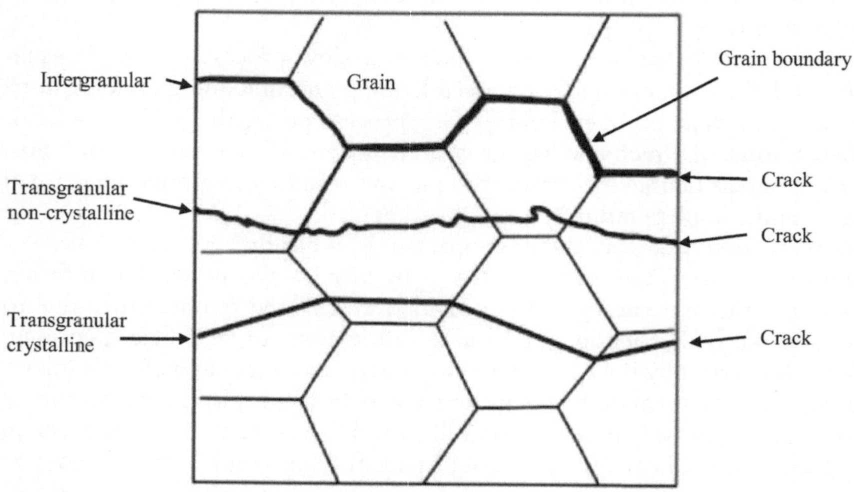

Figure 2.40 Granular or transgranular crack paths.[4]

In TG fractures, the crack penetrates through the grains. In crystalline materials, it follows the crystalline planes, whereas in non-crystalline materials, it does not adhere to these planes. Even ductile materials can experience IG failure due to corrosion or induced material weaknesses.

There are two types of TG failure, which typically occur through the grains.

1. *Crystalline TG failure:* This type of crack propagates along well-defined crystalline planes within the grain. In highly crystalline materials, the fracture follows the crystal structure and crystallographic planes, resulting in cleavage surfaces that are typically found in brittle materials.
2. *Non-crystalline TG failure:* This type does not follow any specific rules. The crack traverses the grains without adhering to crystalline planes, leading to abrupt failure due to impurities, voids, inclusions, or foreign materials in the microstructure.

If the material contains inclusions, the path of crack propagation can change. Instead of following the grain boundaries, the crack will move through the TG regions, which can be either non-crystalline or crystalline. The classification of fractures in polycrystalline materials takes note of this variation in their direction through the microstructure. Fractures in polycrystalline materials can be classified based on the trajectory of the crack within the microstructure. IG fracture happens when a crack runs along the edges of grains that are next to each other. Materials with weak grain boundaries or those susceptible to corrosion are more likely to experience this type of fracture. TG fracture involves crack propagation through individual grains and can further be divided into two subtypes: crystalline TG fracture, where the crack advances along distinct crystalline planes, often observed in materials with high crystallinity; and non-crystalline TG fracture, which occurs in materials lacking well-defined crystalline structures, resulting in more random fracture surfaces. Factors such as microstructure, composition, processing conditions, and applied loading influence the observed fracture type in polycrystalline materials.

Figure 2.41 illustrates various patterns of brittle fracture. This figure demonstrates the presence of cracks initiating within the microstructure, cleavage lines, and IG crack lines, leading to brittle failure.

Figure 2.42 illustrates a fracture showcasing both brittle and ductile characteristics. Micro-voids formed by inclusions, grain boundaries, and structural discontinuities cause fracture. With increasing load and plastic deformation, these micro-voids elongate, diminishing the available load-sustaining area and leading to cleavage fractures. In regions of pronounced stress concentration, cleavage fractures may further develop as load and plastic deformation escalate, culminating in surface failure. Materials with

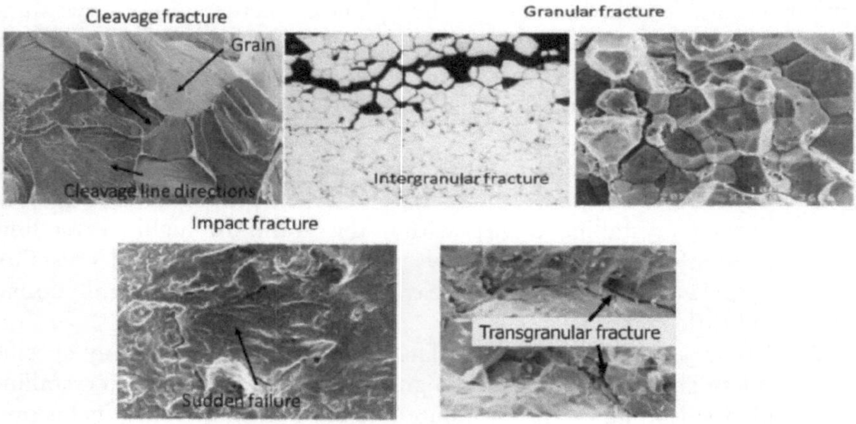

Figure 2.41 Different patterns of brittle fracture.[4]

intricate microstructures often exhibit this blended fracture mechanism under specific loading conditions. It commences with the formation of micro-voids within the material due to localized plastic deformation, occurring at various imperfections like inclusions, grain boundaries, or structural discontinuities, indicative of ductile behavior. As stress increases, the material undergoes plastic deformation, causing elongation and thinning of regions surrounding the micro-voids, allowing it to endure significant strain before eventual failure. At the same time, cleavage fractures can happen in places with a lot of stress. This type of fracture causes the material to separate along certain crystallographic planes, creating smooth, flat fracture surfaces. This brittle behavior is commonly associated with materials that have well-defined crystalline structures, and cracks can propagate rapidly.

Following brittle or ductile fracturing, the newly revealed surface undergoes transformations influenced by environmental factors like UV radiation, temperature shifts, moisture, stressors, or external forces, modifying its characteristics. For instance (Figure 2.43), we observed chloride formation on the fracture surface when it encountered a very mild HCl environment, with an HCL percentage of only 0.025%, deemed negligible and insufficient for failure.

Examining such a surface might lead one to speculate that internal chloride formation from chlorine ingress caused the failure. However, this explanation does not correlate with the immediate exposure to the HCL environment post fracture. Therefore, it is probable that the alteration to this state stemmed from a different source. Therefore, we must prioritize the initial fracture and the subsequent treatment of the exposed surface. Achieving a suitable microstructure for SEM, TEM, or other analyses

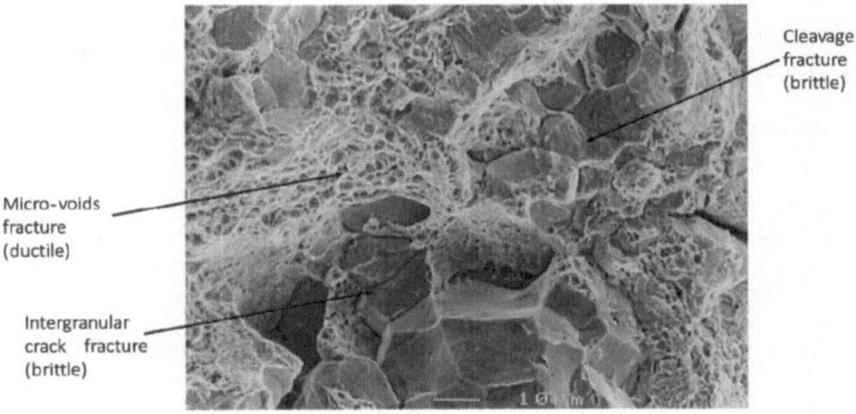

Figure 2.42 Combined brittle and ductile fracture.[4]

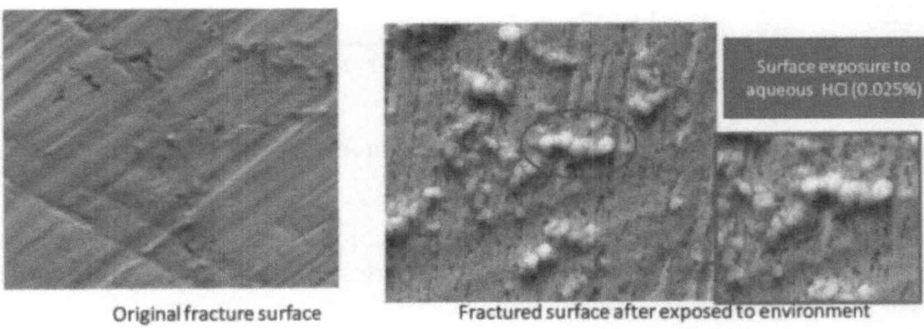

Figure 2.43 Impact of the surrounding environment on fracture.

necessitates specific equipment and lab procedures, which often require significant time and can take up to several days to access. Hence, meticulous care is imperative. Consequently, it is essential to highlight both actions to undertake and those to avoid.

It is absolutely critical to abstain from reassembling fracture surfaces under any circumstances. Attempting to reconnect fractured pieces often introduces additional failures or may even distort the original cause of the fracture. Therefore, we strongly advise against rejoining pieces or marking the cracked surface. In practical scenarios, it is best to avoid inscribing numbers or labels on fractured surfaces. Moreover, it is essential to refrain from removing any pieces, no matter how small, as doing so could obscure crucial evidence of the root cause of the failure. Similarly, using harsh equipment or sharp edges to clear debris from the broken surface is highly

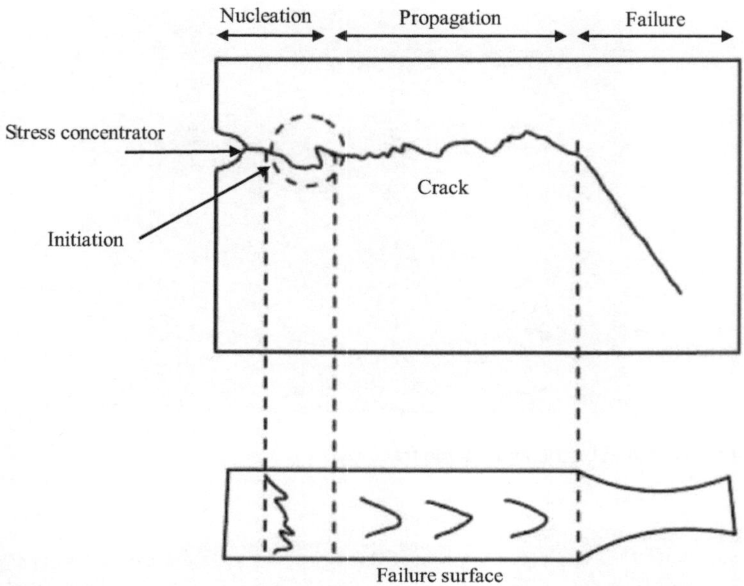

Figure 2.44 **Progression of fracture failure through three stages.**

discouraged. In addition, it is important to avoid touching fractured surfaces with bare hands, as the moisture and particles on our fingers could introduce further failure or alter the evidence of the original failure.

Figure 2.44 illustrates the three essential stages of fracture: nucleation, propagation, and final failure. Nucleation, which we have previously explored in the context of fatigue failure, marks the beginning of a crack. Propagation, occurring even in brittle failures, progresses at a notably faster rate. Finally, the failure stage culminates in either neck formation or complete failure. These stages follow a sequence: initially, there is no crack; then nucleation begins, initiating a crack; and subsequently, propagation leads to the proliferation of cracks. Numerous books and mathematical formulas devote extensive study to these stages.

Crack nucleation often originates from voids, manufacturing defects, high stress, or material impurities like inclusions or microstructural imperfections. These flaws can expedite the crack-initiation process, significantly reducing the material's lifespan. For example, due to these imperfections, what might have taken 10,000 cycles to initiate a crack could occur in just 100 cycles. Once a crack forms, stress concentrations in defective regions can exceed the material's strength, hastening the progression toward final failure.

Within the realm of crack nucleation, imperfections like void inclusions or microstructural flaws are pivotal. At first, we may not consider cracks smaller than a nanometer as significant discontinuities. However, these imperfections, though smaller than microcracks at first, have the potential to grow in size due to stress concentration. Depending on variables such as the defect size and its distance from the point of maximum load, the stress multiplication factor can vary between 1.5 and 2.5. This increased stress concentration causes microcracks to enlarge on a visible scale.

Next, let us discuss crack propagation. Once a microcrack forms, it naturally advances. This process involves the crack enlarging in size with persistent stress concentration near its vicinity, potentially intensifying. As the crack extends, the region resisting failure gradually diminishes. Fracture progression can occur gradually or abruptly, depending on various conditions. Under stable load conditions, fractures occur slowly. However, significant changes in load conditions, temperature fluctuations, or exposure to moisture or corrosive substances like acid can greatly accelerate crack propagation.

The speed of crack extension typically reflects the material's characteristics and stress conditions. Slow propagation often indicates ductile failure, whereas spontaneous progression suggests brittle failure. Finally, failure occurs when the crack reaches a critical length where the remaining material in the cross-sectional area is inadequate to support the applied load. This critical length varies based on design, material composition, and structural shape.

Understanding this critical length is vital as it signifies the loss of load-bearing capacity, which can occur suddenly or gradually, depending on the type of failure. This loss significantly compromises the structural integrity, leading to the separation of fracture parts and a substantial reduction in load-bearing capacity. To model these fracture phenomena, mathematical models and empirical rules are available.

Several common examples illustrate these principles; let us cover a case study from literature[5] published in 2023, illustrating the significant role of temperature in changing failure characteristics. This case study focuses on the fracture of a brazed diamond abrasive grinding wheel. A company used a brazing process to make the grinding wheel (shown in Figure 2.45) with a nickel–chromium–phosphorus–silica active alloy powder as the link between a steel base and synthetic diamond. The company created the wheel, referred to as a diamond wheel for cutting purposes, using the filler alloy, synthetic diamond grit, and other materials. The cutting face was made using a laser welding process between the steel surface and the brazed surface. The wheel's diameter is approximately 355 mm. The filler alloy and synthetic diamond grit were cleaned using ultrasonic cleaning with acetone and ethanol. This cleaning process is common in manufacturing. The cutting tool, or brazing diamond cutting saw, was made by following the diamond

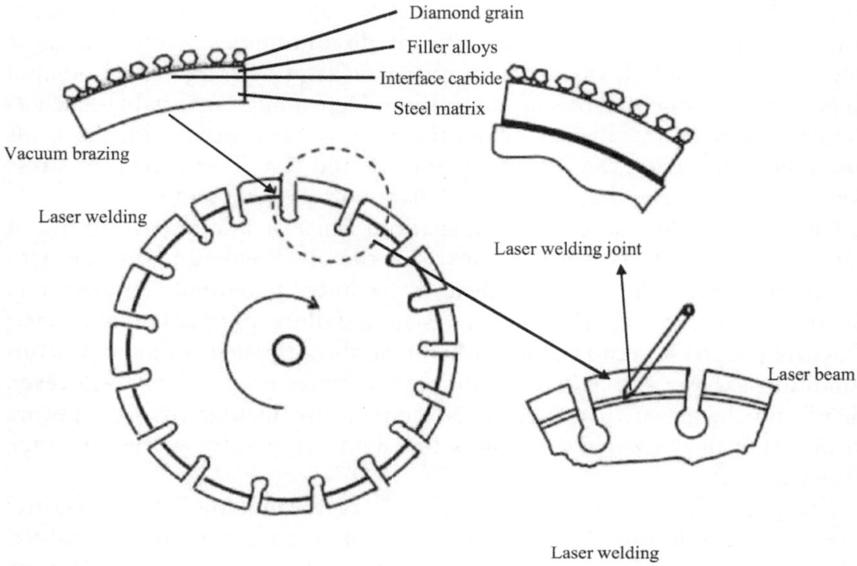

Figure 2.45 Fracture of brazed diamond abrasive grinding wheel using Ni–Cr–P–Si active powder.[5]

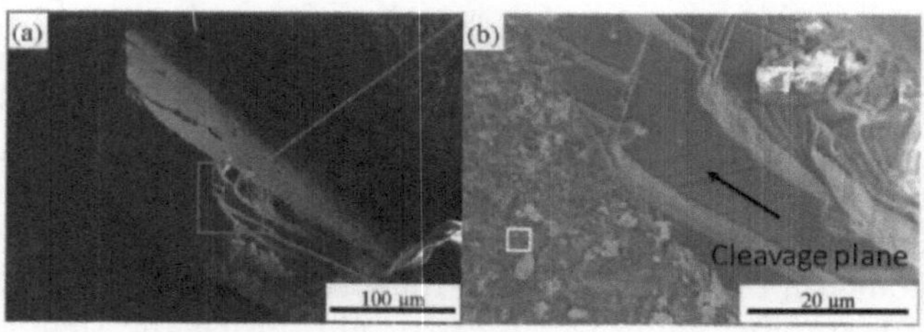

Figure 2.46 Cleavage plane indicating fracture of brazed diamond abrasive grinding wheel using Ni–Cr–P–Si active powder.[5]

brazing process and kept in a vacuum furnace. The brazing temperature was around 1026°C, maintained for 15 minutes.

To evaluate the diamond wheel's performance, the trials involved cutting cast iron pipes with a thickness of 18 mm to evaluate the performance of the diamond wheel. A key aspect of the test was the exclusion of cooling agents or lubricants due to their cost and sustainability concerns. During the test, visible cracks appeared on the diamond, along with markings and cleavage planes (Figure 2.46), indicating brittle failure. This suggests that the diamond wheel could not withstand the high temperatures generated during cutting without lubrication, which can reach up to 1000°C—a temperature diamond is typically able to endure.

The findings imply that the 18-mm pipe thickness led to a high rate of heat generation, causing the diamond structure to soften. This transformation resulted in the SP3 carbon bonds of the diamond converting to SP2 bonds. The change in structure from SP3 to SP2 bonds caused the abrasive particles to break and fracture. Contrary to initial assumptions, these fractured pieces did not detach from the alloy filler but remained in place, indicating a structural change in the diamond itself.

Figure 2.46(a) illustrates the fracture profile of a diamond abrasive particle. The presence of SP2 bonds was confirmed by microstructural analysis. This supported the observation that the high temperatures during the cutting process had softened the diamond and changed its structure.

2.9 INFLUENCE OF HEAT AND RADIATION ON SURFACE DEGRADATION

The impact of radiation and heat on materials is extensive, ranging from basic thermal effects to radiation-induced degradation. Heat triggers alterations in the physical and chemical properties of materials, while radiation can induce flaws in their atomic and molecular structures, gradually accumulating and eventually leading to material failure. Temperature fluctuations and radiation exposure significantly affect most materials, with a few exceptions. This investigation aims to understand the various mechanisms through which materials deteriorate when subjected to heat or radiation. The goal is to introduce the topic of heat and radiation, as well as their effects on surface degradation. Numerous articles and books are available on this topic. If you are interested, you can also delve into or refer to those resources.

Figure 2.47 illustrates the deterioration of mechanical properties due to high-temperature exposure. This degradation can result from grain growth, leading to a more flexible structure, or recrystallization, introducing flaws such as vacancies and dislocations. Phase changes may also occur, making the material softer and weaker. These flaws compromise the material's mechanical properties and reliability, and cavity formation further degrades the surface. Some materials become brittle, and metals exposed to radiation experience atom migration from within, reducing ductility and toughness, which can lead to cracking. In some cases, radiation displaces and rearranges atoms, causing swelling and surface inflation, as seen in the frequent failure of nuclear fuel rods. Melting is one of the most severe types of surface damage and should be avoided. Establishing effective cooling systems is necessary to limit temperature rise.

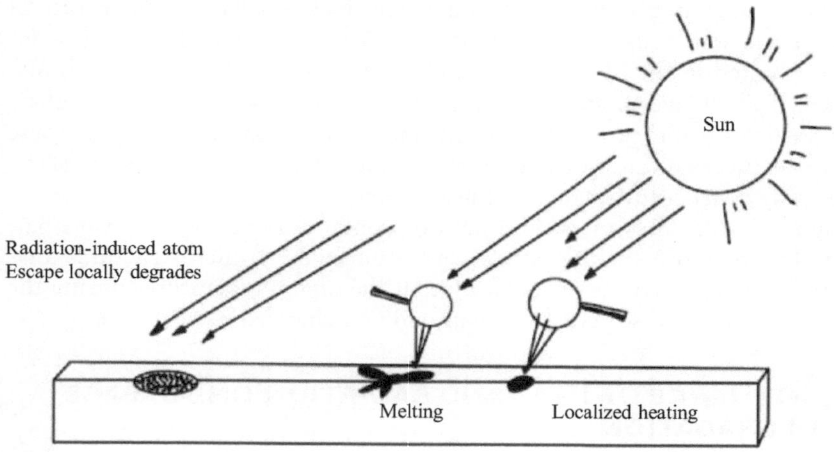

Figure 2.47 Surface degradation caused by heat and radiation.

The impact of temperature on microstructure and the resulting increase in brittleness due to dislocations can be observed through experimental analysis of microstructural changes. For instance, conducting experiments on steel by subjecting it to different cooling methods after austenitizing, such as normalizing (air cooling or AC) and quenching (water cooling or WC), and then applying the same tempering procedure for both cases, allows for a comprehensive analysis. The resulting microstructures can then be examined using techniques like OM and SEM, as depicted in Figure 2.48, to gain the necessary insights. Following initial cooling, the identical tempering process is applied to both cases for further comparison and analysis.

The microstructure of the as-normalized sample is primarily composed of granular bainite (GB) with a minor component of martensite, accounting for about 5% by volume (Figure 2.48(a)). The formation of martensite occurs when certain alloying elements are segregated during the rolling process. Small martensite–austenite (MA) constituents (<1 μm) are uniformly dispersed in parallel within the bainitic ferrite matrix, while larger banded or blocky MA constituents form along the prior austenite grain boundaries (PAGBs). No significant carbides are observed (Figure 2.48(b)). The as-quenched sample, on the other hand, has a fully martensitic structure (Figure 2.48(c)), with three to four martensite packets inside each austenite grain (Figure 2.48(d)).

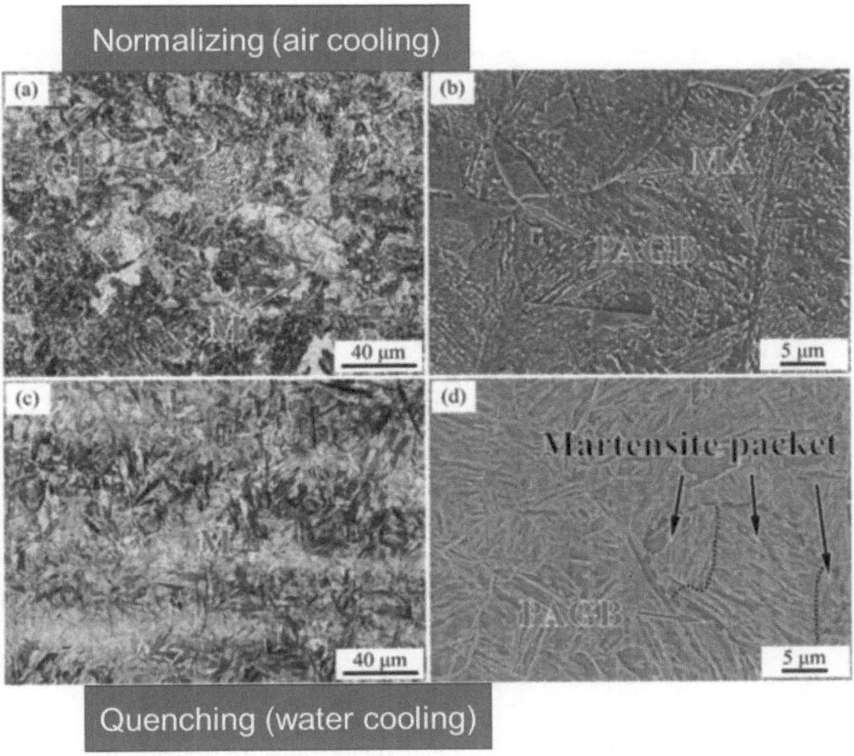

Figure 2.48 Impact of two cooling methods post-austenitizing of steel.[6]

Figure 2.49 plots the sample's continuous cooling transformation (CCT) curve to analyze cooling behavior and the effects of different cooling methods, such as WC and AC. Comparison of the WC and AC curves in Figure 2.49 reveals distinct outcomes: WC produces a fully martensitic microstructure, indicating an exceptionally high cooling rate exceeding 100°C/s, whereas AC produces a GB structure. The martensite resulting from quenching consists primarily of martensite laths with notable dislocation substructures. Conversely, the bainite structure observed after normalization consists of a ferrite matrix and some MA constituents. Notably, both normalizing and quenching treatments, conducted at similar austenitization temperatures, did not significantly alter the prior austenite grain size. In the case of water-quenched samples, carbide is absent, and the matrix is entirely martensitic with a slightly reduced grain size of approximately 24 μm. The PAGB consists of three to four martensite packets. The CCT curve also shows the difference in cooling rates between AC (about 10°C/s) and WC

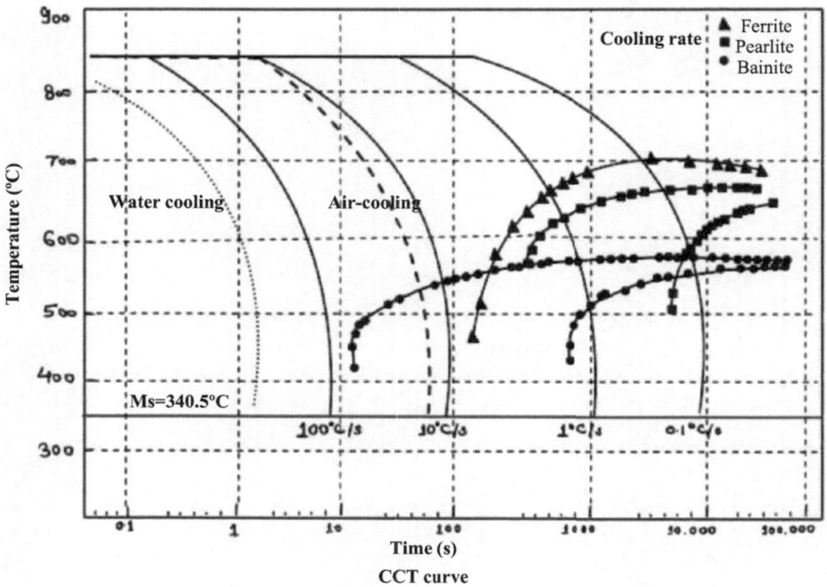

Figure 2.49 Analysis of cooling behavior and the impact of cooling methods.

(about 340°C/s). This shows that WC has a fully martensitic microstructure because it cools quickly, while AC has a GB structure. The presence of several dislocations in the martensite formed after quenching increases the susceptibility to brittle fracture.

Photochemical degradation refers to the process of material deterioration triggered by exposure to light, particularly UV light. UV radiation can absorb photons with sufficient energy from a material's molecules, thereby initiating chemical reactions that can alter its properties. In polymers, photochemical degradation often involves bond breakage or crosslinking of the polymer chains, leading to a decrease in the polymer's molecular weight and a subsequent decline in mechanical properties such as strength and toughness. Crosslinking may also occur, potentially increasing the material's modulus and brittleness. Other materials, including dyes and pigments, are also susceptible to photochemical degradation. Exposure to UV light can cause color fading or alteration by disrupting chromophore groups or inducing other chemical transformations. Occasionally, this degradation may be reversible, as seen in photochromic materials capable of transitioning between two distinct forms under light exposure.

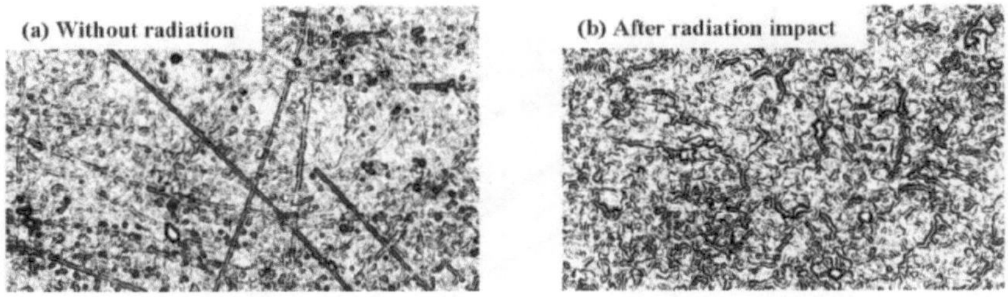

Figure 2.50 Impact of radiation on copper.

Photoionization occurs when an atom absorbs a photon with energy exceeding the binding energy of one of its electrons, causing the electron to shift to a higher energy level or be ejected from the atom. This process can alter the atomic configuration and potentially displace the atom from its original position, leading to changes in material properties such as electrical conductivity, optical characteristics, and mechanical attributes. The extent of photoinduced dislocation is determined by factors such as light intensity and wavelength, the amount of energy required to overcome the atom's binding energy, and interactions among neighboring atoms in the lattice. High-energy photons, such as those from UV and X-rays, are particularly effective at inducing photoinduced dislocation due to their ability to penetrate deep into materials and cause ionization. In such cases, metals can experience significant changes in their mechanical properties, such as increased brittleness, reduced ductility, and altered strength. Non-metals' susceptibility to radiation damage varies based on their composition and structure; for example, ceramics and glasses may exhibit increased resistance to radiation-induced damage, while polymers and organic materials may undergo severe degradation when exposed to radiation.

Radiation exposure can induce several forms of damage in copper, such as the displacement of copper atoms from their lattice sites, the creation of vacancies, and interstitials. These defects can adversely affect the mechanical properties of the metal, leading to reduced ductility and increased brittleness (Figure 2.50). In addition, copper is prone to radiation-induced embrittlement, especially under high-energy neutron exposure. This can trigger the formation of hydrogen gas within the metal, leading to swelling and cracking, ultimately resulting in material failure.

One of the most common forms of radiation affecting polymers (Figure 2.51) is ionizing radiation, which includes gamma rays, X-rays, and electron beams. These types of radiation can cause chain scission, where the polymer chain breaks into smaller segments, resulting in a decrease in

Figure 2.51 Polymer degradation upon radiation exposure.

molecular weight and changes in mechanical and thermal properties. In addition, radiation can induce crosslinking, forming a three-dimensional network structure. Radiation exposure can form discontinuities and remove branches from a polymer (Figure 2.51). In the presence of another monomer, new structures, known as grafted chains (Figure 2.51), can form, significantly altering the polymer's microstructure. This process can improve the mechanical and thermal properties of the polymer, making it more resistant to heat and chemicals. However, excessive crosslinking may lead to embrittlement and reduced flexibility.

UV radiation-induced degradation can have significant effects on various mechanical properties of materials (Table 2.1). First, it can lead to a decrease in tensile strength by introducing new chemical or functional groups on the material's surface, weakening intermolecular bonds, and promoting the formation of microcracks or crazing within the material. These defects and structural changes contribute to the reduction in tensile strength. Similarly, UV radiation can diminish the Izod impact strength of materials by rendering molecular chains more brittle, reducing the material's ability to absorb energy during impact loading. Degradation-induced microcracks and defects also act as stress concentrators, making it easier for cracks to start and spread and lowering the resistance to impact even more.

Table 2.1 Impact of UV radiation on epoxy surface

Sample	*Tensile strength (MPa)*	*Izod impact strength (kJ/m²)*	*Pull off strength (MPa)*
A1 (0 h)	29.12	26.24	4.42
A2 (250 h)	27.87	24.87	4.33
A3 (500 h)	25.39	23.14	4.04
A4 (800 h)	20.25	20.33	3.52

The results listed in Table 2.1 indicate that the considerable damage inflicted on the epoxy, which possesses a relatively low melting temperature, arises from radiation-induced displacement of atoms, resulting in surface damage and alterations in surface behavior. Initially, the epoxy exhibited a tensile strength of approximately 29 MPa. After 250 hours, a decrease of approximately 5% was observed, with a further decline noted after 500 hours, although the bulk material still retained around two-thirds of its original strength. The toughness, initially measured at approximately 26.24, declined to approximately 20. This reduction in toughness and shear strength, also referred to as pull-off strength, underscores the overall decrease in properties over time due to radiation exposure.

Furthermore, UV radiation exposure can affect the pull-off strength, which represents the adhesive strength between a coating or surface layer and the substrate. UV radiation can induce the degradation of the coating material, altering its physical and chemical properties by inducing photo-oxidation and the formation of weakening chemical groups. In addition, UV radiation can cause surface degradation, such as roughening or cracking, which compromises the interfacial bond strength and ultimately reduces the pull-off strength.

Figure 2.52 shows the fracture morphologies observed in Charpy impact specimens aged at 475°C for various periods, as well as the relationship between embrittlement and the microstructural changes of α-ferrite in duplex stainless steel. Initially, during the early stages of aging, a prevalent ductile fracture mechanism is evident, as shown in Figure 2.52(a) and 2.52(b), displaying numerous deep dimples.

After 8 hours of aging, a noticeable shift occurs in the fracture appearance (Figure 2.52(c)). Smaller and less defined facets become linked by ridges or shallow dimples. Within these facets, a faint river pattern emerges, extending from the center of each. As aging progresses to 16 hours (Figure 2.52(d)), this characteristic river pattern becomes more pronounced on the fracture surface.

In specimens aged at 475°C for 32 hours (Figure 2.52(e)), predominant quasi-cleavage areas emerge, indicating the prevalence of a brittle fracture mechanism. This indicates material embrittlement, with α-ferrite cleavage

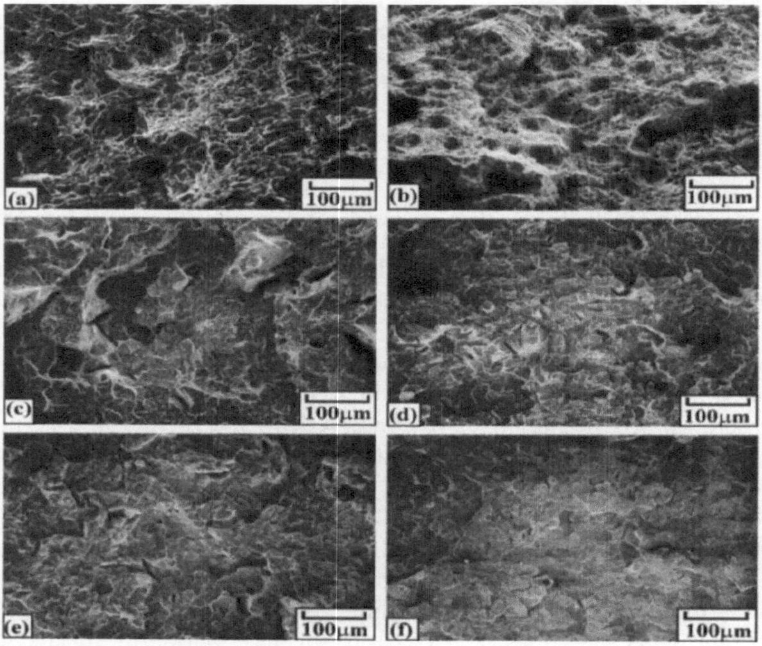

Figure 2.52 Embrittlement of stainless steel over time.[4]

playing a significant role in the loss of toughness. With age, the α-ferrite phase becomes notably more brittle, leading to a transition in the austenite phase's fracture mode from dimple rupture to brittle rupture.

During aging, the hardness of α-ferrite substantially increases, while the hardness of austenite remains relatively stable. This consistent observation suggests that the embrittlement phenomenon in the examined duplex stainless steel is primarily associated with microstructural changes occurring in α-ferrite.

Figure 2.53 demonstrates the relationship between Rockwell C hardness (HRC) and aging time at different aging temperatures. It reveals that hardness generally increases with aging time across all temperatures, except during the initial stage of aging at 400°C. This trend suggests a correlation between embrittlement and increased hardness. The explanation provided is that aging significantly enhances the hardness of ferrite, which is present in stainless steel but not in austenite. While the hardness of austenite remains relatively stable, the hardness of ferrite steadily increases with aging. This continuous increase in ferrite hardness correlates with increased embrittlement, leading to a higher susceptibility to brittle failure. Over time, the ferrite phase becomes increasingly fragile, predisposing the material to

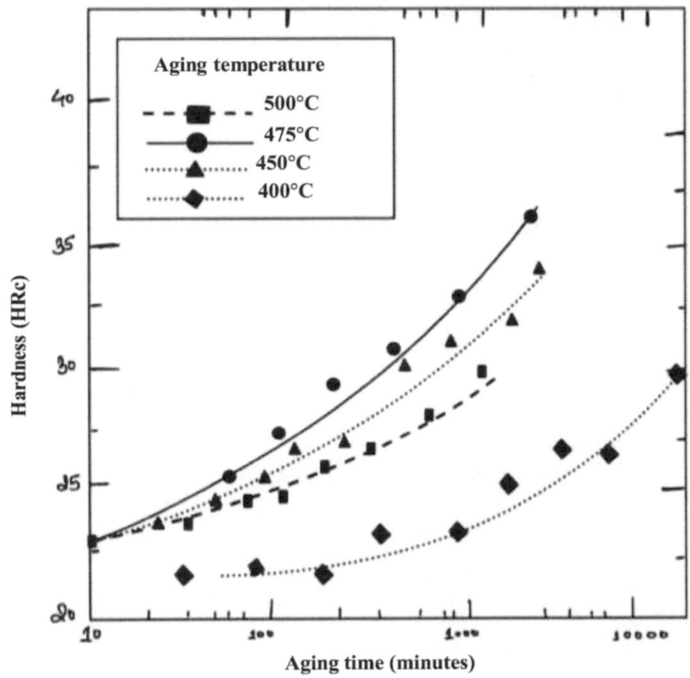

Figure 2.53 Variation of HRC with respect to aging time at various temperatures.

brittle failure. As noted, the decrease in dimples observed on the surface demonstrates this shift in microstructure, indicating a transition from ductile to brittle rupture.

2.10 ROLE OF FATIGUE AND FRACTURE MECHANISMS IN WEAR DEBRIS FORMATION

The current book focuses on identifying surface faults to prevent failures. Our primary concern is the failure caused by wear particles emerging from the surface. The goal is to understand the process of particle separation from the surface, which is associated with fatigue and fracture. Thus, we believe that fatigue and fracture are responsible for generating wear debris.

Fatigue and fracture mechanisms provide valuable insights into wear phenomena by effectively separating wear debris from the parent surface, underscoring their significance. Fatigue, often induced by repetitive cycles, leads to increased brittleness in materials and the gradual accumulation

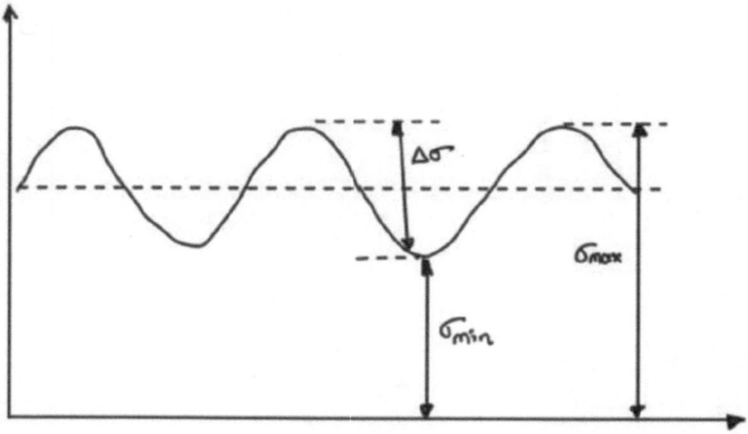

Figure 2.54 Reversible stress fluctuations in stress diagrams.

of localized damage over successive loading and unloading cycles, eventually resulting in fracture and wear debris. Surface irregularities account for approximately 70% to 80% of fatigue incidents. Stress undergoes reversible fluctuations between maximum and minimum values, illustrated in stress diagrams (Figure 2.54) with the range denoted by $\Delta\sigma$. Furthermore, we can describe stress as sliding instead of just the load, implying that contact during sliding triggers the load, allowing us to depict fatigue phenomena in terms of reciprocal or unidirectional sliding.

Sliding contact occurs primarily at the asperities, where individual roughness peaks load onto each other or onto the material surface itself. During sliding, there is a continuous cycle of loading and unloading, meaning both processes happen simultaneously. If we know the exact load magnitude, we can analyze load-related factors; otherwise, we consider sliding and relative motion. In this book, we highlight how localized degradation, stemming from these asperity contacts, can ultimately lead to mechanical unit failure. Nearly all mechanical systems harbor inherent cracks, and fracture mechanics delves into how these cracks propagate, effectively linking fatigue and fracture phenomena in our discussions. When it comes to the creation of wear debris, it can be envisioned as crack initiation followed by detachment from the surface. Thus, I perceive fracture as intrinsically intertwined with wear debris, where the separation of debris from the surface is indicative of a fracture event. Understanding these mechanisms is pivotal for averting wear debris generation and curbing surface deterioration to prolong component lifespan. In this pursuit, striving for ultra-mild wear is a paramount objective.

Fracture mechanics specifically addresses the initiation and propagation of cracks. From my standpoint, cracks are ubiquitous in materials, arising from various factors such as boundary conditions, crystallization processes, or material alloying. The crucial question arises: Does the system possess sufficient energy, load, and velocity to facilitate crack propagation or enlargement? The study of fatigue and fracture mechanisms, which underlie wear debris generation, provides an answer to this inquiry.

In addition to the widely acknowledged S–N diagram, or S–N criteria, which is commonly used to analyze fatigue phenomena, attention is also directed toward the strain–life cycle, which encompasses strain and the number of cycles until fracture occurs. According to fracture mechanics, fatigue contributes to the development of brittleness in a material, thereby initiating or propagating cracks. As a result, it is critical to approach fatigue analysis through a fracture mechanics lens. To address the inherent uncertainties in fatigue behavior, a probabilistic approach is frequently utilized. This entails integrating various fatigue design philosophies within engineering practices, each of which offers distinct methodologies and considerations:

- *Stress–life (S–N) approach:* Also known as the S–N curve approach, this method relies on the relationship between the applied stress range and the number of cycles to failure. S—N curves are established through laboratory testing, where specimens endure various stress amplitudes at a constant stress ratio until failure occurs. These curves help estimate a component's fatigue life based on the stress range it experiences.
- *Strain–life (ε–N) approach:* This approach considers the relationship between strain amplitude and the number of cycles to failure. Strain–life curves are created through experimental testing, similar to the S–N approach. It is particularly useful for materials exhibiting significant cyclic plastic deformation.
- *Fracture mechanics approach:* This approach evaluates crack propagation and its influence on a component's fatigue life. Principles of fracture mechanics, such as stress intensity factors and crack growth rates, are utilized to assess crack behavior and predict remaining fatigue life. It is commonly applied to structures with existing cracks or to evaluate the impact of flaws on fatigue life.
- *Damage tolerance approach:* Focusing on a structure or component's ability to withstand defects or damage without catastrophic failure, this approach assesses the size and location of potential flaws. Fracture mechanics principles are used to determine their effect on fatigue life. It is crucial in industries like aerospace, where small defects are inevitable but must not lead to catastrophic failure.
- *Probabilistic methods:* These methods analyze variations in material properties, loading conditions, and other factors to assess reliability

against fatigue failure, offering a thorough grasp of fatigue behavior by addressing uncertainties in design and manufacturing. In a probabilistic approach, we assume dynamic, reversible processes due to continuous loading and unloading. This method is fundamentally statistical, as it accounts for the likelihood of various problems in a material and the probabilistic nature of loading.

The selection of a fatigue design philosophy hinges on a variety of factors, including material properties, loading conditions, component geometry, and industry standards. Engineers often integrate these philosophies to devise designs that are resilient to fatigue while meeting safety and performance criteria. Inherent cracks exist in most mechanical systems, prompting a fracture mechanics approach to comprehend crack propagation and its impact on fatigue life. Understanding the science behind debris creation is crucial, considering the progressive damage from fracture initiation. Fracture mechanics principles, encompassing stress intensity factors and crack growth rates, aid in assessing crack behavior and predicting remaining fatigue life.

Moreover, probabilistic methodologies assess the reliability and safety of components or structures against fatigue failure by accounting for statistical variations in material properties, loading conditions, surface profiles, and other influential factors. Therefore, adopting a probabilistic approach is advisable. In this context, it is essential to acknowledge the statistical fluctuations inherent in material properties. This recognition leads to the establishment of lower and upper limits for material strength, along with mean values and deviations, which are crucial for accurately representing material properties. Similarly, loading conditions exhibit statistical distributions, with maximum and minimum values, variations, mean values, and associated deviations. These considerations underscore the fatigue-related significance of incorporating probabilistic assessments into fatigue analysis methodologies.

By incorporating the aforementioned theories, one can model the removal of wear debris from the surface. The model must assume certain patterns, conduct experiments, and integrate the experimental results back into the model. Models cannot be generalized and must work in tandem with experimental setups. This approach is crucial for preventing corrosion and surface degradation, so it is important to keep these considerations in mind. As detailed modeling can be challenging, we often use a simpler approach to understand material behavior under fatigue loading, which we categorize into three types, as illustrated in Figure 2.55.

The first category involves plastic strain accumulation, known as plastic ratcheting. After a number of cycles, we observe patterns of fractures, such as surface cleavages and beach marks, caused by high stress levels exceeding

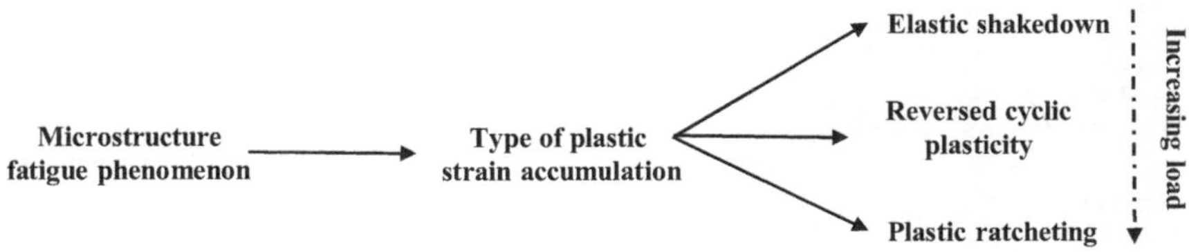

Figure 2.55 Fatigue loading behavior.

the plastic limit. Continuous plastic strain buildup on the surface eventually leads to a transition from ductile to brittle fracture, resulting in failure.

The second category is reversible cyclic plasticity. Here, the load is lower, resulting in a pattern with a much longer life and more cycles compared to plastic ratcheting.

If we further reduce the loads, we achieve an elastic shakedown. In other words, elastic shakedown refers to stress or strain levels below which cyclic plasticity stops. When a material goes through reversed plastic straining during cycling without a net directional accumulation of plastic deformation, this is called reversed cyclic plasticity. It is also known as plastic shakedown. Plastic ratcheting describes the condition where the material undergoes a net directional plastic strain increment during each cycle.

Elastic shakedown does not imply that the stress level never reaches the plastic limit; it does. Initially, asperities are in contact, leading to plastic deformation. Imperfections change shape, increasing the contact area. When the contact area increases, the maximum stress decreases, potentially falling just below the yield limit. When this happens, the material is considered mature enough to sustain the load, reaching an endurance limit. This phase, known as the fatigue limit phase, occurs during elastic shakedown and is where the term endurance limit applies.

We can relate these three stages to the previously discussed bathtub curve (Figure 2.17). From A to B, or B to A, wear is at its maximum. This happens because, at level A, a few sharp asperities come into contact, resulting in high stress levels. As they slide, these asperities change shape. Initially, sharp peaks undergo plastic deformation, developing contours at the bottom (Figure 2.56). These contours increase the projection or cone angle, thereby reducing the wear rate.

Following this initial wear phase, the material enters a stage of elastic shakedown, represented by BC on the curve. During this phase, the material can sustain the endurance limit, and the number of cycles increases. This means that the material can better resist further deformation and wear, as the stress levels have adjusted to be more manageable.

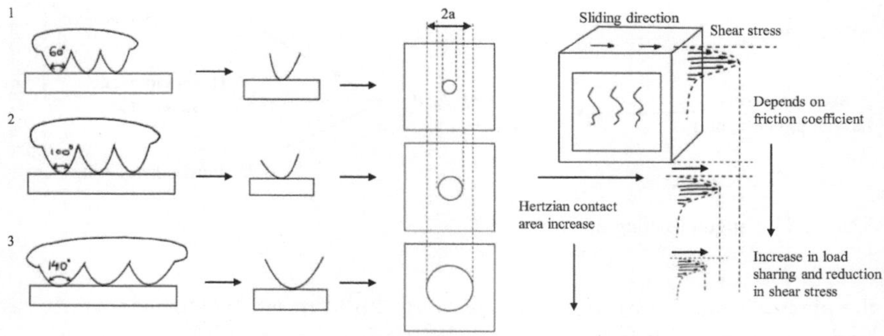

Figure 2.56 Plastic deformation contours increase Hertzian contact radium and reduce wear rate.

Finally, in the phase represented by line CD, the contact area decreases again due to an increase in the number of cracks. As these cracks propagate, the stress rises once more, leading to failure. This final stage links fatigue and fracture phenomena with wear, as the material ultimately succumbs to the cumulative damage incurred during its lifecycle. This demonstrates the interconnected nature of wear, fatigue, and fracture in material degradation.

Sometimes, people say these phenomena are unconnected. Elastic shakedown leaves residual stress on the material surface, improving wear resistance through micro-hardening. This process can result in rounded edges instead of sharp peaks, which reduces wear. As the asperities round off, load redistribution occurs. This redistribution lowers the maximum contact pressure, leading to a more efficient distribution of contact pressure. More rounded asperities come into contact, reducing friction and significantly lowering wear. This increases the chances of elastic shakedown, allowing the material to undergo fatigue fracture rather than immediate failure. The acceptable wear coefficient is generally less than 10^{-7}, meaning the material can endure many cycles of contact without significant wear. This wear coefficient helps relate to the probability of failure or wear debris generation.

Understanding the connection between fatigue, fracture, and wear is crucial. In my opinion, they are interrelated. To understand fracture, we must consider the separation of debris particles from a surface. It is unnecessary to assume that a complete fracture or separation of the material throughout its entire thickness is the only possibility. Instead, we should focus on where and how the debris separates.

2.11 CONCLUSIONS

Understanding surface degradation is critical for ensuring the longevity and reliability of components subjected to various wear mechanisms. This chapter delves into the intricate processes by which wear phenomena—such as adhesive, abrasive, and fatigue wear—interact with material properties and external conditions to accelerate degradation. Each type of wear presents unique challenges. Factors including load, sliding speed, temperature, and environmental conditions like moisture or corrosive agents further influence the interplay of these wear mechanisms. The present chapter provides a comprehensive understanding of these variables. The chapter emphasizes the importance of proactive measures to combat wear, including the selection of materials with high wear resistance and the use of lubrication to minimize friction and heat generation. In addition, regular maintenance practices, such as inspecting and replacing worn components, ensure that systems continue to perform efficiently, reducing the risk of unexpected failures. The chapter concludes that a thorough grasp of degradation mechanisms and the application of tailored engineering solutions are foundational to enhancing product performance and sustainability.

REFERENCES

1. Wang, F., Zhou, D., Zhou, X., Xiao, N., & Guo, C. (2020). Rock breaking performance of TBM disc cutter assisted by high-pressure water jet. *Applied Sciences*, *10*, 6294. https://doi.org/10.3390/app10186294
2. Lim, S. C., & Ashby, M. F. (1987). Overview no. 55 wear-mechanism maps. *Acta Metallurgica*, *35*(1), 1–24.
3. Bartz, W. J., Chatterley, P., & Moore, A. J. (1993). *Lubrication of gearing: lubricants and their properties: design of gears: Practical gear lubrication: failure analysis*. Mechanical Engineering Publications.
4. FRACTURE, a project report, https://esm.rkriz.net/classes/ESM5344/ESM5344_NoteBook/crcd/lectures/fracture/contents.html
5. Liu, S., Liu, M., Liu, T., Xiao, H., & Xiao, B. (2023). Wear characteristics of brazing diamond abrasive wheel on high efficiency grinding ferrous metals. *Wear*, *514*, 204580.
6. Tong, Z., Zhou, G., Zheng, W., Zhang, H., Zhou, H., & Sun, X. (2022). Effects of heat treatment on the microstructure and mechanical properties of a novel H-grade sucker rod steel. *Metals*, *12*, 294. https://doi.org/10.3390/met12020294

Chapter 3

Corrosion

The preceding chapter examined the influence of mechanical forces on surface degradation and how enhanced hardness can mitigate this effect. It is essential to understand the effects of surface degradation caused by chemical interactions, known as corrosion, with ambient gases, moisture, and pollutants. A multitude of students commenced their study of "corrosion" in the sixth grade, signifying a progressive enhancement of their foundational comprehension of the subject. Given our prior study of corrosion, what necessitates an additional chapter on its effects?

Corrosion occurs via electrochemical reactions, wherein reduction (electron gain) transpires at the anode and oxidation (electron loss) takes place at the cathode. Excessive oxidation of metals leads to detrimental degradation. Corrosion may occur together with mechanical wear. Corrosion can induce void formation, whereas mechanical loads may cause cracks to propagate perpendicularly to the surface, resulting in sudden system collapse. Understanding these issues is essential, as the design protocols for alleviating surface deterioration caused by external conditions may evolve.

The ramifications of corrosion are substantial, including diminished performance, heightened maintenance expenses, and jeopardized safety resulting from the degradation of structural integrity and strength. Annually, the degradation of diverse constructions such as pipelines, tanks, mechanical components, ships, bridges, and marine infrastructure incurs a total expense beyond INR 1.5 trillion. Moreover, safety issues emerge as corrosion-related failures can result in disastrous outcomes in essential apparatus such as pressure vessels, boilers, hazardous chemical containers, turbine blades, bridges, aircraft components, automotive systems, nuclear power plant equipment, and nuclear waste management systems. Efficient corrosion management and surveillance are essential for reducing economic and safety-related losses attributed to corrosion.

Surface treatments, material selection, and environmental adjustments are some significant techniques to mitigate corrosion. Choosing

DOI: 10.1201/9781003545019-3

corrosion-resistant materials, such as aluminum, stainless steel, or specialty alloys, mitigates the risk of corrosion based on operational and environmental conditions. In industrial systems, inhibitors are chemicals that mitigate corrosion by forming protective layers or neutralizing detrimental elements. Paints, polymers, and galvanization are surface treatments that create a barrier to safeguard things from corrosive oxygen and moisture. Specialized coatings, including ceramic or metallic layers, provide enhanced protection in vital sectors such as maritime and aerospace industries. By integrating these methodologies, engineers can establish robust protections against corrosion, enhance material longevity, diminish maintenance expenditures, and augment safety. This mitigates the financial and safety issues resulting from rusting.

3.1 INTRODUCTION

Corrosion occurs when a material chemically interacts with its environment, resulting in the formation of corrosion products. These products frequently produce a thin, cohesive coating, generally measuring between 1 and 10 nm in thickness. This is advantageous as it does not significantly alter the material's size. This slender barrier reduces the likelihood of delamination by maintaining a strong adhesion to the surface and exhibiting insolubility or resistance to dissolution. The layer serves as an impermeable coating, effectively safeguarding the underlying material and preventing further corrosion. This protective effect is particularly evident in certain metals that inherently resist further corrosion by developing a sacrificial oxide layer. These slender, impermeable films can regenerate if the operating parameters remain within specific limits, despite potential degradation under sliding and moderate pressure conditions. The ability to rapidly regenerate following degradation is an essential attribute, ensuring ongoing protection. These metals are ideal for regular usage due to their durable, self-renewing protective covering. Nonetheless, it is crucial to understand how they may fail under specific conditions. As the oxide layer thickens due to repeated contact, it may become brittle and fracture, resulting in the formation of wear particles. If these particles are abrasive, they accelerate wear, necessitating human involvement to mitigate damage. Conversely, if the trash is pliable and yielding, it may function as a lubricant, safeguarding the surface.

The electrochemical potential of a material (as listed in Figure 3.1) serves as a fundamental parameter in assessing a material's susceptibility to corrosion. The standard electrode potentials (E°), measured in volts (V), indicate how likely a species is to gain electrons and undergo reduction under standard conditions (25°C, 1 atm pressure, and a concentration of 1 Molar, M). The positive E° value indicates a favorable reduction reaction, meaning the species has a strong tendency to gain electrons and reduce

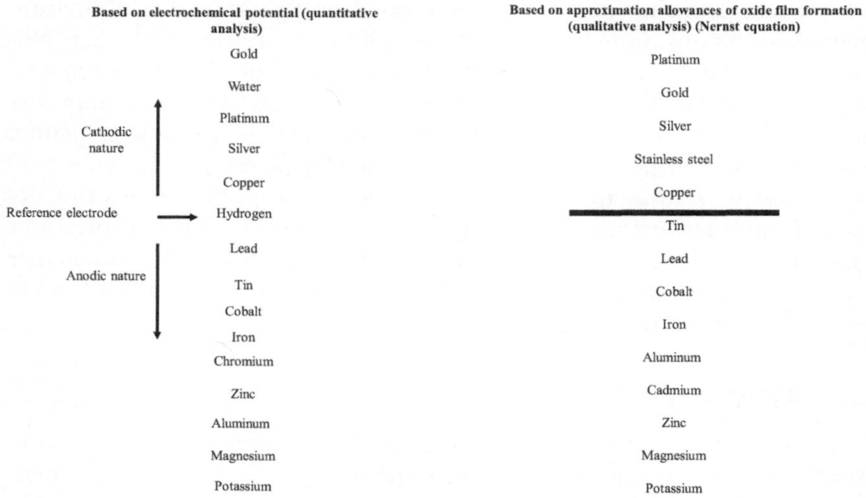

Figure 3.1 Series of electrochemical potentials.

easily. In contrast, a negative E° value suggests that the species does not readily gain electrons, making reduction less likely. For example, gold (Au^+) has a high positive potential of +1.69 V, indicating that it is easily reduced to metallic gold, which explains why gold is resistant to oxidation and corrosion. Platinum ($Pt^{2+} + 2e^- \rightarrow Pt$) has a positive electrode potential (+ 1.18 V), indicating a strong tendency of corrosion resistance. Silver ($Ag^+ + e^- \rightarrow Ag$) has a moderately positive potential (+0.799 V), also making it favorable for reduction. The reduction of hydrogen ions to hydrogen gas ($2H^+ + 2e^- \rightarrow H_2$) serves as the reference point for all standard electrode potentials, with a value of 0.000 V. On the other hand, metals like zinc (Zn^{2+}, −0.762 V) and aluminum (Al^{3+}, −1.68 V) have negative potentials, meaning they are more prone to remain in their oxidized forms, making them more susceptible to corrosion.

The corrosion rate of a substance is significantly influenced by its electrochemical potential. Materials exhibiting a larger negative potential are typically more reactive and susceptible to corrosion, whereas those with positive potential, such as gold and platinum, have more resistance. Metals like aluminum and chromium, while regarded as corrosion-resistant due to their protective oxide coatings, possess negative E° values, indicating their chemical reactivity. This underscores the necessity of comprehending the distinct features of each material and their interactions with the environment when choosing materials and formulating corrosion control measures. Materials having larger negative potentials exhibit more susceptibility to corrosion, whereas noble metals such as gold demonstrate significant resistance. Reactive metals such as sodium, however, exhibit vigorous reactions with water and are typically seen in compound form rather than in their elemental state.

Non-metals such as polymers, ceramics, and composites may also undergo "corrosion degradation" in specific situations. Elevated temperatures can increase a material's activation energy, resulting in enhanced corrosion wear. Certain polymers in ceramics may expand, soften, or dissolve upon exposure to acids. To select the appropriate material for an application, it is essential to evaluate these criteria and employ suitable protective measures. Understanding this facilitates improved material selection.

Sliding over corroded surfaces can result in severe wear, necessitating methods to mitigate the damage. Bearing corrosion, seen in Figure 3.2, results in numerous issues, including diminished performance, increased maintenance expenses, safety hazards for equipment and personnel, and the necessity for bearing replacement. Corrosion can result in bearing failure, causing unanticipated downtime, diminished production, missed deadlines, and financial losses. Early detection allows for maintenance steps such as cleaning, lubrication, and treatments to mitigate or halt corrosion; nevertheless, these measures incur expenses related to manpower and materials. In certain instances, bearing corrosion may pose safety risks, resulting in accidents, injuries, or damage to adjacent components. Numerous techniques exist to establish barriers that proficiently withstand corrosion. One way is "chemical passivation," which safeguards materials from corrosion. This entails the application of protective coatings, such as oxides or hydroxides, to the surface to avert more deterioration. Stainless steel can be passivated with nitric or citric acid solutions. These acids facilitate the removal of free iron and other contaminants from the surface, enabling the formation of a chromium-rich oxide layer. This oxide layer enhances the corrosion resistance of stainless steel, increasing its durability in adverse situations.

An alternative approach is exemplified by conversion coatings. This procedure entails the chemical treatment of the metal surface to convert it into a layer that exhibits enhanced corrosion resistance. This layer

Figure 3.2 Corrosion of "inner ring of roller bearing" and "connecting rod bearing."

comprises compounds like phosphates, chromates, or oxides that function as a protective barrier against corrosion. Phosphating is a prevalent industrial technique for adding conversion coatings to steel, while anodizing is typically employed for aluminum and its alloys. Anodic films produced through anodization typically demonstrate increased thickness, strength, and corrosion resistance relative to natural oxide layers. Coating systems utilized in aviation typically comprise three layers: a conversion coating/anodization layer, a primer layer that provides corrosion resistance, and a topcoat that serves as the primary barrier.

Moreover, physical barriers can function as efficient methods for protecting metal surfaces from corrosion. Coatings like paints and polymer films protect metals from corrosive environments, preserving their structural integrity and extending their lifespan.

Protection can be further augmented using sacrificial anodes, which corrode instead of the primary metal surface. Sacrificial coatings, such as zinc galvanizing, typically demonstrate superior thickness and endurance. Zinc galvanizing is the deposition of a zinc layer onto steel, which oxidizes to safeguard the steel from undesirable corrosion. Rusting specifically refers to the corrosion of iron or iron-based alloys, leading to the development of hydrous ferric oxides.

3.2 CORROSION FUNDAMENTALS

Corrosion transpires when materials, particularly metals, interact with their surroundings. This process generally entails electrochemical reactions, wherein metal atoms undergo electron loss to form ions (oxidation process as delineated in Eq. (3.1)), and subsequently react with oxygen (as specified in Eq. (3.2)) or other substances (as demonstrated in Eq. (3.3)) to generate corrosion products (such as rust), resulting in material degradation and loss of structural integrity.

$$\mathrm{Fe} \rightarrow \mathrm{Fe}^{2+} + 2\mathrm{e}^- \tag{3.1}$$

$$4\mathrm{Fe} + 3\mathrm{O}_2 \rightarrow 2\mathrm{Fe}_2\mathrm{O}_3 \tag{3.2}$$

$$\mathrm{Fe}_2\mathrm{O}_3 + x\mathrm{H}_2\mathrm{O} \rightarrow \mathrm{Fe}_2\mathrm{O}_3.x\mathrm{H}_2\mathrm{O} \tag{3.3}$$

Corrosion presents in multiple forms, each exhibiting unique traits and causes. The six primary types of corrosion are: general (uniform) corrosion, which uniformly affects the entire surface; galvanic corrosion, caused by dissimilar metals; localized corrosion, which occurs in specific areas; microbial-induced corrosion (MIC), instigated by microbial and fungal organisms;

stress corrosion cracking (SCC), resulting from the interplay of corrosion and mechanical stress; and high-temperature oxidation, occurring in reactive gas or molten salt environments. MIC transpires when microorganisms, including bacteria and fungus, expedite corrosion by releasing corrosive byproducts or modifying the chemical milieu. The high-temperature oxidation process accelerates metal oxidation rates, compromises material integrity, and may result in structural failure, particularly in high-temperature applications such as industrial furnaces or jet engines. This chapter will not address MIC. Temperature significantly influences all forms of corrosion; therefore, high-temperature corrosion will not be addressed independently.

Before detailing the types of corrosion, it is worth considering the material's ability to corrode from the interaction between a metal and its surroundings. To understand it, let us consider "Experiment Setup by Evans[1]":

- Two samples of iron are connected using a galvanometer. Both the iron samples are immersed in two solutions of potassium chloride, separated by a porous membrane. An electric current circulates between the two iron samples.
- Air is bubbled through one of the solutions, and this aerated sample becomes the cathode. As air, which contains oxygen, is bubbled through the solution, oxygen dissolves into the liquid, increasing its concentration. The dissolved oxygen promotes a reduction reaction, as oxygen tends to accept electrons. In an electrochemical cell, the cathode is the site of the reduction reaction (where electrons are gained). As the aerated solution now has more oxygen available for reduction, it acts as the cathode in this setup.
- The non-aerated sample becomes the "anode" and starts to corrode.

In another experiment, iron is connected to another metal (like copper, zinc, or magnesium) instead of another iron sample:

- An electric current is observed between the two metals.
- "Iron becomes the anode" and corrodes when connected to "copper."
- "Zinc or magnesium" becomes the anode and corrode instead, providing protection to iron.

Both chemical and electrochemical mechanisms must be considered to analyze corrosion reactions. The chemical process emphasizes reactions that do not entail the transfer of electric charges, such as oxidation or reduction. These reactions adhere to the "law of mass action," which posits that the pace of a chemical reaction is contingent upon the concentrations or quantities of the participating components, and the "law of solubility product," which pertains to the dissolution and reactivity of substances in a solution, influenced by parameters such as ion concentration. The electrochemical process pertains to reactions involving electric charges or electrons. In these reactions, electrons are either acquired or relinquished, as shown in oxidation–reduction reactions. While chemical reactions are influenced by factors such as concentration and pressure, electrochemical reactions furthermore entail the transfer of electric charges, introducing an additional layer of complexity.

A common way to understand corrosion reactions is through a "Pourbaix diagram," which is a graphical tool used to predict how a material reacts in different environments, particularly how it might corrode or remain stable. The "x-axis" of the diagram shows the "pH" of the solution, which ranges from acidic to alkaline. The "y-axis" shows the "electrode potential (Eh)," which indicates whether a substance tends to lose or gain electrons. Using a Pourbaix diagram, one can predict:

- Corrosion zones, where the material is likely to dissolve and lose ions.
- Non-corrosion zones, where the material remains stable.
- Passivation zones, where a protective oxide layer forms on the surface of the material, preventing corrosion.

Pourbaix diagrams assist in determining the zones where metals exhibit stability, corrosion, or passivation (protection by an oxide layer). For example, they illustrate the formation of protective oxide layers on iron or aluminum, enabling engineers to select appropriate conditions for "cathodic protection," "anodic protection," or the application of inhibitors. Nevertheless, these predictions require empirical validation, as the formation of a protective coating does not invariably ensure total protection.[1]

Constructing corrosion-resistant materials and buildings, as well as devising effective strategies to prevent and minimize corrosion, necessitates comprehension of fundamental corrosion concepts. The utilization of corrosion inhibitors, protective coatings, appropriate material selection based on environmental conditions, and effective maintenance practices are all viable techniques in this context. It is essential to reevaluate the classification of corrosion. This book will exclude microbiological and fungal deterioration. Nonetheless, there exist five primary categories of corrosion in a broader context: general corrosion, SCC, crevice corrosion, pitting corrosion, and galvanic corrosion.

3.3 GENERAL CORROSION

This corrosion, termed "uniform corrosion," uniformly impacts the whole surface of a material, resulting in slow degradation. It generally transpires in settings where corrosive substances are uniformly distributed. A typical instance is "aqueous corrosion," which initiates at the anodic regions of a material (where oxidation occurs) and results in the development of a protective layer that safeguards the material from further deterioration.

For example, "aluminum alloys," known for their resistance to corrosion, form a protective oxide layer (Al_2O_3) on their surface. This oxide layer acts as a barrier, preventing more corrosion or oxidation of the metal underneath. If the oxide layer gets damaged, aluminum alloys can "self-heal" by reforming the protective layer in the presence of oxygen, continuing to protect against corrosion.

During corrosion, aluminum alloys experience polarity shifts, meaning the corroded areas become more electrochemically stable (noble) than the surrounding metal. This causes the surrounding material to corrode, forming a protective film. A real-life example of this is when an aluminum sample is partially submerged in a 10% hydrofluoric acid solution, which causes the metal to corrode and develop an etched appearance, showing uniform corrosion.

Aqueous corrosion occurs due to electrochemical processes. Water molecules undergo ionization upon contact with the metal surface, resulting in the release of metal ions. These ions subsequently react with hydroxide ions, resulting in the formation of insoluble metal hydroxides. Environmental factors such as variations in temperature, acidity, or the presence of contaminants might alter the kinetics of this corrosion process. When a metallic surface interacts with an aqueous environment, the surface metal atoms may ionize and dissolve into the solution. Ionization primarily transpires at adatoms, which are atoms located on the peripheries of the metal surface, or along terraces, which are planar areas. Adatoms and atoms situated along terraces exhibit greater susceptibility to ionization due to their bonding with fewer neighboring atoms in contrast to subsurface atoms.

Ionization entails the transfer of electrons from metal atoms to the aqueous solution, leading to the generation of metal ions and the liberation of electrons. Various factors can affect this process, including the solution's pH, the presence of dissolved oxygen or other reactive species, and the properties of the metal surface. In addition to ionization, metal surfaces in aqueous settings may also experience additional types of corrosion, including oxidation and dissolution. Impurities or pollutants in the solution can expedite these processes by functioning as catalysts for the reaction. Figure 3.3 depicts a chain and jaw coupling with a dark film of iron oxide, commonly referred to as rust. Rusting, as illustrated in Figure 3.3 and defined by Eq. (3.3), constitutes a form of "aqueous corrosion" that impacts metals such as iron, steel, and other ferrous substances. Iron and steel are more susceptible to rust than other metals due to their comparatively weaker protective coatings.

As per a few researchers, the topic of "Rusting" and "Aqueous Corrosion" should be described in different headings. Rusting is an oxidation reaction

Figure 3.3 Rusting of chain and jaw-coupling.

where iron reacts with oxygen to form iron oxide. During this process, iron atoms lose electrons, generating iron (II) ions (Fe^{2+}), which combine with water to form iron (II) hydroxide [$Fe(OH)_2$]. Further oxidation produces iron (III) ions (Fe^{3+}), which eventually form rust, a hydrated form of iron (III) oxide [$Fe_2O_3 \cdot xH_2O$].

On the other hand, "aqueous corrosion" involves the electrochemical corrosion of metals like aluminum, copper, and zinc in water environments. This process operates through electrochemical reactions:

- Anode (–): The site where the metal loses electrons (e.g., zinc).
- Cathode (+): The receptor of electrons (e.g., copper).

Aqueous corrosion necessitates the interaction of an electrolyte, such as water containing dissolved ions, with the material's surface. Variations in potential over the metal's surface result in the establishment of anodes (sites of metal dissolution) and cathodes (sites of reduction processes). Metal atoms at the anode undergo oxidation, relinquishing electrons and dissolving into the electrolyte as metal ions. At the cathode, reduction reactions, including the reduction of dissolved oxygen or hydrogen ions, facilitate the rectification of charge imbalance. When the metal ions generated at the anode interact with other environmental components, a variety of corrosion products are produced. Rust can serve as a barrier to impede further corrosion, functioning as a sort of "passive corrosion protection" by creating a layer that obstructs additional corrosion. In certain instances, rusting escalates to a severe condition known as "active corrosion," wherein rust proliferates uncontrollably, as illustrated in Figure 3.4. In such instances, rust can induce the metal to grow, frequently by more than mere nanometers or microns. A further issue with rust is its propensity to flake off due to its porous nature and inadequate adhesion to surfaces. This is particularly problematic when components operate at elevated velocities, as rust particles can detach more readily.

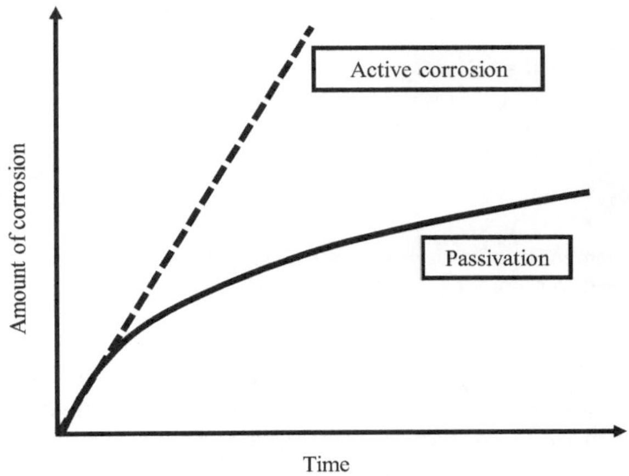

Figure 3.4 Passive versus active corrosion.

A protective coating serves as a barrier against more corrosion. Nonetheless, if this layer fails to form, "active corrosion" may persist, ultimately leading to the entire degradation of the material. In active corrosion, the deterioration exacerbates with time, potentially advancing at a consistent or erratic pace, contingent upon the prevailing conditions. The velocity and intensity of "aqueous corrosion" are affected by variables such as temperature, precipitation, solar radiation, surface inclination, sulfur dioxide concentrations, and the presence of salt or dust in the atmosphere.

A fundamental principle is that specific corrosion films stop forming after reaching a given thickness, thereby passivating the surface and restricting future oxygen exposure to the metal. This process is referred to as "chemical passivation," which establishes barriers and successfully ceases the corrosion process. It is imperative to refrain from hastily removing discolored or reddish surfaces, as this may compromise the protective oxide layer and intensify corrosion. A comprehensive comprehension of passivation mechanisms is important. In certain cases, factors like diminished moisture or pressure on a component might trigger natural self-passivation, therefore preventing further corrosion without external intervention. Nonetheless, if the corrosion layer is fragile or porous (as depicted in Figures 3.5 and 3.6), oxygen may infiltrate further into the material, thereby hastening degradation.

Figure 3.6 illustrates the expedited rusting process triggered by air oxygen, water, and pollution, with variables such as temperature and sulfur compounds additionally influencing the corrosion rate. Atmospheric

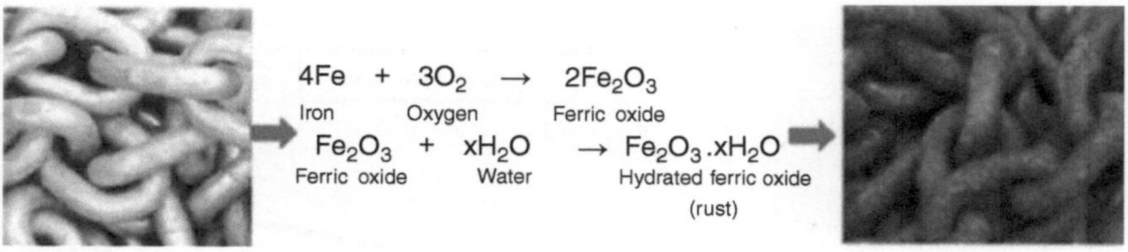

Figure 3.5 Development of outer layers on the chain, primarily composed of hydrated iron oxide, which commonly exhibits a reddish-brown hue.

pollutants promote rusting by introducing corrosive agents. A coating of iron hydroxide develops over metal pits, resulting in a concentrated solution that facilitates more corrosion via electrochemical processes. This initiates a perpetual cycle of rust development, which exacerbates as the process advances.

To mitigate rusting and aqueous corrosion, it is necessary to employ chemical treatments such as nitric acid or citric acid on the surface. Although these compounds may initially induce modest corrosion, they subsequently create a passive coating that prevents further deterioration. Certain compounds are occasionally used in lubricants to create a protective coating on operational surfaces, thereby inhibiting corrosion. Utilizing phosphate or solid lubricants on steel, or anodizing aluminum can improve corrosion resistance. Anodizing aluminum produces a thicker and more resilient protective layer compared to the natural oxide layer, significantly enhancing its corrosion resistance. Although aluminum's inherent oxide layer provides limited resistance under mild conditions, anodizing is advisable for harsher settings, including contact with specific acids, to enhance its protective properties. Anodizing is a conversion coating procedure that establishes a barrier to inhibit aluminum corrosion. Effective corrosion prevention relies on the accurate equilibrium of chemical concentrations and application methods, highlighting the necessity for continuous research to enhance corrosion resistance approaches.

Employing physical barriers such as paint or a polymer coating on a material's surface serves as an alternative method for corrosion protection. These procedures provide a protective coating that safeguards the substance. A layer that establishes a robust chemical bond with the substance is referred to as a "chemical layer." A layer deposited superficially without adhesion is termed a "physical barrier." To assess a material's corrosion resistance, it is essential to evaluate not only the material itself but also variables such as temperature and ambient conditions. Collectively, these elements influence the material's overall capacity to withstand corrosion.

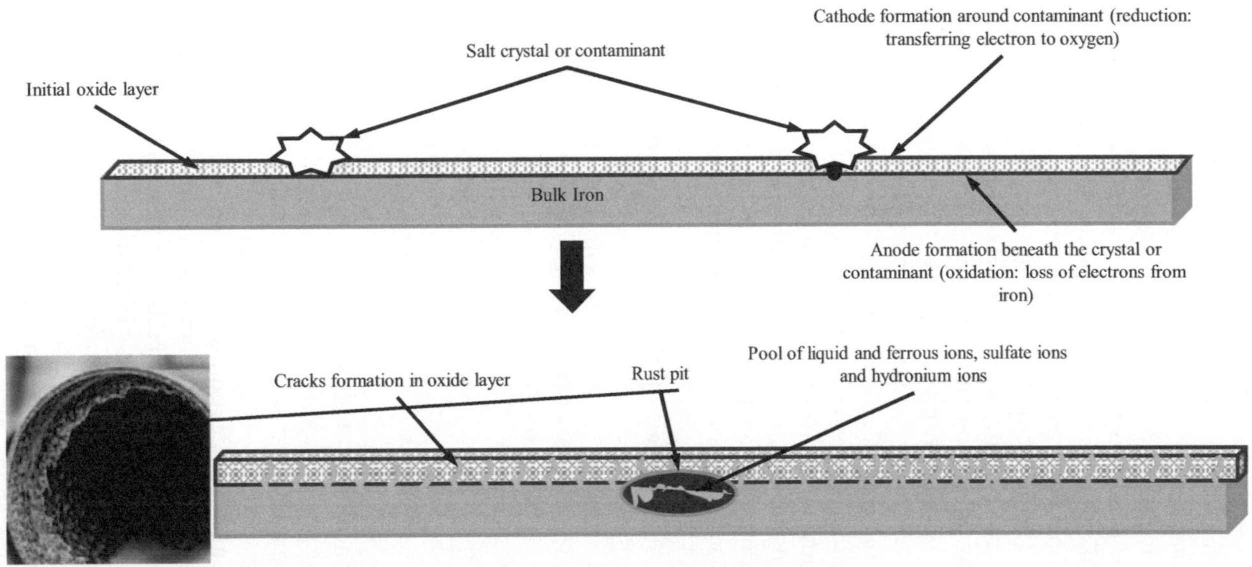

Figure 3.6 Influence of aqueous solutions on the acceleration of electrochemical corrosion.

3.4 GALVANIC CORROSION

Galvanic corrosion transpires when two dissimilar metals or alloys are electrically connected and immersed in an electrolyte, such as water or an aqueous solution. Corrosion happens more frequently when there is a significant disparity in the electrode potential of two metals or when the electrolyte increases ionic conductivity. The more reactive metal undergoes preferred corrosion, while the less reactive metal is safeguarded. One metal functions as an anode and undergoes oxidation (corrosion), while the other serves as a cathode. The anode metal emits electrons that traverse to the cathode metal through the conductive medium, resulting in accelerated corrosion of the anode. This phenomenon is frequently observed in constructions made of several metals, such as pipelines or electrical systems.

Galvanic corrosion requires the presence of two distinct metals or alloys, whereas general corrosion can occur in the absence of a second metal. General corrosion results from the chemical characteristics of the environment, while galvanic corrosion arises from the electrochemical potential disparity between metals and the existence of an electrolyte.

Diverse techniques can be utilized to prioritize materials according to their electrochemical potential. A prevalent method is a galvanic series (see Figure 3.1), which organizes materials according to their respective electrochemical potentials. In this series, the material exhibiting the most negative potential is deemed the most anodic and hence the most prone to corrosion, whereas the material with the most positive potential is classified as the most cathodic and thus more resistant to corrosion. The galvanic series is based on galvanic corrosion, which transpires when two dissimilar metals contact one another in the presence of an electrolyte. In this context, the material with the more negative electrochemical potential serves as the anode and undergoes corrosion, while the material with the more positive potential functions as the cathode and is protected from corrosion.

Electrochemical potential alone does not solely dictate a material's corrosion resistance. Additional elements, including chemical composition, microstructure, and ambient circumstances, significantly influence the corrosion of a material. The microstructure of a metal's surface influences the initiation and propagation of corrosion, with surface defects such as scratches, fissures, and roughness exacerbating localized corrosion. The relative surface areas of two metals in contact substantially influence the rate of galvanic corrosion, as illustrated in Figure 3.7.

When there is a substantial anode area and a small cathode area, corrosion is distributed over a larger surface, resulting in a less concentrated attack on the anode. This results in "minimal corrosion" distributed across an extensive region of the anode, akin to a metal sheet. The larger anode facilitates electron flow, hence lowering the total corrosion rate.

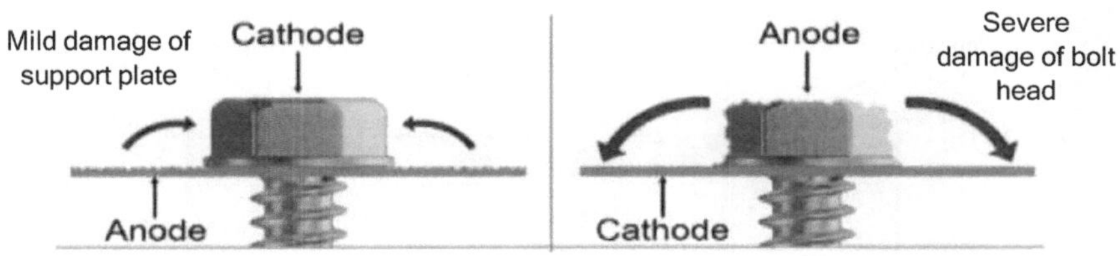

Figure 3.7 Impact of relative surface areas of anode and cathode metals on galvanic corrosion damage.

Conversely, when there exists a "large cathode area and a small anode area," the diminutive anode is required to provide all the electrons to the expansive cathode. This induces a "highly concentrated flow of electrons" at the anode, leading to "rapid and severe corrosion" at the anode, exemplified by the head of a fastener (as depicted in Figure 3.7). The smaller anode cannot distribute the electron flow across a broad area, resulting in accelerated and concentrated damage.

Choosing the appropriate combination of metals and surface areas is crucial to mitigate rapid corrosion. To understand, let us examine the example depicted in Figure 3.7, where "aluminum," being more susceptible to corrosion, functions as the anode, while "stainless steel," exhibiting greater resistance, serves as the cathode. Consequently, "aluminum frames (large anode)" must be utilized with "stainless steel fasteners (small cathode)" to mitigate galvanic corrosion damage.

3.5 LOCALIZED CORROSION

In contrast to general corrosion, which generally impacts a surface, localized corrosion concentrates on select regions, preserving the majority of the material. Localized corrosion denotes a sort of corrosion that manifests in certain regions on a metal surface, resulting in the development of pits, crevices, or fissures. Corrosion often begins at minor surface flaws in the metal, such as scratches or pre-existing corrosion sites, which act as nucleation places for more corrosion. This phenomenon can be exceedingly harmful as it frequently advances swiftly, resulting in considerable degradation of material strength.

Intergranular corrosion (IGC), as depicted in Figure 3.8, is a type of localized corrosion that transpires along the grain boundaries of polycrystalline materials such as metals and alloys. Grain borders, which delineate individual crystals (or grains), possess distinct chemical and structural properties compared to the core of the grains. The disparities render grain boundaries more vulnerable to corrosion, and the ensuing localized assault

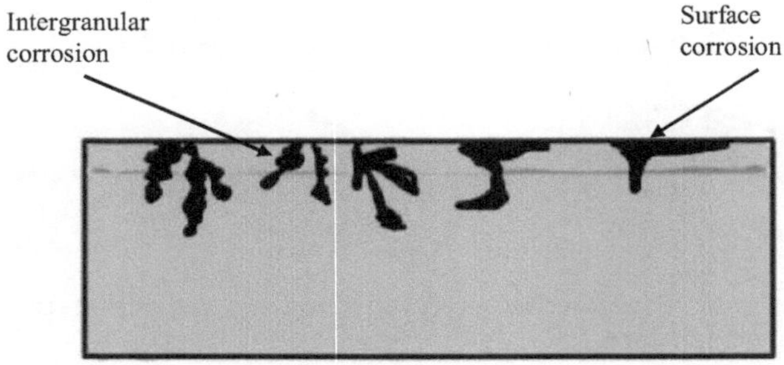

Figure 3.8 Localized corrosion.

can considerably compromise the material's integrity. In stainless steel, corrosion resistance depends on a protective chromium oxide layer; nevertheless, this layer may be damaged at the grain boundaries, especially when chromium-depleted zones develop during heat treatments or welding. This renders these locations exceedingly susceptible to IGC.

IGC frequently arises when secondary phases, such as carbides, precipitate at grain boundaries during the cooling of the alloy from elevated temperatures. This diminishes chromium at the grain boundaries, hence impairing corrosion resistance in those regions. Elevated temperatures can intensify corrosion, rendering grain boundaries more chemically reactive and resulting in pits or cracks in these areas. This sort of localized corrosion is especially detrimental in sectors such as oil and gas, chemical processing, and maritime operations, where harsh conditions may result in cracks or other types of material failure. As IGC frequently remains undetected until substantial damage has transpired, it presents a concealed yet grave risk to the structural integrity of components.

Pitting corrosion results in the formation of microscopic holes or craters on a material's surface, whereas crevice corrosion transpires in confined places, such as gaps or cracks, where stagnant liquids collect. Both categories represent forms of localized corrosion, indicating that the damage is concentrated in specific locations rather than being uniformly distributed across the surface. Some researchers also classify galvanic corrosion, occurring when two dissimilar metals are in contact, as a distinct form of localized corrosion.

Crevice corrosion generally develops in regions with confined or stagnant solutions, such as joints or beneath deposits, while pitting manifests on the exposed surface of the metal. Crevice corrosion is frequently initiated by corrosive agents becoming ensnared in narrow fissures, whereas pitting results from localized chemical or electrochemical reactions on the surface. Pitting and crevice corrosion are detrimental as they can penetrate deeply into the metal, undermining its structural integrity and damaging its strength.

3.5.1 Pitting corrosion

Pitting corrosion, defined by the formation of small pits or cavities on the metal surface, is the most prevalent type. Pitting corrosion transpires when a protective surface coating deteriorates, revealing localized regions of the metal surface to unregulated corrosion. Upon the formation of a pit, it becomes anodic in relation to the adjacent surface, leading to accelerated corrosion within the pit. This corrosion can exacerbate and enlarge the pit, resulting in significant subsurface corrosion, despite the modest surface aperture of the pit.

Pitting corrosion, illustrated in Figure 3.9, initiates when the protective layer, or passivating coating, on a metal surface, deteriorates, thereby exposing localized regions to direct corrosion. The exposed region functions as an anode, while the adjacent metal serves as a cathode, resulting in

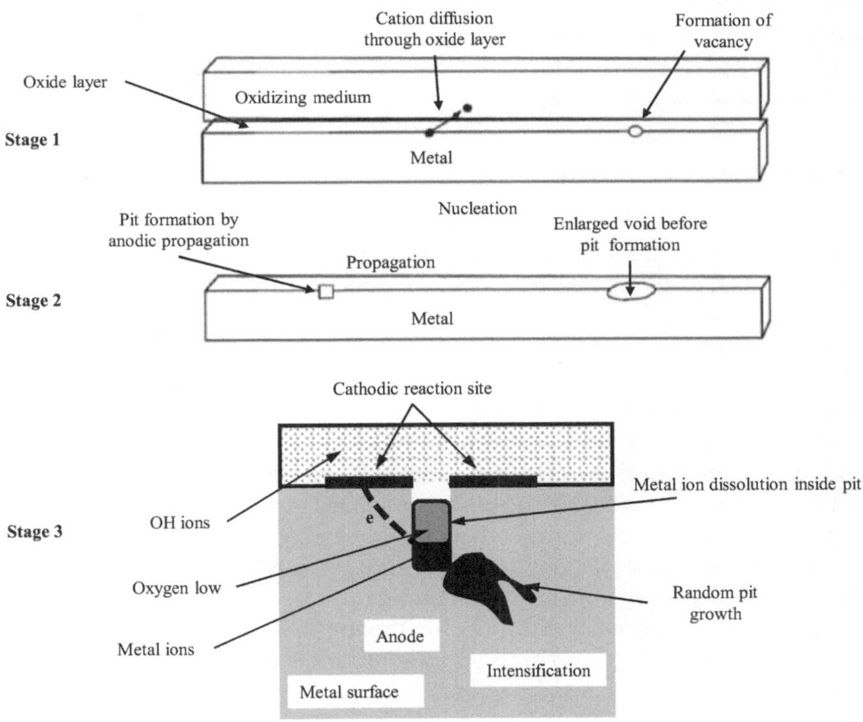

Figure 3.9 Various stages of pitting corrosion.

accelerated corrosion within the pit. Although the surface aperture of the pit may be minimal, the corrosion might penetrate deeper and disseminate beneath the surface. Nucleation, the initial development of the pit, occurs gradually, succeeded by a more rapid phase of growth, known as propagation. A theory posits that pit nucleation arises from the accumulation of crystalline vacancies at the interface between the metal and its protective oxide layer. Over time, these vacancies result in voids, leading to the failure of the protective layer and exposing the metal to corrosion. In the nucleation phase, the metal surface may seem unchanged, referred to as the induction period, until the voids expand to a critical size.

The electrochemical potential of the metal surface, affected by the environment, is crucial in pitting corrosion. Corrosion initiates when the potential falls below a specific threshold, referred to as the pitting potential. A nucleation potential exists, representing the minimum threshold required to begin a pit. Upon the initiation of pitting corrosion, the upper region of the pit functions as the cathode, facilitating the reaction of oxygen and water, whereas the lower region operates as the anode, where metal oxidation occurs. Electrons traverse from the anode to the cathode, so completing an electrochemical circuit. During this process, oxygen is reduced at the cathode to generate hydroxide ions, while metal ions are created at the anode. The metal ions subsequently react with hydroxide in the electrolyte to generate corrosion products, hence exacerbating the ongoing development and deterioration of the pits.

3.5.2 Crevice corrosion

Crevice corrosion is a form of localized corrosion that occurs in confined places, known as crevices, on a metallic surface. Crevices may develop in locations such as joints (see Figure 3.10), gaskets, threads, behind deposits, or as a result of surface flaws.

This form of corrosion arises due to the disparity in chemical conditions between the interior of the fissure and the external environment. These variations may arise from alterations in oxygen concentrations, pH levels, moisture content, stagnant water, or the presence of corrosive agents such as chloride ions. Chlorides and other corrosive substances can accelerate the corrosion process, rendering these areas increasingly susceptible.

In crevice corrosion, the metal within the crevice serves as the anode (site of corrosion), while the adjacent solution functions as the cathode. The deficiency of oxygen and the accumulation of corrosive agents in the crevice establish optimal conditions for the initiation and advancement of corrosion. Crevice corrosion presents a significant issue due to its occurrence in inaccessible or concealed locations, complicating detection and prevention efforts. Consequently, non-destructive testing and proactive strategies

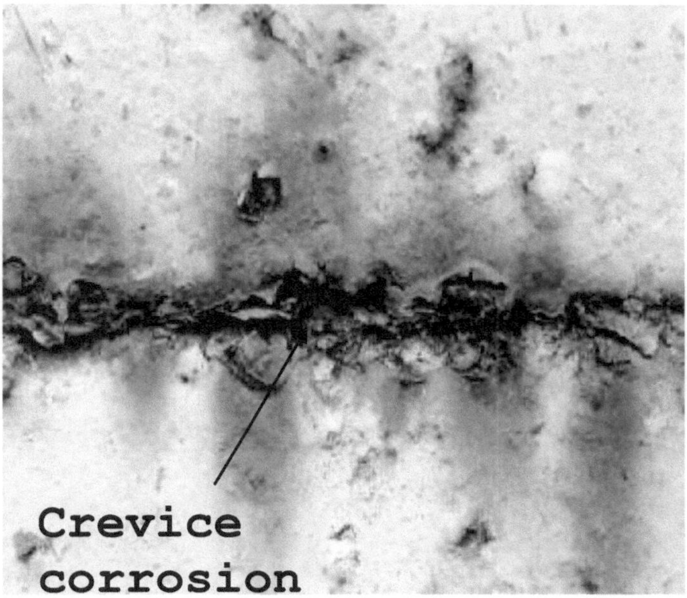

Figure 3.10 Crevice corrosion in welded joint.

are essential to detect and alleviate crevice corrosion, thereby preserving the integrity of metallic structures.

To understand crevice corrosion, we can refer to the ASTM G78 standard. Wang et al.[2] conducted a case study on stainless steel alloys, particularly SS904L, and developed a method for testing crevice corrosion according to these guidelines.

Prior to the crevice corrosion testing, each specimen was meticulously prepared by abrading it with wet 2000 grit SiC paper, degreasing it with acetone and alcohol, and permitting it to air-dry. The specimens were subsequently assembled utilizing a titanium bolt, nut, and washer, which were secured with a torque tool calibrated to 0.1 nm. The technique produced 20 minor crevice sites on either side of the specimen. PTFE tape was applied around the titanium bolt to insulate it from electrical contact with the stainless steel.

The crevice corrosion assessment involved tests utilizing artificial saltwater at twice the standard concentration, with the particular chemical compositions outlined in an accompanying table. The testing temperature was meticulously regulated at 70°C. The results are shown in Figure 3.11.

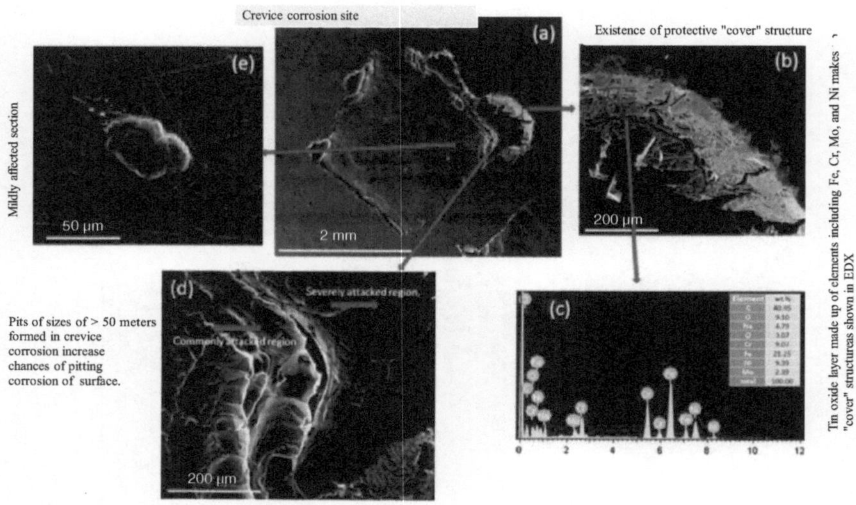

Figure 3.11 **Investigating crevice corrosion in stainless steel. (Adapted figure with copyright permission.[2]**

Figure 3.11(a) illustrates the characteristic form of a crevice corrosion site observed in SS904L, as reported by Wang et al.[2] This illustration delineates three unique areas: a passive region (Figure 3.11(e)), an active region (Figure 3.11(b)), and a variable region (Figure 3.11(d)). A protective "cover" structure is observable at the upper right corner of the crevice corrosion site, positioned above the edge of the active region. Figure 3.11(b) offers a detailed analysis of this "cover" structure.

Energy dispersive X-ray spectroscopy (EDS) study of the red region in Figure 3.11(b) indicated that the "cover" structure comprises a thin oxide layer including components such as iron (Fe), chromium (Cr), molybdenum (Mo), and nickel (Ni) (see Figure 3.11(c)). Figure 3.11(e) depicts the corrosion morphology of a minimally affected region, where subtle scratches from sandpaper remain discernible, signifying only slight corrosion. The data and descriptions presented provide a comprehensive overview of the distinct traits and characteristics of a typical crevice corrosion site in SS904L.

Proactive measures are necessary to prevent crevice corrosion in goods and systems. Select corrosion-resistant materials, such as certain stainless steels or titanium, for components used in corrosive environments. Designs must facilitate unobstructed access to all surfaces, thus aiding in inspection and maintenance, and simplifying the identification and remediation of crevice corrosion, particularly in inaccessible regions. Reduce joints, gaskets, and other elements that form confined areas conducive to moisture and corrosive agent accumulation. Should joints be required, contemplate designs that facilitate adequate drainage and ventilation.

Applying protective coatings or sealants in regions susceptible to crevice corrosion can diminish moisture infiltration and safeguard metal surfaces from corrosive conditions. Perform comprehensive testing in simulated environmental settings to evaluate the risks of crevice corrosion, which must be included in the product validation process. Aim for minimalist designs with reduced components to minimize the likelihood of creating concealed recesses. By taking these elements into account during the design phase, designers may markedly diminish the risk of crevice corrosion, hence improving the lifetime and reliability of their products.

3.6 STRESS CORROSION CRACKING

SCC transpires when a material, including titanium alloys, steels, or stainless steels, is exposed to simultaneous tensile stress and a corrosive environment (as depicted in Figure 3.12). Tensile stress may originate from multiple sources, including applied tensile loads, elevated temperatures, and residual strains from manufacturing, assembly, or heat treatment procedures. Additionally, materials may be subjected to corrosive conditions, including saline, acidic, or alkaline solutions, as well as corrosive gases.

A significant concern associated with SCC is hydrogen embrittlement, which poses risks in industries like aerospace and oil and gas. The mechanisms of SCC are not fully understood, but the process is believed to involve several key stages:

- *Initiation:* Cracks or defects can develop on the material's surface or within its structure (as shown in the sketch of Figure 3.12), providing sites where SCC can commence.
- *Propagation:* Once initiated, these cracks tend to propagate along the material's grain boundaries, typically perpendicular to the applied tensile stress.

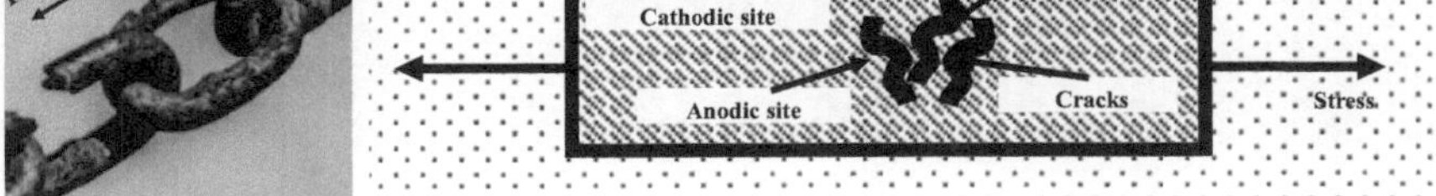

Figure 3.12 Stress corrosion cracking in chain.

- *Fracture:* If left unaddressed, this propagation can lead to catastrophic failure of the material.

Consequently, the selection of suitable materials is essential to prevent SCC. Duplex stainless steel, characterized by its distinctive amalgamation of austenitic and ferritic crystal structures, exemplifies super stainless steel. Its exceptional strength-to-corrosion resistance ratio renders it appropriate for rigorous applications, including offshore oil and gas exploration, chemical processing equipment, and the pulp and paper sectors.

Similarly, super-austenitic stainless steel, fortified with elevated concentrations of nickel, chromium, and molybdenum, provides outstanding resistance to diverse corrosive conditions, such as acidic chloride solutions. This substance is essential in various industries, including chemical processing, desalination, and pulp and paper production. In product development, understanding the properties and behaviors of these materials under stress and corrosive situations is crucial for ensuring reliability and durability in final applications.

Wu et al.[3] conducted SCC tests in accordance with the ASTM G36 standard. The specimens were milled and subsequently exposed to a boiling magnesium chloride solution at 155.0°C ± 1.0°C for 2 hours. Scanning electron microscopy was employed to analyze the specimens, and crack density was assessed by measuring the lengths of cracks in specified areas. Figure 3.13 depicts the morphologies of SCC cracks on the milled surfaces

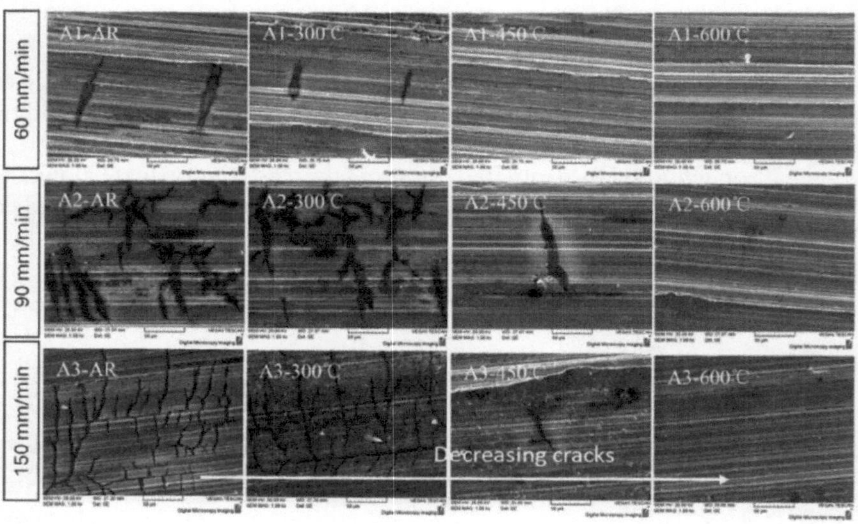

Figure 3.13 Morphologies of SCC on A1 to A3 specimens before and after annealing at different temperatures. (Adapted figure with copyright permission.)[3]

of the A1 to A3 specimens, both before and after annealing at various temperatures. The specimens milled at a speed of 60 mm/min are designated as A1, while those milled at 150 mm/min are designated as A3.

A consistent pattern of diminishing crack densities was noted in all specimens as annealing temperatures increased, suggesting that elevated temperatures facilitated a drop in crack density. Specimens annealed at 300°C exhibited negligible reduction in crack start, indicating that this temperature was less efficacious in preventing the onset of cracks. Conversely, substantial enhancements were observed when specimens were annealed at 600°C, with no crack start identified, indicating that this temperature effectively inhibited crack formation in the majority of specimens. The results demonstrate empirical support for the advantageous impact of increased annealing temperatures in diminishing crack density and preventing crack formation in the examined specimens. The scientists determined that utilizing elevated annealing temperatures, especially at 600°C, markedly improves the specimens' resistance to SCC.

3.7 CONCLUSIONS

This chapter underscores that effective corrosion prevention hinges on appropriate material selection, protective coatings, and design adjustments to minimize exposure to corrosive environments. Corrosion, caused by electrochemical reactions, leads to structural degradation, diminished performance, and safety risks, often compounded by mechanical wear. The annual cost of corrosion in various sectors exceeds INR 1.5 trillion, highlighting the need for efficient management and surveillance to mitigate economic and safety impacts. By addressing both chemical and electrochemical mechanisms, the chapter illustrates how specific treatments, like anodizing and passivation, enhance resistance to degradation. Proactive corrosion control not only extends material longevity but also reduces maintenance costs and enhances safety. These insights emphasize the importance of a systematic approach to managing corrosion in engineering design, ensuring the reliability and durability of critical infrastructure.

REFERENCES

1. Winston Revie, R., ed. (2011). *Uhlig's Corrosion Handbook*. John Wiley & Sons.
2. Wang, C. G., Li, X. F., Wei, J., Wei, X., Xue, F., Ma, R. Y., Dong, J. H., & Ke, W. (2018). Crevice corrosion of several supper stainless steels in the simulated LT-MED environment. *Acta Metallurgica Sinica (English Letters)*, *31*, 1137–1147.

3. Wu, H., Li, C., Fang, K., Zhang, W., Xue, F., Zhang, G., & Wang, X. (2018). Effect of residual stress on the stress corrosion cracking in boiling magnesium chloride solution of austenite stainless steel. *Materials and Corrosion*, *69*(11), 1572–1583. https://doi.org/10.1002/maco.201810201

Chapter 4

Corrosion and surface deterioration and fractography

Understanding how different degradation mechanisms interact synergistically is crucial because their combined effects often result in more severe surface degradation than any single mechanism acting independently. This raises a vital question: why is it important to study the interplay between multiple failure mechanisms? Does such a synergistic effect genuinely occur, and if so, how does it amplify the degradation process?

Based on available knowledge, there is indeed significant synergy between various failure mechanisms. The overall rate of degradation and the progression of damage are heavily influenced by the interaction of different surface degradation processes, such as mechanical, chemical, thermal, or radiation effects.

Mechanical actions, such as rubbing or abrasion, can compromise the protective layers on a material's surface, exposing the underlying material to chemical attack. Once this protective barrier is disrupted, the newly exposed surfaces become more vulnerable to corrosion. The interaction between mechanical and chemical degradation mechanisms can significantly accelerate material failure.

Additionally, relative motion at mechanical interfaces, especially during high-speed sliding or rubbing, generates frictional heat. This heat can speed up thermal degradation processes like thermal cracking, further weakening the material's structural integrity. The elevated temperatures also enhance the diffusion of reactive species, intensifying chemical reactions and accelerating the breakdown of the material. Understanding the combined effects of these degradation mechanisms and their synergy is critical for preventing product failure.

Ionizing radiation adds another level of complexity to material degradation by triggering or speeding up chemical reactions. Radiation can generate reactive species that attack materials, worsening the effects of mechanical, chemical, and thermal stresses. These degradation processes often work together. For example, heat and mechanical wear expose surfaces, making

DOI: 10.1201/9781003545019-4

them more vulnerable to chemical and radiation damage, which creates a cycle of accelerated deterioration.

Understanding these combined interactions is essential for developing strategies to protect surfaces and extend product life. Radiation can initiate chemical reactions that release reactive particles, further damaging the material. This creates a chain reaction where one form of damage leads to another, making it necessary to identify the root cause of failure. Pinpointing the primary factor allows for more targeted treatments to prevent surface damage effectively.

4.1 SYNERGISTICALLY COMBINED DEGRADATION MECHANISMS

The interplay between surface degradation mechanisms—mechanical, chemical, and thermal or radiation actions—can profoundly influence both the rate of degradation and the extent of damage progression. This interaction is often visualized as three overlapping bubbles (Figure 4.1), where each bubble represents one mechanism: mechanical action, chemical action, and thermal or radiation effects. The overlaps illustrate the synergistic relationships between these forces, emphasizing their combined impact on material degradation.

Figure 4.1 illustrates the synergistic interactions among surface degradation mechanisms—mechanical, chemical, and thermal/radiation. A comprehensive framework for understanding material degradation must consider the interplay between material properties, contact conditions, service environments, and failure mechanisms. The behavior of degradation mechanisms is significantly influenced by the inherent properties of the material and the conditions under which it operates. Key material properties, such as composition, microstructure, and mechanical and thermal characteristics, directly impact how a material responds to mechanical wear, chemical reactions, and thermal or radiation exposure, emphasizing their importance in designing more durable systems.

- *Material properties:* To prevent surface degradation effectively, it is vital to review material composition, microstructure, and mechanical and thermal properties. Key characteristics include mechanical properties like strength, stiffness, toughness, and fatigue resistance; chemical properties such as corrosion resistance, reactivity, and solubility; and thermal properties like thermal conductivity, expansion, and melting point. Material compositions that are designed to withstand mechanical wear may not be able to handle chemical attacks. Surface deterioration is influenced by particle size, grain boundaries, and phase distribution.

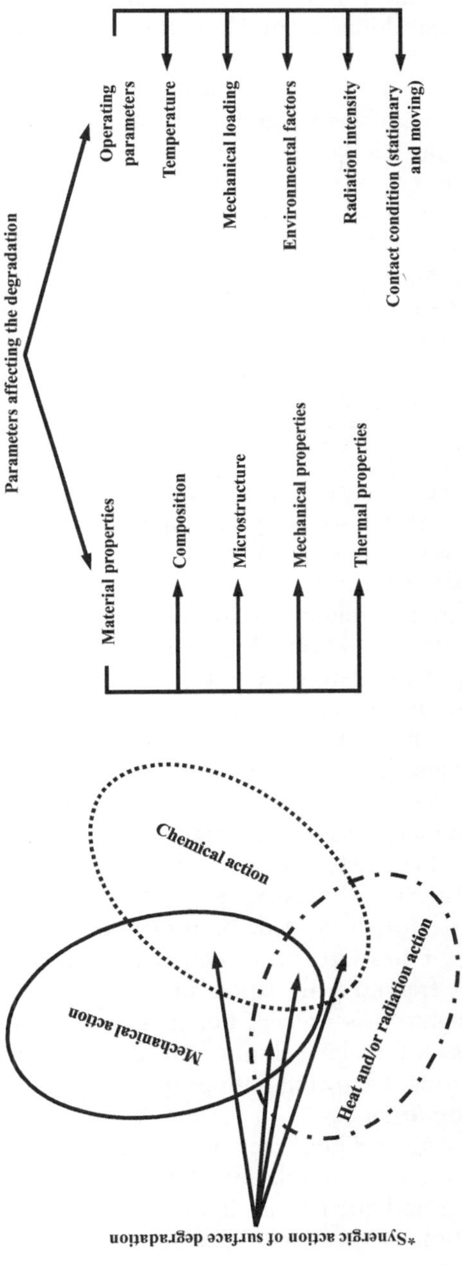

Figure 4.1 Parameters influencing degradation and potential synergistic effects.

- *Composition:* The material's composition influences its susceptibility to different degradation mechanisms, determining which mechanisms dominate and how they interact.
- *Microstructure:* Factors like grain size and phase distribution impact a material's resistance to degradation. For example, materials with smaller grains are often more resistant to mechanical wear. However, having too many grain boundaries can make the material more vulnerable to chemical attack, leading to corrosion and possible rupture.
- *Thermal properties:* To ensure effective heat dissipation and prevent localized hotspots—which can cause wear, corrosion, or fractures—the material must have good thermal conductivity. Materials with a high specific heat capacity and a high melting point are ideal for reducing hotspots and minimizing related damage. As operating temperatures approach 30% to 70% of a material's melting point, creep becomes a major concern, and elevated temperatures can trigger various degradation processes.

However, in some cases, thermal insulation may be necessary to prevent unwanted heat transfer, so the specific application and conditions must be carefully assessed. When applying coatings or layers, it is crucial to consider thermal expansion mismatches between contacting materials. Significant differences might require material changes or intermediate steps to reduce stress. Additionally, applying compressive forces during assembly can generate tension due to thermal expansion mismatches, which may result in fractures or other structural issues.

Ultimately, a detailed analysis of thermal conductivity, thermal expansion, and insulation is essential for designing robust solutions.

- *Surface coatings or treatments:* Surface coatings or treatments modify a material's surface, making it more resistant to damage. High hardness helps prevent plastic deformation and wear, but it can also raise the risk of erosion and brittle fractures. While hardness is often needed for protection, increasing it through alloying or heat treatment can sometimes lead to cracks and corrosive fractures. To handle shock or impact loads effectively, high toughness is crucial, but it usually comes at the cost of hardness. A possible solution is to use a tough substrate with a hard coating, balancing both properties.

- *Operating conditions:*
  - *Temperature:* Higher temperatures speed up degradation processes such as oxidation, fatigue, and chemical reactions. The range and fluctuations in operating temperatures also affect degradation mechanisms. When a material is exposed to heat, its structure can be weakened due to thermal expansion, bond disruption, or oxidation.

- *Environmental factors:* Environmental conditions such as humidity, moisture exposure, corrosive agents (e.g., reactive gases, liquids, or contaminants), pressure, vibrations, and pH levels have a significant impact on the interaction of degradation mechanisms. These factors can accelerate or modify the way materials deteriorate over time, making it essential to account for the service environment. Humidity, for instance, can act as a corrosive medium, while pressure may cause cracks to expand or propagate. Temperature fluctuations can induce thermal expansion, further accelerating failure, especially at higher temperatures.
- In acidic environments with a pH below 7—commonly around 4 or 5 in various liquids—the corrosive potential must be carefully evaluated. Contaminants like sulfur dioxide (SO_2) and nitric oxide can also compromise material integrity. Even seemingly less harmful gases, such as carbon dioxide, contribute to corrosion dynamics. While oxygen can have beneficial effects under certain conditions, it can also play a role in material degradation. Understanding the combined influence of these factors is crucial to predicting and mitigating material failure.
- *Mechanical loading:* The nature of the load—whether static or dynamic—significantly affects degradation, especially in challenging environments such as dusty or humid conditions. While static loading is often mentioned, it is important to recognize that all loading is dynamic to some extent, with inherent fluctuations. The frequency of cyclic loading and dwell times also contributes to this process. Factors such as the number and duration of operational cycles, along with the magnitude, frequency, type (tensile, compressive, shear), and distribution of applied mechanical loads, are key in determining the combined effects of mechanical and chemical degradation. Impact loading, caused by substantial dynamic or cyclic forces, is particularly critical for erosion, as particles—whether solid or fluid—striking a surface significantly contribute to wear.
- *Radiation intensity:* Ultraviolet (UV) damage refers to the harmful effects that UV radiation can have on materials and surfaces exposed to it. UV radiation is a component of the electromagnetic spectrum, emitted by both the sun and certain artificial sources. This type of damage primarily impacts organic materials, including polymers, paints, coatings, plastics, and textiles. UV radiation possesses sufficient energy to break chemical bonds in organic molecules, leading to a process known as photodegradation or photochemical degradation. This deterioration often involves the cleavage of carbon–carbon bonds,

resulting in chain scission, a reduction in molecular weight, and a loss of mechanical strength and integrity. The type and intensity of radiation—such as ionizing or UV radiation—play significant roles in affecting chemical reactions and the generation of reactive species, which enhance the interaction between radiation and chemical degradation. Additionally, radiation can induce structural changes, such as bond breaking or the generation of free radicals, which accelerate the aging of materials.

- *Surface contact:* The type and intensity of contact with other surfaces—whether sliding, rolling, or abrasive—are key factors in determining wear and degradation. In contact scenarios, it is important to consider macrocyclic loading and motion, such as fretting corrosion, which can occur at amplitudes as small as 1 to 2 μm, even when the movement appears static. Cyclic loading, along with sliding and rolling contact, must be considered in analyses. Surface roughness also plays a significant role in how surfaces interact and wear over time, as it directly influences wear patterns and contact mechanics.

Synergistic effects occur when different failure mechanisms, such as stress-assisted corrosion and radiation-enhanced wear, work together to speed up material degradation. To effectively prevent surface degradation, it is essential to evaluate a material's composition, microstructure, and mechanical and thermal properties. Cumulative damage refers to the gradual impact of these degradation processes over time, leading to a decline in the material's properties and performance. Additionally, feedback mechanisms can develop when various failure actions interact; for example, cracks created by mechanical stress may serve as points for accelerated corrosion.

In understanding surface degradation, it is important to consider the service environment, which includes factors like pollutants, gas reactions, corrosive substances, pressure, vibrations, and the type and intensity of radiation (such as UV or ionizing radiation). When multiple failure actions are present, they can have synergistic effects, meaning that one failure action can amplify the effects of others, leading to quicker degradation. For instance, mechanical stress can create fissures that become sites for rapid corrosion. Similarly, the combination of radiation-enhanced wear, stress-assisted corrosion, and the thermal activation of chemical reactions under mechanical stress can create a cycle where the degradation from one mechanism accelerates the degradation from another. Understanding these interactions helps us develop better strategies for mitigating surface degradation.

Figure 4.2 shows a framework for understanding synergistic failure degradation by looking at how material properties, contact conditions, and service environments interact. Surface roughness is important because it affects wear and how surfaces contact each other. The service environment

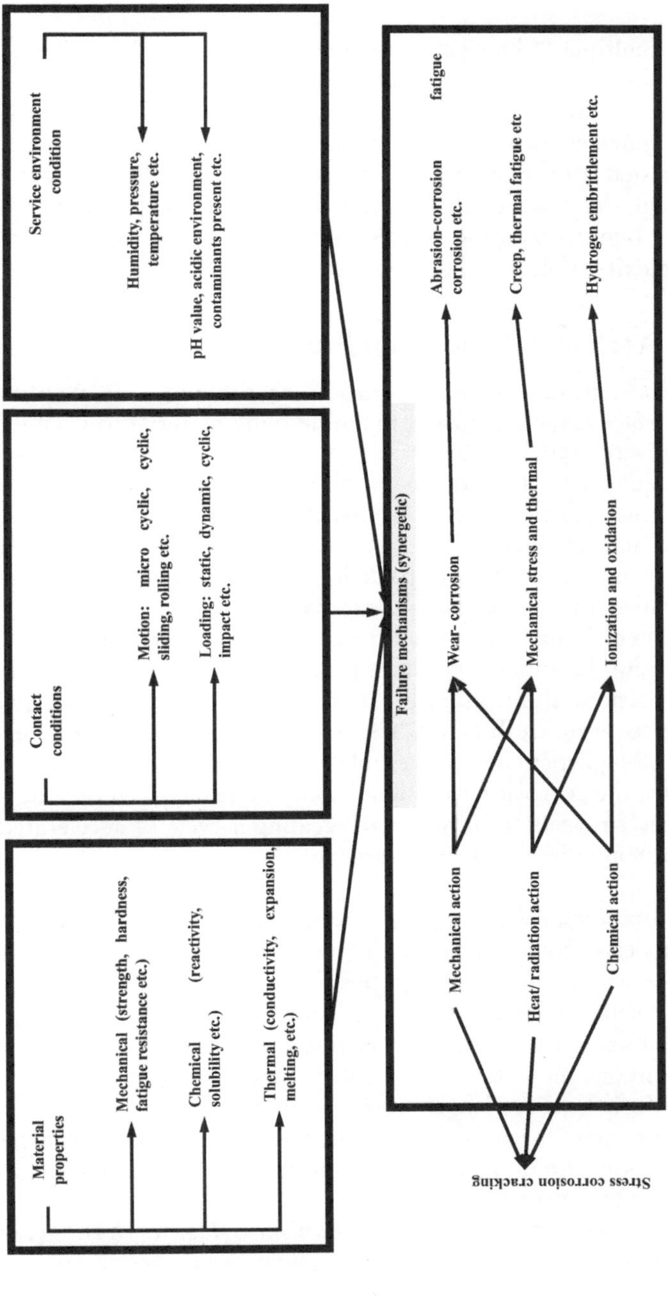

Figure 4.2 Framework for studying the combined effects.

includes factors like pollutants, gas reactions, corrosive substances, pressure, vibrations, and the type and intensity of radiation, such as UV or ionizing radiation.

When multiple failure actions occur together, they can have synergistic effects, meaning one action can increase the impact of others, leading to faster degradation. For example, cracks from mechanical stress can become areas for quick corrosion. Other examples include radiation-enhanced wear, stress-assisted corrosion, and thermal activation of chemical reactions under mechanical stress. Overall, the way different failure mechanisms interact can cause one degradation process to speed up another, thus emphasizing the complexity of material degradation.

4.1.1 Mechanical and chemical actions

Mechanical actions, such as abrasion or rubbing, can strip away protective surface layers, exposing the underlying material to chemical agents. This exposure increases the reactive surface area, which accelerates chemical reactions and hastens material degradation. Similarly, mechanical wear removes surface material, revealing fresh, reactive metal surfaces. These surfaces are highly susceptible to corrosion, a process that generates byproducts acting as abrasive particles. This feedback loop intensifies wear and amplifies the overall rate of degradation.

When steel is exposed to sulfuric acid, the interaction between wear and corrosion becomes even more pronounced. Mechanical wear, whether through friction, abrasion, or erosion, removes oxidized or passive surface layers, uncovering fresh metal. These newly exposed surfaces, being highly reactive, are particularly vulnerable to corrosive environments. In such conditions, the corrosion rate significantly increases, and the ongoing wear continuously reveals fresh material, creating a cycle of accelerated degradation that magnifies the overall damage.

- *Corrosive and sliding wear mechanisms:* Figure 4.3 illustrates the modes of corrosive and sliding wear in steel exposed to sulfuric acid. These modes involve the exposure of unoxidized metal and the plastic deformation of surface layers, contributing to corrosion. As the concentration of sulfuric acid changes, the steel wear mechanisms shift from one mode to another. Interestingly, very high concentrations of sulfuric acid are less corrosive to metals than lower concentrations. This occurs because concentrated sulfuric acid lacks sufficient water to dissociate into an anionic radical and two hydronium ions, which drive the corrosive process.
- *Wear-assisted corrosion mechanisms (Figure 4.4):* Wear-assisted corrosion occurs when mechanical wear accelerates corrosion,

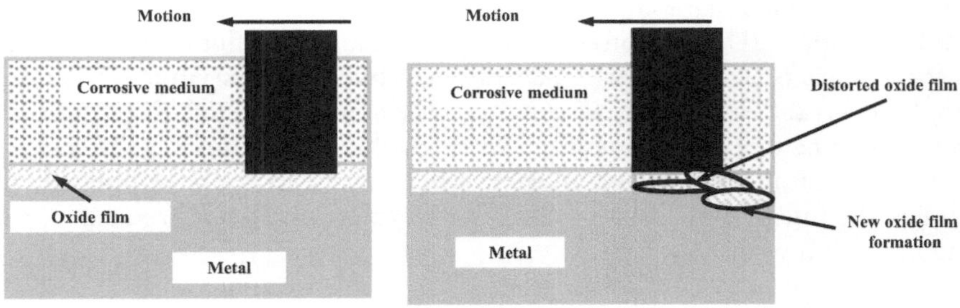

Figure 4.3 Corrosion and sliding wear of steel in a sulfuric acid (SA) environment.

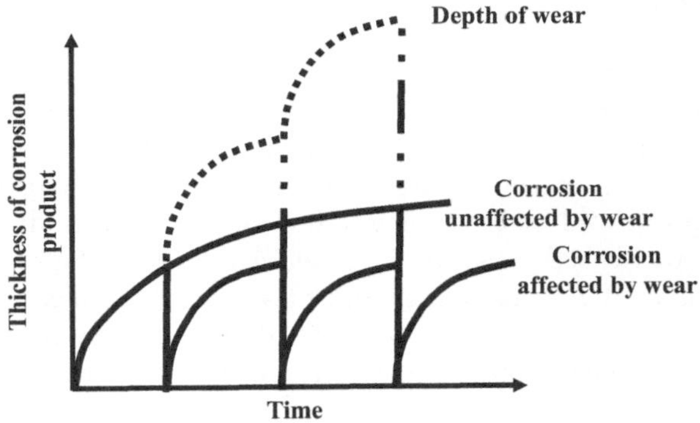

Figure 4.4 Wear-assisted corrosion.

leading to faster material degradation. Figure 4.4 illustrates this with a graph showing the relationship between time and the thickness of corrosion products, highlighting wear's impact. The graph presents two curves: one representing normal corrosion, where the thickness of corrosion products increases gradually over time, and the other showing wear-affected corrosion, characterized by a cyclical pattern. In the wear-affected scenario, corrosion products build up but are repeatedly removed as wear exposes fresh material, restarting the corrosion process. A dashed line marks the "depth of wear," showing material loss over time, and when this thickness falls below a critical level, replacement is necessary. Freshly exposed surfaces are much more reactive and prone to corrosion, leading to increased material loss as cycles progress. By the third cycle, material loss has tripled, demonstrating that wear combined with corrosion causes significantly more damage than corrosion alone.

Several mechanisms contribute to the interaction between wear (abrasive wear: surface material is removed by hard particles (3D abrasion) or hard asperities (2D abrasion); erosion wear: caused by fluid flow or solid particle impacts; fretting wear: occurs when two surfaces vibrate at small amplitudes) and corrosion, leading to accelerated material degradation. This phenomenon can be analyzed through three mechanisms: Corrosion plays a significant role across all these mechanisms, as we have previously studied the interactions between abrasive wear, erosion wear, fretting wear, and corrosion wear:

- Abrasive wear-assisted corrosion occurs when mechanical wear removes material due to hard particles or asperities, exposing fresh metal surfaces. These surfaces are highly reactive and more prone to corrosion. The surrounding corrosive environment, often acting as an electrolyte, accelerates electrochemical reactions at the wear sites, intensifying the corrosion process.
- Erosion-corrosion is a mechanism where fluid flow or solid particles mechanically erode protective films or coatings from the metal surface. This action exposes bare metal to the corrosive medium, leading to rapid material degradation. Xu et al.[1] conducted a study on ferrous-based steels to investigate this degradation, using two different configurations to examine flow-accelerated corrosion (FAC) and erosion-corrosion. The primary difference between these setups lies in the positioning of the test specimen. In the first setup, the linear velocity on the sample surface was controlled by adjusting the rotational speed, making it easier to manage the flow velocity. In the second setup; however, the slurry flow rate on the sample surface could not be directly correlated with the disc's linear velocity due to edge effects, leading to a more uneven corrosion pattern. In the first setup, the angle between the sample and the flow direction remained fixed, causing erosion and metal loss to occur mainly in the tangential direction at the start of the test. In contrast, the second setup allowed the angle between the sample and flow direction to be adjusted from 0 to 90 degrees, providing more flexibility for studying the effects of sand impact angles on erosion-corrosion. The aim was to monitor erosion-corrosion while preserving the electrochemical environment and local reactions in the flowing electrolyte. The researchers examined the surface of API X100 steel after several cycles of erosion-corrosion and found that corrosion worsened erosion through various factors. These included increased surface roughness, removal of hardened layers, faster extraction of embedded sand particles, and subsurface crack development during corrosion. Erosion disrupted the cementite network (a phase in steel), which promoted pitting corrosion and further increased surface roughness,

accelerating the corrosion process. By analyzing potential and galvanic current distributions, the researchers tracked dynamic changes in anodic and cathodic sites, which helped them understand the interaction between localized erosion and corrosion. This allowed them to quantify metal loss due to erosion (W_a), corrosion (W_c), and the synergistic effects of both (W_s).

- Fretting corrosion occurs when two surfaces undergo small, repetitive oscillations under load, leading to surface damage, removal of the oxide film, and the generation of wear debris. This debris increases the likelihood of both two-body and three-body abrasion. The exposed surfaces become vulnerable to localized corrosion, further accelerating material wear. As the surfaces vibrate under applied force, the protective layer is stripped away, exposing the underlying metal to corrosive elements. Despite increased peripheral forces, the debris remains, causing significant degradation of the protective layer. The opening of connections often reveals substantial debris, indicating considerable material degradation. This outcome emphasizes the detrimental effects of wear-assisted corrosion.
- Sliding wear-assisted corrosion results from the relative motion between two surfaces in high friction contact. This motion disrupts or removes passive films, leaving fresh metal surfaces unprotected.

In corrosive environments, these newly exposed surfaces undergo accelerated corrosion due to the absence of protective layers, increasing overall material degradation.

4.1.2 Mechanical and heat/radiation actions

High-speed rubbing or sliding generates frictional heat, raising surface temperatures and accelerating thermal degradation processes such as thermal oxidation and cracking. These elevated temperatures can weaken material properties, hastening surface deterioration. Similarly, repeated mechanical loading and unloading can cause localized heating and cooling, leading to cyclic temperature changes. These thermal cycles, coupled with mechanical stress, promote thermal fatigue, which manifests as surface cracking and eventual failure. The combined effects of frictional heating and mechanical stress create a dynamic environment where materials degrade more rapidly than under isolated conditions.

When mechanical stress interacts with thermal changes, materials are prone to failures like creep and thermal fatigue. Creep occurs when a material undergoes slow, permanent deformation under constant stress at high temperatures, as seen in jet engine turbine blades subjected to both high rotational forces and extreme heat. Over time, creep can reduce efficiency or lead to catastrophic failure if unaddressed. Thermal fatigue arises from repeated heating and cooling cycles, causing materials to expand and contract, introducing internal stresses that lead to cracking. Components such as exhaust manifolds in internal combustion engines experience this phenomenon, with small cracks gradually growing into significant failures. Together, mechanical stress and thermal effects synergistically degrade materials in ways that exceed the impact of either factor alone.

To combat these issues, Akhtar et al.[2] studied the T6 heat treatment cycle applied to an AC series aluminum alloy. This cycle includes "solution treatment," "quenching," and "aging hardening" processes that enhance the alloy's strength and durability. "Solution treatment" involves heating the aluminum alloy to high temperatures, allowing the magnesium and silicon in the alloy to distribute more evenly. "Quenching" is the rapid cooling of the alloy, usually in hot water, to trap the dissolved elements in a supersaturated state, which increases strength during the aging process. "Aging" is done at lower temperatures to form tiny, strong precipitates within the alloy, further enhancing its strength. During the aging process, the alloy was heated between 100°C and 200°C in 25°C intervals for 1 hour. This allowed the material to develop optimal hardness and strength through the formation of fine precipitates. Tensile and impact tests were performed on the treated samples, following ASTM standards, to confirm the improvements in strength.

The aluminum alloy used had a specific composition: 11.1% to 11.2% silicon (Si), 0.6% to 0.8% iron (Fe), 4.3% to 4.5% copper (Cu), 0.1% to 0.2% manganese (Mn), 0.5% to 0.8% magnesium (Mg), 0.4% to 0.6% nickel (Ni), ≤0.1% zinc (Zn), ≤0.08% tin (Sn), ≤0.2% titanium (Ti), ≤0.1% lead (Pb), and ≤0.1% chromium (Cr). This composition emphasizes the significant amounts of silicon and copper, with smaller amounts of other elements such as manganese, magnesium, and nickel. Akhtar et al.[2] demonstrated that thermal treatment of these alloys enhances their strength. To assess the effects, tensile tests (ASTM E8) were conducted to measure strength, and impact tests (ASTM E23) evaluated the material's toughness under stress.

4.1.3 Chemical and heat/radiation synergy

Elevated temperatures significantly accelerate chemical reactions, including oxidation, by increasing the diffusion rates of reactive species. This intensifies chemical processes and fosters aggressive chemical attacks, leading to material degradation. Similarly, ionizing radiation, such as UV rays or ion beam irradiation, can trigger or catalyze chemical reactions within materials. Radiation-induced reactions create reactive species that further degrade the material through intensified chemical activity.

Embrittlement, a reduction in material ductility and toughness, often results from the combined effects of chemical, thermal, and radiation factors. Hydrogen embrittlement occurs when hydrogen atoms penetrate the material, interact with the metal lattice, and cause cracks or fractures. Radiation-induced embrittlement arises from the displacement of atoms and the accumulation of defects in the crystal lattice due to exposure to ionizing radiation, compromising mechanical properties. Thermal embrittlement occurs when elevated temperatures alter the material's microstructure, leading to increased brittleness and susceptibility to failure under stress.

The synergy of chemical, thermal, and radiation effects creates degradation mechanisms that are greater than the sum of their individual impacts. Heat accelerates reaction rates by enhancing molecular movement, while radiation provides energy to break molecular bonds or generate free radicals, intensifying chemical reactions. For example, in polymer chemistry, heat and UV light are often combined to initiate polymerization, with heat increasing molecular activity and UV light activating specific reaction pathways. In space environments, materials subjected to solar radiation and thermal cycling experience rapid deterioration, as radiation weakens molecular bonds and thermal fluctuations exacerbate oxidation and stress fatigue, leading to faster material failure.

Another field that utilizes the synergy of chemical, thermal, or radiation effects is catalysis. Heat provides the activation energy for chemical reactions in many catalytic processes, while a catalyst lowers the energy barrier for reaction pathways. Radiation can further enhance this process by providing additional energy to the reactants or directly activating the catalyst surface. For instance, photocatalysis relies heavily on the synergistic effect of light and heat to drive chemical reactions on a catalyst's surface. Photocatalytic processes, like water splitting for hydrogen production, benefit from heat that enhances molecular movement, increasing reaction rates. When combined with light energy, this leads to significantly higher efficiency in energy conversion than either heat or radiation alone.

The hydrogen embrittlement phenomenon is another example of the synergistic effects of chemical action—often in the form of hydrogen—and external factors like heat or radiation. The interaction between these elements can significantly accelerate material degradation, leading to catastrophic failure in engineering components. Hydrogen embrittlement exemplifies the powerful synergy between chemical action (hydrogen interaction with metals) and heat or radiation. The combination of these factors leads to accelerated diffusion of hydrogen into metal, increases the probability of defect formation, and enhances the material's susceptibility to cracking and failure. This synergetic interaction is of critical importance in the design, maintenance, and safety management of components in industries ranging from aerospace to nuclear energy. Understanding and mitigating hydrogen embrittlement requires addressing the complex interplay of chemical, thermal, and radiation effects to ensure the reliability of high-performance materials.

Hydrogen embrittlement, especially common in marine environments, occurs when hydrogen infiltrates materials, increasing their brittleness and leading to failures. Hydrogen degradation manifests primarily in two forms (as sketched in Figure 4.5): blistering, which appears as surface-level defects, and embrittlement, which affects internal or subsurface layers. The mechanism involves hydrogen atoms migrating within the material, which can cause surface blistering and promote cracking or pitting in environments rich in hydrogen, such as aqueous solutions or high-pressure gas.

Hydrogen-induced material degradation occurs when hydrogen atoms penetrate and diffuse through a material, accumulating at the surface to form gas pockets or blisters, particularly in hydrogen-rich environments like aqueous solutions, high-pressure hydrogen gas, or hydrogen sulfide. This can result in surface blistering, cracking, pitting, or delamination. Additionally, hydrogen atoms can migrate into the material's lattice structure, increasing brittleness and the risk of fracturing, especially in steels and titanium alloys. Common sources of hydrogen ingress include manufacturing processes such as electroplating and welding or exposure to hydrogen-rich environments such as water, oil, or gas pipelines. Higher temperatures accelerate the hydrogen diffusion process, allowing atoms to accumulate in areas of stress concentration, such as grain boundaries. Once inside the material, hydrogen weakens atomic bonds, leading to embrittlement, reduced ductility, and toughness, which can cause microcracks and sudden catastrophic failure under mechanical stress.

Hydrogen embrittlement demonstrates how chemical interactions and heat work together to accelerate material failure. Elevated temperatures play a crucial role in this process by increasing the rate of hydrogen diffusion within the metal lattice, allowing hydrogen atoms to quickly accumulate in vulnerable areas like grain boundaries or regions of high stress. This accelerates embrittlement compared to lower-temperature exposure. Additionally, heat enables hydrogen to overcome energy barriers for diffusion, especially in hydrogen-rich environments like welding or heat treatment, where thermal energy speeds up hydrogen uptake. The combination of stress and elevated temperatures further intensifies embrittlement, as heat promotes hydrogen diffusion while mechanical stress provides sites for hydrogen to concentrate, increasing the likelihood of cracking and material failure.

Radiation plays a significant role in intensifying hydrogen embrittlement by creating additional defects in the material structure, particularly in environments like nuclear reactors. Radiation generates vacancies, dislocations, and other lattice defects that act as traps for hydrogen atoms, promoting their accumulation in localized areas. This trapped hydrogen can cause cracking at lower stress levels than would typically be required. Additionally, radiation accelerates the diffusion of hydrogen atoms, much like heat, further enhancing embrittlement. In nuclear environments, where radiation, heat, and hydrogen exposure coexist, these factors synergistically increase

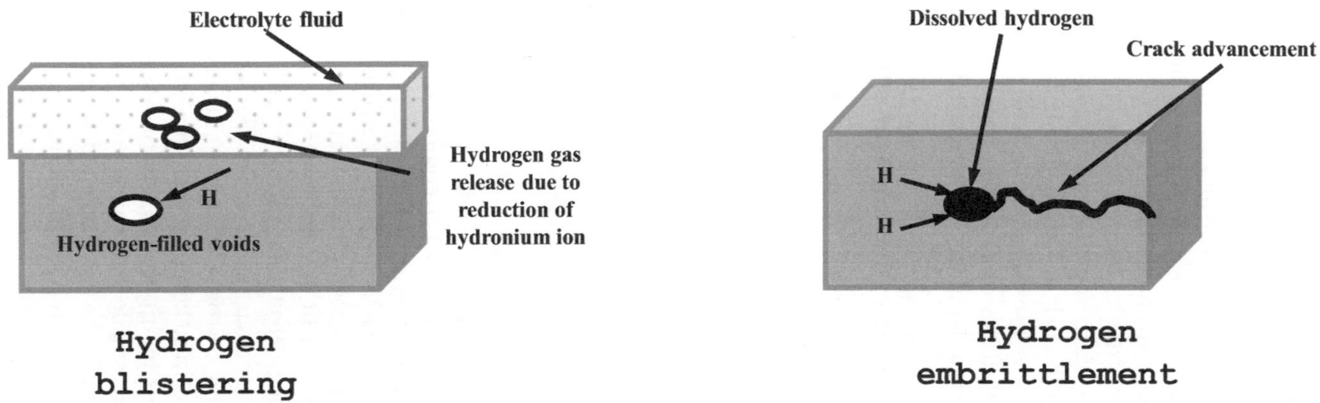

Figure 4.5 Types of material degradation resulting from hydrogen diffusion.

hydrogen diffusion and damage, leading to rapid degradation of structural components and heightened susceptibility to embrittlement.

Various industries that expose materials to hydrogen, high temperatures, or radiation have experienced numerous failures due to hydrogen embrittlement. In the oil and gas sector, pipelines transporting hydrogen-rich fuels are especially vulnerable, with embrittlement-induced cracks causing leaks or ruptures that pose environmental and safety risks. In aerospace, high-strength steel components like fasteners, landing gear, and turbine blades are susceptible to hydrogen embrittlement in high-stress, high-temperature conditions, increasing the risk of failure. Similarly, in nuclear reactors, critical components such as pressure vessels and piping face embrittlement due to the combined effects of radiation, heat, and hydrogen exposure, threatening the safety and longevity of nuclear infrastructure.

4.1.4 Mechanical, heat/radiation and chemical actions

Interaction of mechanical, thermal (or radiative), and chemical actions leads to severe damage in materials, with specific emphasis on industries such as aerospace, oil and gas, and nuclear energy. The combination of mechanical stress, heat or radiation, and chemical action significantly accelerates material degradation. Whether in hydrogen embrittlement, stress corrosion cracking (SCC), or radiation damage, the combined effects of these factors lead to a rapid decline in material performance and increase the risk of catastrophic failures. By understanding these synergistic interactions, engineers can design more robust materials and systems that mitigate the risks of degradation, ensuring the safety and longevity of critical infrastructure across various sectors.

Mechanical stress significantly contributes to material degradation, particularly in high-load environments like aerospace, pipelines, and power plants, but its impact is magnified when combined with chemical and thermal factors. In hydrogen embrittlement, for instance, mechanical stress drives the propagation of microcracks initiated by hydrogen diffusion, with crack growth accelerating as hydrogen accumulates at stress points, increasing the risk of brittle fracture. In nuclear environments, radiation exacerbates degradation by creating defects like vacancies and dislocations, which trap hydrogen atoms and weaken the material. The combination of radiation, high temperatures, and corrosive chemicals in reactors leads to embrittlement and cracking in critical components. This synergy between stress, heat, and chemical reactions multiplies the rate of degradation, as seen in SCC, where materials under tensile stress in corrosive environments experience accelerated crack growth, further hastened by elevated temperatures that increase reaction rates and atomic diffusion.

Nuclear reactors, which simultaneously expose structural components to hydrogen, elevated temperatures, and radiation, provide a prominent example of hydrogen embrittlement. The mechanical stress from the operational load creates vulnerable areas in the material, such as grain boundaries or dislocations. Heat increases the mobility of hydrogen atoms, while radiation further enhances the diffusion of these atoms by generating defects in the material's structure. As hydrogen accumulates in these stress concentration zones, microcracks form and propagate, leading to sudden and catastrophic failure of critical components.

The aerospace industry uses high-strength materials such as titanium alloys and steel in turbine blades, landing gear, and fasteners. These components experience extreme mechanical stress during operation. At the same time, high temperatures from engine operation accelerate the diffusion of impurities or hydrogen into the material. Additionally, environmental factors like corrosive chemicals from jet fuel can initiate surface degradation. The synergy between stress, heat, and chemical exposure often leads to embrittlement, cracking, or fatigue failure.

Similarly, in the oil and gas industry, pipelines transporting hydrogen-rich fuels are subject to mechanical loads from internal pressure, thermal fluctuations from operating conditions, and chemical exposure from the transported fluids. The combination of these factors accelerates hydrogen diffusion into the pipeline materials, promoting embrittlement and increasing the likelihood of failure through cracks, leaks, or ruptures.

Mitigating synergistic degradation requires an in-depth understanding of the combined effects of mechanical, thermal, and chemical actions on materials. In various industries, strategies are developed to reduce these factors and slow degradation, such as using coatings or treatments to prevent hydrogen ingress in hydrogen-rich environments or applying thermal barriers and cooling systems in high-temperature applications. Advanced materials, like specially engineered alloys that resist hydrogen embrittlement, corrosion, or radiation damage, are increasingly employed in sectors like aerospace and nuclear energy. Additionally, stress-relief techniques, such as annealing and surface treatments during manufacturing, help minimize defects and stress concentrators, enhancing material resilience against synergistic degradation.

Xu et al.[3] conducted an experimental study on hydrogen embrittlement in AISI 4135 steel under high-pressure hydrogen exposure, focusing on its corrosion, hydrogen permeation, and SCC behaviors in a tidal zone. Steel samples were exposed to hydrogen gas at 120 MPa and 200°C for 160 hours in an autoclave, followed by uniaxial tensile testing. The results showed that hydrogen-exposed samples (labeled HC1 and HC2) exhibited brittle failures, with encapsulated hydrogen proving difficult to eliminate even after polishing. The study revealed that hydrogen embrittlement significantly accelerates crack propagation, increasing the likelihood of fractures, especially when combined with corrosive media. Gas chromatograph thermal desorption analysis confirmed that hydrogen content remained in the samples, indicating no substantial hydrogen loss during transportation or testing. These findings underscore the challenges of mitigating hydrogen embrittlement and its critical role in material degradation under harsh environmental conditions.

4.2 FRACTOGRAPHY

In earlier chapters, we introduced fractography, focusing on fractographs and key features like cleavage and dimples in materials. This section builds on that foundation to offer a deeper understanding of fractography.

The scanning electron microscope (SEM) uses electrons instead of light to create images. Its narrow electron beam allows for high-depth, three-dimensional images that are particularly helpful in examining surface structures. SEM can reveal various features, including compositional differences and surface topography (such as shape, inclination, and edges), making it a valuable tool for studying material microstructure.

Fractography is the detailed study of fractured materials, focusing on how they break and analyzing the patterns and characteristics of those fractures. Fractures occur when a material is subjected to stress—such as fatigue, tension, or bending—causing it to break into two or more pieces. Regardless of the type of stress, the outcome is the separation of the material.

In this section, the analysis will go beyond fractures to include topics like wear and fatigue. Earlier chapters, which covered the fracture process, crack initiation, and propagation, serve as a foundation. All structures have natural discontinuities, including differences in grain size and boundaries due to their polycrystalline nature. Small cracks may not propagate unless they have enough energy, so understanding crack initiation and propagation is essential in preventing material failure.

The new concepts in this section will focus on building upon previous knowledge while introducing new details. Magnified images will be used to examine features that are often as small as 5 to 10 μm—too small to be seen with the naked eye. Earlier discussions noted that anything under 50 μm is difficult to observe. This section will utilize SEM and optical microscopy to study fracture surfaces in detail. These tools help identify important features like fracture initiation sites, propagation paths, secondary cracks, and patterns such as dimples, beach marks, striations, and cleavages. Instruments such as SEMs, optical microscopes, and transmission electron microscopes (TEMs) are essential for analyzing the topography and morphology of these fractures. Additionally, chemical analysis may be required to study related phenomena like wear processes.

Figure 4.6 depicts an SEM image of a fractured surface showing two key features: dimple microstructure and cleavage fractures. The dimple microstructure consists of tiny depressions, commonly found in metals and alloys subjected to fatigue loading. These dimples form as small cracks in the

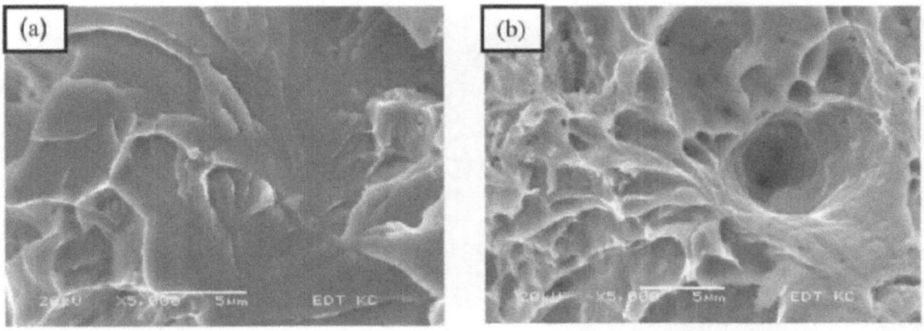

Cleavage at X5,000

Dimple rupture at X5,000

Figure 4.6 SEM image of a fractured surface.[4]

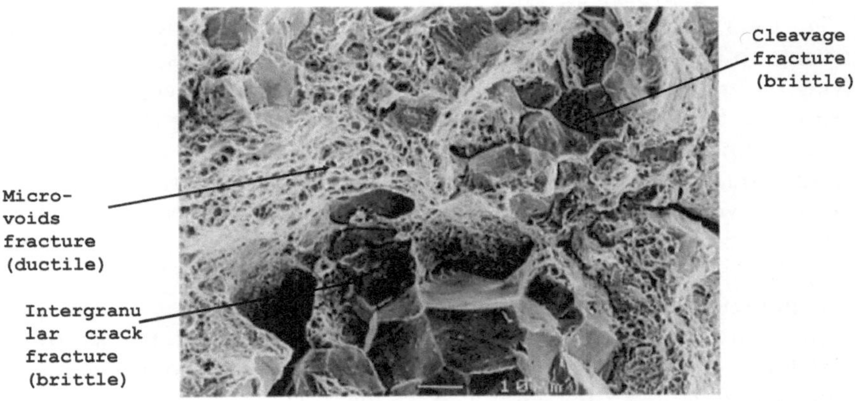

Figure 4.7 Multimode fracture.

material concentrate stress and propagate under cyclic loading. In contrast, cleavage fractures occur along weak planes or grain boundaries, producing smooth, flat surfaces. These fractures follow pre-existing cracks and are influenced by grain boundaries in polycrystalline materials, often showing ridge steps, which are elongated, narrow protrusions requiring more energy than cleavage fractures.

Figure 4.7 depicts a widely available image from public web sources. To enhance clarity and facilitate academic use, appropriate labels have been added to illustrate the multimode fracture observed in many ductile materials subjected to variable loading. This multimode fracture demonstrates the coexistence of various fracture mechanisms on a single surface, including micro-voids, cleavage, and intergranular cracks. Micro-voids are characteristic of

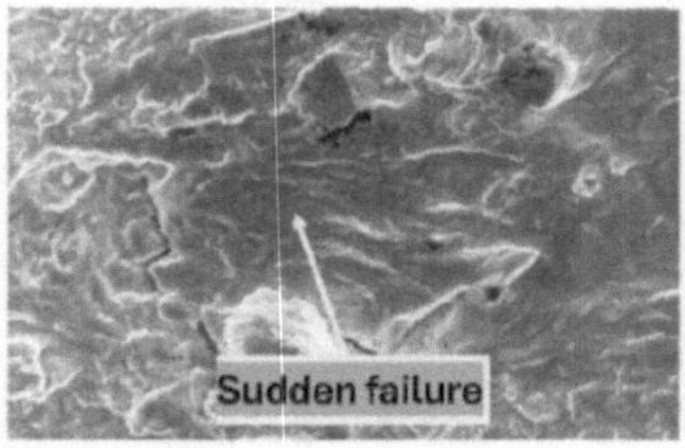

Figure 4.8 Impact fracture.

ductile materials, while cleavage fractures are commonly associated with brittle materials. The image underscores the intricate nature of fractures, where both ductile micro-voids and brittle cleavage features can coexist on the same fractured surface.

Figure 4.8 presents a commonly available image sourced from the public domain. To improve clarity and support academic purposes, a label identifying the crater has been added to highlight an impact fracture featuring deep voids. Such fractures are typically observed in ductile materials like copper, aluminum, and steel. These materials exhibit substantial plastic deformation, where tensile stress causes initial discontinuities to develop into micro-voids. With continued loading, these voids coalesce, ultimately resulting in a sudden fracture.

Figures 4.6 to 4.8 provide a comprehensive view of various fracture types, highlighting the differences between micro-voids, dimples, and cleavage fractures. Micro-voids, typically seen in ductile materials, form under lower stress levels and coalesce to cause fracture, as shown in both multimode and impact fractures. Dimples, which are microscopic depressions, emerge from fatigue loading and resemble smaller versions of craters, which indicate more sudden and energy-intensive failures. Cleavage fractures, often occurring in brittle materials, follow weak planes or grain boundaries and require higher stress levels to propagate. These images collectively demonstrate how multiple fracture modes can coexist on a single surface, contributing to material failure in different ways.

Figure 4.9 presents fractured surfaces after crack tip opening displacement (CTOD) testing under in situ hydrogen cathodic charging at 20 mA/cm^2, showing features linked to hydrogen embrittlement. These features include (a) river pattern morphologies, (b) stair-like embrittled surfaces, (c) distinct bands and fracture surfaces, (d) micro-void coalescence, (e) quasi-cleavage facets with teardrop shapes, and (f) voids with fish-eye morphology and stepwise micro-cracking.

Previous discussions covered "hydrogen embrittlement," a process where increased hydrogen content leads to swelling and crack growth on material

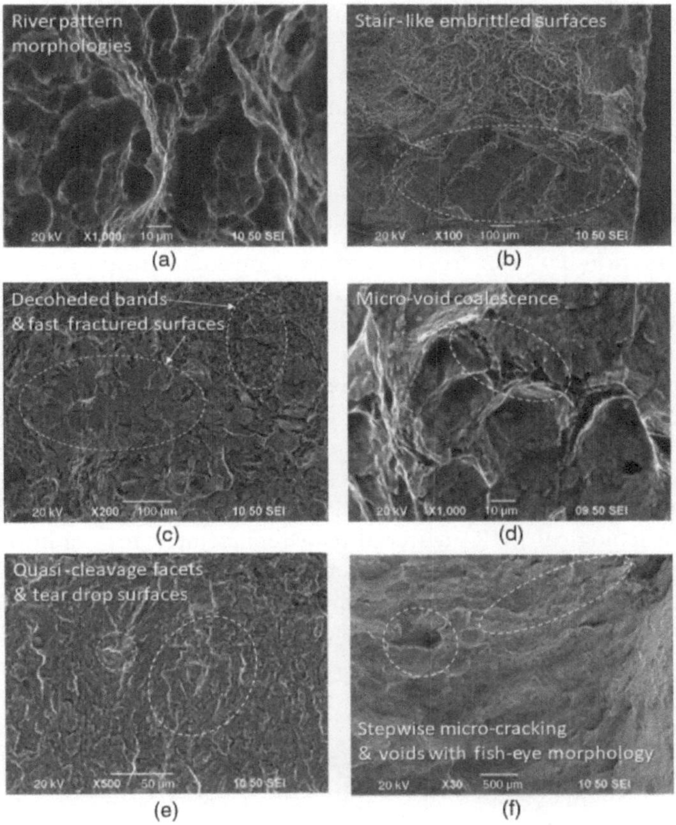

Figure 4.9 Fracture morphologies from CTOD testing under hydrogen embrittlement.[5]

surfaces. When examining fracture surfaces in hydrogen-affected materials, discontinuities, and microstructural changes are observed, which can promote crack propagation.

Figures 4.6 to 4.9 demonstrate that fractures in materials often involve multiple modes, not strictly brittle or ductile. Instead, multimode fractures, where different fracture types coexist on the same surface, are common. For example, micro-voids, cleavage fractures, and intergranular cracks may all appear together, making it challenging to categorize the fracture by just one mechanism.

Assessing multimode fractures requires a more refined and probabilistic approach. Instead of a binary classification, fracture surfaces must be evaluated by identifying and measuring various features, like dimples (indicative of ductile failure) and cleavage planes (suggestive of brittle failure).

By embracing this probabilistic method, engineers and researchers can better understand the factors driving fracture initiation and propagation, leading to improved predictions of material behavior. This approach enables the design of more resilient materials and structures, addressing the combined effects of multiple fracture mechanisms rather than focusing on a single type of failure.

4.2.1 From erosion and wear mechanisms to fractography

The interaction between erosion and wear mechanisms plays a crucial role in determining fracture characteristics. Variations in impingement angles lead to different wear behaviors: at lower angles (e.g., 30 degrees), cutting and abrasive wear dominate, whereas higher angles (e.g., 90 degrees) are associated with fatigue or impact wear. By analyzing material properties and wear responses (as illustrated in Figure 4.10), one can better predict fracture patterns and implement strategies to prevent wear-induced failures.

In a previous chapter, we discussed the use of a steel disc rotating against magnetorheological (MR) material as a replacement for traditional brake shoes. Fractographic analysis can be applied to study disc wear and identify multimode failures. This analysis emphasizes the need to enhance surface hardness to prevent particle embedment and minimize two-body abrasion.

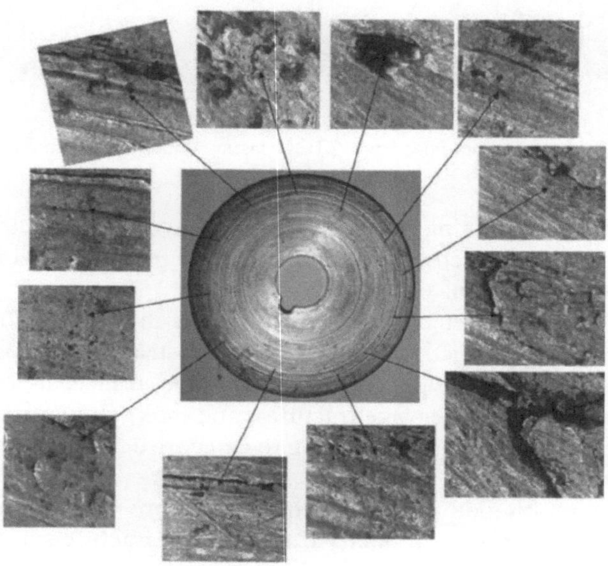

Figure 4.10 Disc of MR brake: SEM images of worn surfaces.

Figure 4.10 depicts a steel disc at the center. These magnified regions highlight different surface features and damage patterns, which provide important insights into the wear mechanisms at play. The central steel disc shows wear marks, resulting from its interaction with MR material. The radial grooves suggest that rotational movement occurred, and the close-up views show different types of surface damage, including abrasion, scuffing, and possible material transfer or particle embedment.

In the magnified sections, there are visible signs of two-body abrasion, where hard particles or asperities from the counter surface may have ploughed through the disc, creating grooves. Some areas show microcracking or fatigue, indicating possible material weakening due to repeated stresses. The close-up analysis suggests multimode wear mechanisms—abrasion, particle embedment, and possibly fatigue—all contributing to the overall wear of the disc. This type of fractographic analysis can help in diagnosing the causes of failure and improving surface treatments or material choices to enhance the disc's durability.

Figure 4.11 showcases a collection of widely recognized fracture patterns commonly observed in engineering failure analysis, sourced from public domain references. To enhance academic utility and improve understanding, labels and directional arrows have been added to clearly distinguish the various fracture modes. This figure highlights seven distinct fracture line patterns critical for identifying fracture origins and deciphering primary failure mechanisms. The following patterns are illustrated:

1. *Chevron lines:* Typically indicative of high-energy impacts, these patterns suggest rapid crack propagation originating from a concentrated stress point.

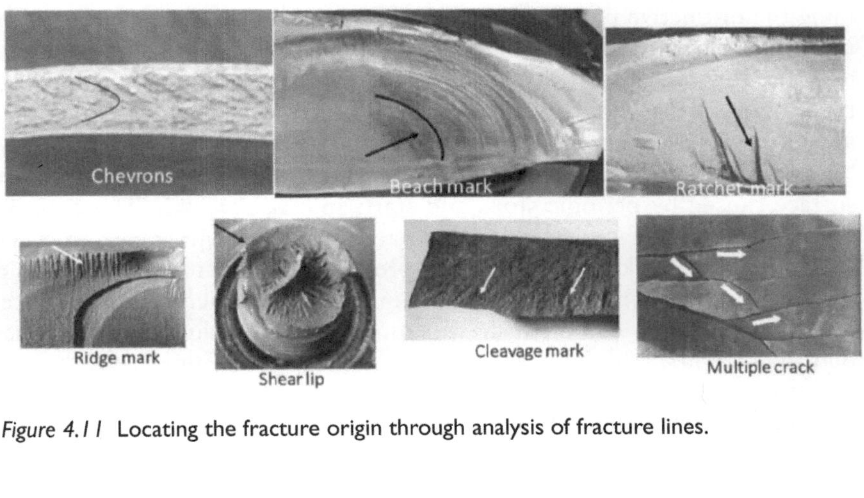

Figure 4.11 Locating the fracture origin through analysis of fracture lines.

2. *Beach marks:* Characteristic of fatigue failure, these concentric, wave-like patterns are formed due to repeated stress cycles and progressive crack growth.
3. *Ratchet marks:* These asymmetrical, gear-like patterns are associated with fatigue cracks that originate at multiple points and eventually merge under cyclic loading.
4. *Ridge marks:* Resembling erosion patterns, ridge marks form due to microstructural features such as inclusions or impurities influencing crack growth pathways.
5. *Cleavage marks:* Smooth, planar features typical of brittle materials, indicating crack propagation along specific crystallographic planes under low-energy conditions.
6. *Shear lips:* Indicators of ductile failure, these patterns result from significant plastic deformation near the crack edges during fracture.
7. *Multiple cracks:* These signify complex fracture events involving simultaneous crack initiation and propagation, often requiring advanced analysis to identify the dominant failure mechanism.

These fracture patterns provide essential information about the material's response to external forces and environmental conditions. For instance, fatigue-induced features like ratchet and beach marks point to cyclic loading, while chevron and cleavage lines highlight the role of impact or brittle failure mechanisms. By carefully analyzing these patterns, engineers can better understand failure origins, mitigate material degradation, and develop strategies to enhance surface performance and reliability.

4.2.2 Understanding fracture and wear debris using fractography

Fractography allows us to uncover the underlying causes of failure by examining the distinctive patterns left behind on fractured or worn surfaces. By using fractography, one can identify the mode of fracture, the material's response to stress, and how wear and wear debris contribute to system degradation. "Chevron patterns," for example, are V-shaped markings that appear on the fracture surface due to high-energy impact or explosive stress. "Beach marks," on the other hand, are concentric rings on the fracture surface caused by periodic stops and starts in crack growth. These marks are indicative of "fatigue failure" and are associated with repeated cyclic loading. Using this kind of knowledge provides valuable information about the "microscopic" or "nanoscale" mechanisms behind fracture. For instance, a "ductile fracture" will often show features like "shear lips"—curved lines that indicate material deformation under shear stress. Conversely, "brittle fractures" are typically associated with "cleavage marks" that form along crystallographic planes.

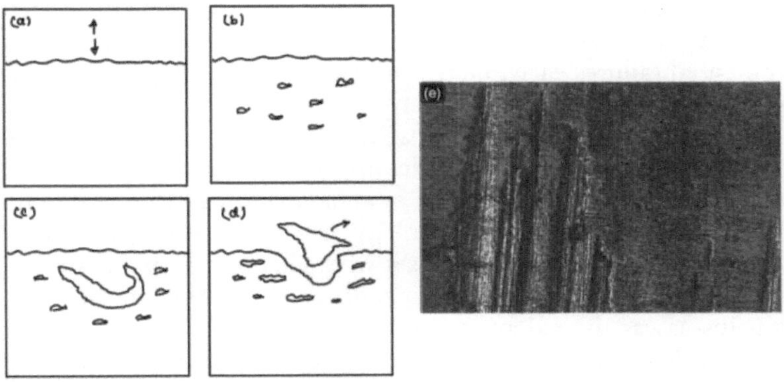

Figure 4.12 Formation of "wear debris" and "surface pit."

While fracture analysis focuses on the complete failure of a component or system, "wear debris" acts as an early warning sign of impending failure, as illustrated in Figure 4.12. In Figure 4.12(a), the diagram shows the initial surface contact where external forces cause stress concentration. Figure 4.12(b) depicts the formation of subsurface cracks due to repeated stress cycles, leading to micro-fractures below the surface. In Figure 4.12(c), crack propagation continues, and small fragments of material (wear debris) begin to detach. Figure 4.12(d) depicts the formation of a surface pit as debris is ejected, creating surface irregularities that accelerate wear. Finally, Figure 4.12(e) presents a fractographic microscope image showing real-life surface pitting with visible grooves and pits caused by debris formation.

Wear occurs when two surfaces slide against each other, leading to the gradual removal of material. "Wear debris" refers to the small particles generated during this process as material is worn away from one or both surfaces. While wear is normal in many mechanical systems, excessive wear or the buildup of wear debris can signal deeper problems, which, if not addressed, could lead to fractures or even catastrophic failure.

Unlike fractures, which clearly indicate system failure, wear debris can provide valuable information much earlier before major issues arise. The debris forms as materials in contact break down over time, and its size, shape, and composition can reveal important details about the system's condition. By monitoring and analyzing wear debris, engineers can detect early signs of excessive wear, misalignment, material incompatibility, or lubrication problems—issues that, if left unchecked, could eventually cause failure or fracture.

The presence of wear debris acts as an early warning system, allowing for proactive maintenance. For example, a sudden increase in debris might point to abnormal stresses or the wear of a part, indicating the need for corrective actions such as replacing a component, realigning the system, or improving lubrication. By monitoring wear debris, maintenance teams can catch potential failures early, increasing system reliability and reducing the risk of unexpected downtime.

As material loss accumulates, wear debris can be analyzed using fractography to better understand the material's degradation. Wear debris serves as an early warning sign of system malfunction or material degradation. By examining the morphology and composition of this debris, engineers can identify the root causes of wear, such as poor lubrication, material incompatibility, or misalignment. Fractographic analysis of wear debris helps assess the extent, severity, and nature of the wear, allowing engineers to detect specific issues before they lead to major failures. For example, "adhesive wear" occurs when material transfers from one surface to another due to friction, generating wear debris with distinctive shapes and characteristics that offer insight into the wear mechanism at play.

In Figure 4.13, two images highlight key phenomena related to adhesive wear. The first image shows the transfer of material from the carbon graphite seal onto the steel surface. Adhesive wear occurs when two surfaces in contact slide against each other, causing material from one to adhere to and transfer onto the other. Here, carbon graphite from the seal is deposited onto the steel shaft due to the frictional forces during operation. The second image refers to the small particles or fragments that have broken away from either the deposited carbon graphite on the steel shaft or the carbon graphite seal itself as a result of adhesive wear. These debris particles, generated by adhesive wear, can intensify wear by circulating between contacting surfaces and causing further abrasive wear.

Figure 4.13 illustrates the adhesive wear process between a stainless-steel shaft and a carbon graphite seal, highlighting both material transfer and

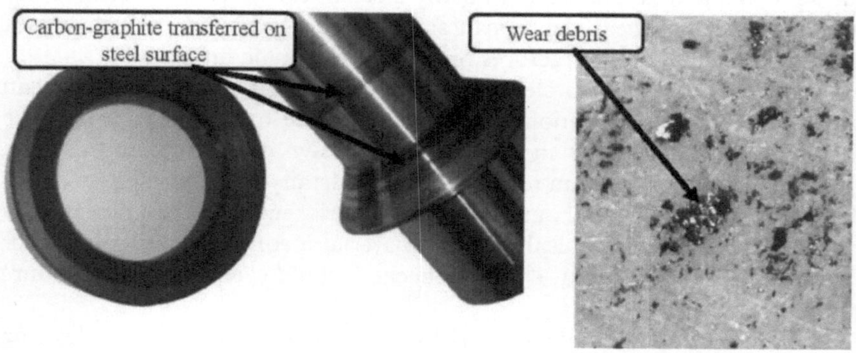

Figure 4.13 Adhesive wear between stainless steel shaft and carbon graphite seal.

the formation of wear debris. A similar process is shown in Figure 4.14. In adhesive wear, molecular attraction causes cold welding between surfaces, leading to material transfer from the softer surface to the harder one, potentially inducing work hardening and further promoting material transfer. Over time, this transferred material accumulates and reaches a critical size, causing detachment, often from weaker sections of the material, resulting in fracture. The resulting plate-like debris is a characteristic sign of adhesive wear, which can eventually transition into abrasive wear.

In Figure 4.14, two surfaces come into contact and begin sliding relative to each other, as indicated by the arrow showing the direction of motion. At specific localized points, surface asperities (microscopic high points) make contact, and adhesion forms due to the pressure and friction between the surfaces. As the sliding continues, these adhered points are subjected to shearing forces, causing stress concentrations. Small cracks may form at the interface, weakening the bond.

With ongoing sliding motion, these bonds break, leading to the detachment of small pieces of material (wear debris) from one or both surfaces. As more material is removed, debris accumulates in the contact zone. This process illustrates how adhesive wear evolves from initial contact to the

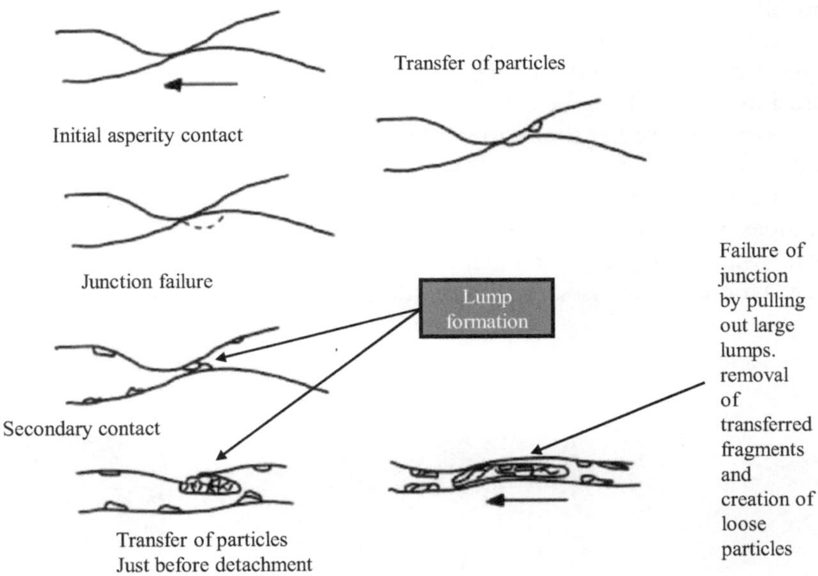

Figure 4.14 Progressive sequence of events related to adhesive wear.

formation of debris, which can then lead to abrasive wear. Abrasive wear occurs when hard particles, such as accumulated debris, scrape against the surfaces, leaving behind characteristic scratch marks and grooves.

Wear mechanisms are typically classified into adhesive and abrasive wear, which together account for 70% to 80% of all wear-related phenomena. Surface fatigue also plays a significant role in wear, often initiating from material defects that act as starting points for crack formation. These cracks can propagate and eventually lead to larger wear debris. As the cracks expand, they may merge and reach the surface, causing the detachment of small particles, generally ranging from 15 to 20 μm in size. While this debris does not necessarily conform to traditional fracture definitions, it is still a crucial indicator of wear in the system. By analyzing the morphology and chemical composition of wear debris, valuable insights into the system's condition can be obtained, helping to identify the underlying wear mechanisms without the need for physical disassembly. Even small amounts of debris can reveal signs of potential issues, such as corrosion or material degradation. Quantifying factors like particle density and wear coefficients allows for more accurate predictions of the remaining lifespan of components, helping to prevent further surface degradation.

Some types of wear debris are shown in Figure 4.15. Analyzing these debris involves examining their microstructure, shape, and composition to gain insights into system performance and detect issues early. As depicted in Figure 4.15, wear debris typically appears irregular, lacking a defined shape, and often resembles a three-dimensional structure. Some debris may take on a spherical form (as shown in Figure 4.15, with a diameter of less than 2 μm), which is generally linked to rolling fatigue phenomena. The type of abrasion, caused by spherical debris, is less harmful because the particles roll more freely rather than slide, resulting in a less aggressive wear process. Additionally, the presence of contamination or oxide formation on wear debris can serve as a key indicator of potential system malfunctions. For example, wear debris approaching 1 mm in size often signals significant wear, while particles greater than 100 μm typically indicate rapid surface degradation, often associated with severe adhesive wear. The appearance

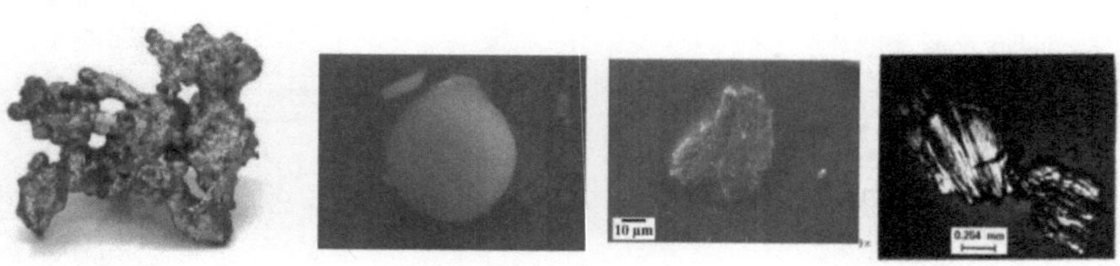

Figure 4.15 Different types of wear debris.

of plate-like stripes, measuring 254 μm in the figure, is a clear indicator of adhesive wear.

By examining wear debris under a microscope/SEM, fractographers can identify early indicators of material fatigue, microcrack formation, and other forms of microscopic damage, which can eventually lead to fracture. Fractographic analysis can identify these wear-induced cracks and differentiate them from other fracture mechanisms. This understanding allows engineers to take preventive measures, such as improving material selection, altering operating conditions, or enhancing lubrication, to mitigate the effects of wear and prevent subsequent fracture. Furthermore, examining the "composition" of wear debris can help identify the role of "surface contamination" or "foreign particles" that may exacerbate wear or fracture. For example, abrasive particles from external sources can cause "excessive wear," leading to material weakening and crack initiation. By understanding the correlation between wear and fracture, engineers can design systems that better resist both wear and catastrophic failure, leading to longer-lasting, more reliable materials. Fractography reveals the underlying causes of fractures and helps to identify the early stages of wear, enabling proactive measures to prevent catastrophic system failures. By linking wear debris to fracture, fractographic analysis provides a comprehensive understanding of the material's behavior under stress and wear, aiding in the development of more durable, reliable materials and mechanical systems. The ability to detect wear at the micro or nanoscale before it leads to fracture is crucial for improving system reliability, preventing unplanned downtime, and ensuring safe, efficient operations. Fracture is often regarded as a catastrophic, higher-level failure within mechanical systems. It occurs when a material or structure experiences a complete separation due to applied stress that exceeds its intrinsic strength. Fracture is typically the result of progressive damage that accumulates over time, often signifying a critical and irreversible failure that may disrupt the system's function entirely. Examples of this include the breaking of a mechanical part, leading to total system failure or a significant safety risk. The process leading to fracture is often accompanied by observable warning signs, such as crack propagation or deformations, but it is generally viewed as the endpoint in the failure process.

4.3 CASE STUDY 1

A multi-station Hip Simulator is a valuable tool for investigating surface deterioration and fractography in medical hip implants. This study focuses on acetabular cups made from various materials, including ultra-high molecular weight polyethylene (UHMWPE) (Figure 4.16), and tests them under simulated hip joint conditions that include physiological cyclic loads and bi-axial rocking motions against a cobalt–chromium (CoCr) femoral

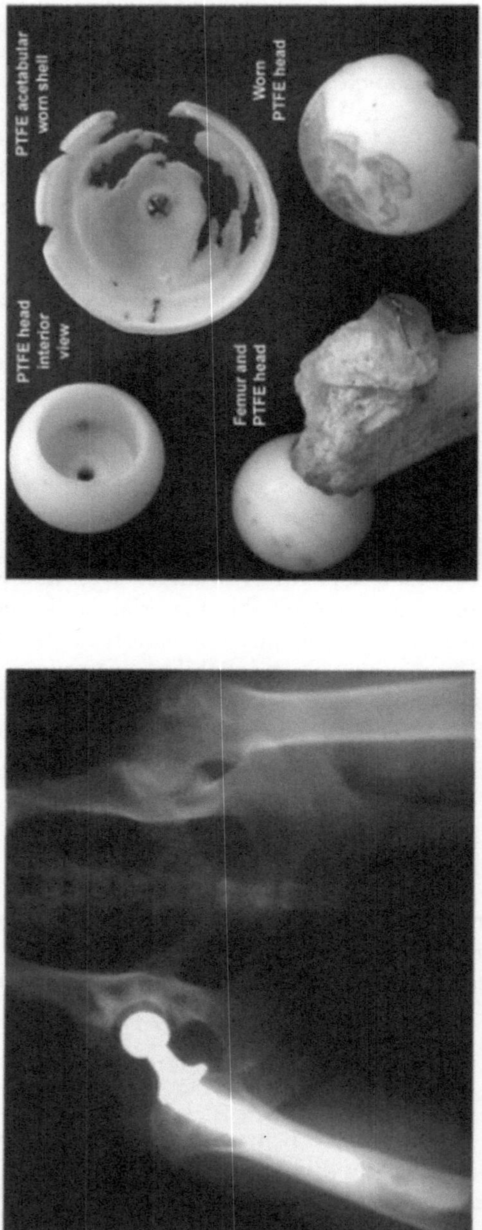

Figure 4.16 Position of the hip joint cup and potential failure modes.[6]

head. The cups undergo different surface treatments, such as being untreated, sterilized with ethylene oxide (EtO), sterilized with gamma radiation, and pre-soaked in bovine calf serum, to closely mimic real-life conditions. Usually, sodium azide is added to the bovine calf serum to prevent bacterial degradation. This experimental setup enables researchers to effectively observe wear and deformation patterns on the surfaces of the cups.

Wang et al.[7] conducted a study on various sterilization methods for acetabular cups and found that gamma-irradiated cups, due to cross-linking, had a significantly lower wear rate of 43 ± 3 mm^3 per million cycles. In comparison, cups sterilized with EtO showed no cross-linking and exhibited increased surface fibrillation. Untreated cups displayed a surface morphology like that of the EtO sterilized cups but had a higher wear rate. Both gamma-irradiated and EtO-sterilized cups exhibited micro-voids and ripple formations, indicating material fatigue from cyclic loading. SEM images revealed rippled surfaces and micro-voids in the worn areas, with gamma-irradiated cups demonstrating better wear resistance but typical brittle fracture patterns. Conversely, EtO-sterilized cups showed extensive fibrillation, indicating ductile failure. This study emphasizes the significant impact of different sterilization methods on the wear and durability of implant materials, offering valuable insights into the longevity and safety of medical implants in real-world scenarios.

Shah et al.[6] studied fretting fatigue behavior of polytetrafluoroethylene (PTFE) as a material for medical implants. They performed fractography analysis of fractured PTFE samples and observed that a small amount of debris is present at the fractured surface of collar, indicating that the polymer underwent a tearing action at the neck.

Figure 4.17 depicts the formation of debris caused by fatigue phenomena and the presence of striation on the fractured surface. Striation consists of linear grooves or markings that appear as fine lines running in a specific direction across the surface. These observations suggest that PTFE demonstrates both brittle and ductile fracture behaviors. Ductile fractures lead to the formation of a slanted lip, whereas brittle fractures result in a flat, grooved surface.

Current understanding indicates that typical ductile failure is characterized by micro-void coalescence, whereas brittle fractures exhibit a cleavage pattern. Many alloy materials display mixed-mode failure, where the matrix region experiences ductile failure while the precipitates exhibit intergranular failure. Temperature plays a crucial role in influencing the number and size of dimples observed in these fractures. As temperature increases, variations occur in yield strength, hardness, density of micro-dimples, and striation patterns. Therefore, it is essential to analyze failure by considering all operating conditions, alongside the material and surface properties. Additionally, it is important to note that fractures subjected to sudden, impact-type loading often demonstrate more brittle characteristics compared to those

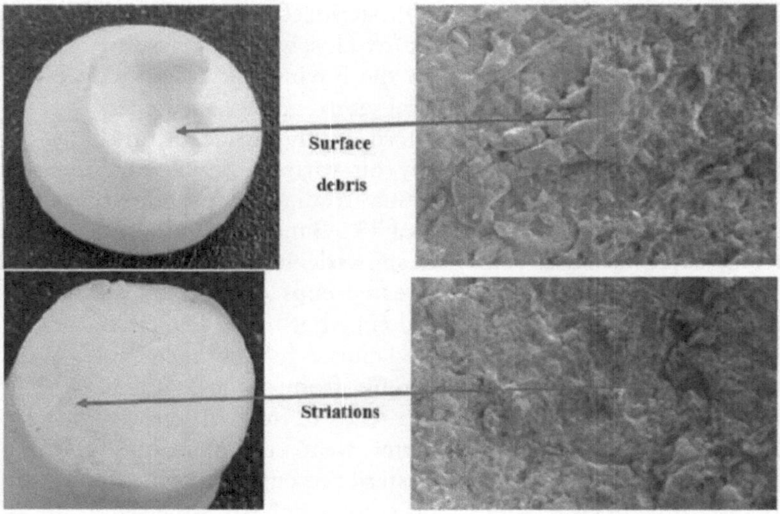

Figure 4.17 "Surface debris" and "striation" aspects of the wear process in PTFE.[6]

experiencing gradual tensile loading. This difference highlights the need for a comprehensive evaluation of the material's behavior under varying loading conditions to accurately predict failure modes.

4.4 CASE STUDY 2

The demand for pipeline steels has surged in recent decades due to their efficient and economical ability to transport oil and natural gas over long distances. However, two major damage modes, SCC and hydrogen-induced cracking (HIC), pose significant challenges to these materials. SCC arises from the combined effects of a corrosive environment and mechanical stress, with various factors influencing the initiation and propagation of SCC cracks, including the steel's microstructure, the distribution of inclusions and precipitates, residual stresses, chemical composition, applied loads, alternating current (AC) effects, and the pH levels of both the transported fluids and the surrounding soil. Additionally, crystallographic texture plays a crucial role in crack propagation. Generally, hard and brittle phases facilitate the advancement of SCC cracks. More explanation of these factors has been reviewed by Mohtadi-Bonab.[8]

Investigating the behaviors and mechanisms of corrosion, hydrogen permeation, and SCC in steel can lead to effective strategies for mitigating degradation in marine engineering applications. Enhancing SCC resistance may require modifications to the steel's microstructure and improved control over its texture and grain boundary characteristics. Several innovative methods can enhance the performance of pipeline steel against SCC. Crystallographic texture and grain boundary engineering are critical in improving SCC resistance in these materials. Arafin et al.[9] investigated the influence of texture on X65 pipeline steel and found that a deeper understanding of texture and grain boundary characteristics could facilitate the development of pipeline steels with enhanced intergranular SCC resistance. Their research indicated that a higher concentration of coincidence site lattice (CSL) boundaries and low-angle grain boundaries (LAGBs) could effectively inhibit the propagation of intergranular SCC cracks, achievable through modifications in surface texture.

The CSL concept describes a lattice structure where atoms in two adjacent crystals or grains are not perfectly aligned, yet a certain fraction of lattice points from both grains coincide. This alignment reduces the energy of the grain boundary compared to non-coincident boundaries, leading to increased stability and better mechanical properties, such as enhanced resistance to grain boundary sliding and improved ductility. The presence of CSL boundaries can enhance resistance to SCC and HIC by providing paths that are less conducive to crack propagation. Engineers can manipulate processing conditions to promote the formation of CSL boundaries.

LAGBs are characterized by a small misorientation, typically less than 5 degrees, between adjacent grains. Formed by arrays of edge dislocations, LAGBs accommodate misalignment and allow some degree of strain accommodation across the boundary. These boundaries exhibit lower energy than high-angle grain boundaries, contributing to enhanced stability and reduced susceptibility to crack propagation in certain materials. Engineers can adjust processing conditions, such as temperature and strain rate, to promote LAGBs, thereby improving the properties of materials subjected to harsh environments.

Additionally, the role of inclusions in SCC crack initiation is significant. Producing pipeline steel with fewer inclusions or a uniform distribution of inclusions is essential for achieving high SCC resistance. Stress concentration at corrosion pits tends to induce and propagate cracks, and the geometry of the corrosion pit, particularly its depth, plays a crucial role in crack initiation.

To further optimize the performance and lifespan of pipeline steel, the integration of digital twin technology (will be discussed in subsequent chapters) offers a promising solution. A digital twin acts as a virtual replica of the physical pipeline system, allowing engineers to monitor and analyze real-time data from various sources, such as sensors and Internet of Things (IoT) devices. By correlating the wear and degradation patterns observed in the digital twin with the characteristics of wear debris and the condition of the actual pipeline, maintenance strategies can be developed to early address potential failures. This capability enhances decision-making processes related to maintenance, as the digital twin can simulate various conditions and predict wear patterns based on actual operating data.

Moreover, the insights gained from the digital twin can inform the design modifications needed to improve the steel's microstructure, such as promoting the formation of CSL and LAGB boundaries, as well as reducing the number of harmful inclusions. By utilizing digital twins in conjunction with advanced materials engineering techniques, the industry can achieve a more resilient and efficient pipeline infrastructure, ultimately reducing the risks associated with SCC and HIC.

4.5 CONCLUSIONS

This chapter underscores the necessity of understanding the synergistic interactions among mechanical, chemical, thermal, and radiative degradation mechanisms to effectively prevent surface deterioration. Such understanding enables the development of robust materials and surface engineering strategies that address the compounded effects of multiple degradation factors. As demonstrated, combinations like mechanical wear with corrosion or radiation-accelerated chemical degradation can accelerate failure processes, making it crucial to approach material design with an integrated perspective on degradation pathways. The chapter concludes by advocating comprehensive material testing, utilizing advanced fractographic techniques to analyze failure modes. Enhanced predictive models that account for synergistic degradation can guide material selection, processing, and treatment techniques, leading to improved reliability and lifespan in high-stress, high-risk environments.

REFERENCES

1. Xu, Y., Liu, L., Zhou, Q., Wang, X., Tan, M. Y., & Huang, Y. (2020). An overview of major experimental methods and apparatus for measuring and investigating erosion-corrosion of ferrous-based steels. *Metals*, *10*(2), 180.
2. Akhtar, M., Qamar, S. Z., Muhammad, M., & Nadeem, A. (2018). Optimum heat treatment of aluminum alloy used in manufacturing of automotive piston components. *Materials and Manufacturing Processes*, *33*(16), 1874–1880.
3. Xu, Y., Huang, Y., Cai, F., Lu, D., Wang, Z., & Wang, X. (2022). Study on corrosion, hydrogen permeation, and stress corrosion cracking behaviours of AISI 4135 steel in the tidal zone. *npj Materials Degradation*, *6*(1), 96.
4. Richard, L., Ellsworth, P. E., Chris, S., & Spies, P.E. (2019). Blog published on 09.01.2019, www.edtengineers.com/blog-post/fractures-part-1
5. Kyriakopoulou, H. P., Karmiris-Obratański, P., Tazedakis, A. S., Daniolos, N. M., Dourdounis, E. C., Manolakos, D. E., & Pantelis, D. (2020). Investigation of hydrogen embrittlement susceptibility and fracture toughness drop after in situ hydrogen cathodic charging for an X65 pipeline steel. *Micromachines*, *11*, 430. https://doi.org/10.3390/mi11040430

6. Shah, Q. M. Z., Kowser, M. A., Chowdhury, M. A., Chani, M. T. S., Alamry, K. A., Hossain, N., & Rahman, M. M. (2022). Modeling fracture formation, behavior and mechanics of polymeric materials: A biomedical implant perspective." *Journal of Composites Science*, *6*(1), 31. https://doi.org/10.3390/jcs6010031
7. Wang, A., Stark, C., & Dumbleton, J. H. (1996). Mechanistic and morphological origins of ultra-high molecular weight polyethylene wear debris in total joint replacement prostheses. *Proceedings of the Institution of Mechanical Engineers, Part H: Journal of Engineering in Medicine*, *210*(3), 141–155 https://doi.org/10.1243/PIME_PROC_1996_210_407_02
8. Mohtadi-Bonab, M. A. (2019). Effects of different parameters on initiation and propagation of stress corrosion cracks in pipeline steels: A review. *Metals*, *9*(5), 590. https://doi.org/10.3390/met9050590
9. Arafin, M. A., & Szpunar, J. A. (2011). Effect of bainitic microstructure on the susceptibility of pipeline steels to hydrogen induced cracking. *Materials Science and Engineering: A*, *528*, 4927–4940. https://doi.org/10.1016/j.msea.2011.03.036

Chapter 5

Environmentally assisted surface deterioration

Surface degradation is primarily caused by corrosion and various environmental factors. As discussed in previous chapters, environmental conditions significantly impact surface wear and weakening. This chapter explores how these factors contribute to surface damage and examines strategies to prevent or mitigate it. Key factors include moisture, extreme temperatures, chemical exposure (such as acids and toxic gases), and electromagnetic radiation such as ultraviolet (UV) light, which can impair performance and cause surface degradation. For instance, UV radiation can cause swelling in polymers, as previously noted in earlier chapters. Additionally, abrasive particles such as soot, dust, and sand further accelerate surface wear.

Corrosion significantly impacts everyday infrastructure, particularly steel and iron pipes used for transporting drinking water. As these pipes corrode, they can release debris and contaminants into the water supply, creating health risks that necessitate immediate replacement. Similarly, industrial fluids can become contaminated, which affects performance and emphasizes the need to address atmospheric corrosion and other environmental factors that lead to failures in systems, components, or elements.

The interaction between corrosion and wear is especially crucial, as it substantially impacts various types of industrial machinery, such as slurry pumps in the oil sands sector and mining equipment. In one of my earlier lectures, I discussed how corrosion, wear, and fracture interact and whether they exacerbate damage significantly. This interplay is vital in affecting many industrial machines. For instance, in slurry pumps, the presence of abrasive particles leads to erosion, which exposes new surfaces to corrosion, accelerating deterioration.

Likewise, mining equipment faces numerous erosion risks, with surfaces becoming exposed and the presence of water further contributing to corrosion. Therefore, corrosion, wear, and erosion are continuous processes that must be considered. These factors are crucial when selecting materials, designing products, defining component lifespans, or enhancing the reliability and serviceability of equipment.

DOI: 10.1201/9781003545019-5

This chapter revisits the crucial relationship between corrosion and wear, as previously mentioned. By carefully evaluating each parameter, we can enhance safety and reliability while reducing the risk of future incidents. It also underscores how hidden chemical reactions in buried pipelines can cause unexpected corrosion when moisture infiltrates, gradually compromising the structural integrity. As highlighted earlier, atmospheric corrosion represents a serious safety risk. If corrosion is neglected over time, it can reduce a structure's factor of safety to unacceptable levels, increasing the likelihood of structural failure. Additionally, corrosion can alter the fatigue behavior of materials such as steel and iron, making them more brittle and susceptible to sudden failures.

Throughout history, various incidents[1] have demonstrated the need for corrective actions in science to address these challenges. One example, a direct quote from the references,[1] is the incident on April 28, 1988, involving Aloha Airlines flight 243, a Boeing 737. During the flight, the aircraft experienced explosive decompression, causing a part of the fuselage to detach. While the plane landed safely, the incident resulted in one fatality. Cusick et al.[1] provided numerous case studies. The case studies pertain to airplanes or craft, detailing the problems encountered and the evolution of technologies over time. This incident has taught us a valuable lesson, as it involved a single door that malfunctioned, resulting in corrosion-assisted damage. This example illustrates how corrosion can reduce the fatigue limit from its initial lifespan to as low as 60% to 70% of that time.

Satellites in low Earth orbit (LEO) encounter environment-assisted degradation. One of the primary threats is atomic oxygen (AO), a highly reactive form of oxygen that rapidly erodes satellite materials. In addition to AO, these satellites must contend with UV radiation, charged particles from solar wind and cosmic rays, micrometeoroids, and space debris. Each of these elements contributes to the potential degradation of satellite components, making their design and longevity crucial considerations. Understanding the challenges posed by the LEO environment is vital for ensuring the effective operation of satellites. Engineers must consider these factors in their designs to maintain satellite functionality throughout their operational lifespans. To further protect satellite surfaces, special coatings—such as temperature-control layers and AO-resistant materials—are employed to minimize damage. Innovations like nanocoatings, which improve hardness and reduce wear on components, also play a crucial role in safeguarding satellites against environmental exposure. This proactive approach not only slows down aging but also extends the service life of various structures.

Additionally, insights gained from studying degradation mechanisms can be applied to the dismantling of large vessels approaching the end of their operational life. By utilizing efficient dismantling methods, we can significantly reduce the labor-intensive processes currently in use.

5.1 ENVIRONMENTAL FACTORS CONTRIBUTING TO SURFACE DEGRADATION

5.1.1 Water and humidity

Water plays a key role in atmospheric corrosion, even at low moisture levels from 100 to 10,000 ppm. Water on metal surfaces helps transport ions, forms electrolytes, and triggers localized corrosion. When water vapor in the air reacts with gases like sulfur dioxide (SO_2) and nitrogen oxides (NO_x), it creates acids, leading to acid rain that corrodes structures like bridges, buildings, and monuments. Even small amounts of water vapor can encourage electrolyte formation, causing localized corrosion events and speeding up the breakdown of materials.

5.1.2 Temperature variations

Fluctuating temperatures, from extreme highs to lows, can expand and contract materials, creating microcracks and accelerating surface degradation. Temperature impacts chemical reaction rates, further influencing how pollutants and environmental factors degrade surfaces.

5.1.3 UV radiation

UV radiation, primarily from sunlight, accelerates surface degradation, particularly for polymers and paints, by breaking down chemical bonds. UV-induced damage, combined with environmental pollutants, exacerbates surface wear and requires additional protection strategies.

5.1.4 Chemical exposure

Surfaces wear down faster when exposed to acids, gases, and other reactive substances. Gases such as sulfur dioxide and nitrogen oxides react with water to create sulfuric and nitric acid, which cause acid rain, while chlorine compounds can form hydrochloric and hypochlorous acid, both of which are highly corrosive to metals. The damage depends on factors such as concentration, exposure time, temperature, humidity, and material type. UV radiation, mechanical stress, and chemical reactions also play a role. Sunlight, combined with stress from strong winds and particles, along with chemical reactions from gases and acids, further speeds up surface damage.

5.1.5 Abrasive particles

Dust, sand, and debris in the environment wear down surfaces as they are blown by the wind, especially when moisture or rain is present. These particles can damage the exteriors of outdoor tools, vehicles, and structures.

Moisture carried by the wind can speed up this surface damage and even cause corrosion. This is especially true for structures made of different metals, where differences in how they expand can make the damage worse.

Atmospheric factors such as water and humidity, temperature changes, UV radiation, chemical exposure, and abrasive particles often work together, leading to more severe damage than any one factor alone. For instance, while water alone does not cause rapid corrosion, dissolved ions like oxygen, chlorides, and sulfates combine with other factors—especially fluctuating temperatures—to speed up the process. Heat increases chemical activity, making gases and ions more reactive, which raises corrosion rates in humid environments. Additionally, UV radiation, along with high temperatures, can break down protective layers on materials. Industrial emissions, such as sulfur and nitrogen oxides, mix with water to create highly corrosive acids, adding another layer of damage to exposed surfaces.

Soil and atmospheric factors also interact to intensify corrosion. Metals buried in sand experience subsurface corrosion due to soil moisture, chemicals, and abrasive particles, which wear down surfaces. Soil pH and the presence of chlorides and sulfates further impact the rate of degradation, particularly in buried structures. Above ground, dust and airborne abrasive particles can damage surfaces, especially when combined with trapped moisture. Temperature fluctuations lead to condensation, forming electrochemical cells that drive corrosion, while UV radiation and chemical exposure further stress materials, creating a powerful synergy that speeds up the degradation process.

5.2 SIGNS OF SURFACE DEGRADATION

Atmospheric corrosion is the gradual breakdown of surfaces due to environmental factors such as rain, sunlight, and temperature changes. These conditions, along with pollutants from vehicles, factories, and construction sites, introduce gases, particles, and water vapor that accelerate corrosion on materials like metals, polymers, ceramics, and composites. Rain, oil spills, and poor waste disposal can also worsen the effects, weakening materials and making them brittle and prone to fractures.

In critical structures and machinery, corrosion can reduce load-bearing capacity over time. For instance, bridge support beams exposed to de-icing salts can corrode, weakening structural integrity. In high-stress environments, like jet engines, constant stress combined with a corrosive environment can lead to cracks that expose fresh metal surfaces, accelerating the corrosion cycle. Wear from particles can further aggravate this damage, as abrasive corrosion products grind down surfaces under tangential, or sideways, loads.

Environmental factors like humidity, pH, temperature, pollutants, and mechanical stress all contribute to visible signs of surface damage, including

scratches, pitting, grooves, and cracks. Acid rain can create pits on vehicle surfaces that become entry points for more corrosion. In marine environments, saltwater corrosion on ship hulls can lead to pitting, which, if unchecked, expands into larger voids. Mechanical stress can amplify this damage, spreading cracks that open up surfaces to electrochemical reactions, further speeding up degradation.
Common signs of environmentally induced surface degradation include:

5.2.1 Erosion and material removal

Over time, riverbeds, bridges, and building facades are worn down by wind, water, and particulate erosion. For example, sand and wind in desert climates can visibly erode the surface of buildings and monuments, removing material layer by layer.

5.2.2 Peeling, delamination, or cracking

Paint layers on vehicles or pipelines in highly humid areas may peel or delaminate, exposing the underlying surface to more severe corrosion. For instance, cracks in road surfaces caused by freeze–thaw cycles open up, worsening degradation.

5.2.3 Discoloration and corrosion

Bridges often exhibit discoloration from rust, which forms due to prolonged exposure to air and moisture. On statues and building exteriors, the green patina that forms on copper or bronze surfaces, like the Statue of Liberty, indicates exposure to environmental oxidation over time.

These signs of degradation reveal underlying wear and structural vulnerabilities, pointing to the need for protective measures and timely maintenance.

5.3 HISTORICAL PERSPECTIVES ON ENVIRONMENTAL DEGRADATION

Since ancient times, people have recognized that natural elements such as water, air, and temperature fluctuations contribute to material wear. Greek philosophy identified four essential elements—water, air, earth, and fire—each relevant to the corrosion processes of metals and alloys (as shown in Figure 5.1). In this discussion, metal corrosion in water and air, metal oxidation at elevated temperatures, and metal degradation in soil have been explored. Water has consistently played a major role in material degradation, often accelerating the process in combination with air and temperature changes.

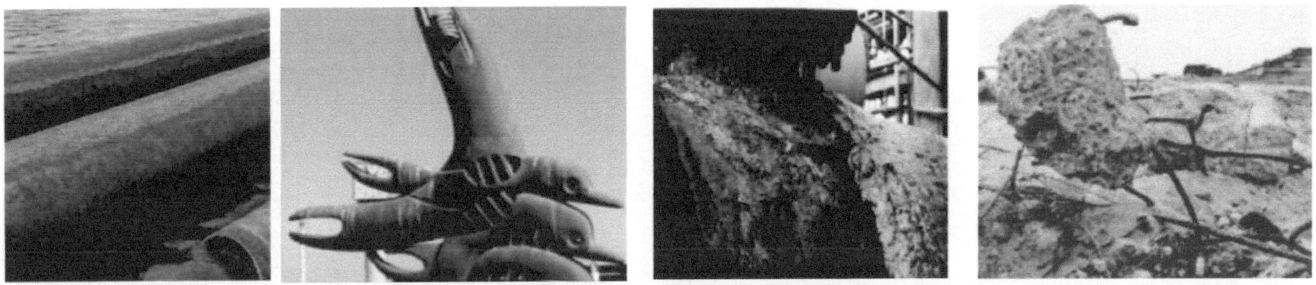

Corrosion of steel tubes in water environment | Natural patina formation on bronze statue due to airborne contaminants | Corrosion of a heat exchanger in a fire-exposed environment | Corrosion of concrete in soil environments

Figure 5.1 Elemental causes of metal and alloy degradation.

The images in Figure 5.1, sourced from the public domain, illustrate the impact of environmental factors on various materials. As previously discussed and depicted in Figure 5.1, Greek philosophy identified four fundamental elements—water, air, earth, and fire—each playing a critical role in the corrosion processes affecting metals and alloys. Atmospheric corrosion, defined as the rate of material degradation due to exposure to environmental conditions, varies significantly based on local atmospheric factors and surroundings. For example, metals exposed to water are particularly susceptible to corrosion processes like electrochemical reactions, where dissolved metal ions in water create electrolytes that intensify degradation. Key factors influencing corrosion rates in water include oxygen levels, flow velocity, water chemistry, and temperature. The presence of dissolved oxygen, chlorides, and sulfates can further speed up corrosion, especially in conductive environments, and high temperatures increase chemical reaction rates, heightening metal deterioration.

In the air environment, metals encounter moisture, oxygen, and pollutants, each of which plays a role in corrosion. Humidity provides moisture for corrosion to occur, with high humidity accelerating this process. Oxygen exposure leads to the formation of oxides or hydroxides on metal surfaces, further degrading the material, while pollutants such as sulfur dioxide and nitrogen oxides from industrial and vehicular emissions intensify corrosion, particularly in urban areas. Shifts in temperature and humidity can destabilize protective oxide layers, allowing corrosion to advance more quickly.

Fire environment or high temperatures, like those found in furnaces, introduce oxidation or "fire corrosion," where oxidation rates increase with temperature if ample oxygen is present. In contrast, soil environments present unique corrosion challenges for metals, as soil moisture, chemistry, and resistivity significantly influence electrochemical corrosion. Moist soils rich in corrosive ions can foster rapid deterioration, with soil pH and aeration further impacting corrosion behavior. By understanding these environmental factors, material selection and protective measures can better address corrosion risks.

Several common examples demonstrate environmental corrosion processes: carbon steel tubes (as shown in the first image of Figure 5.1) corrode over time when exposed to water, while bronze statues (second image in Figure 5.1) gradually develop a patina due to atmospheric pollutants, leading to surface degradation. Subsurface degradation also affects structures buried in sand (fourth image in Figure 5.1), where moisture and abrasive particles erode foundational materials. In high-temperature environments, such as boilers, elevated temperatures accelerate chemical reactions, resulting in faster material degradation. Finally, electrochemical corrosion underscores the role of water as an electrolyte that dissolves metal ions, enhancing conductivity and accelerating corrosion in the presence of dissolved oxygen, chlorides, and sulfates.

This environmental variability is further impacted by *wind and rain*, which exhibit a dual influence on corrosion. While wind can help evaporate moisture, reducing corrosion risks, it can also carry abrasive particles that strike and damage surfaces, creating pits. These pits often trap water, forming localized *electrochemical cells* that accelerate corrosion. Airborne pollutants—including sulfur dioxide (SO_2), nitrogen dioxide (NO_2), nitrogen oxides (NO_x), carbon dioxide (CO_2), and chlorides—interact with moisture on surfaces, producing *corrosive acids* that significantly increase the corrosion rate, particularly in urban and industrial environments. Thus, corrosion is not only driven by the elemental factors described but also by dynamic environmental interactions, highlighting the need for comprehensive corrosion management strategies across diverse settings.

5.4 INFLUENCE OF ENVIRONMENTAL CONDITIONS ON CORROSION RATES

Atmospheric corrosion rates, which indicate the pace at which materials degrade upon exposure to the atmosphere, are influenced by various local conditions. To effectively analyze this phenomenon, atmospheric environments are classified into major categories based on typical exposure conditions. These classifications include rural, urban, industrial, marine, and indoor environments, each with unique factors contributing to the corrosion process.

5.4.1 Rural environment

In rural settings, atmospheric corrosion rates are influenced by relatively low levels of air pollution and minimal human activity. The atmosphere mainly consists of natural elements such as oxygen, nitrogen, and carbon dioxide. Factors such as vegetation, soil, and limited industrial activities contribute to the atmospheric composition. However, corrosion rates may vary based on geographical location, climate, and proximity to agricultural activities.

Rural environments typically experience a higher degree of humidity and moisture retention due to vegetation, which can contribute to corrosion. In such settings, it is crucial to pay special attention to agricultural applications, where exposure to elements like fertilizers, pesticides, and water can introduce corrosion risks. Selecting materials capable of withstanding these conditions while maintaining longevity is essential for effective rural design.

5.4.2 Urban environment

Urban areas experience significantly higher corrosion risks compared to rural settings due to increased anthropogenic activities and pollutants from various sources, such as vehicles, industries, and construction sites. Urban atmospheres contain elevated levels of pollutants, including sulfur dioxide (SO_2), nitrogen oxides (NO_x), particulate matter (PM), and volatile organic compounds (VOCs). When combined with moisture in the atmosphere, these pollutants accelerate corrosion on exposed materials.

In urban settings, engineers and designers must adopt conservative design approaches, utilizing materials and coatings specifically formulated to resist the harsher conditions presented by urban pollution. This proactive planning is crucial to ensure the longevity and durability of infrastructure. The interaction of pollutants with moisture creates corrosive environments that necessitate comprehensive strategies for corrosion prevention and control.

5.4.3 Industrial environment

Industrial environments pose significant corrosion risks, primarily due to the presence of corrosive gases, vapors, and PM. Industries release corrosive agents like sulfuric acid (H_2SO_4), hydrochloric acid (HCl), chlorine (Cl_2), and various metal oxides. Even at low concentrations, these substances can substantially increase the corrosion rates of exposed materials.

The combination of high humidity levels and temperature fluctuations typical in industrial settings exacerbates material degradation. Therefore, understanding the specific chemical composition of the environment is critical for developing effective corrosion mitigation strategies. Regular monitoring and maintenance, along with the use of advanced protective coatings, can significantly extend the lifespan of equipment and structures in these environments.

5.4.4 Marine environment

Marine environments present unique challenges for corrosion due to the presence of salt-laden air, high humidity levels, and dissolved ions in seawater. These factors lead to elevated rates of atmospheric corrosion that can severely affect the integrity of materials used in coastal structures. Continuous exposure to marine conditions, combined with salt and moisture, accelerates corrosion.

Design considerations for marine environments must account for the effects of salt exposure and salinity, which can corrode metals more rapidly than in other settings. Protective coatings, cathodic protection systems, and the use of corrosion-resistant materials are essential strategies for mitigating the risks associated with marine corrosion. Proper design and regular maintenance of structures in these environments are critical for ensuring long-term durability and performance.

5.4.5 Indoor environment

While indoor environments are often perceived as safer than outdoor ones, they can still exhibit significant corrosion risks. Factors such as temperature, relative humidity, air quality, and the presence of corrosive substances from cleaning agents or indoor emissions can impact the corrosion process. Indoor environments may have higher concentrations of ammonia and formic acid compared to outdoor settings, indicating that indoor contamination should not be overlooked when assessing corrosion risks.

In certain cases, indoor environments experience relatively low corrosion rates compared to outdoors, especially when humidity is controlled and corrosive agents are minimized. However, the presence of cleaning agents can release gases that contribute to corrosion, necessitating careful consideration in indoor design.

Mitigating corrosion through effective design strategies involves several critical components. The initial condition of a surface, particularly the presence of protective layers or coatings, plays a pivotal role in determining the rate of atmospheric corrosion; any compromise of these protective layers due to wear or environmental factors can significantly increase corrosion rates, especially when excessive rain strips away these coatings, emphasizing the need for regular maintenance and monitoring. Additionally, in complex systems, a thorough understanding of the environmental sensitivities of each component is essential for designing more durable products; implementing sheltering techniques and maintaining protective coatings can extend the service life of structures, while environmental considerations such as wind velocity and moisture exposure must be integrated into the design processes to ensure long-term performance. Furthermore, engineers and scientists should focus on developing targeted corrosion-mitigation strategies that are tailored to the specific environmental factors present in various settings, recognizing that solutions effective in urban areas may not be suitable for industrial or marine environments. By understanding local environmental conditions, more effective prevention measures can be crafted, leading to enhanced durability and resilience of materials and structures in diverse atmospheric conditions.

5.5 FACTORS AFFECTING ATMOSPHERIC CORROSION

Numerous factors significantly influence atmospheric corrosion, including wetness duration, precipitation (rain, dew, fog), dust particles, temperature, initial exposure conditions, sheltering, wind velocity, and pollutant presence. Each factor contributes differently to the corrosion process, either accelerating or decelerating metal deterioration in diverse environments.

5.5.1 Wetness duration

The duration that a metal surface remains wet plays a crucial role in corrosion. Prolonged wetness provides ample opportunity for moisture to react with the metal, leading to accelerated corrosion rates. Atmospheric conditions that prolong wetness, such as high humidity or frequent rainfall, can significantly enhance corrosion processes.

5.5.2 Precipitation

Precipitation in the form of rain, dew, and fog introduces moisture and contaminants to metal surfaces, initiating corrosion processes. Rainwater, particularly, may contain dissolved gases and pollutants that further promote corrosion. The flow of rainwater can also erode protective coatings, exposing the underlying metal to corrosive agents.

5.5.3 Dust particles

Dust and airborne particles settle on metal surfaces, forming a barrier that traps moisture and corrosive agents. This layer can accelerate corrosion by extending the duration of wetness and facilitating electrochemical reactions. The accumulation of dust not only promotes corrosion but also complicates maintenance efforts.

5.5.4 Temperature

Temperature influences corrosion rates, with higher temperatures enhancing the activity of corrosive agents, while lower temperatures may reduce corrosion rates. Temperature fluctuations can lead to condensation formation, which can further contribute to corrosion. Understanding the thermal dynamics in different environments is critical for developing effective corrosion management strategies.

5.5.5 Initial exposure conditions

The initial condition of the metal surface, along with any preexisting corrosion products, influences the rate and extent of further corrosion. Oxide layers or protective coatings may act as barriers against corrosive elements or serve as sources for localized corrosion processes. Proper surface preparation and coating application are essential for optimizing corrosion resistance.

5.5.6 Sheltering

Sheltering from environmental elements can significantly affect corrosion rates. Sheltered areas often experience lower corrosion rates due to reduced exposure to moisture, rain, and wind-borne contaminants. Designers must consider sheltering techniques to enhance the durability of structures in various environments.

5.5.7 Wind velocity

Wind can influence corrosion by promoting faster moisture evaporation from metal surfaces or carrying corrosive agents and particles to the surface. Higher wind velocities can also cause mechanical damage to protective coatings, exposing the underlying metal to corrosion. Understanding wind patterns and their potential impact on corrosion is essential for effective design.

5.5.8 Pollutants

Atmospheric pollutants such as sulfur dioxide (SO_2), nitrogen oxides (NO_x), carbon dioxide (CO_2), and chlorides accelerate corrosion by forming acidic solutions or corrosive compounds when reacting with moisture. These pollutants can significantly increase corrosion rates, necessitating robust protective measures in exposed materials.

5.6 PRIMARY ATMOSPHERIC CONTAMINANTS AFFECTING CORROSION

The atmosphere comprises several key components, including "nitrogen (N_2)," "oxygen (O_2)," "water vapor," "sulfur compounds," "chlorine compounds," and "carbon dioxide (CO_2)," each playing distinct roles in the corrosion process. While "nitrogen" is largely inert, "oxygen" is a critical factor driving surface reactions that lead to corrosion. In other words, key corrosive agents in the environment are:

1. *Oxygen:* This gas is a primary contributor to corrosion, acting as an "electron acceptor" in "cathodic reactions," which facilitates the oxidation of metals.
2. *Carbon dioxide:* Carbon dioxide, frequently overlooked, is found in our atmosphere at around 330 ppm. When it reacts with water, it forms carbonic acid, which plays a crucial role in acidification by decreasing pH levels. This decline in pH fosters the creation of corrosive species that can severely harm metal surfaces, significantly accelerating the corrosion process.
3. *Water vapor:* Water vapor is crucial in the corrosion mechanism, as it helps create an environment conducive to the formation of corrosive agents and facilitates electrochemical reactions on metal surfaces.

4. *Sulfur compounds:* Naturally occurring sulfur compounds, such as "sulfur dioxide (SO_2)" and "hydrogen sulfide (H_2S)," are recognized as major contributors to atmospheric corrosion. These compounds can lead to the formation of "sulfuric acid (H_2SO_4)," which is particularly corrosive to metals. The presence of sulfuric acid can cause extensive damage to metal surfaces, thereby increasing the overall corrosion rate.
5. *Nitrogen oxides:* Although nitrogen itself is relatively inert, "nitrogen oxides (NO_x)" are significant atmospheric pollutants primarily produced by automobile emissions. These nitrogen oxides contribute to corrosion by forming "nitric acid (HNO_3)," which can accelerate the degradation of metal structures and components.
6. *Chlorine compounds:* Chlorine compounds in the environment can generate either "hydrochloric acid (HCl)" or "hypochlorous acid (HOCl)," both of which are highly corrosive to metals. The rate of corrosion is influenced by the concentration levels of these corrosive species. There exists a threshold concentration; below this level, these gases or elements do not significantly impact exposed metals. However, once this threshold is surpassed, corrosion rates can escalate rapidly.

Temperature significantly impacts the concentration of corrosive species present in the environment. Additionally, the duration of exposure—ranging from minutes to hours—can impact the severity of corrosion. The interplay of factors such as gas concentration, temperature, and exposure time collectively determines the overall impact of corrosion, highlighting the importance of monitoring and mitigating these environmental conditions to protect metal structures and materials effectively.

5.7 UNDERSTANDING ENVIRONMENTAL VARIATIONS FOR EFFECTIVE DESIGN

To create resilient designs, it is essential to consider various environmental factors, including solar radiation, contaminants, humidity, gases, high temperatures, and acids. These factors dramatically influence the corrosion processes that materials undergo. Additionally, geographical elements such as latitude and longitude contribute to environmental variations that must be understood and accounted for in the design phase.

1. *Universal versus location-specific approaches:* While it is challenging to create designs tailored to every specific location, a universal approach can provide flexibility in adjusting maintenance schedules, conducting failure analyses, and implementing remedial measures as needed. Understanding general corrosion risks associated with

various environments allows designers to create durable and resilient designs that can withstand a wide range of conditions.

2. *Indoor versus outdoor specific approaches:* Although indoor environments are frequently perceived as safer than outdoor ones, they can still present significant corrosion risks due to the presence of cleaning agents and other chemicals that may emit gases contributing to corrosion. This necessitates careful consideration during indoor design. Furthermore, indoor spaces can often exhibit higher concentrations of ammonia and formic acid compared to outdoor settings, highlighting that indoor contamination should not be ignored when evaluating corrosion risks.

In summary, atmospheric corrosion is influenced by a myriad of factors across diverse environments, including rural, urban, industrial, marine, and indoor settings. Understanding these environmental classifications, along with the specific factors that influence corrosion rates, is crucial for engineers and materials scientists in developing effective corrosion prevention and protection strategies. By implementing appropriate design techniques, selecting suitable materials, and maintaining protective measures, it is possible to enhance the lifespan and durability of structures and materials exposed to atmospheric conditions.

5.8 CASE STUDIES

This section explores environmental factors causing surface degradation across various case studies, including lead corrosion in humid environments, satellite erosion in LEO, and aircraft and building degradation. Case studies highlight key contributors like carbonyl compounds, AO, and atmospheric pollutants such as chloride ions and ozone. For instance, Tétreault et al.[2] studied the effects of formic, acetic, and formaldehyde on lead and copper artifacts, finding acetic acid to be the most corrosive. Similarly, AO in LEO accelerates material erosion, necessitating protective coatings like multilayered nanocoatings for satellites. Aircraft face erosion from rain, hail, and pollutants, while chloride ions cause severe pitting in coastal structures like bridges. Studies on hydrogen embrittlement and temperature-assisted oxidation emphasize the interplay between environmental stressors and material degradation. For example, Hu et al. demonstrate that thermal cycling in turbine blades significantly reduces material fatigue life. Chloride-induced crevice corrosion was shown to weaken protective oxide layers, increasing structural vulnerabilities. The degradation cycle, fueled by wear, corrosion, and fracture, accelerates material loss and debris generation. Friction and tangential loads exacerbate these effects, highlighting the need for robust surface engineering. By understanding these mechanisms, engineers can design durable materials and implement preventive measures to extend product life and reduce environmental impacts.

5.8.1 Case study 1

This case study explores the effects of carbonyl compounds on lead in a humid, 75% relative humidity environment, examining interactions with formic acid (FA), formaldehyde (F), and acetic acid (AA). Tétreault et al.[2] examined the corrosion behavior of copper and lead under controlled conditions with varying concentrations of formaldehyde (methanal), formic (methanoic) acid, and acetic (ethanoic) acid, relative humidity (RH), and exposure duration. A total of six exposure scenarios were explored: clean conditions, formic acid alone, a combination of formic acid and formaldehyde, formic and acetic acids together, all three chemicals combined, and acetic acid alone.

It is interesting to note that in museum indoor environments, "formaldehyde, formic acid, and acetic acid" carbonyl compounds are commonly encountered as airborne pollutants, originating mainly from materials used in building construction, exhibit displays, storage units, and even some preservation materials. These impact the museum indoor air:

- *Formaldehyde (methanal):* It is released from a variety of construction and display materials, such as particleboard, fiberboard, plywood, and other engineered wood products. Additionally, paints, varnishes, adhesives, and some textiles used in museum displays emit formaldehyde. Due to its high volatility, formaldehyde quickly diffuses into the air, often reaching higher concentrations in enclosed or poorly ventilated spaces. While it can pose a health hazard to humans, its impact on artifacts is generally indirect; however, prolonged exposure can alter the stability of organic materials, including paper, textiles, and natural dyes.
- *Formic acid (methanoic acid):* It is often a byproduct of the degradation of formaldehyde and other organic compounds. It is also emitted directly from some materials used in museum storage and display, such as certain types of wood and adhesives. For example, wood and paper products can emit low levels of formic acid, especially as they age. In museum conditions, formic acid can react with metal artifacts, especially under conditions of elevated humidity. The corrosion of metals like copper and lead is accelerated in the presence of formic acid, leading to visible tarnishing and potential structural damage. Given its role as both a primary and secondary indoor pollutant, managing formic acid levels is important for the conservation of metallic and organic artifacts alike.

- *Acetic acid (ethanoic acid):* It is a common emission from materials like cellulose acetate film, adhesives, and some wood products, especially oak and other hardwoods. It is also present in some coatings, paints, and preservation materials. Acetic acid is relatively volatile, making it a notable component of "off-gassing" within enclosed display cases or storage areas. It is particularly harmful to metals, where it can initiate and accelerate corrosion. Lead and copper are especially susceptible, developing surface corrosion that can lead to pitting and surface degradation. For organic materials, high levels of acetic acid can cause brittleness, yellowing, and degradation of cellulose-based materials.

Tétreault et al.[2] studied the distinct corrosion behaviors of copper and lead when exposed to mixed vapors of formaldehyde, formic acid, and acetic acid. Results showed that acetic acid induces the most severe corrosion on lead, especially in combination with formic acid, while formaldehyde alone has minimal impact. Formic acid can form a protective layer on lead, composed of lead formate or lead formate hydroxide, which helps reduce further oxidation. The study establishes the no observed adverse effect level (NOAEL) for formic acid around 0.1 ppmv in 54% to 75% humidity conditions, where corrosion remains low below this threshold. However, mechanical disruptions, like scratches, compromise this layer, exposing fresh lead and increasing susceptibility to corrosion.

Copper, on the other hand, primarily reacts with formic acid, while lead undergoes more complex interactions. On lead surfaces, acetic acid typically forms a thick, white corrosion layer, whereas formic acid creates a thinner, darker film of lead formate compounds that can inhibit further acetic acid reactions. When this stable formate layer is intact, it reduces overall corrosion risk, though repeated surface damage can re-expose vulnerable lead. To preserve lead artifacts effectively, minimizing acid emissions in display and storage materials is critical, as wood-based cases with finishes that undergo oxidative polymerization can elevate corrosion risk. Using low-emission materials in museum settings can help mitigate this effect and protect valuable artifacts.

In addition to the preventive measures proposed by Tétreault et al.,[2] adequate ventilation and air filtration are crucial for preventing the buildup of harmful carbonyl compounds in display and storage areas. Regular monitoring of indoor air quality enables museums to detect any rise in these compounds early and take timely steps to protect sensitive artifacts. While formaldehyde typically does not oxidize into formic acid under standard conditions, high levels of peroxide may form within an enclosed space, particularly if a new polymer coating is still curing. This peroxide presence heightens corrosion risk, so it is essential to allow coatings to cure completely and avoid sealing artifacts in environments where reactive compounds are present.

5.8.2 Case study 2

Satellites in LEO, located between 200 and 2000 km above Earth, face tough environmental conditions that impact their performance and lifespan. Orbiting at high speeds, these satellites complete an orbit in just 88 to 128 minutes and provide essential services like broadband internet, mobile communications, and remote sensing. To remain in orbit, they must maintain their speed; any slowdown could disrupt their trajectory and require significant design changes.

In this high-speed environment, satellites are constantly bombarded by ions, leading to oxidation that can wear away protective coatings and damage materials. They are also exposed to atmospheric gases such as oxygen, ozone, and water vapor, which contribute to surface degradation. Additional threats include radiation from photons and particles, extreme temperature changes, thermal cycling, impacts from micrometeoroids and debris, contamination, and exposure to AO. The degree of material degradation varies based on the satellite's orbit, mission length, solar activity, and orientation relative to the Sun. This complex mix of risks underscores the need for effective protective strategies and careful material selection.

A significant factor in satellite degradation is AO. In the LEO environment, especially at altitudes between 200 and 700 km, AO is present in considerable amounts. This results in a significantly accelerated reaction rate, leading to a greater impact on satellites compared to those in orbits above 700 km. It is important to note that LEO extends from 200 to 2000 km, meaning that the upper range of 1300 km has a lower abundance of AO.

Furthermore, when molecular oxygen is present in the environment and interacts with UV light, it can trigger photochemical reactions that generate AO as a byproduct. Therefore, for satellites located between 700 and 2000 km, it is crucial to evaluate the actual levels of molecular oxygen and determine whether these photochemical reactions are occurring. This assessment is vital for understanding the potential production of AO and its subsequent effects on satellite materials.

Satellites traveling at high speeds collide with AO, typically experiencing collision energies around 4 to 5 eV, which is sufficient to break chemical bonds in various materials, resulting in substantial wear.

The reactivity of AO leads to the erosion of protective coatings designed to shield satellite surfaces from environmental damage. Over time, continuous exposure to AO can degrade these coatings, leaving underlying materials vulnerable. This erosion can compromise critical components like solar panels and thermal coatings, which are vital for satellite functionality.

The durability of protective coatings is crucial for satellite performance, as they must withstand the harsh conditions of space. Depending on their effectiveness, these coatings may last anywhere from 2 to 10 months before the satellite needs to be decommissioned or removed from orbit. In addition to AO, UV radiation accelerates surface degradation, compounding the challenges for these satellites.

In other words, LEO satellites encounter various environmental threats, including mechanical, chemical, and thermal risks. While AO is a major concern, other harmful agents, such as photon radiation and charged particles, also contribute to surface degradation. The constant bombardment by high-speed particles results in erosion characterized by grooves and cuts, with the collision angle affecting the severity of corrosion. Additionally, high contact pressures can cause abrasion and adhesion issues, negatively impacting the satellite's operational efficiency.

To mitigate the harmful effects of AO and other environmental factors, engineers employ several strategies in satellite design. Protective coatings that can withstand AO attack significantly extend a satellite's operational life. Furthermore, incorporating systems for regular maintenance and upgrades helps manage wear and tear over time, improving reliability and functionality.

5.8.2.1 Design considerations for LEO satellites

LEO presents unique challenges in satellite design and operation. Engineers must address issues like high-speed erosion, material degradation from AO, extreme temperature variations, and the threat of orbital debris (OD) to ensure satellite durability and functionality. Satellite designs for LEO require careful consideration of factors such as mission duration, solar events, satellite velocity, and orientation. During sunlit periods, one side of the satellite faces the Sun while the other side faces cold space, leading to a significant temperature gradient. In non-sunlit periods, both sides face cold space.

Thermal control systems, categorized as either passive or active, are crucial for maintaining satellite precision and efficiency. Passive thermal control systems, commonly used in LEO satellites, rely on specialized thermal control coatings with specific solar absorptance and thermal emittance properties to keep temperatures within safe operational limits. These coatings help regulate heat by reducing solar gain on the sun-facing side and retaining warmth on the shaded side. Research shows that solar absorptance tends to increase in most materials after exposure to UV and particles, while emittance values generally remain stable.

Satellites in LEO face a significant risk from OD, much of which originates from rocket fragments left behind after launches. Solid rocket boosters and spent propellant contribute to this issue, with abandoned rocket stages alone comprising approximately 50% of the total debris mass in LEO.

These debris particles can remain in orbit for years, increasing the likelihood of collisions that could damage operational satellites.

Since the 1980s, a process called passivation has been implemented to minimize debris. This procedure involves venting any remaining fuel from rocket bodies to prevent explosions that would generate additional debris. If fuel is not vented, it can cause catastrophic explosions, worsening the debris problem. However, while passivation helps control debris, it can also release AO and other reactive ions, which can corrode satellite surfaces.

Passivation, though essential for managing OD, introduces new challenges due to the release of AO—a highly reactive gas that can erode and damage satellite surfaces. To enhance satellite durability and lifespan, engineers must carefully consider how gases like AO interact with materials in LEO. Designing materials resilient to LEO's specific atmospheric conditions and contaminants is essential to address these complex environmental risks effectively.

Liu et al.[3] introduced a degradation modeling approach for thermal control coatings on satellites in LEO, focusing on materials like S781 white paint and F46 second-surface mirrors. Their research examines various factors, including solar radiation, infrared radiation from Earth, and reflected solar radiation. The goal was to optimize the thermal control system, essential for maintaining safe operating temperatures for satellites. To achieve effective temperature regulation, the researchers utilized a dual-layer coating system. The first layer is S781 white paint, a conventional coating known for its reliability. The second layer, referred to as F46, functions as a second-surface mirror, often enhanced with silver to reflect harmful radiation, thereby protecting the underlying thermal control coating. This focus underscores the critical role of thermal control coatings in ensuring that satellites operate within safe temperature limits without incurring damage.

If the thermal control system fails, other environmental factors, such as AO, can significantly degrade the satellite's surface, leading to potential failure of the entire system. Additionally, even the best coatings experience performance degradation over time, indicating that no single coating can be relied upon for long-term durability. Consequently, understanding the quality and performance of coatings is essential. Key performance metrics include low solar absorption and high infrared emittance, with effective coatings exhibiting minimal solar absorption and optimal emittance values.

Liu et al.[3] highlighted the use of zinc oxide as a pigment in the S781 coating, chosen for its chemical stability, non-toxicity, and cost-effectiveness. In S781, SUV radiation alters ZnO pigments, changing the material's color and increasing solar absorptance. While affordability is a consideration, the primary focus in spacecraft applications is on consistent performance; failures can incur significant costs. In competitive situations, however, cost may be a factor. The research also utilized an organic binder made from silicon rubber, silicon resin, and epoxy resin to create a thin, high-quality coating that enhances adhesion and minimizes radiation exposure.

For the F46 second-surface mirror, fluorinated ethylene propylene (FEP) serves as the surface attachment, with a silver backing to reflect solar radiation while allowing necessary energy to pass through. To ensure optimal performance, a silicon oxide binder was employed for its adhesion properties. Researchers aimed to minimize air gaps between layers, with a conductive layer of indium tin oxide (ITO) applied to prevent thickness growth. This layer, thicker than previous discussions at around 100 nm, protects against erosion from AO.

The study indicated that this new composite coating outperforms traditional single-layer coatings. While the F46 semi-solid metal (SSM) is considered superior to the zinc oxide coating, the researchers acknowledge the potential for further improvement and innovation in this area. Their findings suggest that employing a multilayer coating system, where each layer serves a distinct function, could enhance overall performance and yield better results.

Liu et al.[3] formulated a comprehensive equation to explore potential enhancements in coatings and structures, aiming to improve performance compared to existing materials. Utilizing in-orbit telemetry temperature data, their approach employs an energy balance equation to assess the solar absorptance of coatings on various LEO satellites. Experimental investigations into the S781 and F46 SSM coatings revealed that their degradation mechanisms are primarily influenced by AO and solar ultraviolet (SUV) radiation. The modeling of their degradation pathways involves Wiener processes, with the F46 SSM coating demonstrating greater reliability than the S781 under comparable conditions.

Kapton and palladium are commonly used materials in the construction and protection of satellites operating in LEO. Kapton is renowned for its exceptional thermal stability, maintaining its properties over a broad temperature range from –269°C to +400°C. It is an effective electrical insulator and resists various chemicals, including corrosive agents often found in the space environment. Its lightweight nature and thermal characteristics make Kapton an ideal substrate for thermal control coatings and insulation, serving as a thermal barrier that helps manage heat within satellites, particularly during direct sunlight exposure and when exposed to AO. As a 25 μm-thick polymer foil, Kapton is integral to satellite construction, ensuring structural integrity and thermal stability under extreme conditions. However, it is susceptible to erosion from AO in LEO, which can gradually degrade its surface and structural integrity. This surface damage increases with exposure to AO, especially when combined with particle impacts.

While Kapton offers essential thermal and electrical properties, applying a palladium coating enhances protection against erosion from AO and impact damage. Known for its excellent corrosion resistance and high thermal conductivity, palladium is an ideal choice for protective coatings in harsh environments like LEO. By adding a thin layer (10 to 100 nm) over Kapton, it effectively protects against erosion and other environmental problems. This greatly increases the useful life of satellite parts by reducing the effects of AO and other reactive species. Interestingly, the effectiveness of palladium coatings is influenced by their thickness, with thinner coatings potentially experiencing greater erosion. Palladium-coated Kapton demonstrates better performance against hypervelocity impacts than uncoated materials, maintaining functionality and structural integrity in demanding space conditions. But if the palladium coating isn't thick enough, the impact can cause a lot of mass loss and damage to the surface, because the erosion pits that form let AO get through and completely wear away the Kapton underneath. The erosion of the impacted foil is significantly more severe than that of foils solely exposed to AO, resulting in numerous erosion pits caused by AO ingress through the impact holes.

In this case study, it can be concluded that multilayer coatings—composed of several thin layers, each designed for a specific function such as reflecting harmful radiation, absorbing heat, resisting erosion caused by AO, or providing thermal insulation—are essential for enhancing performance. This layered structure enables tailored properties that single-layer coatings cannot achieve. The versatility of multilayer coatings allows for customization to meet specific environmental challenges, making them applicable across various industries, including aerospace, electronics, and optics. Furthermore, the field of multilayer coatings is poised for innovation, with ongoing research focused on developing new materials and configurations that can further improve performance and address emerging challenges in technology and materials science.

5.8.3 Case study 3: What caused the Titanic to strike the iceberg?

In this case study, let us examine the infamous sinking of the *Titanic*. Many people are familiar with the story, largely popularized by films, which have spurred various theories and discussions. This event is an important part of history, highlighting how the advancements of science and technology can sometimes introduce unforeseen challenges. I will present my perspective on the *Titanic*'s failure, inspired by DiLisi,[4] who examined the *Titanic*'s sinking in the context of thermal inversion and atmospheric refraction phenomena.

What are thermal inversion and refraction phenomena? The *Titanic* disaster occurred on April 14, 1912—more than 110 years ago—and the primary cause was the ship's collision with an iceberg. While the iceberg was a solid, cold mass, the impact it had on the *Titanic* was catastrophic, resulting in extensive damage to the ship. Notably, the *Titanic* sank much more quickly than anticipated following the collision, indicating that something went awry, possibly in the contingency measures or safety protocols in place. This resulted in a tragic loss of life.

DiLisi[4] analyzes the *Titanic* disaster and its collision with an iceberg, highlighting the environmental conditions that influenced the event. The intersection of the cold Labrador Current and the warmer Gulf Stream created a temperature mix that likely impaired visibility. As illustrated in Figure 5.2, the Labrador Current, a large surface current in the North Atlantic, originates in the Atlantic Ocean, carrying frigid water from the Arctic to the Grand Banks region and along the western side of the Labrador Sea. This flow introduces extremely cold, nearly icy water alongside warmer currents, creating a significant temperature contrast between the two streams. This situation resembles optical illusions, where perception does not match reality. For instance, a shimmering effect that looks like water may appear on roads due to light refraction. Similarly, in deserts, distant mirages can create the illusion of pools of water, leading to mistaken perceptions of the environment.

The significant temperature difference between the cold Labrador Current and the warmer Gulf Stream may have initiated a natural thermal inversion process. What does this thermal inversion entail, and what might occur if the captain encountered a false horizon? A false horizon could cause the captain to perceive an iceberg as much smaller in size (as illustrated in Figure 5.3), while a larger iceberg was actually present but not fully visible due to poor visibility at the time. The real iceberg could have been two to three times taller than what was observed.

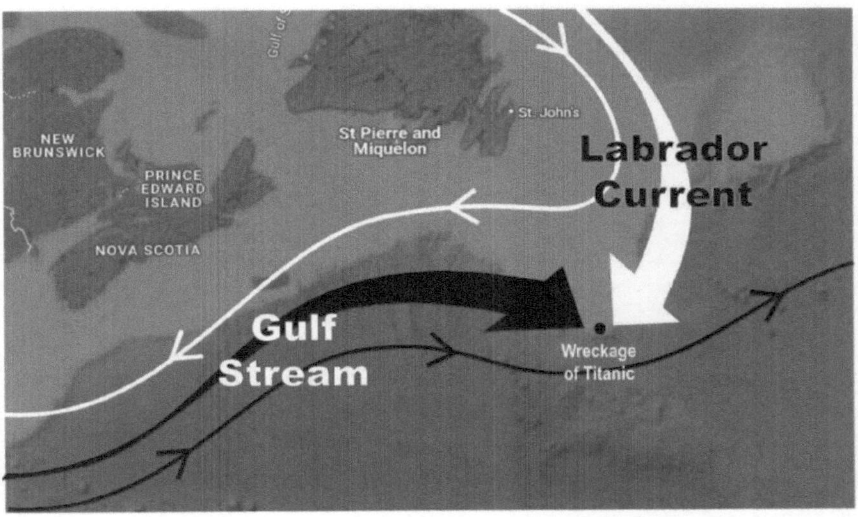

Figure 5.2 *Titanic* wreckage location.

Thermal inversion significantly influences the formation of looming, superior, and inferior mirages by affecting light refraction in the atmosphere. When light transitions from a colder medium (cool air) to a warmer one (warm air), it bends downward due to refraction. This bending occurs because light travels faster in warmer air compared to cooler air. When distant objects, typically hidden by the horizon, emit light, this light can curve around the Earth's curvature. This phenomenon allows observers to see these faraway objects, which appear elevated or displaced from their actual position, a phenomenon known as looming.

Superior mirages occur when light is refracted downward through a thermal inversion, giving the impression of objects positioned above their true location. In this case, a warm layer of air sits atop cooler air, creating a gradient that bends light downwards. Consequently, distant objects, usually obscured by the horizon, may appear to float or be situated higher in the sky, potentially misleading observers into seeing objects that aren't physically present in that location.

Conversely, inferior mirages arise when light is refracted upward, often observed on hot days over warm surfaces like roads or deserts. Unlike superior mirages, inferior mirages develop when a layer of warm air rests above cooler air, causing light rays to bend upwards. This bending creates the illusion of pools of water or reflections on the ground, deceiving observers into believing they see water or objects that are actually below the horizon or far away. All these phenomena are interconnected through the principles of light refraction influenced by the temperature gradients created during thermal inversion.

In one notable instance, the actual horizon was distorted due to this visual bending, leading to the false impression of an iceberg. This optical illusion might have concealed the tip of the iceberg or rendered it entirely invisible. As illustrated in Figure 5.3, the upper solid line represents a false horizon

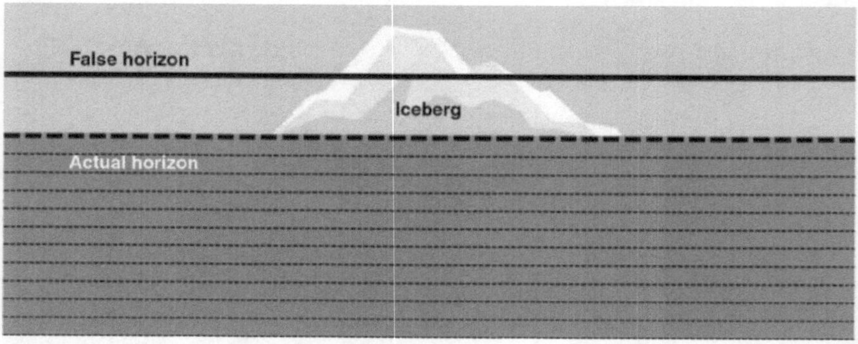

Figure 5.3 Discrepancy between actual and false horizon, reducing iceberg height to half or one-third.

that obscured the iceberg's surface, resulting in a direct collision with the submerged mass. This situation was attributed to thermal inversion, also referred to as the refractive index or refraction. Consequently, if the captain was unable to see the large iceberg, the subsequent events may appear natural in hindsight. Despite receiving multiple warnings—at least six—about nearby icebergs, the captain was unable to spot them and could not determine whether the warnings pertained to his ship or another vessel.

Thermal inversion is a meteorological phenomenon characterized by a layer of warm air trapping cooler air near the Earth's surface. In typical conditions, air temperature decreases with altitude; however, during thermal inversion, the warm air layer prevents cold air from rising. This creates a sort of "lid," keeping the cooler air trapped and leading to unusual temperature gradients. As a result, this phenomenon can produce misleading effects on visibility and weather conditions.

DiLisi[4] suggests that the thermal conditions likely impaired the crew's visibility of the iceberg, making it difficult for the captain and crew to detect it in time. Refractive phenomena, such as superior or inferior mirages, can distort distance perception, which may have led them to misjudge how close the iceberg was. Despite having lookouts and visual aids, the crew might have believed there was enough distance for evasive action, only to realize too late that the iceberg was dangerously nearby.

Now, let's examine the scene from a different angle, focusing on the ship and the nearby massive iceberg. If the ship were to collide with the iceberg, a significant impact would inevitably occur. In such situations, it is crucial to reduce speed upon receiving a warning—like how a driver would slow down when faced with reduced visibility. For example, traveling at 40 or 50 km/h, sudden visibility limitations might necessitate slowing to as little as 10% of the original speed. However, the *Titanic*'s captain maintained a speed of about 22 knots, just under the maximum design speed of 23 knots. This type of ship could reach that top speed, and maintaining such a high speed in the presence of a large mass like an iceberg can lead to immediate and catastrophic consequences. In any significant collision, it is essential to consider the energy involved. When a collision occurs, the kinetic energy is related to the square of the velocity, which means even small increases in speed can lead to large increases in energy. Additionally, the mass of the object striking the surface must be carefully accounted for.

Another important point is that there is often no clear direction on how evacuation should proceed during such events. This lack of organization in the evacuation process likely contributed to the high number of casualties. While we do not intend to assign blame, this serves as a critical lesson about the need for clear planning and thorough documentation of evacuation procedures.

It is vital to have a solid plan for responding to emergencies. Human nature often leads individuals to prioritize their own safety over that of others. This instinct was evident during the *Titanic* disaster, where the lifeboats available could not accommodate all passengers. Many were confused about the situation, and some felt a strong emotional attachment to the ship, making them reluctant to leave.

Regardless of the chosen approach, defensive measures may not be sufficient; they can only save a limited number of individuals, as seen in the significant collisions between ships and icebergs. The captain was likely misled by the optical illusion caused by thermal inversion, which distorted his perception of the iceberg's height. Relying on the ship's technology, he navigated confidently past what appeared to be a minor obstacle. Thus, it is reasonable to consider how such challenges might be anticipated and addressed.

Reflecting on the advances in science since 1912, one can see that thermal inversions are more prevalent during winter, where cold air is trapped beneath warmer layers, often leading to fog and reduced visibility. This trapped fog can create significant transportation challenges, as visibility decreases and necessitates a reduction in speed for safety. Although modern vehicles are equipped with better visibility technologies, the natural phenomenon of thermal inversion can still obscure vision. The process of thermal inversion can have serious implications for visibility and environmental health. It can trap pollutants and contribute to an unhealthy atmosphere, making it essential to understand and mitigate these occurrences whenever possible.

Why discuss this topic in this book? In my view, thermal inversion is a critical factor affecting air quality and atmospheric conditions. This book emphasizes the importance of reducing atmospheric pollutants, as these can intensify thermal inversions and contribute to corrosion. During a thermal inversion, a layer of warmer air traps contaminants near the surface, creating an artificial barrier that prevents pollutants from dispersing naturally. This trapped layer degrades air quality, posing risks to the degradation of the products. During an inversion, particularly in winter, stagnant air conditions can lead to smog formation and significant reductions in air quality. To mitigate pollution, measures such as water spraying are often employed, though this practice itself is a major contributor to corrosion.

From a technical perspective, the *Titanic*'s failure was influenced by various material and design flaws. For example, the rudder was undersized, limiting effective maneuverability, and although there was confidence in the watertight compartments' ability to prevent flooding, a number of these compartments also flooded. This highlights the importance of robust design in both maritime and environmental contexts.

Another factor contributing to the *Titanic*'s failure was the use of lower-grade steel, unlike the superior-grade steels available today that could have better withstood such impacts. The lower-grade steel used in the hull led to more severe fracturing than would be expected by modern standards. Additionally, the hull fractured immediately upon collision, without deflecting, leaving no time for the captain to take corrective measures. Reports also indicate that riveted heads may have been broken off or removed, weakening the hull assembly at certain points. Numerous technical challenges—spanning design, manufacturing, and corrosion-related issues—were present in the *Titanic*'s construction. However, it is worth noting that the designers and engineers at the time may not have anticipated a collision with an iceberg of this nature. This case study underscores the importance of thoroughly considering all potential scenarios in modern design processes.

However, when examining the situation as a whole, it becomes evident that human error was a factor, along with issues stemming from faulty construction due to various design flaws. While it may be possible to identify these mistakes now, they might not have been apparent at the time. Thus, human error, structural deficiencies, and environmental conditions all contributed to the *Titanic*'s damage, regardless of the speed at which the collision occurred.

I use the term "human error" because timely warnings could have prompted a reduction in speed or significantly mitigated the impact. After receiving at least six warnings, the ship could have made necessary repairs, potentially lessening the damage. There was a great deal of confusion and a lack of clear directions during the disaster, leading to numerous problems. These issues stemmed not only from technical and human errors but also from environmental factors. This compounded effect has been well documented in history. In this case study, we highlight that atmospheric corrosion or environmental hazards can create significant problems, so it is essential to consider many factors when aiming to reduce surface damage.

Particularly regarding ship hulls, it is important to recognize that damage should be categorized into various sections. Dividing the hull into sections increases the surface area, which could help with buoyancy. However, it was discovered that some sections were not watertight. If water leaks into these areas, the ship can sink. The design faced various difficulties and shortcomings, and the bulk of the vessel was not safe. Additionally, using two engines or adding extra engines can increase the risk of atmospheric damage.

Historically, ships were designed with a single hull, but regulations have since changed to require double hulls for better protection. Today, aircraft are built with safety norms in mind from the outset. When atmospheric hazards occur, ensuring one's safety is not solely an individual responsibility; it is crucial for everyone involved. This underscores the importance of educating individuals about safety precautions in case of disasters.

For example, when traveling by air, safety briefings are provided at the beginning of the flight. Often, passengers may overlook this information, but safety drills are vital. It is essential to acknowledge that while such incidents should not happen, we must be prepared for them. Understanding how to manage emergencies effectively, prevent chaos, and avoid further incidents is critical.

Communication issues also arose during the *Titanic* disaster, as radio transmissions were not used effectively due to anxiety and confusion. This case study offers valuable insights into product design and damage minimization. Recognizing the various atmospheric conditions, considering every detail, and creating comprehensive documentation are vital steps in implementing safety measures effectively. While we have discussed the *Titanic*, similar incidents have occurred throughout history.

5.8.4 Case study 4: Sinking of the Erika[5]

A similar incident to the sinking of the *Titanic* occurred in 1999, although human involvement was less of a factor. Following this event, the social and environmental impacts were severe and lasting. On December 12, 1999, the vessel *Erika* sank, resulting in a significant oil spill. Amid severe weather off the coast of Brittany, France, the *Erika* tanker fractured, releasing around 19,800 tons of heavy fuel oil at the time, out of a total capacity of approximately 31,000 metric tons. The ship broke in two, leading to a major oil spill into the sea.

In this instance, corrosion was identified as the main cause of the structural failure. Corrosion, as a progressive issue, typically provides ample warning signs before catastrophic failure occurs, though it often goes overlooked due to inadequate maintenance practices. Corrosion does not lead to failure overnight; it advances gradually, giving clues over months or even years. However, ignoring these early signs can lead to sudden breakdowns.

When moving a ship from one location to another, corrosion-related degradation continues to progress. This is not a static failure but rather a continuously evolving one. Without intervention, even a small crack can grow significantly, with the rate of growth potentially accelerating. Understanding the sensitivity of the structure and accurately predicting when failure might occur could improve the operation and maintenance of such vessels, enhancing safety and longevity.

Corrosion issues on the *Erika* had been identified as early as 1994, with records available to port state control authorities and potential charterers. These inspections revealed concerns, including U.S. Coast Guard records noting deficiencies in the tanker's firefighting and inert gas systems, which posed a potential explosion risk. Frequent inspections at U.S. ports also highlighted issues such as holes in the main deck coaming, degraded inert gas system risers, and firefighting equipment in poor condition, often receiving temporary patches instead of adequate repairs.

Despite numerous warnings, no immediate action was taken to address *Erika's* deteriorating condition, raising serious questions about whether the ensuing oil spill could have been prevented. In March 1999, the tanker's certificate of financial responsibility for trading in U.S. waters expired and was left unrenewed. This lack of accountability and maintenance led to a catastrophic oil spill when the ship broke in two, spilling around 20,000 metric tons of oil into the ocean, with devastating environmental impacts on marine life, local economies, and tourism.

This event underscores the critical role of proactive maintenance and adherence to regulatory standards. Corrosion is a progressive issue, providing ample warning signs and ample time for intervention before leading to a failure as severe as the *Erika's*. Modern scientific methods, such as simulations and thorough documentation, could help operators better predict and manage structural risks, improving safety and extending the lifespan of vessels. After this incident, stricter international marine standards were implemented, including age limits on vessels and mandatory double-hull designs for oil tankers to minimize the risk of similar disasters.

Ultimately, the *Erika* incident underscored the repercussions of neglecting maintenance and regulatory standards. The economic impact—manifested through losses in tourism, fishing revenue, and restrictions on seafood trade—significantly affected the local community. This situation illustrates that failure to promptly address structural issues can lead to extensive consequences. The lessons learned from the *Erika* disaster emphasize the necessity of prioritizing scientific rigor and transparent recordkeeping to avert such failures, equipping future generations with the knowledge and tools to develop safer and more reliable systems.

When a specific efficiency target is established, a product that does not achieve that standard is considered a failure. This designation does not necessarily imply that parts are broken or have failed in large numbers; rather, it may suggest that the product is not functioning as intended. Nanocoatings can protect surfaces from wear, corrosion, and fracture. While microcoatings also exist, advancements in technology have directed our attention toward nanocoatings, which should maintain a controlled thickness. Following the removal of these coatings, refurbishment or reconditioning of the products may be necessary. In our earlier discussion regarding LEO satellites, we highlighted the considerable benefits of employing nanocoating to effectively manage corrosion and fracture phenomena.

Nanocoating consists of nanocrystalline deposits that can be either crystalline or amorphous. This coating enhances the surface hardness and reduces the wear rate, thereby providing greater resistance to corrosive and chemical attacks and decreasing the likelihood of fracture. Additionally, if the nanocoating exhibits low friction properties, most particles will remain embedded below the surface, minimizing their exposure to corrosive environments, ultimately extending the lifespan of the product. Consequently, we recommend mitigating corrosion by passivating surfaces and isolating contact points. This surface engineering layer acts as a barrier between the product and its environment, representing the forefront of current technology.

Understanding the various types of nanocoatings is essential, and we will explore specific coatings in greater detail when we begin our lecture on coating and surface engineering. Coating and surface modifications are among the most prevalent and effective tools in surface engineering, and these topics will be revisited during our discussions.

Furthermore, in cases involving large metal structures, we can leverage our understanding of these phenomena. For instance, if we design a ship with an anticipated lifespan of 25 years, it is essential to consider that, after this period, the ship may need to be dismantled. We have observed that large units are frequently deconstructed for repurposing. The process of remelting and recycling these materials requires substantial effort and resources.

What if we could control the chemical degradation of a product within a specific timeframe—say, 10, 15, or even 20 hours? Instead of merely acknowledging the negative effects of corrosion, we could harness this scientific understanding to transform it into a productive and efficient process. This perspective highlights that maintaining large metal structures, such as ships nearing the end of their operational life, can be economically beneficial.

I propose the idea of programming these products for controlled self-destruction. If a product is designed to self-destruct, we could expose it to a controlled environment that accelerates corrosion, facilitating its breakdown. While we typically advise against exposing products to atmospheric conditions, there are instances where it becomes necessary to dismantle or decommission a unit. By implementing a corrosive environment, we could effectively manage the disassembly of the system into multiple parts.

In recent years, many countries, including India, have resorted to manual labor for dismantling old ships, often relying on low-cost labor. However, with the scientific advancements at our disposal, we could achieve this process more efficiently. If we receive a product in a highly functional state, we can carry out reconditioning to remove any nonfunctional components, potentially allowing us to remelt and reuse materials.

Thus, understanding the underlying science enables us to optimize product lifecycle management. Preventing corrosion and failure is essential, but equally important is the application of scientific modeling to maximize benefits.

5.8.5 Case study 5: Impact of corrosion and cracks on steel bridges

In this case study, we examine a familiar example: the steel bridge. Many of us have seen steel bridges (as shown in Figure 5.4) showing signs of discoloration, pitting, and corrosion. Similar to the stresses that affected

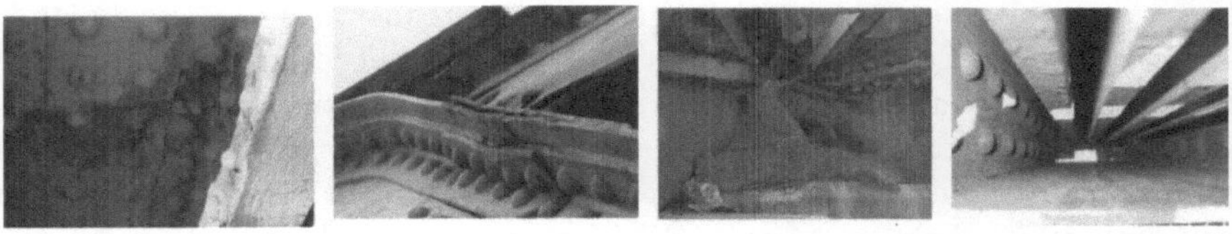

Figure 5.4 Corrosion and cracks in historical steel bridges. (From public domain references.)

the *Titanic*, these issues can lead to misalignment and even structural separation over time. Numerous historical steel bridges have developed defects, including severe corrosion and material fatigue, due to various contributing factors.

Even in "well-designed" bridges, irregularities caused by improper fabrication or inadequate maintenance can create stress concentration points where cracks may begin and spread. Corrosion significantly undermines the structural integrity of steel bridges, as exposure to environmental factors like moisture and chemicals leads to rust formation. If untreated, this rust results in cracks that progressively weaken the structure. For instance, corrosion at a single joint (e.g., bolt or rivet joint) can disrupt load distribution and shift stress concentration. Without timely intervention, repeated cycles of loading and unloading cause microscopic cracks to accumulate, expanding into serious structural issues over time. Initially, the bridge may appear flawless, but even one loose joint can eventually lead to misalignment and compounding problems. While eliminating joints could reduce such risks, the associated increase in construction costs makes this approach impractical.

Assembly imperfections are often unavoidable, given the large number of components involved, resulting in dynamic increases in stress levels over time. Population growth and rising vehicle traffic further push bridges beyond their original design limits, with bridges initially intended for lighter loads eventually becoming overburdened, leading to unexpected stress concentrations. A bridge that may start with a safety factor of 2.5 could, without maintenance, see this reduced to 1.2 within 5 years and even to 1 in 10 years, highlighting the importance of regular inspections. This decline in safety emphasizes the need for a comprehensive maintenance plan and strengthened maintenance strategies.

High-quality fabrication is essential for ensuring bridge durability. Although joints are integral to bridge construction, they require careful attention, as improper welding/joining can compromise connection strength. Many workers learn welding on the job without formal training in techniques such as face alignment and point-specific welding, which can result in weak joints. Although a bridge may initially perform well, accumulated rainwater in gaps and crevices can lead to corrosion and misalignment over time.

Regular maintenance is crucial to address these vulnerabilities. To reduce the risk of collapse, an essential strategy is to repair and seal cracks and crevices to prevent further corrosion. Corrosion-induced cracks and reduction in material fatigue-strength over time are common causes of structural deterioration, both of which significantly contribute to steel bridge failure. Steel bridge structural failures often stem from several key issues, including weak connections caused by improper joint design and welding, which reduce component lifespan. Corrosion, frequently due to insufficient preventive maintenance, gradually diminishes structural capacity, while poor welding quality, lacking in rigorous joint design and quality control, weakens connections and accelerates deterioration. If left unaddressed, these factors collectively compromise bridge durability and structural integrity, underscoring the importance of precise design, regular inspections, preventive maintenance, and skilled fabrication. These practices are essential for extending bridge lifespan and ensuring safety.

Zhang et al.[6] conducted a review on average daily truck traffic (ADTT) over bridges, analyzing its impact on the fatigue-related crack growth rate and the resulting reduction in the service life. They noted that uncoated weathering steel (UWS) bridges are preferred for their cost-effectiveness; however, the corrosion-fatigue (C-F) issue associated with UWS presents a significant concern. The authors developed a simulation model for the C-F process, treating the stages of crack initiation and propagation independently. Their findings indicate that during the crack initiation stage, the pit, gap, or crevice acts as an initial flaw that grows due to corrosion, and once this flaw reaches a critical size, fatigue effects begin to play a role in propagation, ultimately becoming the dominant growth driver leading to failure. They recommend the use of high-performance weathering steel (HPS 70W) to improve the service life of bridges.

One major challenge is understanding hidden corrosion mechanisms like crevice and pitting corrosion, commonly observed in bridges. In these cases, cracks and chemical infiltration points are often unknown, forming concealed electrochemical cells that promote corrosion. For example, structures like bridges contain numerous hidden electrochemical cells. If we can effectively prevent these, we can extend the product's life to resemble high-cycle fatigue. However, without intervention, hidden electrochemical cells shorten lifespan, leading to low-cycle fatigue. Recognizing and managing these hidden cells are essential for improving the durability and longevity of engineered structures. If we do not fully understand the formation of electrochemical cells on or just beneath the surface, we cannot recommend effective surface engineering solutions to prevent corrosion. Preventing structural collapse, damage, or rupture is essential, and we must assess the extent of corrosion over time to understand its impact on material strength. Initially, we might assume a certain corrosion factor, but this decreases with time. For instance, if a material starts with 100% strength, we could estimate it to retain 90%

strength after 5 years, 80% after 10 years, and so on. This provides a threshold, below which equipment should be decommissioned and replaced with a new system. Accurate lifecycle analysis depends on understanding these factors and systems.

To mitigate structural collapse, it is crucial to account for the gradual reduction in strength of components exposed to corrosive environments. This topic is fundamental to our focus, as shown in earlier studies of satellites in LEO. We learned two types of protective coatings: one for temperature control and another for resistance to AO, which helps prevent surface degradation and maintain operational temperatures. As previously discussed, LEO environments contain numerous pollutants, with AO posing a significant threat.

In high-speed applications, such as satellites or ships like the *Titanic* (which sustained significant damage partly due to its high speed of 22 knots upon colliding with an iceberg), the sensitivity to velocity plays a critical role in determining the extent of damage. Therefore, it is essential to develop high-performance coatings capable of withstanding such conditions. Researchers recommend nanocoatings due to their minimal weight and precise application, which helps prevent issues such as delamination and internal bubbles. Applying these coatings effectively requires advanced technologies, some of which we will explore in the upcoming chapters.

5.8.6 Case study 6: Atmospheric-assisted surface degradation of aircraft

In this case study, we will examine a critical aspect: the atmospheric-assisted surface degradation of aircraft. While clouds might appear familiar, they contain countless particles and water droplets. When an aircraft flies through a cloud, its relative speed is significantly high, creating a situation akin to a ship navigating through icebergs, albeit with a smaller mass and much higher velocity. The surfaces of aircraft exhibit patterns of small and large dots, indicating potential erosion areas. Regardless of the size of these dots, erosion on the aircraft's surface remains a significant concern. For instance, a collision between an aircraft and a flying bird can cause substantial damage, despite the bird's relatively small mass compared to the aircraft, primarily due to the high relative speed at which they meet. Consequently, the coatings and surface protections on aircraft require careful attention, leading manufacturers to invest years in developing effective solutions.

In this context, larger dots generally signify medium- to high-speed impacts, while smaller dots indicate lower speeds. Due to the approach of aircraft at relatively high speeds toward objects, significant failures are likely to occur at the front end of the aircraft, although other areas may also exhibit signs of damage. It is crucial to closely monitor any pit formation that could lead to corrosion. There is a need to continuously monitor the health of aircraft and track impacts from birds and insects carefully. When these collide with an aircraft, it is termed foreign object damage (FOD).

Additionally, weather conditions and airborne particles such as rain, hail, ice, sand, volcanic ash, and dust can induce pits and corrosion. Rainwater contains impurities and gases that exacerbate corrosion. Hail and ice also pose significant threats to aircraft structures; large hailstones can cause dents, abrasions, and structural deformations, while ice formation disrupts airflow, negatively affecting aerodynamics and performance.

When examining atmospheric-assisted surface degradation of aircraft, it is essential to consider airborne pollutants such as rain, ice, small particles, sand, volcanic ash, and various types of dust. These substances possess abrasive properties that can erode exterior surfaces, compromise protective coatings, and expose metal components to corrosion. Notably, certain particles, including volcanic ash and specific dust varieties, are especially damaging due to their high abrasiveness, posing a risk to structural integrity. Additionally, airborne materials like sand and ash can be more abrasive than steel, significantly increasing the likelihood of erosion. Consequently, when these airborne particles collide with the aircraft, they contribute to its atmospheric-assisted surface deterioration.

The initial impact from an object underscores the importance of distinguishing between erosion and impact. Erosion refers to the removal of material, often described as erosive wear, whereas impingement is simply the act of striking a surface. Although impacts can cause material removal, resulting in erosion, we often use the terms "impingement" and "erosion" interchangeably. Research on erosive wear indicates that a smaller angle of attack increases cutting wear, while higher angles (such as 80 to 90 degrees) may lead to fatigue failure. Erosion can be categorized into two types: abrasive wear and fatigue wear. While impingement might not always result in material removal, it can still lead to the formation of pits or surface deformations, causing short-term physical degradation. Additionally, the presence of water particles can exacerbate the situation, leading to long-term corrosion damage.

Hailstone impacts often result in the most significant damage to aircraft. When larger particles strike the surface, they can cause substantial harm by creating larger dents and removing material, which leads to surface abrasion. In severe cases, this can result in the overall deformation of the aircraft structure. Incidents, such as the damage to a Sabre aircraft caused by rain,[7] underscore the adverse weather hazards that aircraft encounter. Rain and other environmental factors can cause physical damage, lead to structural problems, and create safety risks. This highlights the critical need to understand and address these dangers in aircraft design, operation, and maintenance, particularly during high-speed flight scenarios. Moreover, such impacts can remove protective coatings or paint, heightening the risk of corrosion. Once a sample pit forms, particularly in areas with indentations or impingements, and rainwater is present, the rainwater can dissolve gases from the environment. This process creates an ideal environment for electrochemical reactions, thereby accelerating corrosion.

To illustrate these concerns, consider two examples.[7] The first example is the SABRE aircraft, specifically the OOL 8, which suffered damage to its radome after rain exposure in 1959. This incident occurred at a high altitude of approximately 5000 feet and at a speed of around Mach 0.76. The erosion observed on the aircraft's front highlighted the need for design modifications, adjustments to operational procedures, and an emphasis on effective maintenance strategies.

The second example examines tests conducted in 1973 to assess erosion effects on a modified Phantom YF-4M aircraft. Operating at lower altitudes and reaching speeds close to Mach 0.99, the aircraft experienced significant rain damage on the leading edges, radome, paintwork, and pitot probes. These observations led to suggested design improvements and maintenance practices to reduce the impact of erosion on aircraft structures. Findings from both cases emphasize the critical need for continuous aircraft health monitoring and regular maintenance.

In 2011, a new aircraft design underwent testing to enhance performance, underscoring the importance of rigorous research, protective measures, and proactive maintenance strategies to maintain structural integrity and extend the lifespan of aircraft.

It is also important to note that while cost is often a critical consideration in system design, for aircraft, safety and reliability take precedence due to the severe socioeconomic impacts of even minor failures. In such systems, atmospheric pollutants and contaminants can exacerbate surface degradation, necessitating robust protection and maintenance.

Moreover, not all systems show visible signs of failure. In certain cases, internal components may fragment, altering the structure significantly. Even in the absence of visible wear, if a system's efficiency—typically expected around 95%—drops below acceptable levels, the product may become ineffective. For instance, a motor unable to deliver the required torque, "x" Nm, will render the unit nonfunctional, necessitating replacement even in the absence of full structural failure. Thus, degradation goes beyond fractures; a reduction in operational efficiency to levels below 95%, 90%, or even 85% is sufficient to mark a system as ineffective.

This concept demonstrates how environmental factors accelerate machine aging, impacting not only structural durability but also efficiency, which is crucial for maintaining operational reliability in critical systems.

5.8.7 Case study 7: Building degradation

In earlier case studies, we emphasized the critical role of environmental factors in surface damage. As previously outlined, the interplay between wear, corrosion, and fracture can collectively result in significant degradation. Wear alone can cause surface defects such as scratches, pitting, and grooves, creating initial points of weakness. While some issues may arise from fabrication, most surface defects stem from wear-related deformation, sliding, or similar processes.

Once these defects are present, corrosion plays a key role by exploiting pits or grooves to form electrochemical cells. This allows cracks to initiate on these vulnerable surfaces, facilitating the propagation of secondary cracks that ultimately lead to fractures, breaking the surface into segments.

In this cycle, wear, corrosion, and fracture operate sequentially. Presenting them in this way helps illustrate how they amplify one another. Environmental factors such as acidic conditions (pH below 7), increased temperature, and humidity intensify these effects, raising the wear rate. Structural crevices or joints—like those found in bridges—pose additional risks. Imperfections in the fabrication process often result in small gaps, allowing electrochemical cells to form. While environmental severity varies, joints remain a significant concern. Ideally, a joint-free product design would eliminate this issue, though it would come at a high fabrication cost. Thus, ensuring cost-effectiveness remains a crucial consideration.

Let's examine how environmental factors contribute to the deterioration of buildings. The formation of an electrochemical cell can lead to ongoing corrosion, potentially resulting in structural collapse or discoloration. The degradation resulting from atmospheric pollutants threatens the structural integrity and aesthetic value of these buildings while posing potential health risks due to harmful substances on their facades. Therefore, preserving monuments, particularly those important for tourism, is essential. Without regular maintenance, the materials of these structures can deteriorate, compromising their integrity and aesthetic appeal. It is also noteworthy that materials such as aluminum and titanium may not have been commonly used 200 years ago, which can impact the longevity and preservation of these historical buildings.

India, like many other countries, is home to numerous monuments, several of which are designated as UNESCO World Heritage Sites. Many of these monuments contain banned metals that pose risks to human health, with their surfaces serving as reservoirs for these harmful substances.

To illustrate this, consider the example of brake shoes. The composition of brake shoes has evolved over time, and older brake particles can become airborne and settle on these monuments. If regular maintenance is neglected, these particles can accumulate on the surfaces. Although these contaminants may not present immediate health risks in everyday environments, they can become problematic when visitors interact with the monuments. Touching these surfaces can transfer particles onto our hands or fingers, potentially leading to health issues.

Therefore, when evaluating the overall impact, these considerations become crucial. Oliveira et al.[8] conducted a study on two architectural sites in Bordeaux, France, which attract numerous tourists, revealing contamination across nearly all examined areas. Utilizing techniques such as energy dispersive X-ray spectroscopy (EDX), the researchers analyzed contamination levels on the facades of two UNESCO-listed historic sites: Grosse Cloche and Cathédrale St-André. The study investigates the impact of atmospheric pollutants originating from vehicular traffic, public transit systems, industrial emissions, and agricultural activities. These pollutants contribute to the accumulation of gases and PM on building surfaces, leading to the deterioration of historic structures. Preserving these architectural treasures is essential not only for their historical significance but also for safeguarding human health.

Naturally, these contaminants originate from various sources and settle on the surfaces of monuments, necessitating thorough cleaning and careful consideration of the underlying issues. It is essential to apply scientific insights to develop effective maintenance practices. Many areas of monuments were found to be covered with a black crust, which may indicate a lack of oxygen or sunlight. Such contamination can block sunlight from reaching the surface, leading to cumulative damage. Therefore, immediate removal is crucial, as the presence of a black crust not only detracts from the monument's aesthetic appeal but also perpetuates the degradation of its surface.

Aesthetic considerations are significant, as people are drawn to visually appealing surfaces, like authentic monuments, rather than overlooking layered or obscured structures. Furthermore, neglecting proper maintenance will inevitably lead to further degradation. In their[8] detailed analysis, researchers examined microscale images revealing both spherical and nanoscale particles. They noted that poor air quality in many buildings arises from their porous structure, allowing external spherical particles to settle on the surfaces without causing direct damage. Urban areas, such as Bordeaux, are known for high concentrations of atmospheric pollutants.

The primary challenge faced by Grosse Cloche and Cathédrale St-André is the limited sunlight exposure on certain sections for much of the year. This lack of light hinders the hydration of construction materials and promotes the formation of black crusts caused by moisture-loving microorganisms. To protect these structures, authorities must enhance their efforts by promptly removing black crusts to prevent further deterioration and implementing measures to minimize PM accumulation. Potential strategies include deploying air filtration systems, improving urban planning to reduce pollution sources, and enforcing regulations on vehicle and industrial emissions.

Analysis of samples collected from both sites indicates similar conditions marked by porous spherical materials influenced by deteriorated urban air quality. These particles may include agglomerated nanoparticles (NPs) containing hazardous elements such as arsenic (As), lead (Pb), cadmium (Cd), and chromium (Cr), which pose significant threats to environmental quality and human health. The agglomerated NPs are particularly alarming, as they tend to attract other particles, resulting in increased accumulation if not properly controlled. This buildup may encompass toxic substances like arsenic, cadmium, and chromium, which have been banned for years yet still persist on building surfaces today. Furthermore, the presence of ultrafine particles (UFPs) around the structures' pores suggests ineffective adsorption, underscoring the urgent need to mitigate the presence of these hazardous elements and UFPs to preserve the sites effectively. The existence of these NPs raises serious concerns regarding the potential release of toxic substances, which could lead to various health issues.

Samples collected from monuments near train tracks showed metallic particles likely originating from the train's brakes and tracks, mixed with soil and settled dust from construction materials. Energy dispersive spectroscopy (EDS) analysis revealed elements such as aluminum (Al), silicon (Si), calcium (Ca), potassium (K), sodium (Na), chromium (Cr), iron (Fe), antimony (Sb), molybdenum (Mo), and other significant metallic materials. The absence of similar particles in samples taken from more distant locations indicates that these emissions are specifically linked to train operations. These particles are likely the result of wear from train brakes and tracks, combined with soil and settled dust from building materials. Essentially, the source of these metallic particles is the trains themselves, rather than the degradation of soil or building materials.

Preventive measures are crucial to protect the well-being of train crew members and regular passengers. Although tourists may face a lower level of risk due to their shorter exposure times, local government officials should consider strategies to mitigate the impact on residents and workers in the area. Initiatives such as mandating the use of brake pads free from potentially harmful components and implementing routine street cleaning could prove beneficial. This highlights the importance of protective measures to address harmful substances and PM, ensuring the effective preservation of Grosse Cloche and Cathédrale St-André while safeguarding both environmental quality and human health.

5.8.8 Case study 8: Surface degradation due to ozone

This case study investigates the well-documented harmful effects of ozone (O_3), a highly reactive gas present naturally in Earth's atmosphere and also generated by human activities, such as industrial operations and pollution. It is widely recognized that ozone significantly impacts carbon-based and organic materials, which will be discussed further.

The question arises: How does ozone contribute to surface degradation? The process operates through a cyclic mechanism, where ozone molecules continuously interact with the double bonds in organic polymer chains, particularly those containing carbon. Oxygen plays a crucial role, as its high reactivity transforms double bonds into single bonds, creating new functional groups that enhance the molecules' chemical reactivity. This leads to further degradation as the functional groups initiate additional reactions with ozone. Organic materials—particularly rubber, plastics, and certain paints—are especially vulnerable to this degradation. In rubber, for example, ozone can cause a network of surface cracks known as ozone cracking, which manifests as symptoms like surface fissures, crazing, discoloration, loss of flexibility, and significant reductions in strength, ultimately rendering the material nearly ineffective. To address these issues, it is essential to investigate the source of ozone exposure and conduct thorough chemical analyses of the affected surfaces using techniques such as scanning electron microscopy (SEM) or EDX to assess the extent of damage and determine appropriate interventions.

Wang et al.[9] investigated the structural changes in alkali lignin (AL) resulting from ozone treatment and assessed their effects on the enzymatic hydrolysis of Avicel, highlighting the significant implications of ozone exposure for material integrity and functionality. The study found that factors such as increased ozone concentration, elevated temperatures, and prolonged exposure times significantly exacerbate the degradation process, emphasizing the importance of closely monitoring environmental conditions. Using commercially available ozone generators, the researchers explored the effects of ozone on this carbon-containing material, illustrating the ongoing research into ozone's impact on various substances and the necessity for effective monitoring and management strategies to mitigate degradation.

The researchers conducted an experiment to evaluate the effects of ozone on AL by exposing the system to ozone at a concentration of 85 mg/L with a flow rate of 3 L/min for 45 minutes, maintaining a temperature of approximately 45°C. Following this treatment, the ozonated lignin was cleaned with deionized water, vacuum-filtered to remove potential inhibitors, and then vacuum-dried. The samples were classified as follows: the original AL, ozone-treated lignin at pH 3 (OL-pH3), ozone-treated lignin at pH 7 (OL-pH7), and ozone-treated lignin at pH 12 (OL-pH12). In other words, to assess the material's response under varying conditions, the researchers varied the pH levels to 3, 7, and 12, analyzing scenarios that included both no ozone exposure (AL) and exposure to ozone in acidic (OL-pH3), neutral (OL-pH7), and alkaline (OL-pH12) environments.

The visual inspection of AL after ozonation showed notable alterations in its surface morphology. Untreated lignin had a smooth appearance with minor debris, whereas ozonated lignin at pH 3 displayed significant damage, including holes, cracks, increased roughness, and wrinkles, indicating serious impairment of its physical structure due to ozone exposure in an acidic environment. The sample with no ozone exposure remained clean, with debris approximately one micron (1 μm) in size. When the pH level was raised to 7, the deformation of the lignin was significantly reduced; however, some debris and brittle fractures were still present, though less severe than in the acidic condition. In alkaline conditions, the original AL showed minimal damage, demonstrating its resilience. This highlights that ozone exposure is particularly harmful in corrosive acidic environments, leading to greater degradation compared to neutral or alkaline conditions, thereby emphasizing the need to avoid such exposures to mitigate material degradation.

5.8.9 Case study 9: Surface degradation due to chloride

Chloride-induced surface degradation is a significant concern, particularly in coastal and industrial areas with high chloride levels. Chloride ions (Cl^-) are highly reactive and can penetrate the protective oxide layer on metals like stainless steel. This oxide layer typically shields metals from corrosive agents, but exposure to chloride ions can compromise it, allowing metal chlorides to form on the surface. This process accelerates localized corrosion, including pitting (small, deep holes) and crevice corrosion (in confined spaces), both of which progressively weaken the metal.

Once the protective layer is breached, the environment shifts from protective to corrosive, increasing the risk of material damage. It becomes essential to determine whether the resulting corrosion is pitting or crevice corrosion, as each type has distinct impacts.

In concrete structures with steel reinforcement, chloride ions can infiltrate through the concrete cover, reaching and corroding the embedded steel. This "chloride-induced reinforcement corrosion (shown in Figure 5.1)" leads to cracking, spalling, and deterioration of the concrete, ultimately compromising structural integrity and heightening the risk of failure. Effective protective measures and regular maintenance are vital to managing chloride-related degradation in such environments.

In this case study, we discuss one research article.[10] Yeh et al. investigated crevice corrosion and its effects on P04L stainless steel, specifically in chloride environments, building upon prior discussions about the significance of distinguishing between types of localized corrosion, such as pitting and crevice corrosion. Crevice corrosion is particularly problematic in chloride-rich environments, like coastal or marine areas, where the concentration of chloride ions can severely degrade surfaces. These ions can penetrate the protective oxide layer on the steel, leading to crevice formation and increased susceptibility to corrosion.

Yeh et al.[10] investigated the use of excess canisters as protective barriers for nuclear fuel, aiming to apply a coating to shield the fuel from environmental exposure. They discovered that these canisters, despite being made of 304 stainless steel—which contains low carbon, high silicon, and elevated levels of chromium, nickel, and manganese—were susceptible to localized corrosion, specifically pitting and crevice corrosion, especially in marine environments with high concentrations of chloride ions.

To evaluate the impact of chloride-induced corrosion, the researchers conducted experiments under controlled conditions. The surfaces of the specimens were polished to meet established standards for uniformity, and they utilized M6 bolts and various polymers to facilitate the formation of crevices, which allowed for increased infiltration of corrosive agents. A torque value of 1.13 Nm was maintained, adhering to standard testing procedures. The specimens were exposed to synthetic seawater, prepared according to ASTM D 1141-98 specifications, with varying chloride concentrations (0.1 and 1 g/m^2) and subjected to a temperature of 45°C with relative humidity levels of 45%, 55%, and 70% for a total of 5000 hours.

The study's findings revealed a significant correlation between humidity levels, chloride concentration, and corrosion damage. At 0.1 g/m^2 chloride concentration, the specimens exhibited the highest damage at 45% humidity, while the least damage occurred at 55% humidity. However, when humidity increased to 70%, there was a resurgence in surface damage. The microstructural analysis, conducted using SEM, showed varying levels of damage, with no cracks observed in the corroded regions at any humidity level, indicating that while crevice corrosion was present, it did not escalate to crack formation.

Overall, the research highlights the complex interplay between chloride concentration, humidity, and temperature in influencing corrosion processes. The authors conclude that a humid environment is crucial for promoting chloride infiltration into crevices, particularly at lower chloride concentrations.

Often, aluminum and its alloys are used to reduce the chances of corrosion. Aluminum is widely utilized in the aerospace and transportation industries for its excellent mechanical properties and corrosion resistance. However, aluminum alloys are prone to corrosion, particularly in a chloride environment. To handle the chloride environment, often anodizing treatment of aluminum is recommended. Anodization is a process that enhances corrosion resistance by forming a protective oxide layer on the aluminum surface. This is achieved by immersing the discs in an electrolytic solution and applying a direct current (DC) to induce controlled oxidation on the surface. Cabral-Miramontes et al.[11] utilized two different electrolyte solutions for the anodization process: a sulfuric acid solution with a concentration of 199.04 g/L and a citric–sulfuric acid solution containing 210.12 g/L citric acid with 5 mL/L of 98% sulfuric acid. The anodization was conducted using a model XLN30052-GL high-power programmable

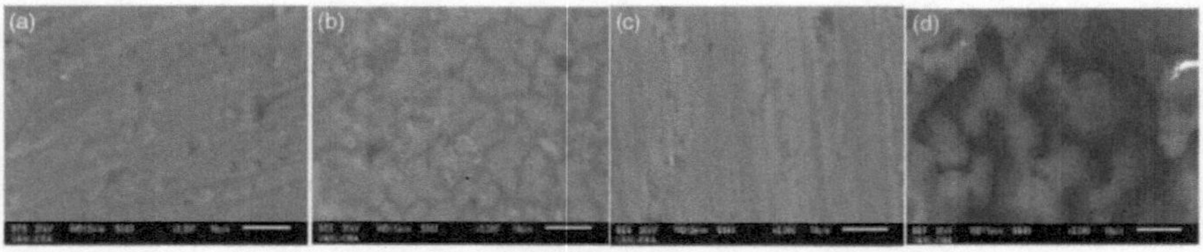

Figure 5.5 SEM micrographs showing surface morphology of samples anodized in: (a) sulfuric acid solution at 1 A/cm², (b) sulfuric acid solution at 7.2 A/cm², (c) citric–sulfuric acid solution at 1 A/cm², and (d) citric–sulfuric acid solution at 7.2 A/cm².[11]

DC power supply, applying current densities of 1 A/cm² and 7.2 A/cm² for 30 minutes.

Cabral-Miramontes et al.[11] anodized AA 6061 samples in both sulfuric acid and citric–sulfuric acid baths and subsequently exposed to a 3.5 wt.% NaCl solution. SEM analysis of anodized samples (shown in Figure 5.5) revealed that surface morphology varied significantly based on current density and electrolyte type. At a higher current density of 7.2 A/cm², samples displayed porosity and superficial cracking due to increased energy input, which likely caused localized stress and structural instability in the oxide layer. In contrast, samples anodized at a lower current density of 1 A/cm² in both sulfuric acid and citric–sulfuric acid solutions exhibited a more homogeneous, smoother surface morphology with fewer defects, enhancing corrosion resistance by limiting pathways for chloride ion penetration. Notably, the sample anodized at 1 A/cm² in the citric–sulfuric acid solution (CA 1 A/cm²) demonstrated the smoothest, most regular surface, indicating superior corrosion protection. These findings suggest that optimizing current density and electrolyte composition is critical for achieving the most effective anodized coating properties, particularly in chloride-rich environments. Among samples anodized with a lower current density of 1 A/cm², the citric–sulfuric acid anodized sample demonstrated the best corrosion resistance, followed by the sulfuric acid anodized sample. The sample anodized in citric–sulfuric acid at a higher current density of 7.2 A/cm² showed the poorest performance.

At the higher current density (7.2 A/cm²), the anodized layer was significantly thicker but exhibited porosity and cracking. In contrast, samples anodized at the lower current density (1 A/cm²) formed a homogeneous and compact anodic film. These findings suggest that both electrolyte composition and current density play crucial roles in determining the corrosion resistance and structural characteristics of anodized AA 6061 surfaces in a chloride environment.

5.8.10 Case study 10: Hydrogen embrittlement

In our previous discussions on corrosion, we delved into hydrogen embrittlement and its more significant effects on materials compared to conventional chemical degradation. Hydrogen embrittlement occurs when atomic hydrogen infiltrates metals, drastically reducing their strength and causing brittleness, much like how rubber becomes less flexible and develops cracks when exposed to ozone. This infiltration results in the loss of critical properties such as elasticity and damping, leading to unexpected fractures that compromise structural integrity. The issue is particularly severe in high-stakes industries like aerospace, oil and gas, and automotive, where structural failures can have dire consequences for human safety and environmental stability. Hydrogen atoms typically arise from common processes like corrosion or electroplating, where reactions between metals (such as iron) and water generate hydrogen ions that penetrate the material's surface. Over time, these hydrogen atoms accumulate near grain boundaries—already vulnerable points within a material—creating microstructural defects and heightening the risk of abrupt fractures. The connection between corrosion and hydrogen embrittlement is crucial, as hydrogen is often produced during corrosion reactions; for instance, when iron corrodes in water, hydrogen ions (H^+) are reduced to atomic hydrogen, which diffuses into the material along internal imperfections. This accumulation can lead to the formation of hydrogen gas bubbles that weaken the metal's lattice structure, rendering it brittle even under low-stress conditions. Thus, effectively addressing hydrogen embrittlement necessitates a comprehensive understanding of its interactions with material structures and the implementation of preventive measures, as it plays a pivotal role in both crack initiation and propagation, ultimately contributing to sudden material failures.

Unlike corrosion, which progresses over time, hydrogen embrittlement advances more quickly, making it a pressing issue for industries that handle hydrogen or hydrogen-enriched environments. While corrosion may transform high-cycle fatigue into low-cycle fatigue over extended periods, hydrogen embrittlement accelerates this process, shifting the failure mode from ductile to brittle and significantly reducing the fatigue life. In tensile tests, samples exposed to hydrogen display brittle fracture surfaces, lacking the typical cup-and-cone shape and micro-voids of ductile failures.

For example, in one study in which specimens were charged with hydrogen showed brittle shear failure at a 45-degree angle was observed, with SEM images revealing quasi-cleavage features, primary and secondary cracks, and intergranular cracking. Even at low temperatures, hydrogen embrittlement leads to brittle failure; however, with increased temperature, the problem worsens. Higher friction or thermal cycling due to environmental variations accelerates degradation through oxidation, corrosion, or embrittlement.

It is worth noting that hydrogen embrittlement, temperature-assisted oxidation, and corrosion are interconnected phenomena that significantly reduce material lifespan in environments with high stress and temperature fluctuations. Effective modeling and mitigation of these effects are essential to ensure safety and durability in high-risk industries.

5.8.11 Case study 11: Thermal mechanical fatigue for turbine blade superalloy: Temperature-assisted oxidation

A case study on turbine blades by Hu et al.[12] demonstrates the significant effects of temperature and oxygen on material degradation. The research emphasizes that the combination of elevated temperatures and the presence of oxygen greatly accelerates the formation of oxide layers, resulting in more severe damage during fatigue loading.

In this study, two temperature conditions were maintained, with one ranging from 500°C to 1000°C, indicating fluctuations, while cyclic variations were kept below 500°C. The findings revealed that the life cycle of the material decreases by half as temperatures rise. For instance, with an initial life cycle estimated at around 4675 cycles, the increased temperature significantly shortens this lifespan. The researchers noted that ductile failure is characterized by secondary cracks identified as fatigue striations. However, at elevated temperatures, more pronounced features such as ridges were observed, indicating a higher likelihood of fracture.

As temperatures increase, the fatigue life diminishes, and material properties deteriorate. This raises the question: Should an increase in temperature make materials softer and more ductile? Contrary to this expectation, thermal cycling at lower temperatures induces a specific type of fatigue, especially at brittle fracture sites. Thus, there is a strong correlation between temperature and oxidation, where an enhanced oxidative environment leads to reduced fatigue life. Consequently, we conclude that high temperatures exacerbate oxidation, resulting in numerous cracks in the material.

This phenomenon suggests that elevated temperatures lead to an increased formation of cracks, potentially causing the superalloy material to disintegrate and exhibit a significantly shorter lifespan than anticipated. The interactions between temperature-assisted oxidation, corrosion, and hydrogen embrittlement underscore the importance of understanding these environmental factors, as they can reinforce one another. Therefore, a comprehensive approach to this topic is essential, which brings us to the conclusion of this presentation.

5.9 FINAL REMARKS

In this and previous chapters, we examined how corrosion accelerates degradation, making materials more brittle and prone to fracture. Corrosion increases the wear coefficient, often making the lifecycle of wear debris—typically marked by high plastic buildup and a nondimensional wear coefficient around 10^{-6}—much less predictable. This change in material properties leads to easier fracturing, which can cause the product to either completely disintegrate or shed debris from its surface.

It is also worth noting that high-cycle fatigue, which occurs in normal environmental conditions and without corrosion, can lead to wear debris generation after many cycles, like 1 million or 10 million. However, corrosion changes the process to resemble low-cycle fatigue, where debris might appear after only 100 or even 1000 cycles. This effect functions almost like a multiplier. Corrosion facilitates the removal of material from the surface, and each newly exposed layer becomes more chemically active in the environment, leading to further corrosion. The combined effects of wear and fracture under corrosive conditions can lead to significant surface damage or even catastrophic failure. With each new surface exposed to the corrosive environment, we see continued wear or fracture, which results in ongoing debris formation. When debris is removed from a surface, a fresh layer becomes exposed to the environment and is prone to further corrosion, significantly accelerating the wear rate. This exposure allows the corrosive process to remove material, which both speeds up corrosion and produces additional byproducts.

The fate of this debris is also important. As the surface deteriorates, particles emerge and can be regarded as corrosion byproducts. These particles, originating from combined corrosion and wear, often act as abrasives. This interaction explains why the wear coefficient, typically around 10^{-9} under normal conditions, can increase to 10^{-6} when both corrosion and wear are present. When corrosion, wear, and fatigue-fracture conditions act together, the wear coefficient may rise further to 0.001. These interconnected phenomena underscore the importance of understanding them.

Another crucial factor is friction. When friction arises between two surfaces, a tangential load is required to move one surface relative to the other, consuming energy that converts into heat. This frictional force increases stress, elevating shear stress and resulting in additional surface damage. Higher friction forces increase the chance of surface damage, while lower friction levels can delay the progression of subsurface cracks to the surface.

High friction increases the likelihood that subsurface cracks (unintended subsurface voids generated during manufacturing) will reach the surface immediately, leading to debris removal. This cycle generates even more debris through ongoing corrosion and associated processes. Thus, we can conclude that tangential loading can significantly amplify the degradation effects of wear, corrosion, and fracture. By accounting for these three factors along with frictional forces and modeling them accurately, we can design products that are durable, resilient, and less susceptible to environmental degradation.

In this chapter, we learned how environmental exposure accelerates surface degradation. Once a product is fully damaged, it contributes to waste material, further impacting environmental health. By integrating "surface engineering" into both the product and its environment, we can effectively mitigate surface degradation, thereby reducing overall environmental impact. We will implement these strategies by enhancing the interface through surface engineering in the next chapters.

In the next chapter, we will shift focus to testing standards and explore the various testing methods available to ensure consistency in reporting results.

5.10 CONCLUSIONS

1. *Corrosion–wear interplay:* The interaction between corrosion and wear significantly impacts the performance and longevity of industrial machinery, infrastructure, and aerospace systems. Understanding this relationship is critical for designing durable materials and systems.
2. *Environmental contributions:* Factors such as water, humidity, temperature, UV radiation, and abrasive particles collectively accelerate surface degradation. The synergistic effects of these factors necessitate comprehensive mitigation strategies tailored to specific environments.
3. *Case study insights:* Real-world examples, including satellite degradation, bridge corrosion, and the *Titanic* disaster, underscore the importance of addressing both material selection and environmental challenges in design processes.
4. *Mitigation strategies:* Advanced protective coatings, like nanocoatings, and regular maintenance are essential for preventing or minimizing degradation. These approaches extend service life and improve safety in critical systems.
5. *Future directions:* Innovations in material science, including multilayer and nanocomposite coatings, hold promise for addressing emerging challenges in corrosion and wear. Additionally, proactive design and lifecycle management strategies are essential for sustainable engineering practices.

These conclusions emphasize the need for multidisciplinary approaches to mitigate surface degradation, ensuring the safety, reliability, and longevity of engineered systems.

REFERENCES

1. Cusick, S. K., Cortés, A. I., & Rodrigues, C. C. (2017). *Commercial Aviation Safety*. McGraw Hill. www.abul.org.br/biblioteca/79.pdf

2. Tétreault, J., Cano, E., van Bommel, M., Scott, D., Dennis, M., Barthés-Labrousse, M. G., Minel, L., & Robbiola, L., (2003). Corrosion of copper and lead by formaldehyde, formic and acetic acid vapours. *Studies in Conservation*, *48*(4), 237–250.
3. Liu, T., Sun, Q., Meng, J., Pan, Z., & Tang, Y. (2016). Degradation modeling of satellite thermal control coatings in a low earth orbit environment. *Solar Energy*, *139*, 467–474.
4. DiLisi, G. A. (2022). The sinking of the RMS Titanic: A case study in thermal inversion and atmospheric refraction phenomena. *The Physics Teacher*, *60*(1), 11–14.
5. Article published on "Learn from the past: Erika oil spill, Europe's environmental disaster", cm-learn-from-the-past-erika-oil-spill-europes-environmental-disaster/
6. Zhang, Y., Zheng, K., Heng, J., & Zhu, J. (2019). Corrosion-fatigue evaluation of uncoated weathering steel bridges. *Applied Sciences*, *9*(17), 3461.
7. Gohardani, O. (2011). Impact of erosion testing aspects on current and future flight conditions. *Progress in Aerospace Sciences*, *47*(4), 280–303.
8. Oliveira, M. L., Neckel, A., Pinto, D., Maculan, L. S., Dotto, G. L., & Silva, L. F. (2021). The impact of air pollutants on the degradation of two historic buildings in Bordeaux, France. *Urban Climate*, *39*, 100927.
9. Wang, H., Zhao, L., Ren, J., & He, B. (2022). Structural changes of alkali lignin under ozone treatment and effect of ozone-oxidized alkali lignin on cellulose digestibility. *Processes*, *10*(3), 559.
10. Yeh, C. P., Tsai, K. C., & Huang, J. Y. (2019). Effects of relative humidity on crevice corrosion behavior of 304L stainless-steel nuclear material in a chloride environment. *Metals*, *9*(11), 1185.
11. Cabral-Miramontes, J., Gaona-Tiburcio, C., Estupinán-López, F., Lara-Banda, M., Zambrano-Robledo, P., Nieves-Mendoza, D., Maldonado-Bandala, E., Chacón-Nava, J., & Almeraya-Calderón, F. (2020). Corrosion resistance of hard coat anodized AA 6061 in citric–sulfuric solutions. *Coatings*, *10*(6), 601.
12. Hu, X., Yang, X., Shi, D., & Yu, H. (2016). Out of phase thermal mechanical fatigue investigation of a directionally solidified superalloy DZ125. *Chinese Journal of Aeronautics*, *29*(1), 257–267.

Chapter 6

Characterization techniques and prevention

6.1 INTRODUCTION

In today's digital age, with advanced computational tools at our disposal, a central question arises: "Is physical testing still necessary?" The answer lies in our commitment to accurately assess structural integrity and performance. Academically, a foundational understanding of physical processes is essential even before attempting to simulate them. This knowledge base equips us to tackle future challenges, particularly when working in contaminated environments. As we build and refine this foundation, we gain opportunities to innovate and improve our methods, which can lead to higher-quality data and better performance outcomes. Each person's unique perspective enriches the scientific community, fostering diverse approaches toward shared goals, such as enhancing system performance. Through continuous collaboration, we can drive meaningful advancements in the field.

Relying solely on historical data is no longer adequate, as contemporary conditions differ markedly from those of the past. The initial properties of materials are essential for product design, as they directly influence resistance strength and the expected service life or cycle. For instance, one case study demonstrated that increased temperatures can significantly shorten a product's lifespan. Conducting rigorous testing provides more reliable results than depending exclusively on modeling or simulation methods.

Moreover, the rise in environmental pollution over the past two decades, characterized by elevated levels of sulfur dioxide and particulate matter, necessitates a reassessment of our testing and evaluation approaches. To ensure optimal performance in this increasingly polluted environment, it is imperative to adapt our testing methodologies. This adaptability is crucial for maintaining competitiveness in the global market, enabling continuous improvements that meet the evolving demands of our industry. Ultimately, effective testing is vital to ensure product reliability, safety, and compliance with current environmental standards.

Accurate evaluations are essential for ensuring the reliability and safety of various components and systems. Numerous American Society for

DOI: 10.1201/9781003545019-6

Testing and Materials (ASTM) standards offer detailed guidelines for these assessments, highlighting the critical importance of testing in maintaining the integrity of operational environments. A key aspect of non-destructive testing (NDT) is that its established techniques are primarily grounded in standards set by various professional societies. Adhering to these standard practices ensures that tests are universally applicable, facilitating the documentation, publication, and sharing of results across different industries.

In addition, the proactive nature of NDT allows for the early identification of potential failures, serving as a cost-effective strategy for detecting issues such as overlooked cracks or inefficiencies in manufacturing processes. By using NDT, organizations can significantly enhance their quality control measures and address potential problems before they develop into serious failures.

As discussed in previous chapters, safety remains a core priority, and while simulations are integral to modern engineering processes, they must be validated through robust testing to be effective. Simulations depend on accurate input data and can reveal correlations among various parameters; however, the synergy between computational models and physical testing is essential to verify predicted behaviors. Thus, engineering tests are indispensable for rigorously evaluating the performance of diverse components and systems.

This chapter explores vital testing methods that are crucial for evaluating the integrity and functionality of engineering systems, ensuring they meet the stringent demands of real-world applications. The techniques discussed here are fundamental for comprehensive material evaluation and will serve as the basis for our in-depth examination of each method later on. Our objective is to elucidate the purpose of these techniques, the key criteria for their classification, and the reasoning behind the distinction between the two primary categories: destructive testing (DT) and NDT.

6.2 ANALYZING SURFACE DEGRADATION

A significant focus in the present book involves understanding "surface degradation"—the process by which material surfaces deteriorate over time. Although traditional failure analysis is well established, the emphasis here is specifically on surface degradation. Key questions in this analysis include:

- What are the primary causes of surface degradation?
- Are they a result of design flaws, material deficiencies, manufacturing processes, lack of maintenance strategy?

Recognizing these challenges, along with implementing suitable maintenance strategies, is crucial as all systems inevitably degrade over time. Furthermore, factors such as contamination, pollution, and physical damage can have substantial impacts on degradation of surfaces of the equipment. Case studies, covered in previous chapter, consistently highlighted the link between engineering system integrity and societal well-being: Any damage to critical equipment can lead to broader adverse effects. Consequently, safety and performance are closely intertwined, making it essential to establish reliable performance metrics to ensure dependability.

6.2.1 Role of testing in analyzing surface degradation

Testing methods are central to understanding and mitigating surface degradation. We have previously reviewed corrosion testing and examined relevant standards. Corrosion testing, in particular, evaluates material stability in controlled environments to simulate specific degradation processes. Key parameters include:

- Hydrochloric acid, sulfuric acid, and citric acid concentrations for simulating acidic conditions.
- Temperature variations (~100°C to 200°C) to assess thermal effects on material surfaces.

6.2.2 Corrosion testing: Assessing weight loss and surface changes

By maintaining controlled testing conditions, the stability of the environment allows for accurate assessments of material corrosion susceptibility and projected lifespan. This data is vital for developing predictive models related to surface degradation, enabling engineers to anticipate material wear and formulate effective protection strategies. Corrosion testing can be divided into two main categories: DT and NDT techniques. In DT, weight loss is typically measured to determine the extent of corrosion. A positive weight loss indicates that material has been removed from the surface, whereas negative weight loss signifies the addition of material. This material addition often occurs in chemical environments where reactions cause substances from the environment to be absorbed onto the surface. Such absorption results in an increase in weight due to the accumulation of external material. It is crucial to consider this weight change. While positive weight loss is generally expected due to wear, corrosion can also lead to negative weight loss, characterized by the swelling of the surface from accumulating external material. Consequently, weight loss measurements serve as significant indicators of corrosion severity. A substantial weight loss suggests a higher degree of corrosion, indicating that the corrosion process is inflicting considerable damage to the surface.

6.2.3 Corrosion mitigation techniques and testing requirements

To combat corrosion, various coating and protection techniques are used, targeting specific forms such as crevice and fatigue corrosion. One of the simplest preventative strategies is to limit excessive electronic transmission in vulnerable areas. However, any mitigation method introduced must be thoroughly tested before implementation to ensure effectiveness.

It is also critical to recognize that corrosive conditions can alter the nature of failure modes. For example, components exposed to corrosive environments may experience a shift from ductile to brittle failure or from high-cycle to low-cycle fatigue. Understanding these transitions is vital for accurate product design and for ensuring long-term performance. Even after materials and coatings are selected and implemented, comprehensive testing is necessary to validate performance under actual corrosive conditions.

6.3 TESTING METHODS IN ENGINEERING

Testing is an essential aspect of engineering practice, providing insights into structural integrity, performance, and durability under various conditions. The following methods highlight critical testing techniques and considerations for evaluating engineering components and materials.

6.3.1 Scaled-down laboratory models

Creating scaled-down laboratory models allows engineers to replicate real equipment on a smaller scale for controlled testing. This approach enables targeted evaluation of material and structural performance under specific loading conditions. By using scaled models, engineers can perform tests like tensile, fatigue, creep, fracture, and corrosion assessments to gather essential data on material behavior. These insights support informed decisions about material selection, design improvements, and reliability assurance. Standard and accelerated testing techniques applied to these models provide critical information on durability, helping engineers optimize design strategies across various applications. This process typically includes applying varied load conditions to assess parameters such as:

- *Friction coefficient*
- *Wear rate*
- *Material testing:* Material scientists rely heavily on testing results to guide the design of structures, assess the safety of existing components, and develop new materials. Testing provides valuable insights into the effectiveness of manufacturing processes, heat treatments, and composite material formulations. For instance, if a manufacturing

process or heat treatment is not yielding the desired material properties, improvements are necessary. Similarly, in composite materials, testing can reveal whether a particular ingredient, orientation, or mix is optimal for the intended performance. Without empirical testing, simulations alone may not provide reliable results, underscoring the importance of combining practical testing with simulation for technological advancements. For example, the tensile test provides insights into:
    - Yield strain
    - Ultimate tensile strength (UTS)
    - Elongation and ductility
- *Crack initiation and propagation:* In fracture testing, undetectable cracks can become visible under stress conditions, accumulating over time and leading to potential material failure. Understanding crack initiation and growth under operational conditions is crucial for assessing material reliability, especially when materials are subjected to stress levels near their limits. Often impact test can be designed for this purpose, which will be able to evaluate the material's resistance to cracking or breaking.
- *Fatigue testing*
- *Creep testing:* Creep tests are conducted to assess material deformation under constant stress over extended periods, particularly at elevated temperatures. Commonly applied in high-temperature applications like turbine blades or power plant components, creep testing reveals a material's long-term deformation characteristics.
- *Accelerated testing:* Accelerated testing methods simulate long-term conditions in a shorter time by applying higher stresses or loads, allowing for efficient lifespan estimation of materials. This approach is valuable for quick product development and quality assurance, as it enables us to replicate years of corrosion and wear in just days. By increasing load, adjusting chemical concentrations, or enhancing environmental conditions like temperature, we can observe material behavior under continuous stress without rest periods, offering insights that would otherwise require extended testing durations. For instance, rather than waiting 15 years to evaluate a light pole's durability in a marine environment, accelerated testing can compress this evaluation into a 30-day period, providing critical data for material selection and design improvements.
- *Fracture toughness:* Fracture toughness is a vital property for materials prone to crack propagation. This parameter measures a material's ability to withstand stress in the presence of flaws. High fracture toughness is crucial for applications with high-stress exposure or cracking risks, ensuring structural integrity in applications like pressure vessels.

- *Evaluating power losses and energy efficiency:* When evaluating power losses and energy efficiency, it is essential to consider real-world applications. Therefore, in addition to laboratory tests, performance tests on actual systems are important. Many of these evaluations can also be conducted during the initial stages of product development, before the launch of a complete or preliminary product. This often involves scaling down models, which may result in some loss of fidelity but allows for significant gains in understanding. Based on the results obtained from these tests, we can make necessary adjustments and modifications to improve the design and functionality of our systems.
- *Wind tunnel testing for aerodynamic behavior:* Wind tunnel testing evaluates aerodynamic forces, essential for designing buildings, bridges, and aircraft exposed to high wind forces. By simulating wind conditions, engineers can measure the impact of aerodynamic loads on structures, contributing to their stability and safety.
- *Corrosion testing:* Corrosion testing is essential for materials exposed to harsh environments, such as marine applications. By simulating corrosive conditions—using synthesized seawater, for example—engineers can evaluate material resilience. Controlled environments enable analysis over various immersion durations, providing insights into:
  - *Weight and visual changes:* Techniques such as weight measurement and visual examination reveal the material's corrosion susceptibility.
  - *Mechanical properties and endurance strength:* Corrosion can reduce a material's endurance strength, impacting its resistance to repeated cycles. Monitoring changes helps predict the material's longevity under corrosive conditions.

6.3.2 Realistic performance evaluations

Laboratory versus real-world testing: While laboratory tests, highlighted in previous subheading, provide valuable data, we must also consider how systems perform in actual operational conditions. This leads us to incorporate performance evaluations in realistic settings, often using methods such as:

- Vibration analysis
- Deflection assessments
- Strain analysis
- Thermal performance evaluations

The importance of real-world testing is vividly exemplified by the *Titanic* disaster, where compromised watertight compartments led to flooding and underscored the critical need for thorough assessment of structural integrity under real conditions. In developing new technologies or materials, rigorous experimental testing is indispensable. While simulations are valuable, they fundamentally rely on data derived from physical tests; exclusive reliance on laboratory methods can sometimes overlook complexities in actual system behavior. Therefore, extensive testing remains essential in the development of resilient engineering systems.

Material testing in corrosive environments, for instance, often yields results that differ markedly from controlled laboratory conditions. These outcomes may also exhibit time dependence, as material performance can degrade over prolonged periods in the field. To fully understand a material's resilience, it must be evaluated in its real operating environment over time.

A crucial aspect of real-world corrosion testing includes electrochemical measurements, which help assess corrosion rates and monitor how rapidly a material may deteriorate. To facilitate these assessments, advanced instruments like potentiostats, dynamic scanning systems, electrochemical impedance spectroscopy, and polarization resistance devices are integral to electrochemical testing. These tools, as recommended in the *ASM Handbook on Corrosion*, provide valuable insights into real-world corrosion studies.

Ultimately, rigorous real-world testing is essential to enhance the durability and reliability of materials, especially for applications that inherently expose materials to corrosive or other challenging environments.

6.3.3 Sustainability considerations

Maintaining and sustaining engineering systems effectively is critical, especially in environments with increasing temperatures, humidity, and pollution. These conditions often speed up corrosion and surface wear, making it essential to identify the root causes—whether in design, material choice, manufacturing, maintenance practices, or environmental factors. In addition, if faults can be identified at start, then failures can be completely avoided. For this purpose, we can examine wear debris—the particles created by wear—rather than total material failure, as catastrophic failure is highly undesirable. By studying wear debris, corrosion, and surface chipping, we gain insights that help prevent severe breakdowns by catching early signs of deterioration. Furthermore, components can be reconditioned, repurposed, and reused.

One fundamental testing method is visual inspection. However, this technique has limits; it may not detect small faults, particularly those under 50 μm in size, which are often early indicators of degradation. Advanced techniques are therefore needed to fully understand the extent and origins of surface damage.

Visual inspection falls under the broader category of NDT, which includes other methods such as radiographic, ultrasonic, and magnetic particle testing

(MT). NDT is crucial throughout the system's lifespan. While inspection begins in production, maintenance demands differ as conditions change, making regular NDT assessments necessary to avoid unexpected failures. For example, grain boundaries in materials act as natural interruptions and are not considered cracks. However, under high stress or corrosive exposure, these can lead to cracks and fractures.

Although NDT testing starts during production, it is equally important to evaluate how products perform under real-world conditions. NDT is vital for inspecting complex structures or assemblies without the necessity of disassembly. This method offers significant advantages, enabling comprehensive assessments while keeping systems intact. By facilitating proactive maintenance and preventing potential failures, NDT plays a crucial role in enhancing durability and cost-effectiveness.

A systematic approach is essential for sustainable testing, and the fishbone diagram (Figure 6.1) from *ASM Handbook Volume 11* provides a valuable tool for this purpose. It helps in identifying potential causes of issues, such as installation errors, operational overloads, and cost-effectiveness considerations. By way of thorough testing and analysis, the fishbone diagram aids in enhancing system safety, reliability, and performance, ultimately benefiting society.

The diagram emphasizes examining various factors, starting with design aspects. Design includes material selection and shape configurations, both of which can significantly impact system performance. It also involves assessing the skills and knowledge of personnel, as their expertise influences outcomes. Selecting appropriate maintenance strategies—whether regular, periodic, or sensor-based—is crucial, especially given that issues like loose nut and bolt connections can lead to frequent maintenance challenges if not addressed promptly.

Clear, documented procedures for operators are essential to ensure consistent performance and minimize failures. When this guidance is lacking, failures become more frequent. Even with a robust design, manufacturing quality plays a critical role. Poorly executed manufacturing or unquantified

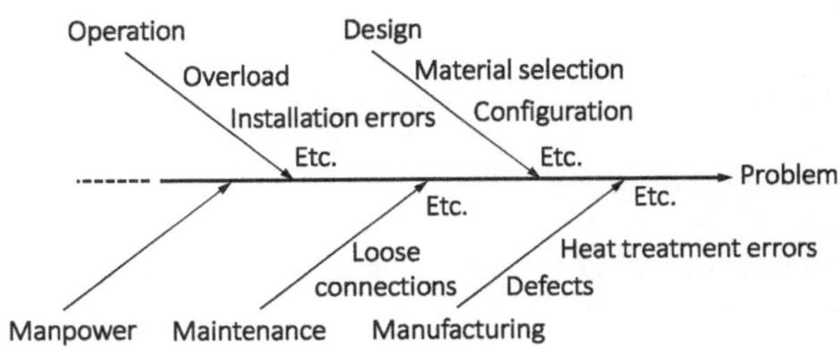

Figure 6.1 Identifying potential factors contributing to the problem.

manufacturing defects can lead to problems, especially in processes like heat treatment. Each process, whether established or novel, requires a comprehensive understanding. Without it, operators may inadvertently contribute to failure.

In the fishbone diagram, each factor branches out to identify specific causes, helping to visualize potential failure modes. This structured approach, combining design, manufacturing, and maintenance factors, enables a holistic understanding of problems and allows teams to address root causes effectively.

6.4 CATEGORIES OF TESTING METHODS

There are various ways to categorize the test methods. The easiest way is to classify as "DT and NDT." This type of classification is based on the objectives of the methods, the potential damage they cause to samples, the types of samples used, and the information they provide to testers. The primary goal of DT is to ascertain the mechanical properties of materials and define their limits. In contrast, NDT aims to identify irregularities and discontinuities within materials, enabling early detection of issues to prevent catastrophic failures and enhance maintenance strategies. DT renders specimens unusable for future testing, whereas NDT permits repeated assessments on the same samples. NDT allows for varying sensor placements and continuous monitoring of material conditions over extended periods, facilitating detailed analyses without damaging the samples or systems.

NDT is particularly advantageous as it does not require specific test samples, enabling in situ or on-site analysis. Conversely, DT typically necessitates a controlled environment and often requires representative models instead of actual components. This can lead to discrepancies, especially when dimensions influence strength; for instance, significant changes in specimen size can affect material properties, making DT results less reliable. Thus, there is a growing emphasis on NDT for its ability to provide insights without the limitations associated with destructive methods.

The limitations of DT are particularly evident in failure analysis, where it is essential to determine whether the properties of the fracture surface have remained stable or undergone significant changes post failure. This evaluation is critical for identifying root causes, such as the shift from ductile to brittle fracture. Subsequent assessments after a failure often focus on the fractured material rather than the original sample. Moreover, DT has constraints regarding specimen size and the comprehensiveness of failure evaluations. For instance, when a large gear fails, it may necessitate removing a section of the gear teeth or analyzing the fractured surface to conduct further investigations. Consequently, DT primarily serves as a vital tool for quality control inspections and maintenance, highlighting the need for increased attention in surface engineering.

Researchers are increasingly working to integrate DT and NDT methodologies to enhance their applications. Although DT is essential for establishing various material limits, NDT focuses on assessing material safety and monitoring the progression of potential failures. DT identifies the upper limits of material performance, recognizing that these limits can fluctuate depending on differing conditions and environments. In contrast, NDT emphasizes the detection of defects and irregularities without compromising the integrity of the sample, which is vital for maintenance and for evaluating the residual life of materials based on specific DT data.

In addition, we have examined the significance of metallographic investigation in supplying critical data for predicting the residual life of a system. This information includes various parameters such as the depth of chemical penetration, the extent of corrosion attack, pit diameter, and the presence of intergranular corrosion, which may indicate severe damage due to high stress levels. Such data is crucial and is obtained through metallographic testing.

This integration underscores the necessity for comprehensive component-level and assembly-level testing. By using both DT and NDT, we can achieve a more holistic understanding of material performance and structural integrity. Component-level testing allows us to evaluate individual parts in isolation, ensuring that each meets required safety and performance standards. Meanwhile, assembly-level testing assesses the interactions between components, revealing how they function together under operational conditions. This dual approach not only enhances our ability to detect potential failures early but also facilitates more effective maintenance strategies, ultimately ensuring the reliability and longevity of engineering systems.

6.4.1 Component-level and assembly-level testing

When developing a system, engineers must ensure that each component is evaluated for its specific performance and durability. Similar to selecting players for a sports team, every component within an assembly must meet certain performance criteria for the entire system to function effectively. For example, each component of a machine, such as bearings, shafts, and gears, has its own set of properties that must be optimized. Individual components need to be rigorously tested to confirm they meet specified performance standards, and then they are tested again at the assembly level to ensure compatibility and durability.

6.4.2 DT versus NDT

Testing methods are broadly classified into two categories: DT and NDT, each serving distinct purposes. DTs focus on identifying performance limits by applying stress or load to a specimen until it fails or permanently deforms. This approach is particularly useful for gaining insights into the material's mechanical properties and understanding its behavior under extreme conditions.

NDT, in contrast, evaluates the integrity and performance of materials or assemblies without causing damage, making it ideal for in situ testing in real-life applications. NDT techniques allow for thorough inspection, assessment, and analysis of actual components, ensuring quality control without compromising the usability of the parts involved.

Both testing approaches play a critical role in confirming that materials meet application requirements. DT provides valuable information on material performance limits and structural boundaries, whereas NDT offers a practical and efficient way to predict potential failures and verify component quality in operational environments. Together, these methods support a comprehensive understanding of material characteristics, aiding in selection and application readiness.

6.4.2.1 *DT: Controlled failure for material assessment*

DT subjects a test specimen to conditions that ultimately lead to fracture or permanent deformation, helping identify the structural and mechanical limits of materials. In DT, materials do not retain their original shape as they are stressed to failure, allowing engineers to gather critical data on how materials behave under extreme conditions.

Purpose of DT: The goal of DT is to identify maximum performance thresholds, such as yield strength or fracture toughness. By incrementally applying stress, engineers observe the point at which a material deforms or fractures, yielding data on energy absorption and resilience. This information is essential for ensuring that components operate safely under extreme conditions.

Evaluating material properties through DT: DT involves pushing materials or components to their failure limits. By testing a specimen to its breaking point, engineers determine the ultimate load, temperature, or speed a material can withstand before failure occurs. Common DT methods include tensile testing, impact testing, and fracture toughness testing. In tensile testing, for example, a universal testing machine (UTM) applies tensile stress until the specimen breaks. This method reveals essential properties, such as yield strength, UTS, and potential failure modes—shear, tensile, brittle, or ductile—offering valuable insights into a material's behavior and limits.

Key structural qualities and mechanical limits: DT gives a detailed investigation of structural properties, which helps us understand how materials respond under stress. Tensile and compressive tests are crucial components of this examination because they offer important information about material performance. Recent research reveals that compressive strength frequently exceeds tensile strength, necessitating separate tensile and compressive testing to accurately characterize material behavior. Bending tests are also helpful for assessing flexibility and fracture resistance. These are frequently used in conjunction with impact testing procedures such as the Charpy test, which evaluates how a material reacts to sudden hits. In addition to other mechanical tests, "Peel-off tests" are important for evaluating coating performance. These assessments work together to provide a full understanding of material behavior under varied settings, allowing for more informed material selection and design decisions.

- *Charpy impact test: Measuring impact resistance:* A standard method in DT, the Charpy impact test assesses a material's resistance to impact and fracture. By striking a notched sample with a hammer, the Charpy test measures the energy absorbed before the material breaks. Low energy absorption indicates a higher likelihood of failure under impact, helping engineers evaluate the toughness of materials used in structural and load-bearing applications.

6.4.2.2 NDT: Ensuring integrity without compromising material

NDT has a major advantage over DT because it enables the examination of materials and assemblies without causing any damage. This is especially important for complete assemblies, as it allows for an assessment of structural integrity while keeping functionality intact. To carry out effective NDT, operators need to be familiar with established regulations and standards. This is why certification at different levels, such as Level 2 or Level 3, is often required for NDT personnel. Following these regulations is crucial to ensure the integrity of the testing process.

Purpose of NDT: The primary objective of NDT is to identify potential issues, such as cracks or surface defects, in critical components without causing any damage. This method is essential in industries where components are expensive or difficult to replace, including aerospace, pipelines, and heavy machinery. In particular, the aerospace sector and pipeline management rely heavily on NDT. Pipelines, which extend over long distances, pose significant risks if failures occur, potentially resulting in hazardous leaks of gases, water, or oil. Pressure vessels also demand careful inspection due to their vital role in safety. NDT focuses on detecting surface as well as subsurface-related failures. For example, defects within a pipeline may not be visible from the outside. As a result, it is critical to monitor the thickness of components and proactively identify any corrosion-related failures. Early detection of issues, especially those near or on the internal surface, is essential. In addition, NDT facilitates routine maintenance by offering valuable insights into the current condition of components and their anticipated lifespan.

Role of NDT in detecting system flaws and ensuring reliability: NDT methods are effective at detecting systemic flaws, abnormalities, and fractures, which are necessary for structural integrity and performance maintenance. Reducing loads, changing temperatures, or applying protective coatings are all examples of preventative actions that can be taken when these issues are identified early on, increasing the reliability of the system. We will explore surface engineering approaches further to promote the development of reliable solutions and decrease the number of flaws that have been detected. NDT is key when evaluating materials and components for quality, defect detection, and servicing frequency calculations.

Applications of NDT across industries: NDT is essential for quality assurance, maintenance, and safety checks in various sectors, as it facilitates the examination of materials and structures in their operational state. For example, in power plants, NDT can be applied directly to active components such as turbine blades, ensuring their safety and performance without causing downtime.

In industries that involve pipelines carrying pressurized fluids or operating at high velocities, NDT methods are essential for inspecting both protective coatings and structural integrity. The integrity of these coatings is particularly critical for preventing corrosion and ensuring long-term functionality. The NDT techniques can be deployed to assess the quality of welds, coatings, and material homogeneity in these applications.

NDT is a relatively recent field of research characterized by ongoing efforts to enhance traditional methods and develop innovative techniques. Conventional approaches, such as visual inspection, can be conducted with the naked eye or augmented with optical instruments like microscopes. In addition to visual inspection, widely utilized NDT methods include ultrasonic testing (UT), radiographic testing, and MT, all of which will be covered in the upcoming slides.

Other essential techniques in the NDT process include eddy current testing (ECT), thermographic testing with thermographic cameras, liquid penetration testing (LPT), and caustic emission testing. I strongly believe that the future of NDT will be shaped by the emergence of innovative start-ups that leverage advanced sensors and sophisticated algorithms. These advancements will facilitate the creation of digital twins—virtual representations of physical systems that accurately model real-world environments.

Such technologies are invaluable for correlating actual conditions with the onset of damage or surface degradation, ultimately providing insights into real performance metrics. I contend that many design processes often concentrate on static conditions, overlooking the dynamic and time-dependent factors that significantly influence performance. Implementing digital twins requires a comprehensive evaluation of both static and dynamic elements. This ongoing comparison between initial design parameters and real-time data will empower us to make informed adjustments to our software designs as needed.

A notable advancement in NDT is the use of digital image correlation (DIC) for stress–strain analysis. This technique captures images of the specimen and correlates them, effectively replacing traditional strain-based sensors. By doing so, it enables the assessment of strength, stress, and elongation without subjecting the specimen to extreme conditions, thereby preserving its integrity during testing.

NDT is specifically designed to maintain the functionality of materials and systems while evaluating their quality and integrity. By identifying weaknesses in subsystems or components and detecting potential issues, NDT facilitates proactive measures to repair, rectify, or mitigate these weaknesses, ultimately enhancing overall product performance. Given these advantages, NDT techniques are fundamental for inspection, monitoring, and quality control throughout the manufacturing process and the entire lifecycle of a product. The significance of NDT cannot be overstated; it should be prioritized whenever possible.

Economics of NDT: The economic aspect of NDT is also critical. It is essential to assess whether the proposed measures will be economically beneficial, ensuring that all activities comply with regulatory requirements. Operators must diligently follow these guidelines to maintain compliance and operational safety.

Furthermore, NDT provides valuable insights into the intricate details of the systems and technologies involved, fostering continuous improvement and innovation. Recent advancements in NDT techniques have led to increasingly innovative methodologies that enhance our understanding and capability in creating superior products.

Three key benefits of NDT are particularly noteworthy: it reduces costs, enhances safety, and ensures compliance with applicable regulations. Most NDT equipment is designed to meet these regulatory standards while preventing damage or deformation of the materials being tested. Consequently, NDT is not only cost-effective and safe but also compliant with all relevant rules and regulations.

Complementary nature of DT and NDT: NDT and DT work together effectively in evaluating materials and systems. NDT is essential for maintaining and ensuring the long-term reliability of active systems, as it identifies potential issues without inflicting damage. In contrast, DT provides crucial information about material limits through controlled failure testing. The data obtained from DT is invaluable for foundational design decisions, including material selection, coating options, and system configurations. When used in conjunction, NDT and DT create a comprehensive approach to assessing integrity across a wide range of applications, balancing the need for insights into both durability and performance. Although NDT is gaining recognition for its advantages, DT remains important for providing critical data points necessary to understand material properties. Thus, both methods are vital for production, quality control, and safety assurance.

It is important to highlight that neither method can be considered superior in all situations; instead, they enhance each other's effectiveness. Current research focuses on advancing NDT techniques to assess properties typically measured through DT, which would represent a noteworthy advancement in the field. As researchers strive to adapt DT methodologies into NDT practices, there is significant potential for improved testing capabilities and more extensive statistical analyses. This transition would facilitate broader data collection beyond the limited sample sizes usually associated with DT. Although DT remains essential for specific applications, such as corrosion testing, ongoing simulation efforts are working to achieve similar assessments through NDT.

- *Surface hardness testing:* Surface hardness testing evaluates material hardness by measuring the depth or width of an indentation or scratch left on the surface after applying a controlled force. In microhardness testing, a fine indentation is created, and the dimensions of the resulting mark reveal important surface hardness properties, which are crucial for assessing wear resistance, as harder surfaces typically exhibit greater durability. This assessment is generally conducted on a representative sample rather than the entire structure, thereby preserving system integrity while determining material properties.

The classification of surface hardness testing is a topic of ongoing discussion among researchers. Some view it as a type of DT, while others see it as NDT. As measuring surface hardness typically does not result in the failure of the sample, many consider it an NDT method. However, it is also important to note that even when testing a representative sample that does not fail, it can still fall under the category of DT.

On the other hand, some researchers argue that NDT is meant to evaluate the quality of materials or systems without causing any failure. This method looks at both the material properties and the overall integrity of the system, identifying any flaws, defects, discontinuities, or anomalies present.

In conclusion, while surface hardness testing shares features of both DT and NDT, its effect on the sample during testing leans more toward being classified as DT. This highlights the need to understand the context in which the testing is performed to determine its appropriate classification.

Figure 6.2 summarizes the testing methods. DT, shown on the left in Figure 6.2, involves evaluating material properties and failure characteristics by causing permanent damage to the specimen. Examples include corrosion testing, fracture and mechanical testing (e.g., tensile test, Charpy impact test, peel test, crush test, fracture test), and fatigue testing. NDT, displayed on the right, is used to identify defects without harming the specimen. Techniques include visual inspection, UT, radiography, MT, ECT, thermographic testing, LPT, acoustic emission (AE) testing, and stress–strain DIC testing. Both methods are essential for quality control, safety assurance, regulatory compliance, and ongoing improvements in production and maintenance processes.

Figure 6.3 presents a comparison of key aspects between DT and NDT methods across several criteria. DT is primarily utilized to assess the mechanical properties, behavior, and performance limits of materials, typically resulting in permanent damage to scaled-down samples or models within controlled environments (ex situ analysis). In contrast, NDT is aimed at detecting defects and discontinuities in actual components or structures without inflicting harm (in situ analysis).

The chart outlines differences in purpose, potential damage, sample types, information obtained, and applications. It emphasizes DT's significance in research and failure analysis, whereas NDT is essential for quality control, inspection, and maintenance. Furthermore, compared to DT, NDT processes are more sustainable. As discussed previously, the importance of sustainable development in manufacturing cannot be overstated, particularly in terms of minimizing waste and maximizing resource efficiency. NDT aids in achieving these objectives by utilizing fewer resources and generating less waste, thus enhancing overall production efficiency.

6.5 CLASSIFICATION OF NDT METHODS

NDT includes a wide array of techniques for inspecting, evaluating, and analyzing materials, components, and structures without causing any damage. These methods are crucial for detecting defects, irregularities, or discontinuities while preserving the integrity of the materials being tested. NDT is widely used across numerous industries for quality assurance, maintenance, and safety inspections.

In the remainder of this chapter, we will provide an overview and continue our discussion on NDT. NDT can be a foundational element in a preventive maintenance strategy by offering valuable data that enables the early detection of potential faults. This proactive approach supports the development of preventive maintenance policies based on early indicators, reducing the need to wait for a complete system failure before taking action.

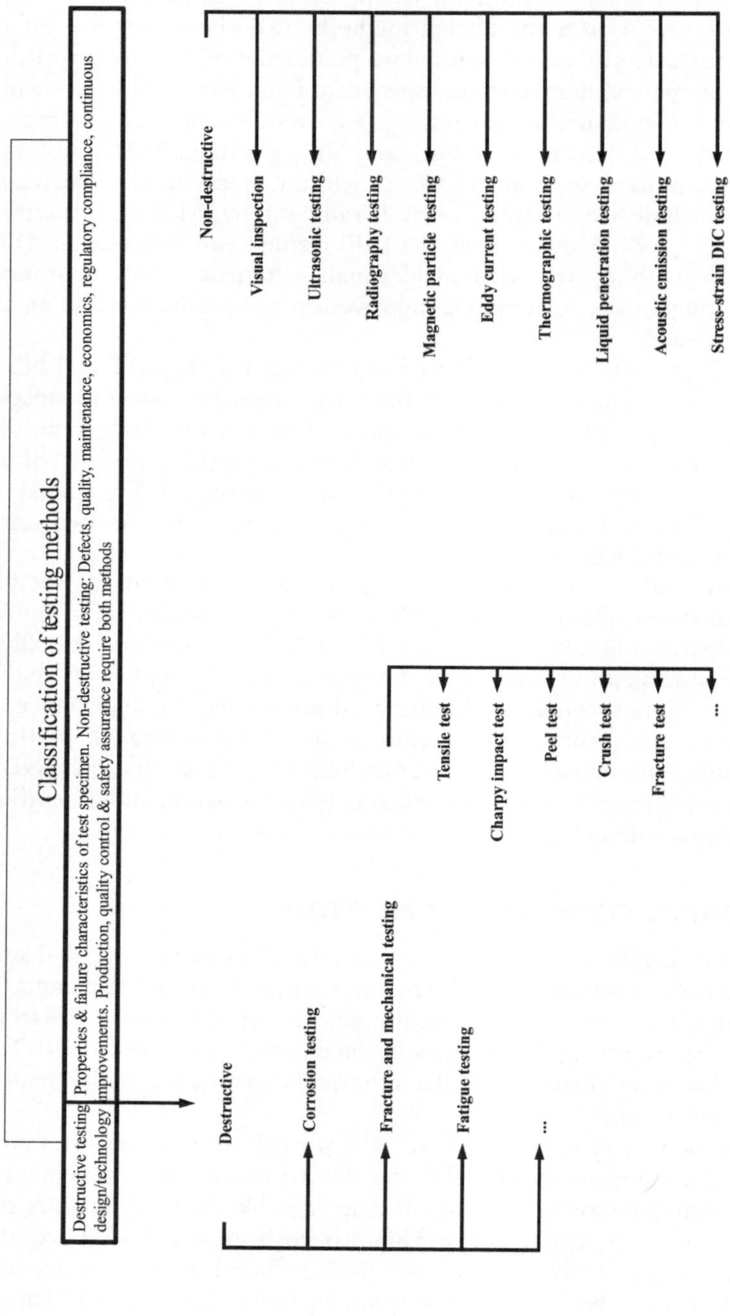

Figure 6.2 Overview of destructive and non-destructive testing methods.

Destructive and non-destructive testing techniques		
Destructive technique		**Non-destructive technique**
To determine the mechanical properties, behavior, and performance limits of materials	Purpose	It focuses on identifying defects, irregularities, or discontinuities that may affect the integrity or performance of the tested object
Cause permanent damage	Damage	Cause no or minimal damage
Scale down representative models of actual machine element (ex-situ analysis)	Test samples	Actual components/material/ structure (in situ analysis)
Detailed information about the material's mechanical properties, behavior under specific conditions, failure mechanisms, and performance limits	Information obtained	Information about the presence, location, size, and severity of defects or irregularities within materials or structures.
Research, material characterization, and failure analysis	Application	Quality control, inspection, and maintenance purposes

Figure 6.3 Comparison of destructive and non-destructive testing techniques.

Through the early detection of potential failures, NDT allows us to mitigate these issues or take corrective actions, thereby preventing catastrophic failures. In addition, it contributes to maximizing and potentially extending the lifespan of components, systems, and materials. In this book, we have covered nine primary NDT techniques; however, many others are continuously being developed as research progresses. New methods and principles emerge regularly, offering potential advancements and alternative options to DT techniques.

This evolution of NDT techniques offers substantial societal benefits. As NDT comprises a variety of methods, it is often beneficial to apply multiple techniques in combination to achieve accurate and reliable results. NDT enables the assessment of material, component, and system quality and integrity without affecting their functionality or structural soundness. This long-standing approach, including methods like visual inspection, has been effective for many years and remains essential across various industries.

By allowing for detailed inspection and evaluation without damaging or altering the item, NDT is particularly well-suited for quality control, maintenance, and safety protocols in sectors such as aerospace, automotive, construction, and manufacturing, where regular, non-invasive testing is crucial.

Key NDT methods (arranged by increasing complexity) are:

- *Visual inspection:* Direct visual examination is used to detect surface irregularities, cracks, or other visible defects. This is often the first step in NDT to identify obvious issues.

- *LPT:* A liquid dye or fluorescent solution is applied to the material's surface, seeping into surface-breaking defects to make them visible. This method is commonly used for detecting cracks on the surface.
- *MT:* Magnetic fields and particles are applied to detect surface and near-surface defects in ferromagnetic materials. This technique is particularly useful for inspecting ferrous metals for cracks or weaknesses near the surface.
- *ECT:* Electromagnetic induction is used to identify surface or near-surface defects and to measure conductivity or coating thickness. This technique is effective on conductive materials and can reveal small surface imperfections.
- *UT:* High-frequency sound waves are used to detect internal flaws, measure wall thickness, and assess material properties. This method is valuable for locating hidden defects deep within the material.
- *Radiographic testing:* X-rays or gamma rays are used to inspect internal structures, identify defects, and evaluate material integrity. This method provides a detailed internal view of the object being tested.
- *Thermographic testing:* Infrared (IR) radiation is used to detect heat variations, which can indicate defects such as cracks, voids, or delaminations. This is useful for identifying hidden defects based on temperature differences.
- *AE testing:* This technique detects and analyzes transient elastic waves generated by rapid energy releases within a material, such as crack growth or fluid leakage. It is useful for monitoring active damage mechanisms in real-time.
- *Stress and strain analysis:* Mechanical behavior under load is evaluated by measuring material deformation and response, providing insights into performance using appropriate software and hardware combination. This analysis is valuable for understanding material durability and resilience.

6.5.1 Visual inspection

Let us begin our exploration of the first technique: visual testing (VT), which is a widely used method. VT is one of the most widely used techniques and is familiar to many individuals. This method allows inspectors to assess defects in materials primarily through visual inspection, relying on sight or tactile feedback, such as using a finger to evaluate surface roughness or identify surface features. While VT can effectively highlight apparent flaws—such as irregularities or excessive surface roughness—it may not detect deeper imperfections that are not immediately visible. The effectiveness of VT largely depends on the skills, training, and experience of the inspector, particularly when direct access to the surface is possible.

To enhance VT, it is crucial to optimize lighting conditions and ensure clear visibility. Magnification aids, such as devices offering 3×, 5×, or even 1000× magnification, are often necessary to enhance clarity. An incident light beam angle of approximately 5 to 45 degrees has been established as ideal for obtaining accurate results. Inspectors must possess clear vision; any blurriness can compromise the reliability of the findings. Thus, excellent illumination, unobstructed vision, and appropriate magnification are essential for achieving acceptable results.

Although surface irregularities can be identified without specialized techniques, deeper defects that lie beneath the surface may evade detection. This limitation necessitates the use of more advanced methods, such as LPT and MT. Moreover, visual inspection is frequently supplemented by other techniques to improve accuracy and reliability. For instance, tools such as borescopes, fiberscopes, and video scopes allow inspectors to examine surfaces that are otherwise inaccessible. These devices enable the assessment of conditions in hard-to-reach areas, providing valuable insights into the material's integrity. Closed-circuit television (CCTV) systems have also become increasingly common for remote monitoring in large industrial settings and buildings where direct access is limited. Such remote VT techniques significantly enhance the effectiveness of visual inspections by facilitating the examination of challenging environments.

The visual inspection technique has been established for some time; however, it continually evolves with the incorporation of advanced features that enhance its utility. A notable advancement is the remote VT technique, which is used to examine areas difficult for human inspectors to access. For example, direct inspection of the interior of a long pipe is impractical. However, by utilizing a videoscope or a camera positioned within the pipe, comprehensive images of that area can be transmitted, providing critical localized information.

Videoscopes are versatile tools that can be used for visual inspections in various settings, including tubes, pipelines, and welds. They yield effective results, even in challenging scenarios such as forgings, where inspections are conducted within confined surfaces. The American Society of Mechanical Engineers (ASME) has established codes and classifications for VT. For instance, VT1 pertains to standard VT, while VT2 specifically involves inspecting the pressurized boundaries of fluid-containing systems to detect leaks. This classification is especially beneficial in in-service situations where visual inspection is essential, and functional testing is conducted under pressurized conditions to pinpoint the sources of leaks, further underscoring the importance of these inspection methodologies in maintaining system integrity.

While VT offers numerous advantages, it is essential to acknowledge a significant drawback: the reliance on the expertise of the individual conducting the examination. This dependency on human skill is a limitation, and while automation of the process would be beneficial, achieving complete automation remains challenging due to various factors.

In conclusion, effective utilization of visual inspection necessitates a certain level of expertise, skills, and substantial experience. Without this foundational knowledge, the efficacy of VT is significantly diminished. Regardless of the technique used, a lack of comprehensive knowledge will hinder the inspection process. Therefore, knowledge, scientific understanding, and analytical skills are paramount, underscoring the importance of this course from an educational perspective.

6.5.2 Liquid penetration testing

LPT, commonly referred to as dye penetrant testing, is a widely used NDT method designed to assess the integrity of non-porous materials. This technique provides a simple and cost-effective way to detect surface defects across various materials. LPT plays a vital role in ensuring quality and safety in manufacturing and maintenance processes by identifying surface-breaking defects that could compromise the structural integrity of components, making it an essential tool for quality assurance, safety evaluation, and preventative maintenance.

- *Purpose and importance:* LPT aims to detect surface flaws, such as cracks, porosity, or inclusions, which could lead to catastrophic failures if left unaddressed. By identifying potential issues early, organizations can implement effective maintenance strategies, reduce the risk of failure, and maximize the lifecycle of components.
- *Methodology:* Implementation of LPT process requires the following key steps:
    1. *Surface preparation:* The surface of the component being tested must be clean and free from contaminants such as grease, dirt, and paint. This step is crucial to ensure that the penetrant can effectively infiltrate any defects.
    2. *Application of liquid penetrant:* A liquid penetrant, which can be either fluorescent or visible dye, is applied to the surface of the test specimen. The penetrant has low viscosity, allowing it to seep into small cracks and crevices through capillary action.
    3. *Dwell time:* After application, the penetrant is allowed to dwell on the surface for a specified period, typically ranging from a few minutes to several hours, depending on the material and the size of potential defects. This soaking period is essential for the penetrant to adequately fill any surface-breaking defects.
    4. *Excess penetrant removal:* After the dwell time, any excess penetrant on the surface is removed. This is typically done with a

solvent or a specific removal agent, which ensures that only the penetrant within the defects remains.

5. *Application of developer:* A developer is then applied to the surface. The developer, usually in the form of a white powder or spray, draws out the penetrant from the defects to the surface, creating a visible indication of any flaws.
6. *Inspection and evaluation:* The technician inspects the surface under appropriate lighting conditions (often ultraviolet (UV) light for fluorescent penetrants) to identify any indications of defects. The presence of colored marks on the surface highlights the locations of cracks or other defects.

The effectiveness of LPT is significantly influenced by the characteristics of the penetrant used, including its viscosity, surface tension, and fluorescence; lower viscosity enhances infiltration into small defects, while lower surface tension promotes capillary action, allowing the penetrant to fill cracks more effectively. In addition, fluorescent penetrants become visible under UV light, facilitating the detection of flaws. LPT is primarily applied to non-porous materials such as metals, ceramics, and certain plastics, as these materials do not permit the penetrant to escape once it has infiltrated a defect. Although recent studies have explored its use on porous materials, LPT remains predominantly suited for non-porous substrates.

Limitations:

1. *Qualitative insights:* LPT primarily provides qualitative information about defects; it does not quantify the size or depth of a defect.
2. *Operator skill:* The success of LPT heavily relies on the expertise of the operator. Experienced technicians can interpret indications effectively and differentiate between true defects and benign surface conditions.
3. *Surface-breaking defects:* LPT is specifically designed to detect surface-breaking defects, which are flaws that extend to the surface of the material being inspected. This limitation arises from the fundamental principles of how LPT operates: The method relies on the ability of the penetrant to seep into cracks, voids, or other discontinuities that break the surface of the material. When a defect is present, the penetrant infiltrates the opening through capillary action and can be drawn out by a developer to highlight its presence.

6.5.3 Magnetic particle testing

MT is an NDT technique specifically designed to detect surface and near-surface defects in ferromagnetic materials like iron, nickel, and cobalt. MT works by applying a magnetic field to the material under inspection and spreading ferromagnetic particles, such as iron or iron oxide, across its surface. When defects such as cracks or discontinuities are present, they disrupt the alignment of magnetic flux lines between the magnetic poles, creating localized disturbances in the magnetic field. These disturbances cause the ferromagnetic particles to cluster in defect zones, providing a visual indication of the flaw.

In practical applications, the magnetic field in MT can be generated using either a permanent magnet or an electromagnet. Electromagnets are often preferred because they allow for adjustable magnetic field intensity by varying the current, giving inspectors more control and flexibility over the testing process. When a defect, like a crack, breaks through the material, it causes a disturbance in the magnetic flux lines, creating new north and south poles around the crack edges. This phenomenon, known as magnetic flux leakage, results in the concentration of magnetic particles around the defect, making it visible to the inspector.

One of MT's unique advantages is its ability to detect both surface and near-surface defects, which is achieved by observing how the magnetic field changes at varying depths. Although MT is highly effective in identifying defect locations, it is primarily a qualitative method and does not precisely quantify the depth or size of the defect. However, the clustering of particles offers insight into the defect's severity and allows for assessment of potential structural concerns.

The choice of magnetic particle type also influences test reliability. Pure iron particles often yield more accurate results than iron oxide (Fe_2O_3) particles, as they have better magnetic retention properties and create a stronger field. For MT to be effective, the tested material must be ferromagnetic, as non-ferromagnetic materials lack the necessary magnetic permeability for particle clustering.

By detecting disruptions in magnetic flux lines, MT effectively highlights cracks, inclusions, and other discontinuities that compromise material integrity. Although MT does not measure defect dimensions with precision, it provides a comprehensive method to ensure the quality of ferromagnetic materials in various industrial applications, complementing other NDT methods for an overall assessment of component integrity.

Li et al.[1] used MT to detect the bearing flaws. They outlined six key steps to achieve accurate detection of surface irregularities, which are:

1. *Preprocessing:* Before testing, the bearing's surface is cleaned and prepared to remove dirt, grease, or other contaminants. This preprocessing step ensures a clear and even distribution of magnetic particles across the surface, enhancing test accuracy.
2. *Magnetization:* Apply a magnetic field and observe disruption or discontinuity of the magnetic field, causing a leakage magnetic field to

form at the defect site. This leakage is essential for defect detection, as it draws magnetic particles to the flaw.

3. *Applying magnetic powder:* Ferromagnetic particles are applied to the surface of the bearing. These particles, which can be in solid form (dry method) or suspended in a liquid (wet method), are attracted to areas where magnetic flux leakage occurs due to surface irregularities. As the particles concentrate around defect sites, visible magnetic traces are formed.
4. *Magnetic trace assessment:* Inspection personnel visually examine magnetic traces to evaluate defect characteristics such as location, size, and shape. This qualitative assessment helps identify critical flaws and determine their potential impact on bearing performance. During this process, inspectors rotate the bearing and inspect it under controlled lighting conditions, especially when using fluorescent magnetic powder. This powder contains dyes that emit a high-contrast yellow-green fluorescence in the 510 to 550 nm spectrum under UV light, making defect traces more visible against the bearing's surface. Conducting inspections in a darkroom further enhances visibility, allowing for quicker and more accurate detection of defects. However, prolonged exposure to UV light and fluorescent glare can lead to eye strain and fatigue, potentially impacting inspection accuracy over time.
5. *Demagnetization:* Once testing is complete, demagnetization is essential to remove any residual magnetism, which could interfere with the bearing's function. An alternating magnetic field is applied, reducing residual magnetism through the magnetic hysteresis loop's diminishing effect, ensuring no leftover magnetic field remains.
6. *Post-processing:* After demagnetization, the bearing's surface is cleaned, and preventive measures like rust protection are applied. This final step ensures that residual particles or chemicals from the testing process do not affect the bearing's future use.

According to Li et al.,[1] MT is crucial for maintaining bearing integrity in critical fields such as automotive, aerospace, and industrial machinery. MT enables the early detection of surface defects that could otherwise lead to premature failure, helping prevent costly breakdowns and enhancing operational safety. The process provides quick, efficient visual feedback and can incorporate semi-automated steps, reducing manual labor while upholding high standards of quality and safety for ferromagnetic bearing components. MT's widespread use stems from its ability to rapidly pinpoint defect locations through visual signals. The following examples further illustrate the MT process and its applications, particularly in industrial settings.

1. *Detecting surface cracks in machinery components:* A steel gear used in heavy machinery is tested using MT to check for cracks that could compromise its integrity under stress. The gear is magnetized using an electromagnet to create a controlled magnetic field around it. Ferromagnetic particles are then applied, and as they settle, clusters of particles reveal a hairline crack on the surface. This crack might not be visible to the naked eye but is clearly indicated by particle buildup.
2. *Pipeline inspection for oil and gas industries:* In oil and gas pipelines, near-surface defects can lead to leaks, environmental hazards, and costly repairs. MT is used on ferromagnetic pipeline sections to find corrosion pits and cracks without causing any damage to the structure itself. Inspectors use an electromagnet for adjustable field strength, allowing flexibility to detect flaws at different depths. When a defect is found, the magnetic flux leakage causes particles to cluster, helping maintenance teams prioritize repairs in critical areas and ensure pipeline integrity.
3. *Automotive industry—quality control in engine blocks:* In automotive manufacturing, engine blocks made of ferromagnetic materials undergo MT as part of the quality control process. The MT process can reveal any inclusions or discontinuities within the casting that might weaken the material. When a magnetic field is applied, any voids or cracks will disturb the magnetic flux, causing particles to form clusters around these areas. This visual indication allows for a rapid assessment, ensuring only defect-free components proceed in the assembly line, contributing to product safety and durability.
4. *Aircraft maintenance and safety assurance:* Aircraft landing gear, which experiences significant stress during takeoff and landing, is periodically tested using MT. As surface and near-surface cracks could compromise the landing gear's safety, inspectors use MT to detect early signs of fatigue and stress cracks. The ability of MT to reveal defects in hard-to-reach areas without dismantling components provides a non-invasive solution that reduces inspection downtime, critical in high-demand sectors like aviation.
5. *Assessing welded joints in structural engineering:* In construction projects, MT is commonly used to inspect the welds of structural steel members. Welds are susceptible to defects, like slag inclusions or cracks, which could weaken the joint and affect the structure's stability. By applying a magnetic field and observing particle clustering around imperfections, inspectors can identify weak points in the weld that need repair. This capability is particularly valuable in high-stress applications, such as bridges and high-rise buildings, where weld integrity is essential.

6. *Railroad inspection for track safety:* Railway tracks endure constant stress from heavy loads, and MT is often used to detect near surface and surface defects in the track material. During inspection, MT can reveal stress fractures or fatigue cracks that occur in high-stress areas, such as at rail joints and curves. By identifying these flaws early, MT helps prevent track failure and enhances safety. The inspection process is streamlined using portable electromagnets, allowing field testing without removing the rail section.

These examples showcase how MT helps maintain safety, quality, and reliability in industries reliant on ferromagnetic materials by identifying potential failure points through visual indications of particle clustering around defects.

6.5.4 Eddy current testing

As we advance in NDT techniques, we focus on innovative methods like ECT. Recognized for its effectiveness in detecting defects in conductive materials, ECT uses widely accessible eddy current sensors. Although ECT offers significant advantages for assessing conductive materials, it is essential to consider potential interferences from thin oxide coatings, oil, or other contaminants, which may necessitate calibration adjustments. With a thorough understanding of the system, ECT can substantially enhance the longevity of the surfaces being evaluated.

ECT is a crucial NDT technique utilized across various industries to evaluate the integrity of conductive materials and identify defects. This report aims to clarify the historical context, development, and fundamental principles of ECT, particularly its reliance on electromagnetic induction, which enables it to detect defects with high sensitivity to discontinuities. Such non-destructive methods are vital for assessing material integrity and revealing subsurface flaws that traditional inspection techniques might overlook.

Eddy currents are oscillating electrical currents induced in conductive materials when subjected to alternating magnetic fields, a phenomenon described by Faraday's law of electromagnetic induction. When an alternating current (AC) flows through a coil or probe, it generates a magnetic field around it. Bringing this coil close to a conductive metallic specimen induces eddy currents within the material, which flow in circular or elliptical paths influenced by the material's properties and the characteristics of the alternating magnetic field. Irregularities in conductivity or geometry disrupt the eddy current flow, resulting in measurable changes in the magnetic field. By analyzing these changes, it becomes possible to detect and evaluate defects or anomalies within the material.

The working principle of ECT involves generating a magnetic field around a coil integrated into a probe. Although AC is most commonly used, direct current (DC) and pulsed current methods are also available, with AC being preferred for its adaptability to real-time testing and analysis. When the magnetic field interacts with the conductive material, it induces eddy currents, with their characteristics—such as intensity and flow path—depending on various factors, including the material's properties and specific testing parameters. Understanding these parameters is essential for accurate simulations and result interpretation.

Early detection of cracks is critical, as timely intervention is necessary to bridge and fill defects, underscoring the importance of ECT in preventing structural failures. This method is also suitable for painted or coated surfaces, enabling the identification of cracks that may lead to crevice formation and increased corrosion risk. The coil induces eddy currents along a continuous path within the examined material; any disruptions, such as those caused by cracks or defects, result in measurable changes in the eddy current flow. Defects can alter both conductivity and geometry, making them detectable. For instance, corrosion can reduce conductivity and produce observable variations in the eddy current pattern. A calibrated sensor can identify subtle changes, including multilayer formations or material loss, indicating advanced degradation. Essentially, any discontinuity—whether from cracks, corrosion, or geometric changes—disrupts eddy current flow, revealing underlying material issues.

Key parameters influencing ECT:

1. *Frequency of the applied current:* ECT typically uses AC ranging from 1 kHz to several megahertz. Higher frequencies are suited for shallow inspections, whereas lower frequencies penetrate deeper. The skin effect causes eddy current density to decrease with depth, making it vital to select an appropriate frequency based on material thickness and defect location for accurate results.
2. *Lift-off distance:* The lift-off distance, or the space between the coil and the material's surface, affects eddy current strength. Closer proximity enhances intensity, while greater distances weaken the signal, making calibration crucial to avoid inaccuracies.
3. *Temperature effects:* Temperature changes can alter the conductivity and magnetic permeability of the material, affecting readings if the temperature differs from calibration conditions. Compensation for temperature fluctuations is necessary for reliable results.
4. *Skin effect and penetration depth:* Eddy currents are concentrated near the surface and diminish exponentially with depth, known as the skin effect. For materials like aluminum, penetration depth

can reach up to 10 mm, although deeper inspections may require adjustments in current frequency and intensity.

ECT process:

1. *Inducing a current into a coil:* An AC through a coil generates a magnetic field that conforms to the coil's shape.
2. *Eddy currents generation:* Positioning the coil over a conductive part induces opposing eddy currents within the material due to the coil's alternating magnetic field. These currents follow paths that align with the magnetic field lines.
3. *Interaction with defects:* Defects, such as cracks or voids, disrupt the flow of eddy currents, altering their paths and characteristics.
4. *Measuring the disturbance:* The disturbances caused by defects can be measured by the same coil generating the alternating magnetic field. As eddy currents encounter defects, their flow is impeded, affecting the surrounding magnetic field.
5. *Detecting variations in voltage or impedance:* Monitoring changes in the magnetic field near the coil reveals variations in voltage or impedance, indicating the presence and characteristics of defects. Measurements can be conducted through voltage analysis across a secondary coil or impedance assessment of the original coil.

ECT is an effective non-destructive method for inspecting conductive materials for surface or near-surface defects, providing valuable insights into material integrity. ECT can utilize both single sensors and sensor arrays. Research indicates that sensor arrays, like those detailed in the study by Xie et al.[2] provide significant advantages. These arrays can include up to 100 individual sensors, allowing for a more thorough surface scan, particularly when paired with laser scanning to improve spatial resolution. In this configuration, the sensors are printed on a flexible strip, enabling them to conform closely to the surface of the material. This design enhances sensitivity and accuracy while requiring minimal current. Their research indicates that sensor arrays, which consist of multiple interconnected sensors, significantly outperform traditional single-element sensors commonly used in NDT. Single sensors often generate delayed signals and may require physical movement to detect flaws, which limits their efficiency and flexibility. In contrast, sensor arrays enhance performance by improving efficiency, resolution, and coverage area—critical factors for comprehensive surface analysis.

Xie et al.[2] emphasized the significant advantages of utilizing compact sensors arranged in an array for rapid surface scanning, which results in quicker assessments compared to the labor-intensive manual movement required by single sensors. This increased speed is particularly beneficial in scenarios where both time efficiency and thoroughness are critical, such as routine inspections in manufacturing and maintenance environments. In these settings, the ability to quickly assess large areas without extensive manual effort can lead to more effective monitoring and maintenance strategies, ultimately improving safety and operational efficiency.

The authors introduced a novel flexible planar eddy current sensor array specifically designed for the inspection of microcracks in critical aircraft components. The sensor array is fabricated using a flexible printed circuit board technique, which entails etching both exciting and sensing coils onto polyimide films. This design approach allows the sensors to conform to the complex geometric structures typically encountered in aerospace applications, enhancing their applicability and effectiveness in real-world scenarios.

To address the demands of condition-based maintenance (CBM), the proposed sensor array comprises 64 individual sensing elements. This configuration not only allows for comprehensive coverage of the inspection area but also achieves a remarkable spatial resolution of only 0.8 mm. Such high resolution is crucial for detecting shallow microcracks that could lead to significant structural failures if left unchecked. Furthermore, the design's sensitivity enables the accurate sizing of fatigue cracks, which is essential for predictive maintenance and ensuring the longevity of aircraft components. The authors reported that the sensor array can accurately size cracks within a tolerance of ±0.2 mm, demonstrating its reliability and precision in defect detection.

Overall, the advancements presented by Xie et al.[2] highlight the critical role that flexible eddy current sensor arrays play in enhancing the efficiency and effectiveness of NDT methods in the aerospace industry. By enabling rapid, thorough inspections with high sensitivity and accuracy, these technologies contribute to improved safety and maintenance practices in aviation.

Let us examine a second research paper focused on assessing early-stage corrosion, which highlights the importance of understanding its effectiveness and efficiency in practical applications. Corrosion can lead to significant structural failures if left undetected, particularly in critical applications such as aerospace, marine, and infrastructure. For instance, when evaluating early-stage corrosion, it is crucial to determine the impact of protective coatings on the performance of steel samples. Research has shown that while coatings can significantly mitigate corrosion, they do not entirely prevent it, especially in aggressive marine environments. Among the various techniques available, ECT has proven to be an effective method for evaluating corrosion, especially in materials like steel and aluminum alloys.

A key consideration in this context is whether coatings applied to steel samples influence their performance against corrosion. Testing has demonstrated that while coatings can reduce the rate of corrosion, they do not completely eliminate it, particularly in marine settings. The authors refer to a study published in 2012 by He et al.[3] that discusses the characterization of steel corrosion through pulsed eddy current systems.

In their experiments, they compared uncoated and coated mild steel samples to assess their susceptibility to atmospheric corrosion over exposure periods of 1, 3, 6, and 10 months. In their research, the authors tested multiple specimens with varying crack widths (0.1 and 0.2 mm) and lengths (3 to 11 mm), using 7075 aluminum alloy with a conductivity of 18.5 mS/m. The careful preparation of specimens, coupled with detailed measurements, demonstrates the meticulous approach necessary for accurate corrosion assessment. It is crucial to recognize that different materials exhibit unique responses to eddy current signals; hence, tailored models that account for variables such as conductivity, permeability, and material thickness are essential for reliable testing.

A comprehensive understanding of the electromagnetic properties of corrosion products is vital for interpreting ECT results. As corrosion progresses, it generates iron oxides and hydroxides, each with distinct characteristics that affect the material's behavior. For example, hematite (α-Fe_2O_3) acts as a semiconductor, while goethite (α-FeOOH) demonstrates lower conductivity. The changes in material properties due to corrosion products necessitate careful analysis, as they can significantly influence the results of eddy current assessments.

The authors' findings highlight the potential of ECT for long-term corrosion monitoring. By establishing correlations between pulse eddy current properties and exposure duration, the authors provided a framework for predicting corrosion rates and identifying early signs of structural degradation. Such predictive capabilities are essential for effective CBM, allowing for timely interventions that can prevent catastrophic failures.

6.5.5 Ultrasonic testing

UT is a widely used NDT technique that uses high-frequency sound waves to detect internal defects in materials. Unlike the human hearing range of 20 Hz to 20 kHz, UT typically operates at frequencies from 20 kHz up to 1 MHz. Ultrasonic waves are generated by a piezoelectric crystal, which vibrates when a voltage is applied, producing the necessary high-frequency sound waves. Modern technology enables precise control over these frequencies, extending beyond the previous 1 MHz limit into the gigahertz spectrum.

The method involves transmitting high-frequency sound waves into the material under examination. These waves propagate linearly until they encounter a barrier, which may be an interface between different materials—such as a coating and its substrate—or a flaw within the material itself. Flaws can manifest as linear, perpendicular, or oblique cracks, voids, and other discontinuities, and interactions between grain boundaries and these flaws can complicate assessments of material integrity.

When ultrasonic waves encounter a discontinuity, their behavior changes markedly. Some waves are transmitted through the material, whereas others are reflected back toward the source. UT uses two types of sensors to facilitate this analysis: one for detecting transmitted waves and another for capturing reflected waves. This dual-sensor approach allows for a thorough evaluation of the material's internal structure and helps identify potential defects.

When ultrasonic waves encounter a material interface, part of the wave reflects back while the remainder continues to propagate through the material. This interaction is fundamental for effective signal detection in UT, which relies on two primary techniques. As the waves encounter surfaces with flaws, their behavior changes—waves may be fully reflected, partially reflected, or transmitted, depending on the flaw's characteristics. These discontinuities suggest potential issues that warrant further investigation. Accurately interpreting these wave interactions is crucial for identifying and characterizing material flaws. Such discontinuities interfere with the ultrasonic pulse, disrupting its path; consequently, only a portion of the wave energy is transmitted, while some is reflected back. This interference modifies the received signals, which the transducer then converts into electrical signals for analysis. The data is subsequently examined using an ultrasonic flaw detector or specialized processing software.

Let us examine various techniques for applying ultrasonic waves in material inspection. The first method, called the transmission technique (illustrated in Figure 6.4), uses two transducers placed on opposite sides of the material under inspection. This setup enables accurate analysis of the section between the transducers. There are two possible scenarios in this method:

1. In a flawless material, most of the ultrasonic energy passes directly from the sending transducer to the receiving transducer, shown by a

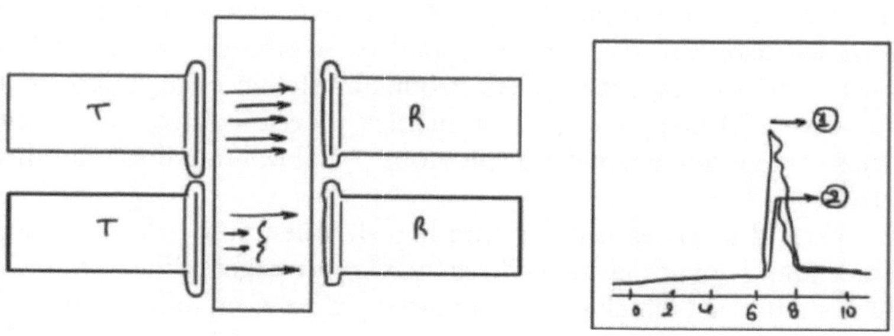

Figure 6.4 Through transmission method (attenuation).[4]

green arrow. This results in a high-amplitude signal, indicating the absence of flaws.

2. When a flaw is present, only part of the energy is transmitted while some is reflected back, resulting in a lower amplitude signal received by the opposite transducer. The reduced amplitude signals the presence of a defect.

This variation in received signal amplitude is critical for flaw detection. As ultrasonic waves experience attenuation as they travel through a medium, calibration is essential to maintain measurement accuracy. Reliable results require the use of suitable curve-fitting methods during calibration. With proper calibration, ultrasonic waves can accurately detect defects, measure material thickness, and assess coating thickness at micron levels. In addition, well-tuned sensors can identify changes at grain boundaries, providing insight into material structure.

As previously mentioned, these measurements are relative, rather than absolute, allowing us to compare conditions, such as a "healthy" versus an "unhealthy" state. In this context, reduced energy received in the second case signals an unhealthy condition relative to the first. The transmission method utilizes two transducers: one as the transmitter and the other as the receiver.

It is important to note that if a sensor has been idle for an extended period, recalibration is essential before reuse. Without regular recalibration, sensors may only provide relative values, which may fluctuate over time rather than yielding consistent, absolute values. Recalibration is thus critical for measurement reliability.

Another key factor to consider is attenuation. When waves are transmitted, energy loss, or attenuation, occurs as they travel, causing the final signal to differ from the initial one due to energy dissipation. Without accurate calibration, this effect could lead to misleading results, potentially causing false defect indications due to energy loss rather than actual material flaws.

Ultrasonic wave behavior changes based on the medium they pass through; for instance, moving from air to a liquid or solid alters the wave's attenuation properties. Even among solid materials like steel, aluminum, and copper, responses can vary significantly. Regular recalibration is therefore essential for maintaining measurement accuracy. Attenuation, or the gradual loss of wave energy as it propagates through a medium, is especially important when measuring from surface to deeper layers. Calibration accuracy generally holds within a specific range—typically between 1 and 3 mm—and measurements beyond this range may be unreliable. Each ultrasonic sensor is designed with defined upper and lower measurement limits, making it essential to operate within these parameters rather than assuming it can measure any distance. Adhering to these operational limits is crucial for achieving accurate measurements.

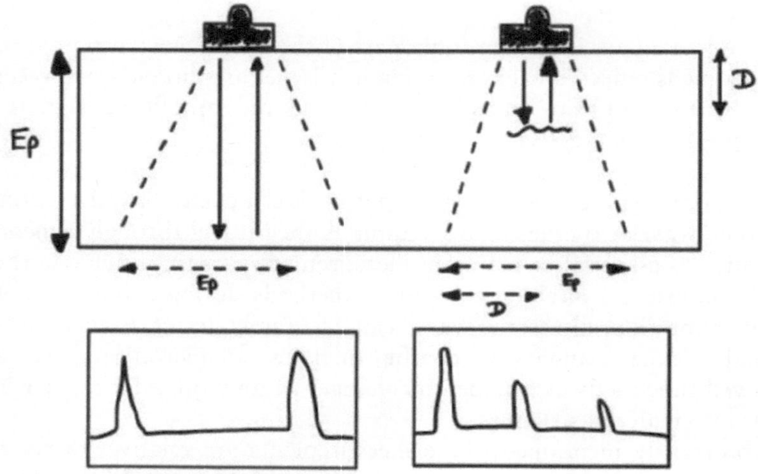

Figure 6.5 Pulse echo method (reflection).[5]

The second related to UT is pulse-echo method, as shown in Figure 6.5. The pulse-echo method is a widely used technique in UT that centers on the reflection of ultrasonic waves to detect flaws within materials. This method utilizes a single transducer, which functions both as a transmitter and a receiver. Here is how it works:

1. *Ultrasonic wave generation:* The transducer is equipped with a piezoelectric crystal. When an electrical signal is applied to this crystal, it vibrates and generates short bursts of ultrasonic pulses. These waves travel into the material being tested.
2. *Propagation and reflection:* As the ultrasonic waves propagate through the material, they may encounter imperfections, such as cracks, voids, or changes in material properties. When these waves hit such discontinuities, some of the energy is reflected back toward the transducer.
3. *Signal reception:* The transducer then acts as a receiver, collecting the reflected waves. The information contained in these echoes can provide insights into the presence and nature of flaws within the material.
4. *Reflection characteristics:* Several factors influence the reflection of ultrasonic waves, including the angle at which the waves hit the discontinuity (the angle of incidence). When the waves strike the defect at a normal angle (90 degrees), the reflection is maximized. In contrast, at other angles, the amount of energy reflected back decreases.

5. *Signal interpretation:* The received signal is often visualized as a series of peaks and valleys on an oscilloscope or a similar device. Each peak corresponds to a reflected wave, and the time it takes for these echoes to return, along with their amplitude, provides valuable information about the defect's characteristics, such as its size and orientation. For example, the method can help determine whether a crack is horizontal, vertical, or at an angle.
6. *Assessment of flaws:* By analyzing the amplitude of the received signals, inspectors can gauge the severity of the detected flaws. Larger amplitudes typically indicate more significant defects.

The pulse-echo method is favored for its simplicity, requiring only a single transducer to perform both transmission and reception of ultrasonic signals. This makes the method easy to implement and efficient for various applications, particularly in industries where the integrity of materials is critical, such as aerospace, automotive, and manufacturing. Its ability to identify subsurface defects and assess material condition is essential for effective quality control and assurance.

Garcia Marquez and Gomez Munoz[6] conducted a study that introduced a novel approach for fault detection, location, and diagnosis through UT of wind turbine blades (WTBs). This research focuses on two structurally identical composite structures featuring a honeycomb core, with one WTB intentionally damaged to induce delamination, while the other serves as a healthy reference for comparative analysis.

The study used macro fiber composite (MFC) transducers, specifically the M2814-P1 model, for transmitting and receiving ultrasonic signals. The transducers were mounted on the outer surface of the WTBs, with one transducer fixed as a transmitter at the tip of the WTB and another as a receiver, which was moved incrementally along the length of the WTB at 100 mm intervals. This setup facilitated data collection from 38 unique measurement points along the WTB's length, allowing the researchers to gather comprehensive information across the top surface of the structure.

The ultrasonic signals were gathered using a "pitch and catch" technique, where the transmitter generated a five-cycle sinusoidal signal modulated by a Hanning window. A sinusoidal waveform was chosen because its smooth oscillations are well-suited to traveling through solid materials and provide consistent, interpretable reflections from within the structure. Hanning windowing technique gradually ramps up and down the amplitude of the signal, reducing any sudden changes at the beginning and end of the waveform. By softening these edges, the Hanning window minimizes spectral leakage—undesired frequencies that can muddy the data. This helps ensure the signal transmitted into the WTB has a clean frequency profile, improving the signal-to-noise ratio (SNR) and enhancing the ability to detect subtle variations caused by faults.

The system conducted a frequency sweep by gradually increasing the frequency from 10 to 100 kHz in 5 kHz increments, chosen to assess the WTB's response across a wide ultrasonic frequency spectrum. Lower frequencies are advantageous for deeper penetration but may be less sensitive to small defects, whereas higher frequencies provide better resolution but experience quicker attenuation within the material. By scanning across this range, the researchers could determine the ideal frequency for both penetration depth and defect detection within the honeycomb structure. Their analysis revealed that 50 kHz offered the best balance, producing the highest SNR for this material. The use of 5 kHz increments allowed for a detailed frequency map, which facilitated a thorough understanding of the material's characteristics and ensured that the most informative frequency was chosen for subsequent defect detection. Frequency scanning is essential in UT, as it helps identify the optimal frequency for achieving clear and accurate results. While 50 kHz proved effective in this study, varying setups may require further frequency assessment to identify the most suitable value. In addition, the WTB's dimensions, with a relatively narrow width compared to its length, could limit the feasibility of using a single sensor, necessitating adjustments in sensor positioning or configuration.

The primary objective of the testing was to scrutinize the ultrasonic signals for indications of defects, particularly delaminations within the WTBs. The first analytical step involved autocorrelation, which compares signals from adjacent measurement points—specifically, distances such as 100 and 200 mm, or 200 and 300 mm. In the absence of defects, the signals from neighboring distances should exhibit similar waveforms and high correlation coefficients. Conversely, the presence of delamination between two adjacent signals alters the ultrasonic wave shape, resulting in a lower correlation, thereby indicating a defect in the WTB.

Following the autocorrelation, cross-correlation was performed between the signals acquired from the healthy WTB and those from the defective WTB. A high correlation between corresponding areas of the two WTBs suggested that the WTB was healthy. In contrast, a low correlation in specific regions of the defective WTB relative to its healthy counterpart pointed toward the presence of a delamination defect in those areas.

The analysis showed that signals from the 200 to 300 mm range exhibited high consistency, indicating uniformity and the absence of defects in that segment. Similarly, a strong correlation was observed between signals from the 700 and 800 mm distances, areas positioned just before the start of the WTB's sandwich region, further verifying structural integrity there. The high correlation coefficients confirmed the composite structure's health in these zones. However, the first flaw detected in the damaged WTB caused a noticeable reduction in signal correlation, pointing to delamination, which altered the ultrasonic wave pattern significantly. Typically, healthy components yield high autocorrelation, whereas unhealthy, damaged, or delaminated areas show markedly lower correlation values. For instance, a correlation factor of 50 may signify a healthy state, while a factor of only 2 suggests damage or delamination, indicating a fault may lie between these two thresholds. For example, a defect might be located between the 2300 and 2400 mm measurements.

This study highlights several key insights for optimizing ultrasonic NDT on WTBs (or any other system), especially for detecting delamination and other structural defects:

1. *Impact of signal attenuation on defect detection:* Ultrasonic waves weakened as they traveled further from the source, amplifying the effects of delamination defects on signal quality. Both natural signal attenuation and structural defects caused a significant drop in SNR, making it more challenging to detect and diagnose faults accurately.
2. *Sensor selection and placement considerations:* Careful sensor selection and precise placement on WTB surfaces, especially thin or noise-prone areas, were critical to ensuring accurate results. Sensors were operated strictly within their calibrated range, as working outside these operational boundaries led to suboptimal performance.
3. *Iterative and flexible analysis approach:* The study emphasized an adaptable approach for signal measurement and correlation threshold setting. Initial measurements allowed for broad localization of defects, with adjustments in measurement intervals and analysis focusing on regions with potential issues. This iterative method improved the effectiveness of fault detection.
4. *External sensor configuration:* All sensors were placed on the WTB's outer surface to maintain NDT standards. No internal sensors were installed to avoid shifting to a DT approach.
5. *Challenges with nanometer-scale thickness:* Testing on very thin materials introduced challenges such as high noise levels, underscoring the need for effective signal filtration and conditioning to obtain clear readings.
6. *Importance of surface preparation:* Clean surfaces were essential for accurate measurements, with any contamination (e.g., dust or oil) removed before testing. Working on very thin or contaminated surfaces risks unreliable data, underscoring the necessity of proper surface preparation.
7. *Signal interpretation framework:* A systematic method was necessary to interpret multiple signal responses accurately. As absolute values could be misleading, the study stressed the importance of relative comparisons between signals from healthy and defective areas.
8. *Reference material and calibration:* Calibration with manufacturer-recommended reference materials and optimal filtration settings was vital for reliable signal readings. Differentiating signals from various

locations and adjusting sensor orientation were also important for accurate detection outcomes.

These findings collectively suggest a robust framework for ultrasonic NDT, with specific emphasis on surface preparation, sensor calibration, signal interpretation, and the use of iterative testing methodologies for precise fault detection in composite structures like WTBs.

6.5.6 Radiography testing

UT, as discussed previously, is a method that uses high-frequency sound waves to detect internal flaws in materials. By contrast, radiographic testing uses radiation, such as X-rays or gamma rays, to identify discontinuities within materials, capturing images of these internal features through digital or film-based methods.

In other words, radiographic testing is an NDT that uses X-rays or gamma rays to examine the internal structure of materials. X-rays are generally preferred for thinner materials or those with lower density, whereas gamma rays can be used for thicker or denser sections. Although gamma rays are less commonly used due to their high energy requirements and associated risks, understanding their applications remains important for a comprehensive understanding of radiographic testing.

Radiographic testing operates by directing X-rays or gamma rays through a material to create an image of its internal structure. A radiation source emits rays toward the object, and by carefully controlling their direction, a clear image of the material's internal composition is obtained. The resulting image, referred to as a radiograph, can be recorded either on traditional film (classical radiography) or on a digital detector (digital radiography). Both methods are effective in revealing internal defects or anomalies within the material.

In radiographic imaging, whether using conventional or digital techniques, the underlying process remains similar. However, the quality of images produced through classical radiography is highly dependent on the technician's skill, who must manage various factors such as exposure settings, film processing, and manual handling. Achieving high-quality images with conventional radiography requires precise manual adjustments to exposure settings, which can vary significantly based on the operator's experience. In addition, the process involves intricate chemical development steps, making it a challenging technique to master.

Digital radiography, in contrast, uses digital X-ray detectors to capture images, converting the X-ray data into electronic signals. These signals are then processed and immediately displayed on a computer screen. This digital method yields high-quality images with real-time feedback, allowing for instant adjustments to optimize image quality. Unlike traditional films, which required days for development, digital radiography provides immediate results, enhancing both efficiency and accuracy in defect detection.

Radiographic testing has several notable advantages. It is applicable to both ferrous and non-ferrous materials, unlike techniques such as MT, which are limited to ferrous materials. It is also effective for both thin and thick structures, with X-rays suitable for thinner sections and gamma rays more effective for thicker ones. Furthermore, radiographic images can be stored on film or digitally, providing valuable records for future reference and trend analysis.

However, radiographic testing comes with significant safety challenges. Strict safety protocols and regulatory measures are essential due to the health risks to personnel and potential environmental hazards from radiation exposure. Radiographic testing is broadly categorized into digital radiography and conventional radiography, with digital radiography now preferred, as conventional methods are increasingly viewed as outdated.

Digital radiography, which uses flat-panel detectors, provides several key advantages over conventional radiography, including increased productivity, enhanced sensitivity, environmental benefits, advanced image processing capabilities, and cost efficiency. This digital approach has gained widespread adoption across industries such as medicine, aerospace, security, and automotive, due to its superior efficiency, image quality, and precision.

One of the main advantages of digital radiography is its reduced radiation requirement, which makes it more environmentally friendly than traditional methods. Furthermore, ongoing advancements in machine learning are continuously improving image processing tools, allowing for more accurate analyses with fewer resources. This heightened efficiency not only improves image sensitivity and quality but also reduces operational costs, making digital radiography the preferred choice for high-precision applications.

In digital radiography, rapid iterations and adjustments can be made, allowing for immediate feedback and faster, more accurate results. Digital images can be readily enhanced and manipulated using a range of software tools, enabling precise analysis. Moreover, digital radiography requires lower radiation doses than conventional radiography, making it a more environmentally friendly option. Its advantages include improved workflow efficiency, advanced image manipulation capabilities, instant image availability, and ultimately, more reliable and detailed results. With each iteration, technicians can refine their technique to produce increasingly accurate images.

While radiography is a valuable technique for inspecting internal structures, it may not be the first choice for detecting surface-level failures; other inspection methods could be more suitable for surface evaluations. However, for examining internal features—such as assessing crack sensitivity, monitoring potential corrosion, or inspecting materials like concrete reinforced with embedded iron where corrosion could lead to failure—radiography is highly effective. In such cases, radiographic testing can provide reliable, detailed insights into internal conditions, making it competitive with other NDT methods. For applications where internal analysis is essential, radiography is recommended for its accuracy and predictive capabilities.

Turning to corrosion testing, particularly for steel reinforcement in concrete, radiography can provide insights into the extent and progression of corrosion. Corrosion often occurs when steel reinforcement is exposed to moisture and oxygen, conditions frequently encountered in construction. Cracks in concrete accelerate corrosion, as they allow corrosive agents to penetrate deeper and come into direct contact with the steel. These cracks can result from multiple factors, such as shrinkage during curing, thermal expansion, contraction, and various loading conditions, including cyclic or excessive loads. Environmental factors also contribute to the formation of cracks, which further facilitate corrosion.

Once cracks allow moisture and oxygen to reach the embedded steel, an electrochemical process initiates, accelerating corrosion. This process creates rust, which has a greater volume than the original steel, thereby increasing internal pressure within the concrete. As rust expands, it exerts tensile stress on the surrounding concrete, potentially causing further internal fractures. Concrete's low tensile strength means it is vulnerable to this stress, hence the need for reinforcement. Over time, this internal pressure compromises structural integrity, as corrosion weakens the bond between the steel reinforcement and the concrete, eventually causing debonding and reducing the load-bearing capacity.

In some cases, a bridging medium, such as epoxy, can be applied to seal cracks within concrete, restoring structural integrity. This can be done through automated methods, using capsules or mechanisms to bridge the gap, or manually by sealing the cracks directly. These are referred to as "automatically healed" or "manually healed" methods, respectively.

Let us examine a case study published by Moreira et al.[7] where radiography was used to inspect pipes for internal and surface-level defects.

The study aimed to detect internal cracks, failures, and surface damage within pipes by applying radiographic testing to sample sections taken from these pipes. Two pipe thicknesses were tested: 12.7 mm (approximately half an inch) and 25.4 mm (about one inch). The pipe material was a low-carbon steel, classified as API 5L X70 grade. For this purpose, test pieces were obtained from pipes manufactured to meet the API 5L 2004 Edition X70 grade specifications, with thicknesses of 0.500 inches (12.7 mm) and 1.000 inches (25.4 mm) as required for X70 grade material. Chemical analyses of the pipes, conducted in accordance with ASTM standard A751, confirmed the composition of the base metal, adhering to established procedures for steel products.

To simulate welded junction conditions and assess material behavior at the welds, six samples were prepared. Each sample measured approximately

250 mm × 250 mm and was taken from longitudinally welded pipe sections with outer diameters of 813 mm (3200) and 1219 mm (4800) and nominal thicknesses of 12.7 mm (0.500 inches) and 25.4 mm (1.000 inches), respectively. Controlled surface cracks were intentionally introduced to specific areas on the samples, enabling a comparative analysis of inspection sensitivity and effectiveness across different techniques.

The study used multiple testing methods, including MT, conventional radiography, and digital radiography, to evaluate the sensitivity and accuracy of each method in identifying defects within steel-grade pipes. Magnetic particle inspection was conducted first, followed by conventional radiography using radiographic films, and finally digital radiography. Conventional radiography involved capturing images on X-ray films, which were then developed and processed with chemicals to produce the final image. Image quality depended on initial exposure settings, film processing conditions, and the technique used, with any necessary adjustments requiring careful control during exposure and chemical processing.

In contrast, digital radiography utilizes flat-panel detectors to convert X-rays into electronic signals, allowing for immediate processing and display on a computer screen. This method produced high-quality images with instantaneous feedback on exposure and technique, enabling digital adjustments to optimize image clarity, such as modifying brightness, contrast, and magnification. The digital format also facilitated the manipulation and enhancement of images using specialized software, providing greater flexibility.

Notably, digital radiography typically requires lower radiation doses than conventional radiography, resulting in reduced radiation exposure. The study's results demonstrated that digital radiography, with its advanced flat-panel detectors, exhibited the highest sensitivity, as reflected by superior image quality indicators (IQIs) and the detection of minor discontinuities. This underscores the effectiveness of digital radiography over conventional methods for accurately identifying and visualizing flaws or defects in welded joints.

Overall, digital radiography offers substantial benefits, including workflow efficiency, enhanced image manipulation capabilities, instant image availability, and reduced radiation exposure, making it the preferred choice for high-precision inspection applications.

A thorough comparison of inspection techniques by Moreira et al.[7] showed that MT was highly effective in detecting surface cracks on ferrous materials, identifying a substantial number of surface flaws. When conventional radiography was applied, it also revealed visible cracks, highlighting its effectiveness as an alternative method. However, MT detected a greater number of cracks than conventional radiography.

With the use of digital radiography, a much more detailed image was obtained, revealing a significantly higher number of cracks, including complex branching patterns that were not visible with conventional radiography.

This finding underscores the superior sensitivity of digital radiography, which allowed for the detection of intricate details such as branching cracks and extended fracture networks, clearly marked in the digital images. This demonstrated that digital radiography surpasses both MT and conventional radiography in capturing fine details of internal flaws.

Unlike magnetic particle inspection, which is limited to ferrous materials, radiography—both conventional and digital—can be utilized on a broader range of materials, including both ferrous and non-ferrous metals, and even polymers. The comparison between conventional and digital radiography demonstrated that digital radiography offers enhanced sensitivity, with a distinct ability to visualize discontinuities in greater detail, including complex fracture patterns. This sensitivity underscores digital radiography as the more effective method for thorough defect detection.

In conventional radiography, images are captured by using X-rays to expose film, which is then chemically developed to produce the final image. This process is like traditional photographic methods, where images are printed on paper and finalized upon development. In contrast, digital radiography captures images via digital X-ray detectors, enabling real-time display, storage, or deletion, reflecting a shift toward more efficient, environmentally friendly, and adaptable digital processes.

In conclusion, digital radiography proves superior in detecting and visualizing internal flaws, especially for identifying fine or branching cracks, while conventional radiography, though effective, lacks this level of detail. For surface inspections on magnetic ferrous materials, MT remains a practical choice; however, for internal or less visible defects, digital radiography is the preferred method due to its heightened sensitivity and flexibility in image analysis.

Van Tittelboom et al.[8] conducted a study using radiography to analyze both longitudinal and orthogonal cracks in concrete structures. Observations were made at intervals of 2 and 30 hours after initiating the exposure process. Their findings indicated chloride ingress in all samples, with varying levels of corrosion noted based on the sealing methods used. This research underscores the potential of radiographic testing for corrosion analysis, effectively identifying both subsurface and surface cracks that can contribute to structural deterioration.

The study by Van Tittelboom et al.[8] evaluated both automated and manual methods for healing orthogonal and longitudinal cracks in concrete structures, revealing interesting insights into the efficacy of these approaches.

- *Healing methods and outcomes:* The research found that manual healing methods consistently produced superior results compared to automated methods. This may be due to the precision and control that manual interventions allow, enabling more tailored applications of the healing material. In the case of orthogonal cracks, significant improvements were also observed, indicating that manual healing can be particularly effective in these scenarios.

- *Role of polyurethane in corrosion mitigation:* The researchers used X-ray radiography to visualize the interface between the rebar (reinforcement bars) and the surrounding cementitious matrix. They specifically assessed the corrosion mitigation achieved through self-healing properties of polyurethane, which was used to encapsulate the fractures. Polyurethane is a versatile polymer that can bond effectively to the concrete, creating a barrier that prevents the ingress of corrosive agents, such as water and chlorides. This action is crucial as it delays the onset of corrosion within the reinforced material, thus prolonging the structural integrity of the concrete.
- Radiographic imaging revealed significant yellowing on the surface of the concrete, a phenomenon that might not be readily visible to the naked eye. This yellowing serves as an indicator of internal damage or the presence of voids that could lead to further structural issues. The ability to visualize these areas through radiography emphasizes its importance in conducting a comprehensive assessment of structural integrity.
- After 70 hours of exposure, additional radiographic examination uncovered further surface cracks, suggesting that continued monitoring is essential to identify potential failures early. This finding underscores the dynamic nature of corrosion processes and the importance of regular assessments to maintain structural safety.
- In the radiographic images, gray areas corresponded to corrosion byproducts, indicating regions within the concrete that were affected by corrosion processes. As these byproducts accumulate, they occupy increasing volumes, leading to the formation of additional cracks. Radiography plays a crucial role in highlighting these yellow areas, which indicate voids and potential cracks that could compromise the structure further.
- To address these issues, remediation strategies may involve:
    1. *Removing corrosive agents:* This could entail physical cleaning or applying barriers to prevent exposure to harmful chemicals.
    2. *Eliminating conductive pathways:* Identifying and mitigating pathways that allow for the passage of corrosive agents through the concrete can also reduce the risk of corrosion.
    3. *Enhancing material properties:* This may involve using additives or coatings that improve the resistance of concrete to corrosion and cracking.

6.5.7 Thermographic non-destructive testing

In this section, we turn our focus to thermography testing (TT), also known as IR thermography. TT is an NDT technique that visualizes heat patterns on an object's surface by detecting IR radiation emitted by objects above absolute zero. By capturing distinctive thermal signatures, TT allows us to identify materials with different thermophysical properties based on thermal contrast.

Using IR cameras, TT captures thermal images for analysis, revealing temperature distribution across the surface of various objects. This data, presented as visible thermal images, provides insights into an object's condition. The importance of TT became especially evident during the COVID-19 pandemic when IR cameras were widely used for body temperature monitoring, highlighting the critical role of high-quality cameras and software for precise thermal analysis. Thermography has a wide range of applications in NDT across various industries. In aerospace and automotive sectors, it is extensively used to detect flaws in composites, metallic components, and structures susceptible to fatigue or crack propagation, thereby ensuring safe operation by identifying potential failure points early. In electrical and electronic components, thermography helps identify overheating, faults, or insulation issues, preventing malfunctions and enhancing safety. In civil engineering and building inspection, it detects structural flaws, moisture, and insulation deficiencies, which improves the integrity of infrastructure. For cultural heritage conservation, thermography enables non-destructive assessment of artworks and historical buildings for structural integrity without causing physical alteration. In addition, in medical applications, it has been used for body temperature monitoring—especially during the COVID-19 pandemic—and for diagnostic purposes related to vascular health.

The focus on temperature distribution arises because the IR radiation emitted by heat sources is invisible to the naked eye. IR cameras can detect this radiation and convert it into visible images through thermal mapping, which allows us to observe temperature variations that provide valuable, though indirect, insights into material integrity. According to established principles, all objects with temperatures above absolute zero emit IR radiation, with the intensity increasing as temperature rises. Hence, the conversion of IR radiation into a visible format is crucial for analysis, producing thermal images that reveal temperature distributions.

Thermography approaches can be categorized into passive and active methods. Passive thermography relies on capturing natural thermal contrasts between an object and its surroundings without the need for external energy input; it is commonly used for applications like monitoring human bodies, electrical components, and building insulation. In contrast, active thermography introduces an external energy source, such as heat or sound, to stimulate the object, thereby creating a thermal contrast between defective and non-defective areas. This approach is particularly useful for detecting voids, delaminations, and other internal defects within materials.

Inspection scenarios in thermography can vary based on configuration, energy transfer mode, and scanning method. "Static configurations" involve a stationary camera, object, and energy source, which is ideal for time-tracked data analysis, whereas "dynamic configurations" use motion, such as real-time scanning on production lines. "Energy transfer modes" include "reflection mode" (detecting near-surface defects from the same side as the energy source), "transmission mode" (inspecting deeper defects by capturing data from the opposite side), and "internal heating" (using ultrasonic or eddy currents to produce heat at defect sites for deeper inspection). "Energy transfer methods" may apply point, line, or surface scanning, depending on whether a specific spot, line, or entire area needs to be stimulated. Accurate thermographic measurements depend on several correction factors, including ambient temperature and the object's emissivity, which varies based on surface treatment. For instance, black paint enhances emissivity, improving the accuracy of thermal readings. In addition, proximity between the camera and the object is essential, as closer distances yield clearer results, although practical constraints often require some separation.

Thermography effectively identifies anomalies, such as irregular heating patterns, misalignments, and damaged components within mechanical systems. For example, bearing misalignment leads to increased friction and localized heating, which TT can detect. Cracks, pores, or voids can also be identified without invasive methods like scanning electron microscopy (SEM). However, thermal imaging only provides data on temperature variations, not the complete structural or geometric profile of the object.

In advanced materials engineering, thermography is gaining importance as industries like automotive and aerospace focus on creating lightweight, efficient components. However, the shift to lighter materials often leads to an increased risk of failure, making predictive maintenance essential. For example, a 2022 case study,[9] examined spur gears with varying rim thicknesses and web ratios to optimize design and reduce weight while maintaining safety. The study revealed that although thinner rims reduce weight, they also accelerate crack propagation, increasing the likelihood of catastrophic failure.

The study demonstrated that cracks initiating from gear teeth, rather than deep within the material, are more manageable. However, if a crack propagates toward the shaft, the likelihood of catastrophic failure increases. The research emphasizes the value of IR thermography for early failure detection by monitoring temperature profiles during fatigue tests.

A fatigue testing machine simulated conditions that could cause gear tooth failure. Researchers painted both standard and thin-rim gears black to enhance emissivity and measured thermal profiles on the tooth root during testing. Results indicated that cracks tended to start at specific surface points, and thermal monitoring allowed crack propagation tracking.

Thermographic techniques encompass several methods for assessing material integrity and defect characterization. Lock-in thermography (LT) applies periodic heating at a controlled frequency to measure thermal response phases, making it particularly effective for determining defect depth. Pulsed thermography (PT) utilizes a brief energy pulse to heat the object, allowing for the observation of varying cooling rates in defective areas compared to intact ones. Square pulse thermography (SPT) uses longer pulses to monitor both heating and cooling, which is ideal for thicker or low-thermal-conductivity materials. Step heating thermography (SHT) concentrates on the heating phase to detect defects, while scanning techniques involve using point or line sources, such as lasers, for detailed surface scanning.

Thermographic testing is also advantageous for field applications, especially in hard-to-reach areas like above-ground tanks, submerged pipes, and storage containers. Portable IR cameras allow quick thermal image capture and storage for analysis, making thermographic non-destructive testing (TNDT) an economical choice for monitoring corrosion in metallic casings. Although UT is sometimes more affordable depending on the equipment, IR inspection is often more cost-effective.

An experimental study[10] published in 2020, on corrosion-related coating testing, used a powerful LED heat source with an imaging camera and a reference sample for thermographic analysis. This study also used a reference tank filled with different materials to evaluate thermographic data under various conditions. Black paint was applied to the tank to increase emissivity, and the camera captured temperature data from approximately 1.1 m away, with the LED heat source at about 0.4 m.

Various thermographic images from tanks of differing thicknesses (2, 4, 6, and 8 mm) showed significant improvements in SNRs after processing. Pulsed thermal testing produced sharper images due to reduced heat diffusion, confirming it as a highly effective approach for identifying defects. Researchers used principal component analysis (PCA) for data processing, which reduces the complexity of variable sets, enabling efficient analysis even when variable relationships are unclear. Advanced signal processing is crucial for improving defect visibility and accuracy in thermography. Key techniques include thermal contrast-based methods, which enhance defect detection through maximum contrast, differential contrast, and thermal tomography. Statistical analysis applies various parameters to refine defect characterization. Matrix factorization, particularly through PCA, reduces data complexity and emphasizes significant defect features. Additionally, transforms such as Fourier and wavelet transforms facilitate frequency-based analysis, notably in pulsed phase thermography (PPT), enhancing the overall effectiveness of thermographic evaluations.

Further insights into the concepts, procedures, and applications of thermography can be found in Ibarra-Castanedo et al.[11]

Thermography offers several advantages and limitations as an NDT method. Among its benefits, it is contactless and safe, unlike X-ray or ultrasound NDT techniques, and allows for fast data acquisition and interpretation, making it particularly useful in industrial applications. It is also capable of inspecting hard-to-reach areas, enhancing its versatility. However, thermography does have some limitations, including susceptibility to non-uniform heating and environmental influences, such as convection losses, which can affect accuracy. It has a limited penetration depth, making it challenging to detect very deep defects, and it requires high emissivity for reliable results, often necessitating surface treatments like black paint. Overall, thermography is a versatile tool for inspecting material integrity across various sectors, enabling effective, rapid, and safe detection of early-stage defects and ensuring the longevity of critical components.

6.5.8 Acoustic emission testing

AE plays a pivotal role in NDT by detecting the elastic or transient elastic waves generated by localized sources within a material, such as flaws, cracks, and surface irregularities. This technique allows for the effective assessment of damage progression by analyzing these emissions. While in previous subheading, discussions have focused primarily on IR emissions, it is important to introduce AE testing, which evaluates the material's response to stress. By correlating emissions with applied loads, AE provides valuable insights into material behavior.

One of the significant advantages of AE is its capability to monitor the initiation and growth of cracks continuously, making it an invaluable tool for applications that require constant surveillance. This continuous monitoring capability positions AE testing as one of the most effective methods for around-the-clock oversight. The data gathered can be utilized not only for immediate assessments but also for long-term analyses, allowing for the development of accurate predictive models. For instance, by forecasting corrosion rates over time, organizations can plan maintenance activities, such as plant shutdowns, at optimal intervals, enhancing operational efficiency.

For the successful implementation of AE testing, it is crucial to ensure that the sensor or transducer is securely attached to the monitored surface or material. The effectiveness of the analysis is heavily dependent on this attachment. Once properly affixed, the sensor captures the elastic waves produced by the material under stress, which can be analyzed using time or frequency methods. A comprehensive database is essential for accurately interpreting these results, as it provides the necessary context for understanding the significance of the detected emissions.

It is important to recognize that AE is not an absolute measurement method; rather, it relies on trend analysis to assess material conditions. To achieve a robust assessment, a well-maintained database, clear interpretation techniques, and advanced analytical methods are essential. In scenarios where a pre-existing database is unavailable, laboratory testing may be required to establish baseline data. However, many organizations opt for complete equipment solutions from suppliers that include extensive databases and libraries, which facilitate accurate predictions of potential failures. When materials undergo strain or deformation, high-frequency elastic waves are generated, and sensors closely attached to the surface can effectively detect and record these waves' characteristics. Through this detection and analysis, underlying issues can be identified, and trends, such as crack growth over time, can be monitored to inform maintenance or overhaul decisions.

In their review, Calabrese and Proverbio[12] highlighted the application of the AE technique in corrosion testing to enable early detection of structural failures caused by corrosion. Their study focused on stress corrosion cracking (SCC), a critical fracture mechanism often triggered by surface defects such as roughness or dents. The authors highlighted the importance of using AE for continuous, real-time monitoring and assessment of SCC. This approach provides valuable insights for managing corrosion-related risks, particularly as SCC is a significant cause of failure in various industries. Monitoring the initiation and propagation of cracks is essential for ensuring safe operation. AE helps evaluate the severity of detected faults, especially when crack propagation accelerates due to activation energy and immediate attention is required, as the damage can rapidly escalate. This offers significant advantages in predicting the timeline for maintenance or replacement, highlighting AE's versatility across multiple sectors.

Figure 6.6 illustrates the progression of SCC through various stages, showing the relationship between pit depth plus crack depth (denoted as "a") and time, under some driving energy/force. It is divided into three primary phases:

- *SCC initiation:*
  - *Stage I:* This is the localized breakdown of the oxide film on the metal surface, leading to pitting corrosion.
  - *Stage II:* Characterized by pit growth, localized concentration of stress, and the nucleation of SCC. This stage includes long-term growth of short cracks.
- *Transition phase:* This marks the shift from an electrochemical to a mechanical dominance in the driving forces of SCC. It is a critical period where the influence of mechanical stress begins to surpass that of the electrochemical factors. Stage II continues in this phase.

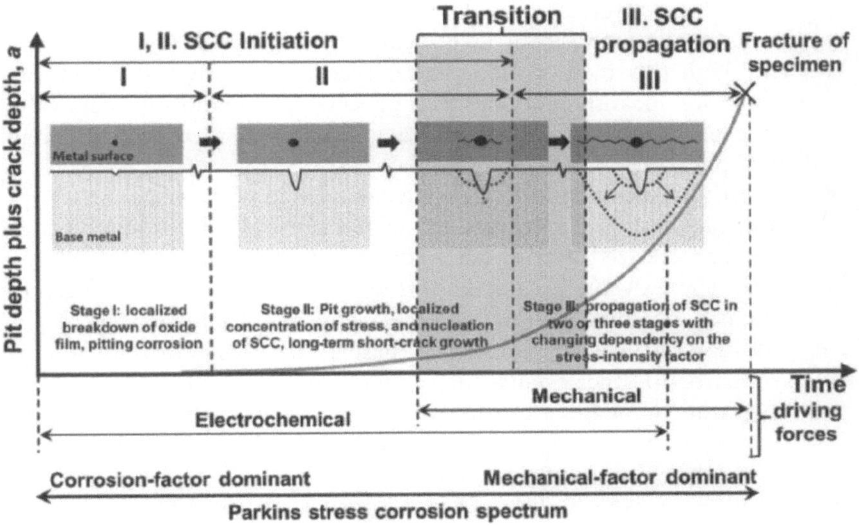

Figure 6.6 Stages and driving forces in the progression of SCC. (From Calabrese et al.)[12]

- *SCC propagation:* SCC propagation occurs in two or three stages, with the growth rate increasingly dependent on the stress-intensity factor. This phase leads to the eventual fracture of the specimen.

The horizontal axis represents time and the shift in driving forces, moving from electrochemical (corrosion-factor dominant) to mechanical (mechanical-factor dominant). The graph suggests that early stages are driven by corrosion processes, while later stages are dominated by mechanical stress, highlighting the importance of understanding both aspects in managing SCC.

Effective AE monitoring relies on accurate sensor placement and signal processing. In a typical setup, sensors are positioned close to the test material. For instance, in a study using a UTM with a dog bone sample, two strategically placed sensors captured continuous data, providing insight into crack behaviors from initiation to propagation through acoustic signals.

Corrosion manifests through various mechanisms—uniform, pitting, and crevice corrosion—each generating distinct AE signals. Accurately associating these signals with specific corrosion types requires an extensive database and advanced software capable of differentiating tribo-corrosion, crevice corrosion, and pitting. The software must discern the signal characteristics of each mechanism to facilitate early warnings. A few key points from the review of Calabrese and Proverbio[12] are:

- *Discriminating SCC mechanisms:* Fast Fourier transform (FFT) can be used to convert AE signals from the time domain to the frequency domain, allowing for the identification of various SCC mechanisms. Although this method necessitates recording the entire waveform, which significantly increases data processing demands, it offers valuable insights into the distinct waveforms associated with different SCC phenomena.
- *Hydrogen gas evolution:* AE events associated with certain SCC phenomena, such as hydrogen gas evolution and anodic dissolution, are characterized by distinct frequency peaks at approximately 50 kHz. These events exhibit high-frequency selectivity, meaning they produce narrow, well-defined peaks in the frequency spectrum. Additionally, the magnitude of these peaks is significantly higher compared to other events, such as passive film rupture. This notable magnitude indicates a greater regularity and intensity in the acoustic waveform, which helps in effectively distinguishing these phenomena from other SCC-related events.
- *Rupture of passive film:* Occurs around 50 kHz with very low amplitude.
- *Metal dissolution:* Metal dissolution during SCC is characterized by a distinct AE frequency peak at approximately 70 kHz. This phenomenon produces a narrow and well-defined frequency peak, which is a clear indicator of its occurrence. The magnitude of this peak is significantly higher—often an order of magnitude greater—than that observed during passive film rupture events. This heightened magnitude reflects the greater energy release and acoustic intensity associated with the dissolution process, making it a prominent feature in the AE spectrum. The distinctiveness of the metal dissolution peak, both in terms of frequency and amplitude, allows for effective identification and differentiation from other SCC-related mechanisms, such as passive film rupture or hydrogen gas evolution.
- *Challenges in AE spectrum interpretation:* Interpreting the AE spectrum during crack propagation presents several challenges due to the presence of numerous secondary AE events, which can complicate the frequency spectrum. The accuracy of the AE spectrum is highly sensitive to deformation and fracture phenomena, making it a reliable tool for detecting such processes. However, this sensitivity also means that distinguishing between various events can be difficult. In addition, the magnitude of AE signals often correlates with the size of defects during the nucleation and growth stages, further complicating the interpretation as different stages of crack development contribute varying signals. These complexities highlight the need for careful analysis and potentially additional experiments to enhance the understanding of AE data during SCC.

Toutountzakis et al.[13] conducted research to evaluate the effectiveness of AE techniques in detecting seeded gear faults within gear systems. Their work aimed to explore how AE can be used to monitor the health of mechanical systems, specifically gearboxes, which are critical components in various machinery.

The researchers designed a test rig featuring a *back-to-back gear arrangement* consisting of two identical gearboxes. This configuration was specifically selected to reduce noise interference during testing, which is crucial for accurately capturing AE signals. They conducted experiments under three distinct load conditions:

- *No load:* This condition only involved the rotation of the gears without any external forces applied.
- *55 Nm:* A moderate load was applied to simulate operational conditions.
- *110 Nm:* A higher load was tested to observe how increased stress affects AE signals.

Sensor placement: To investigate the effect of sensor placement on fault detection, the study used two different setups:

- *Rotating casing sensor:* One sensor was mounted on a rotating casing, requiring a *slip ring* for continuous transmission of signals. This setup allowed the sensor to monitor components that experience motion.
- *Stationary bearing housing sensor:* Another sensor was attached to a stationary bearing housing, providing a stable reference point for capturing AE signals.

This comparative analysis helped determine how sensor location impacts the effectiveness of fault detection.

AE sensors: The AE sensors used in the study demonstrated *linear behavior* across a frequency range of *100 kHz to 1 MHz*. Calibration of these sensors was critical to ensure accurate measurements. In addition, the *gear ratio* of 40:90 was acknowledged as significant for conducting frequency matching analysis, which is essential for identifying specific faults.

Fault localization and challenges: To facilitate fault localization, the researchers introduced controlled faults, such as pits and edge defects, into the gear components. This approach allowed them to test the sensors' capabilities to identify these artificial defects. The study underscored the challenges of diagnosing naturally occurring faults in real-world applications, which can be more complex and varied.

Wear mechanisms and monitoring: The research revealed that wear mechanisms in gearboxes result from *relative motion and load.* During the running-in phase, the wear rate is initially high as surfaces smooth out; however, it decreases to a lower level over time. This emphasizes the need for ongoing monitoring through AE testing to detect potential issues before they escalate.

Running-in period and data capture: A *15-hour running-in period* was part of the study, during which the researchers noted the potential impact of background noise under different load conditions. Capturing reliable data during this period was essential for assessing the effectiveness of the AE techniques.

The researchers used a dual approach using both time-domain and frequency-domain analysis of AE signals, where time-domain analysis was effective in identifying potential issues, and frequency-domain transformation allowed for pinpointing specific faults at higher frequencies, confirming the effectiveness of AE techniques in monitoring gear systems. They found that the performance of AE sensors varied based on their placement, with sensors positioned closer to faults showing superior detection capabilities compared to stationary sensors, thus highlighting the importance of strategic sensor placement for optimal fault detection. In addition, the study emphasized the significance of bearing design on signal response, distinguishing between fully complemented bearings—where all available spaces are occupied by rolling elements—and larger rollers or balls, indicating that the configuration of bearings and sensor positioning significantly influence AE monitoring results and overall gearbox condition assessments. In conclusion, Toutountzakis et al.[13] demonstrated the promising application of AE techniques for fault monitoring and diagnosis in gear systems, emphasizing the critical roles of sensor placement, load conditions, and continuous monitoring in maintaining the reliability and performance of mechanical systems.

6.5.9 Stress–strain DIC testing

Stress–strain analysis using DIC is an innovative NDT technique that enables a comprehensive examination of how materials deform and distribute strain on their surfaces when subjected to external forces. This method involves capturing a series of images before and after deformation, allowing for the tracking of specific marked points (nodes) on the specimen's surface.

DIC is an effective method for analyzing how materials respond mechanically to stress. Stress is defined as the internal force per unit area experienced by a material, while strain describes the deformation or change in shape that occurs as a result. DIC uses high-resolution imaging to monitor surface movements and deformations, allowing for accurate measurement and analysis of stress and strain distributions.

To prepare the surface for analysis, a random speckle pattern is applied, improving the tracking of deformations. High-resolution digital photographs are taken before and after the application of load or deformation using specialized cameras. These images are processed with dedicated software that examines variations in the speckle pattern to compute displacement and deformation fields. These fields allow for the calculation of strain and stress distributions using mathematical techniques, offering insights into the material's behavior under stress.

The DIC software enhances analysis by generating contour plots, vector fields, and graphs that visualize observed displacements, strains, and stresses, which aids in understanding material responses to stress and identifying critical areas of concern. This NDT method is particularly suitable for fragile or expensive items, as it does not require any modifications to the specimen. It offers precise measurements of displacement, strain, and stress, enabling comprehensive studies of material behavior. Unlike conventional strain gauges, DIC analyzes the entire surface of a specimen and records images in real time, making it ideal for dynamic events or transient phenomena. Its versatility allows application across various materials, including metals, composites, polymers, and biological tissues. In summary, DIC uses advanced image processing techniques to measure and analyze deformation, strain, and stress, serving essential roles in research, quality control, and structural integrity testing while providing valuable insights into material performance.

In DIC, we begin with a reference set of images of the specimen at rest and a corresponding set of images taken after the specimen has been deformed. Each point of interest, such as node 1 and node 2, is tracked to ascertain their positions before and after deformation. For instance, initial points can be denoted as $\mathbf{p}$ and $\mathbf{q}$, with their deformed positions represented as $\mathbf{p}'$ and $\mathbf{q}'$. This tracking process allows to compute the displacement of these points, which is essential for calculating strain. A critical aspect of DIC is point matching, which involves identifying corresponding points on the specimen's surface before and after deformation. This comparison facilitates the calculation of transformation matrices and strain values, primarily focusing on 2D displacement data in the x and y coordinates. The process emphasizes relative comparisons to track how deformation evolves, rather than providing absolute measurements. The DIC technique mirrors the principles of the finite element method (FEM), where deformation data are utilized to derive strain values. Unlike traditional methods that often rely on contact sensors placed close to the material, DIC allows for remote observation through high-quality imaging, thereby enabling the analysis of larger surface areas without the need for intrusive measurements.

In DIC, strain calculations typically involve the use of a triangular mesh derived from the captured image data. The main difference between DIC and FEM is how they refine their meshes: FEM relies on progressively finer meshes to achieve accuracy, while DIC directly calculates strain from the image data. To reduce noise and maintain numerical stability, larger triangular elements are preferred for these calculations. In practice, a DIC system is set up with high-resolution cameras positioned strategically to capture images of the specimen throughout the testing process. For instance, in a study on compression testing of granite specimens, a speckle pattern is applied to enhance image contrast, which facilitates accurate tracking of surface deformations as loads are incrementally applied. During the test, images are continuously captured, and detailed strain distributions are analyzed using image correlation techniques. By assessing how the specimen responds under load, researchers can identify areas of increasing strain and potential failure points, even if these regions are not visually noticeable on the surface.

Computational considerations: The analysis of deformation using DIC can be computationally intensive, as it requires processing a vast number of images to accurately assess the changes in the specimen's surface. However, advancements in computational power and the integration of machine learning techniques have made this analysis more efficient and manageable. Noise reduction techniques are also used to enhance the clarity of the results, allowing for more accurate strain measurements.

Application example: Rocket deformation analysis: To illustrate the application of DIC, consider a scenario involving a rocket's surface marked with a grid of points. By tracking the displacement, acceleration, and velocity of the rocket during high-speed impacts, researchers can generate visual representations of the structural behavior through color maps and contour plots. For instance, in a test where a rocket impacts a stationary surface, the DIC method allows for the observation of shape changes and localized deformation without the need for traditional sensors. In this context, the analysis focuses on understanding how the rocket interacts with the target surface during impact. DIC enables researchers to monitor various parameters, including displacement and deformation characteristics, thus providing insights into potential failure zones.

Advantages of DIC: DIC has several advantages over traditional testing methods, especially in its ability to provide optical measurements of deformation. It captures a comprehensive series of images that allow for detailed analysis of how a specimen's surface changes during testing. This is particularly useful for materials with irregular surfaces or complex geometries, where conventional sensors may not offer complete coverage or may struggle to adhere properly. As a powerful non-contact technique, DIC effectively measures deformation and strain distributions in materials. Its capability to deliver detailed optical measurements is especially beneficial for complex shapes and surfaces, making it a valuable complement to traditional testing methods. Although DIC is highly effective on its own, using it alongside other techniques provides a more robust validation of structural integrity, ultimately enhancing the reliability of material analysis findings.

Blikharskyy et al.[14] conducted a comprehensive analysis of the theoretical and experimental findings related to DIC in their review. They emphasized that, through various improvements and modifications, DIC has become a distinct methodology for examining the stress–strain states of structures. This method is versatile, applicable to a wide range of engineering problems and structural materials, providing reliable data at both micro- and macro-scales.

Blikharskyy et al.[14] explored the parameters of the experimental setup (shown in Figure 6.7): camera resolution (c × r, in pixels), the width (w, in mm) and height (h, in mm) of the studied area, the camera's focal length (f, in mm), and the distance from the camera to the sample (d, in mm). The focal length is a variable parameter that allows the researcher to adjust the distance to the image and enhance the sharpness of the displayed area. Camera resolution (c × r) is a technical specification that defines the dimensions of

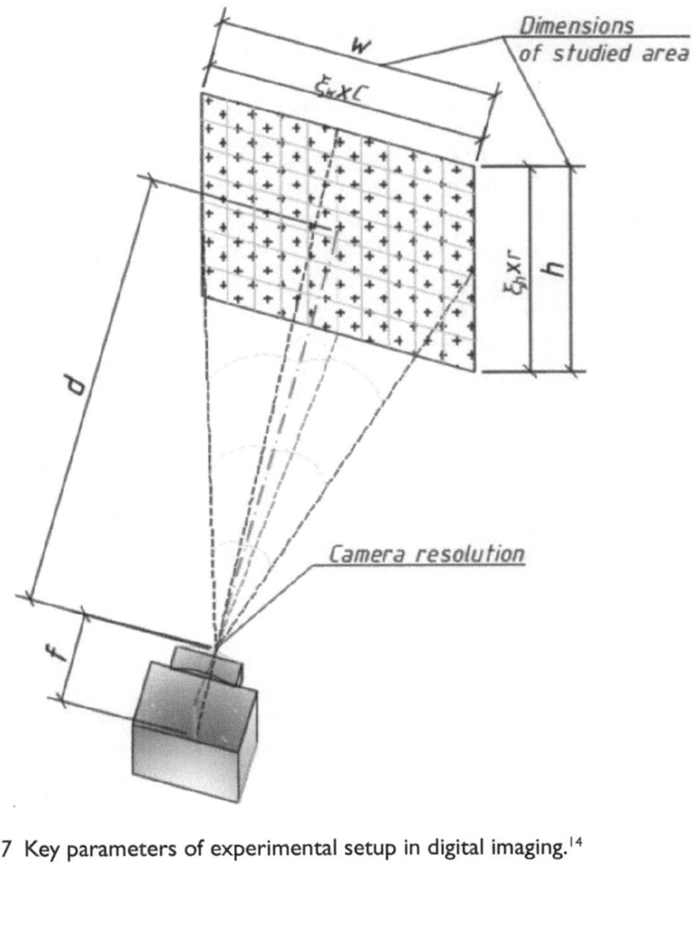

Figure 6.7 Key parameters of experimental setup in digital imaging.[14]

the digital image in terms of columns and rows. By using the predetermined values of c × r and the dimensions of the studied area (h × w), it is possible to calculate the size of the physical surface represented by each pixel in the digital image.

The ongoing development of DIC techniques is expected to enhance the precision of stress–strain measurements and improve the reliability of structural assessments under various loading conditions. Importantly, DIC is applicable not only in laboratory settings but also in field research, where precise measurements and error reduction are critical. The lack of thorough analysis in this area poses challenges for researchers using DIC in site investigations.

In addition, there is no consensus on the optimal specifications and parameters for the speckle pattern that would ensure maximum accuracy. Although the importance of parameters such as dot size, measurement base dimensions, and pattern density is recognized, a clear methodology for defining these values during experimental planning is still lacking. In conclusion, further experimental and theoretical exploration of DIC methodology is an important scientific pursuit that warrants increased attention.

6.6 SELECTION CRITERIA FOR NDT TECHNIQUES

In the previous section, we examined nine NDT methods, emphasizing the importance of selecting the appropriate method based on the types of defects that may arise in components or systems. Defects in materials can greatly influence their mechanical properties and overall performance, making it essential to understand and classify these defects for effective inspection and maintenance. This section will focus on the various defect types, their characteristics, and suitable inspection techniques.

The primary defect types typically evaluated in NDT include linear surface-breaking cracks, volumetric surface-breaking cracks, subsurface linear cracks, subsurface volumetric cracks, near-surface porosity, and internal delamination. Each defect type necessitates careful consideration when selecting inspection techniques to ensure effective detection that is tailored to the specific material and structural requirements. Delamination, for instance, refers to the separation between two bonded surfaces, whereas cracks may manifest as either transverse or longitudinal linear flaws with measurable depths. The orientation of these cracks—whether linear, perpendicular, or parallel to the surface—significantly influences the choice of detection methods. Surface-breaking linear cracks can be identified using techniques such as LPT or MT, which effectively reveal the crack profile in a cost-efficient manner compared to UT. Surface-breaking volumetric cracks are assessable through visual inspection (VT) and may be enhanced by liquid or magnetic testing to detect surface imperfections without the need for more advanced techniques. In contrast, subsurface volumetric cracks impact larger volumes of material beneath the surface, while near-surface porosity can develop, potentially leading to more significant flaws under cyclic loading conditions. Thus, advanced diagnostic methods are essential for the early detection of these subsurface defects to prevent further propagation.

6.6.1 Specific defect characteristics

Each defect type demands a customized approach for effective detection. Using the appropriate NDT methods—from basic visual inspection to advanced techniques—ensures a thorough evaluation and helps maintain structural integrity. Cracks in materials are seldom isolated; they often occur in various orientations and locations. Given the inherent limitations of detection sensors, which cannot inspect the entire thickness, circumference, area, or volume of a material at once, the strategic placement of sensors is essential.

An efficient strategy involves conducting an initial scan to pinpoint critical areas, followed by strategically placing sensors to collect more detailed data. This iterative process enables precise and comprehensive inspection. In scenarios where multiple cracks or faults are present, using a combination of NDT techniques ensures more reliable detection while reducing inconsistencies between methods. Defect types and their characteristics can be categorized into the following six groups:

- *Surface-breaking linear defects:* These flaws penetrate the surface, creating a linear crack that can extend into the underlying material. They include fractures, joints, shrinkage, laminations, leaks, or fissures. If left unchecked, they can propagate, leading to fractures and catastrophic failure. Commonly used NDT techniques include MT, radiography testing, and AE. These methods are particularly adept at identifying open cracks and monitoring defect propagation in real time. ECT and LPT offer moderate suitability for surface-breaking linear flaws.
- *Surface-breaking volumetric defects:* Volumetric defects extend to measurable depths and can coalesce, compromising structural integrity. Examples include holes or initial cracks that affect the material's surface area. Commonly used NDT methods include visual inspection, LPT, MT, UT, radiography, and ECT. These techniques effectively evaluate volumetric flaws at the surface, ensuring comprehensive detection and assessment.
- *Near-surface linear defects:* Found just below the surface, these defects are categorized as normal (perpendicular) or parallel to the surface. Normal defects often result from material inhomogeneities, residual stresses, mechanical loading, or environmental conditions,

whereas parallel defects usually arise from manufacturing processes like grinding and machining. ECT is highly suited for detecting near-surface linear defects due to its sensitivity to surface and near-surface flaws. Thermography and AE techniques offer moderate to high suitability, focusing on thermal patterns and stress wave releases associated with crack propagation.

- *Near-surface volumetric defects:* These non-linear, three-dimensional flaws are located just below the surface. They are critical in cast components or products, as they can significantly impair performance. UT, radiography, ECT, thermography, and AE are highly suitable. Visual inspection and LPT are less effective, while MT offers moderate effectiveness but may face limitations in fully characterizing the defect.
- *Subsurface linear defects:* Subsurface linear defects, typically found at varying depths within the material, are crucial to identify to prevent propagation and failure. These cracks are classified as near-surface but not directly on the surface. UT is particularly effective, providing detailed internal examination. Other methods like radiography can also be used to visualize internal linear cracks. Surface-focused methods like LPT and MT are unsuitable for these deeper flaws.
- *Subsurface volumetric defects:* These deeper flaws, such as voids and inclusions, occur within the material. They are critical for cast components, affecting overall structural integrity and performance. Ultrasonic and radiography testing are highly effective for subsurface volumetric defects. These methods can penetrate deep into the material to provide insights into internal irregularities. Techniques like LPT, MT, and visual inspection are unsuitable due to their surface focus.

6.6.2 Nine NDT techniques: Key insights and limitations

- Visual inspection
  - *Positive points:*
    - Simple and cost-effective.
    - Effective for detecting surface-breaking linear defects.
    - Can be enhanced with optical microscopes for better resolution.
  - *Negative points:*
    - Limited to surface-level defects; not suitable for subsurface flaws.
    - Dependent on the quality of lighting and inspector's experience.
- UT
  - *Positive points:*

    - Ideal for detecting subsurface flaws and measuring material thickness.
    - High sensitivity to small defects.
    - Can provide detailed images of internal structures.
  - *Negative points:*
    - Requires skilled operators and interpretation.
    - Surface condition and geometry can affect accuracy.
- LPT
  - *Positive points:*
    - Cost-effective and straightforward.
    - Excellent for revealing surface cracks and porosity.
  - *Negative points:*
    - Limited to surface defects; cannot detect subsurface or near-surface issues.
    - Time-consuming due to multiple steps (application, dwelling, cleaning, and inspection).
- MT
  - *Positive points:*
    - Effective for detecting surface and near-surface cracks in ferromagnetic materials.
    - Can identify flaws beneath the surface in suitable materials.
  - *Negative points:*
    - Inapplicable to non-ferromagnetic materials.
    - Requires magnetic field generation, which may not be suitable for all environments.
- ECT
  - *Positive points:*
    - Highly effective for detecting surface and near-surface defects in conductive materials.
    - Provides rapid inspection with minimal preparation.
  - *Negative points:*
    - Inapplicable to non-conductive materials.
    - Sensitive to material properties and surface roughness, which can affect accuracy.
- Radiographic testing
  - *Positive points:*
    - Offers detailed internal images, enabling detection of a wide range of defects, including volumetric cracks.
    - Suitable for complex shapes and varying material thicknesses.
  - *Negative points:*
    - Expensive and requires high-energy sources (X-rays or gamma rays).
    - Not practical for in situ applications; safety concerns due to radiation.

- Thermographic testing
  - *Positive points:*
    - Provides a broad overview, detecting surface and near-surface flaws through thermal imaging.
    - Cost-effective and capable of inspecting large areas quickly.
  - *Negative points:*
    - Requires high-quality equipment for detailed analysis.
    - Limited in detecting fine cracks without substantial post-processing.
- AE testing
  - Positive points:
    - Useful for real-time monitoring of crack initiation and propagation.
    - Non-intrusive, can detect defects during normal operation.
  - *Negative points:*
    - Requires sophisticated sensors and data interpretation.
    - Challenging sensor installation, particularly on rotating components.
- Stress–strain DIC
  - *Positive points:*
    - Effective for surface strain analysis and deformation monitoring.
    - Non-contact and applicable to complex geometries.
  - *Negative points:*
    - Limited to surface defects; cannot provide insights into sub-surface conditions.
    - Requires multiple images and post-processing, which can be time-consuming.

Each NDT technique offers unique advantages and faces specific limitations. Understanding these allows for selecting the most suitable method based on the material properties, defect characteristics, and inspection requirements. As technology evolves, the effectiveness of these methods may improve, underscoring the importance of staying updated with advancements in NDT technologies.

To facilitate this, we will use the weighted objective (WO) method, a type of optimization technique. Although many optimization methods exist in the literature, for the purposes of this course, we will concentrate on the WO method.

6.6.3 WO method for selecting NDT techniques

The "WO method" is a systematic approach designed to evaluate and select the most suitable NDT technique by assigning numerical weights to various critical parameters. This method helps balance multiple factors—such as efficiency, quality assurance, cost, and safety—ensuring an objective decision-making process.

6.6.3.1 Key components of the WO method for NDT

- *Defect detection and analysis:* The primary goal is not just to detect flaws but to comprehensively analyze and classify them. Detecting a limited number of flaws may indicate that the method is effective only for larger or more obvious defects, potentially missing smaller, critical ones. The ideal method should capture nearly all defects, meeting specified quality standards and aligning with the defined requirements.
- *Efficiency:* Efficiency in this context can refer to the following: (1) The speed of defect detection and evaluation. (2) The reduction in inspection time while maintaining high detection accuracy.
- *Goal:* The method should offer quick, reliable results without compromising thoroughness, ensuring high output in terms of the number and accuracy of detected flaws.
- *Cost considerations:* It is essential to consider not only the upfront cost of the equipment but also operational costs, including: (1) Training requirements, (2) Staffing costs correlated with inspection time, (3) Maintenance expenses.
- *Operational efficiency:* Equipment that can be used efficiently with minimal training is preferable, reducing long-term operational costs.
- *Safety:* Safety is a critical factor, especially when using methods involving hazardous materials or radiation (e.g., gamma rays, X-rays). Methods must ensure the safety of personnel, making this a paramount consideration in the selection process.
- *Quality assurance:* The method must provide precise and reliable data, which is crucial for maintaining structural integrity and preventing failure. A high-quality assurance level indicates consistent and repeatable results, essential for long-term reliability.
- *Time:* The quicker a method can deliver accurate results, the better, especially in scenarios where time is a critical constraint. Time efficiency directly impacts operational (i.e., measuring parameters and derived parameters) decisions and overall productivity.

6.6.3.2 Process of the WO method

- *Parameter selection and weighting:* Identify relevant parameters: efficiency, quality assurance, cost, safety, operational difficulty, and time. Assign numerical weights (relative weightage) to these parameters based on their relative importance. For example: $\text{Weight}_{\text{Safety}} = 0.3$

(30%); $Weight_{Efficiency}$ = 0.25 (25%); $Weight_{Quality\ Assurance}$ = 0.2 (20%); $Weight_{Cost}$ = 0.15 (15%); and $Weight_{Time}$ = 0.1 (10%). Summation of all the weights must be equal to 1.0.

- *Normalization:* Normalize the values of each parameter to ensure comparability, with scores ranging from 0 to 10. "0" represents the worst outcome (e.g., inapplicability or failure to meet minimum standards). "10" signifies the best outcome (e.g., optimal detection capability and operational ease).
- *Scoring and ranking:* Assign scores to each NDT technique based on its performance in the selected parameters, using data from literature, experimental results, etc. Apply weight factors related to each parameter, multiplying them by the corresponding scores. Calculate an overall score for each technique by summing the weighted scores. Rank the techniques in descending order (from 10 to 0), with the highest overall score typically indicating the preferred method. However, specific conditions may prioritize a technique with lower scores in certain key parameters.

This methodology is versatile, extending beyond NDT to applications such as coating evaluation, tribometer testing, and surface engineering. It systematically assesses each method's effectiveness against predefined criteria, minimizing subjectivity and improving decision-making accuracy.

Let us consider an example of NDT selection, covered by Lee et al.[15] The NDT selection procedure described by Lee et al.[15] involves the following steps:

- *Survey design and objective:* A survey was designed to evaluate the usage of NDT methods in bridge inspections across U.S. state Departments of Transportation (DOTs). The survey aimed to identify commonly used NDT methods for concrete and steel structures, their effectiveness, ease of use, and challenges faced during their application.
- *Selection of NDT methods:* Sixteen NDT methods were selected for concrete and nine for steel structures. These methods included AE, ground-penetrating radar, UT, magnetic particle, and radiographic testing, among others.
- *Survey distribution and response:* The survey targeted bridge inspection unit managers and was distributed to all 52 states and territories, with a total of 40 responses.
- *Evaluation criteria:* Respondents rated NDT methods based on their frequency of use, effectiveness, and difficulty of application. They also identified bridge components commonly inspected using NDT, such as decks, beams, girders, and substructure elements like piers and abutments.

- *Scoring and ranking:* The data was analyzed using scoring systems to determine the preference and effectiveness of each NDT method. Methods were ranked based on their ease of use and effectiveness, with specific scores assigned to reflect the frequency and degree of their application.
- *Findings:* The results highlighted the preference for certain NDT methods, such as penetrant testing and UT, which were found to be both effective and relatively easy to use. The study also noted challenges in using some advanced methods due to their complexity and the need for specialized training.

This study underscores the critical importance of collecting comprehensive data to inform decisions in selecting NDT methods. Relying solely on individual judgment can introduce subjectivity, leading to inconsistent or biased evaluations. By gathering input from multiple sources and experts, the study ensures a more objective, well-rounded assessment of each NDT method's performance. This collective approach helps to mitigate individual biases and enhances the reliability of the findings.

The systematic methodology adopted in this section facilitates a thorough evaluation of various NDT techniques. By analyzing data on effectiveness, ease of use, and practical application, the study provides valuable insights that guide the selection of the most suitable NDT methods for specific inspection requirements. This ensures that the chosen techniques are not only theoretically effective but also practically viable, optimizing inspection outcomes and resource utilization.

Let us consider one example in which one needs to determine some NDT method for evaluating faults in the welded joints. In selecting the most suitable NDT methods for evaluating welded joints, it is essential to adopt a systematic approach that incorporates multiple weighted criteria. This ensures a balanced and objective decision-making process tailored to the specific needs of each inspection scenario. First of all, there is a need to identify key factors:

- *Efficiency:* Efficiency refers to the ability of an NDT method to detect defects within a specified timeframe. This factor is critical in industries, where inspection speed impacts productivity. In the weighted optimization method, this factor could be assigned a high weight to prioritize methods that minimize downtime without compromising the inspection's thoroughness.
- *Quality assurance level:* This factor ensures that the chosen NDT method reliably detects defects to meet stringent quality standards. A higher weight is assigned to this criterion when structural integrity and safety are paramount, ensuring that only methods with proven reliability are considered.

- *Degree of difficulty:* The complexity of implementing an NDT method—considering required expertise, training, and equipment—impacts its practicality. Methods that are technically challenging yet essential may receive moderate weight, balancing feasibility with necessity.
- *Economic criteria:* Economic factors include costs for equipment, personnel, training, and maintenance. Depending on budget constraints, this criterion can be heavily weighted to favor cost-effective solutions that deliver reliable results within financial limits.
- *Safety:* Safety considerations are crucial, especially in environments with potential hazards such as radiation exposure or physical risks. A high weight on safety ensures that methods minimizing risks to personnel and the environment are prioritized.
- *Time:* Timeliness is critical in industries requiring prompt inspections. Assigning a high weight to this factor emphasizes the importance of quick and accurate results, aligning with operational timelines.

Each NDT method is evaluated against these weighted criteria. For example, methods like "UT" and "Radiographic Testing" might score high on efficiency and quality assurance but may have differing scores on economic and safety factors. The overall score for each method is calculated by summing the weighted scores across all criteria. However, the NDT method may change based on the weld defects in one particular component or system. Therefore, there is a need to have comprehensive knowledge and database to make decisions. For example, welding defects may be:

- *Slag inclusion:* Detected using radiographic or ultrasonic methods, prioritized for their efficiency and accuracy in identifying internal flaws.
- *Fatigue cracks:* UT is highly effective for detecting cracks caused by cyclic loading.
- *Setup errors (e.g., porosity, lack of fusion):* Methods like dye penetrant testing or visual inspection, combined with UT, ensure thorough evaluation of surface and subsurface defects.
- *Environmental damage:* Corrosion damage is the primary environmental damage. Techniques such as magnetic particle inspection and ECT are used to assess corrosion and other environmental impacts.

6.6.3.3 Case study: Evaluating NDT techniques for welded joints

In this case study, we will explore various NDT techniques for detecting flaws in welded joints, particularly in the context of manual welding practices. We be using hypothetical data, may not correct from practical point of view, but will provide better understanding of using weight objective method for selecting NDT techniques. My experience indicates that manual welding often leads to numerous defects. For example, in bridge construction, on-site welding frequently results in significant flaws in welded joints.

Importance of flaw detection: If defects are present at the beginning of a bridge's construction, they are likely to worsen over time, potentially leading to structural failures within five to six years. This can significantly increase maintenance costs if these joints are not properly manufactured. Our discussion will focus on the characteristics of welded joints, the types of flaws they may exhibit, and methods for diagnosing those flaws.

Key parameters for evaluation: To conduct our analysis effectively, we will consider the following parameters: efficiency, quality assurance, degree of difficulty, economic factors, safety, and time. Each aspect will be assigned numerical values. According to the literature, five NDT techniques have been widely utilized for evaluating welded joints: UT, magnetic particle inspection, AE, radiography, and ECT. These techniques are essential for diagnosing flaws in welded joints.

6.6.3.3.1 Common flaws in welding

Common welding flaws, particularly in manual processes, often arise from inclusions and other defects. During welding, issues such as cracks, pores, or trapped gas bubbles can occur due to the melting and solidification of materials on-site. Uneven temperature control can also lead to cavities as the weld material fills gaps. One prevalent issue is *incomplete penetration*, where the weld does not reach the necessary depth, resulting in a lack of fusion. This occurs when solidification happens before the molten material adequately fills the intended area. To visualize these issues, I have included Figure 6.8 that illustrates examples of incomplete penetration, where the

```
% List the required parameters
% Efficiency ratio (ER)----High
% Degree of difficulty (DOD) --
-- Low
% Cost (C)---- Low
% Operation Time (OT) --- Low
% Input
% Parameter matrix
% ParMat= [ER DOD C OT]
ParMat=[1.40, 3.11, 3338, 3;
1.99,2.08, 7568,2; 0.88,4.07,
10000, 20; 0.99, 4.31, 3713,5;
1.37,2.68,2000,3];
```

```
ParMat =

   1.0e+04 *

    0.0001    0.0003    0.3338    0.0003
    0.0002    0.0002    0.7568    0.0002
    0.0001    0.0004    1.0000    0.0020
    0.0001    0.0004    0.3713    0.0005
    0.0001    0.0003    0.2000    0.0003
```

Figure 6.8 Common flaws in the manual welding process.

molten metal fails to reach certain areas, leading to inadequate bonding between layers and leaving gaps or trapped inclusions. The presence of inclusions increases the risk of cracking, particularly when combined with potential corrosion. Additionally, small gas particles can form during solidification, contributing to porosity, another significant defect that can lead to delamination. Moreover, we must consider transverse cracks (perpendicular to the weld direction) and longitudinal cracks (parallel to it). These cracks can be detected through visual inspection; however, we will focus on the five aforementioned techniques recognized for their effectiveness in detecting welding flaws.

Uneven cooling during solidification can lead to issues, particularly when liquid is trapped within the weld. Such trapped liquid—referred to as slag or inclusions—compromises the weld's integrity and, consequently, the structural integrity of the entire assembly. NDT methods are essential for evaluating the presence of any slag or other defects.

If defects are identified, we can either remove them or refine our welding procedures to minimize slag formation in future applications. Based on my experience, automated welding processes can be fine-tuned effectively, whereas manual welding often remains subjective, leading to a higher incidence of defects due to improper procedures. Operators in high-temperature environments may struggle to detect issues like porosity, lack of fusion, or undercuts.

6.6.3.3.2 Automation in welding

I advocate for increased automation in welding despite the challenges and costs associated with such systems, which is why manual welding remains prevalent. However, when the complete welding procedure is not adhered to, problems such as inadequate joint preparation, improper edge formation, and misaligned electrodes frequently arise.

In environments prone to corrosion and erosion, even minor defects from the welding process can escalate into significant issues over time. Corrosion and erosion can exacerbate these defects, leading to severe consequences. By using NDT methods, we can promptly evaluate and repair defects, maintaining the integrity of components. However, contaminants such as dirt or oil can infiltrate the weld during the welding process. If not adequately removed, or if high temperatures cause oxide or sulfide formation, these contaminants may become inclusions, leading to additional complications. For instance, insufficient fusion can result in incomplete penetration at the weld junction, creating serious welding defects that prevent proper bonding. This can lead to issues such as delamination or crack formation. The use of gases during welding can also contribute to gas bubble interruptions, causing porosity, which can manifest as blowholes.

If a crack occurs, it can propagate continuously, exacerbating structural issues. A larger crack, particularly one exceeding a certain size, can lead to significant problems. Therefore, it is crucial to ensure that weld junctions are manufactured correctly and that appropriate measures are taken to address any defects.

6.6.3.3.3 Methodology: WO method

We will evaluate five NDT techniques: UT, magnetic particle inspection, AE testing, radiography, and ECT. It is important to clarify that the order of these techniques does not imply effectiveness ranking but reflects a sequence for analysis.

Data for this evaluation can be obtained through experimental work conducted in our laboratory or from existing literature. If the necessary data is not readily available, we can digitize relevant curves for analysis.

Table 6.1 presents a comparative analysis of various NDT techniques. It includes details on the efficiency ratio, degree of difficulty, cost (Rs/ft), and time (in minutes) (excluding setup time) for each method. The parameters presented in Table 6.1 have been compiled from available literature sources. This table provides a comparison that helps in selecting an appropriate NDT technique based on efficiency, cost, difficulty level, and required inspection time.

In this analysis, we have gathered efficiency ratios from the literature: UT has an efficiency ratio of 1.4, MT has a ratio of 1.99, AE testing is slightly lower at 0.88, radiography is at 0.99, and ECT has a ratio of 1.37. Next, we assess the degree of difficulty associated with each method: UT scores 3.11, magnetic particle inspection scores 2.08, AE testing scores 4.07, radiography scores 4.31, and ECT scores 2.68. It is evident that we aim to enhance the efficiency ratios for improved outcomes. However, it is essential to minimize the degree of difficulty associated with these testing methods. Our goal is to achieve higher values for certain parameters while aiming for lower values in others. When focusing solely on the efficiency ratio, MT emerges

Table 6.1 Comparison of key parameters for various NDT techniques

NDT technique	*Efficiency ratio*	*Degree of difficultly*	*Cost (Rs)/ft*	*Time (minutes) (not included setup time)*
Ultrasonic testing	1.40	3.11	3338	3
MT	1.99	2.08	7568	2
Acoustic emission	0.88	4.07	10,000	20
Radiography testing	0.99	4.31	3713	5
Eddy current testing	1.37	2.68	2000	3

as the most favorable option, boasting a value of 1.99. Furthermore, when considering the degree of difficulty—which should ideally be minimized—MT again proves to be the best choice.

In terms of cost per foot, data from a specific source indicates that UT costs approximately Rs. 3338 per foot, while MT is priced at Rs. 3568, AE testing around Rs. 10,000, radiography at Rs. 3730, and ECT is the most economical at Rs. 2000. Thus, from a cost perspective, ECT is the most attractive option due to its lower expenses.

When examining setup time—which does not encompass the complete setup process—MT again demonstrates efficiency, requiring only 2 minutes for setup. In comparison, UT and ECT each take 3 minutes, while AE testing requires a significantly longer 20 minutes. However, advancements in AE technology, such as the development of easier-to-mount sensors, may potentially reduce setup time to between 1 and 3 minutes.

Considering these three parameters—efficiency ratio, degree of difficulty, and setup time—MT generally stands out as the most favorable option. Yet, when evaluating cost, ECT surpasses the others. The challenge remains in determining which parameter is the most critical: cost, time, or degree of difficulty. This dilemma underscores the importance of using the WO method.

To compile this data, we will utilize MATLAB, which is provided in Figure 6.9. We will use MATLAB to create a comprehensive table using a parameter matrix, where "ER" denotes efficiency ratio, "DOD" signifies degree of difficulty, "C" represents cost, and "OT" stands for operating time. This matrix will facilitate effective organization and analysis of the data.

Our analysis will focus on the operating points, utilizing values such as 1.4, 3.11, 3.33, 3.8, and 3 in a 4 × 5 matrix format. This structure will aid in determining which method will be selected based on its ranking. In addition, we must consider how to prioritize time, cost, or degree of difficulty.

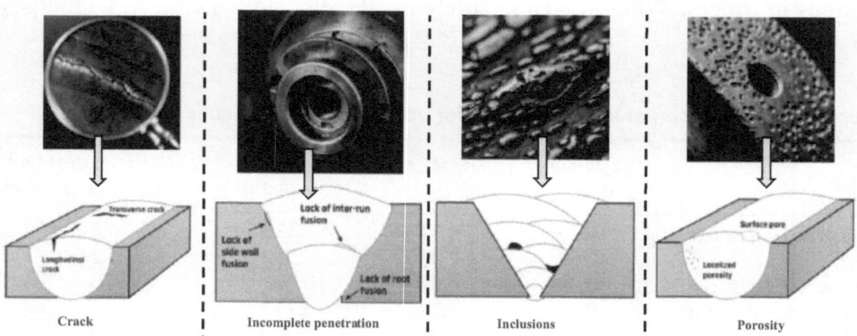

Figure 6.9 MATLAB code and parametric matrix.

Table 6.2 Digital logic method for determining parameter weight factors

Parameters	*1*	*2*	*3*	*4*	*5*	*6*	*7*	*8*	*9*	*10*	*Total*	*Weight of factor*
Efficiency ratio	1	1	1	1							4	0.4
Degree of difficulty	0				0	0	1				1	0.1
Cost		0			1			1	1		3	0.3
Time			0			1		0		1	2	0.2
Dummy				0			0		0	0		

To achieve this, we will use the digit logic method, which involves comparing only two parameters at a time, despite having four parameters overall. It is essential that no parameter is assigned an absolute value of zero, as this would remove it from the algorithm. Thus, a minimum value of at least 1 on the time scale, or 0.1 if normalization is applied, should be maintained for all parameters.

In this digit logic method, we will create a comparison table for the parameters: efficiency ratio, degree of difficulty, cost, and time. A fifth parameter, a dummy variable, will also be introduced, assigned a value of zero, while all other parameters will receive comparative values relative to the dummy (Table 6.2).

To determine the number of columns needed for this table, we calculate based on the number of parameters: multiplying the number of factors by the total number of parameters plus one (to include the dummy parameter) and dividing by two. This results in a total of ten columns, facilitating comparisons among all parameters.

The process will begin by comparing the efficiency ratio with the degree of difficulty, allowing us to assess whether the efficiency ratio should hold more importance. In this case, we establish that the efficiency ratio is indeed more significant than the degree of difficulty, thus assigning it a value of one while the degree of difficulty receives a value of zero. Next, we will compare the efficiency ratio with cost, again determining that the efficiency ratio should be prioritized, so it is assigned a value of one, with cost receiving a value of zero.

When comparing the efficiency ratio against operating time, we conclude that the efficiency ratio is more critical, resulting in another assignment of one to the efficiency ratio. All parameters will score a value of one when compared with the dummy variable.

Moving on to the degree of difficulty versus cost, we conclude that cost is more important, leading to a score of one for cost and zero for degree of difficulty. Similarly, when comparing degree of difficulty against time, we find that time is more important, resulting in a score of one for time and zero for degree of difficulty.

Given that the maximum possible score is 10, we normalize the scores by dividing each by 10:

- Time: 2/10 = 0.2
- Degree of difficulty: 1/10 = 0.1
- Cost: 3/10 = 0.3
- Efficiency ratio: 4/10 = 0.4

This normalization yields weighting factors of 0.4 for efficiency, 0.3 for cost, 0.2 for time, and 0.1 for degree of difficulty, reflecting the relative importance of each parameter, with efficiency holding the highest weight.

This weighting will be documented in the MATLAB code, where the relative importance of the parameters—efficiency ratio (ER), degree of difficulty (DOD), cost (C), and operating time (OT)—is assigned as 0.4, 0.1, 0.3, and 0.2, respectively.

Subsequently, we will establish the utility score for each parameter, as described in Eqs. (6.1) and (6.2).

$$\frac{\text{Objective value } 0.9{*}\text{Minimum value}}{1.1{*}\text{Maximum value } 0.9{*}\text{Minimum value}} {*} 10 \tag{6.1}$$

Establishing utility score for maximization

$$10 - \frac{\text{Objective value } 0.9{*}\text{Minimum value}}{1.1{*}\text{Maximum value } 0.9{*}\text{Minimum value}} {*} 10 \tag{6.2}$$

Establishing utility score for minimization

For instance, we have the following efficiency ratio values from the literature:

- UT: 1.4
- MT: 1.99
- AE: 0.88
- Radiography: 0.99
- ECT: 1.37

These values must be normalized, as they may be measured in different units. We will apply the utility score formula (Eqs. (6.1) and (6.2)), focusing on the objective value for each parameter. For efficiency, the minimum objective value in this case is 0.88, while the maximum is 1.99. It is important to note that, regardless of the method we have chosen—which currently involves five parameters—there remains a possibility for additional techniques to yield lower efficiency ratios than 0.88, or conversely,

Table 6.3 Calculation of overall utility scores

Overall ranking		*Score*
Ultrasonic testing	4.353 × 0.4 + 5.68 * 0.1 + 8.328 * 0.3 + 9.405 × 0.2	6.688
Magnetic particle testing	1.422 × 0.4 + 9.275 × 0.1 + 3.730 × 0.3 + 9.90 × 0.2	7.4570
Acoustic emission	0.629 × 0.4 + 2.339 × 0.1 + 1.086 × 0.3 + 0.99 × 0.2	1.0093
Radiography testing	1.4157 × 0.4 + 1.503 × 0.1 + 7.920 × 0.3 + 8.415 × 0.2	4.775
Eddy current testing	4.132 × 0.4 + 7.184 × 0.1 + 9.782 × 0.3 + 9.405 × 0.2	7.1868

to identify methods with efficiency ratios exceeding 1.99. Thus, we remain open to future developments in this area.

To address our analysis, we apply a method that involves multiplying the maximum value by 1.1 and the minimum value by 0.9. This ensures that we maintain a range that avoids extreme values of 0 or 10, allowing our results to always fall within the interval of 0 to 10. This approach enhances the scope of utility, which we aim to maximize. In cases where minimization is required, we will calculate the score by subtracting it from 10 (Eqs. (6.1) and (6.2)).

Once we compute the utility score of NDT techniques for each parameter, overall scores can be calculated as shown in Table 6.3. After evaluating the utility scores across the four factors for each technique, we found that MT is the most suitable method for welding inspections.

Final rankings and considerations: MT leads in effectiveness, followed by ECT, UT, radiography, and AE testing. However, it is essential to consider material-specific implications. For instance, while MT is effective for ferromagnetic materials, its applicability in aluminum welding requires further examination. In summary, this case study underscores the need for appropriate NDT techniques in evaluating welded joints, particularly in sectors like bridge construction. Using MT ensures effective flaw detection and proper maintenance of structural integrity over time. By integrating advanced NDT methods, we can significantly enhance the quality of welded components and reduce long-term maintenance costs.

6.7 CONCLUSIONS

This chapter highlights the essential role of testing in engineering to identify faults that could lead to surface failure and compromise system integrity if not addressed in time. It emphasizes the need for a balanced approach between DT and NDT. DT provides critical insights into material properties, whereas NDT allows for continuous monitoring and preventive maintenance without causing damage. Advanced techniques like UT, ECT, radiography, AE, IR thermography, and stress–strain DIC testing improve the ability to detect early signs of degradation. By using these methods, engineers can enhance durability, ensure safety, and meet modern standards, contributing to sustainable development and innovation in materials science and system design.

REFERENCES

1. Li, L., Liu, Z., Zhao, H., Xue, L., & Wu, J. (2024). The bearing surface defect detection method combining magnetic particle testing and deep learning. *Applied Sciences*, *14*, 1747. https://doi.org/10.3390/app14051747
2. Xie, R., Chen, D., Pan, M., Tian, W., Wu, X., Zhou, W., & Tang, Y. (2015). Fatigue crack length sizing using a novel flexible eddy current sensor array. *Sensors*, *15*(12), 32138–32151.
3. He, Y., Tian, G., Zhang, H., Alamin, M., Simm, A., & Jackson, P. (2012). Steel corrosion characterization using pulsed eddy current systems. *IEEE Sensors Journal*, *12*(6), 2113–2120.
4. Article published on Ultrasonic Testing, www.unitedgamma-ndt.com/ultrasonic-testing/
5. Common literature available on Ultrasonic testing, https://en.wikipedia.org/wiki/Ultrasonic_testing
6. Garcia Marquez, F. P., & Gomez Munoz, C. Q. (2020). A new approach for fault detection, location and diagnosis by ultrasonic testing. *Energies*, *13*(5), 1192.
7. Moreira, E. V., Simões, H. R., Rabello, J. M. B., Camargo, J. R. D., & Pereira, M. D. S. (2010). Digital radiography for the inspection of weld seams of pipelines—better sensitivity. *Welding International*, *24*(4), 249–257.
8. Van Tittelboom, K., Van Belleghem, B., Boone, M. N., Van Hoorebeke, L., & De Belie, N. (2018). X-ray radiography to visualize the rebar–cementitious matrix interface and judge the delay in corrosion through self-repair by encapsulated polyurethane. *Advanced Materials Interfaces*, *5*(3), 1701021.
9. Curà, F., Mura, A., & Corsaro, L. (2022). Experimental investigation on crack propagation paths in spur gears. In *IOP Conference Series: Materials Science and Engineering* (Vol. 1214, No. 1, p. 012029). IOP Publishing.
10. Simonov, D., Vavilov, V., & Chulkov, A. (2020). Infrared thermographic detector of hidden corrosion. *Sensor Review*, *40*(3), 283–289.
11. Ibarra-Castanedo, C., Tarpani, J. R., & Maldague, X. P. V. (2013). Nondestructive testing with thermography. *European Journal of Physics*, *34*(6), S91.
12. Calabrese, L., & Proverbio, E. (2021). A review on the applications of acoustic emission technique in the study of stress corrosion cracking. *Corrosion and Materials Degradation*, *2*(1), 1–30. https://doi.org/10.3390/cmd2010001
13. Toutountzakis, T., Tan, C. K., & Mba, D. (2005). Application of acoustic emission to seeded gear fault detection. *NDT & E International*, *38*(1), 27–36.

14. Blikharskyy, Y., Kopiika, N., Khmil, R., Selejdak, J., & Blikharskyy, Z. (2022). Review of development and application of digital image correlation method for study of stress–strain state of RC Structures. *Applied Sciences*, *12*(19), 10157. https://doi.org/10.3390/app121910157
15. Lee, S., & Kalos, N. (2015) Bridge inspection practices using non-destructive testing methods. *Journal of Civil Engineering and Management*, *21*(5), 654–665.

Chapter 7

Digitization of established curves and data-driven approaches

The main emphasis of this chapter is on the digitization of known curves and the use of data-driven approaches, a newly developed field of interest that attracted a lot of attention during the previous 3 to 4 years. Unlike traditional approaches, which typically depend on well-defined interactions between variables inside the design domain, this topic remains less investigated in the practical/in-service domain. For example, whereas sophisticated methods such as boundary element method (BEM) and finite element method (FEM) show great efficiency during the design stage, their applicability usually decreases throughout the operating or maintenance stages. In these latter phases, the data-driven methods provide more dependable and pragmatic results.

This chapter will investigate the idea of digitization—especially of existing curves—and go over its importance. Commonly found in engineering and financial literature, these curves have been used for decades to graphically depict data. We will look at how switching from these conventional graphical depictions to a digital, data-driven paradigm might improve research and practice-based analysis and decision-making. We will discuss this. First, let us define digitization and talk about the function of existing curves in various fields of engineering.

7.1 IMPORTANCE OF DIGITIZATION AND DATA-DRIVEN APPROACHES IN ENGINEERING

Within the framework of this book, "digitization" is the process of digitalizing graphs, tables, data, knowledge, or actual objects. Computers can rapidly access, preserve, and evaluate analog materials turned into digital forms. Digitization allows one to evaluate, correct, and apply data using computer tools, therefore enabling advanced techniques such as predictive analytics, simulations, and data-driven modeling. Often data-driven modeling requires "Trend lines." A trend line in Excel is a graphical representation that shows the general pattern or direction of data points in a chart.

DOI: 10.1201/9781003545019-7

It is often used to analyze trends over time and to make predictions. Excel offers different types of trend lines, such as linear, exponential, and polynomial. Users can also customize the trend line to display the equation and the R-squared value, which indicates the line's accuracy. In MATLAB, a trend line can be plotted using built-in functions to analyze data patterns and trends. This is typically achieved through linear regression or other fitting techniques. MATLAB allows users to plot and customize trend lines using commands like "polyfit" for polynomial regression or "fit" for more complex models. MATLAB provides flexibility in adjusting the trend line properties and in automating data analysis using scripts.

One of the most straightforward ways to convert graphs into data is through the use of "Origin Software." Origin is a powerful data analysis and graphing tool widely employed in scientific and engineering disciplines, designed to handle large datasets and perform complex statistical analyses. It is particularly effective for visualizing and analyzing data trends, such as corrosion behavior, using a variety of graph types and customizable plots. One version of this software is free and can be installed on any laptop or desktop computer. To get started, launch Origin, create a new project, and import the graph by selecting "File" > "Open" from the menu bar, then choose the image file and click "Open." The next step is image calibration, where a known reference point or object within the image, with a specified length or dimension, is identified to calibrate the scale. Using Origin's measurement tool, the reference object is measured, and its actual value is entered. After calibration, the digitization process continues with image tracing, where features or data points are manually traced for extraction. Once these features are traced, they are converted into numerical data points. Depending on the data's complexity, tools like "Digitize" or "Data Extraction" are used to extract these points, facilitating the acquisition of x–y coordinates from the traced curves or lines with just a few clicks. More information can be found on website.[1]

Digitizing non-destructive testing (NDT) techniques is turning manually collected inspection data into easily shareable and analyzed digital representations. For ultrasonic or radiographic testing data, for instance, digital storage allows machine learning algorithms to be applied to forecast defect characteristics and suggest the most efficient inspection techniques. This change not only helps to create predictive maintenance plans but also improves safety by tackling possible breakdowns before they start.

Digitization and data-driven approaches reflect a major change in NDT techniques—examined in the last chapter. Whether for surface-breaking, near-surface, or subsurface defect discovery, the choice of NDT methods has always depended on accepted criteria and empirical information. However, digitization lets engineers methodically examine large amounts of data from many NDT checks. Advanced analytics can find trends, improve technique selection, and increase problem detection dependability by grouping these results into a centralized digital database.

Additionally, established engineering safety tools, such as safety factor (FoS) curves, benefit from digitization. Integrating real-time NDT data enables these curves to dynamically update safety assessments based on operational conditions. For instance, acoustic emission data can be used to adjust the FoS for components under varying stress conditions, leading to more accurate and responsive safety evaluations. This synergy between digitization and data-driven insights ultimately advances both the reliability and safety of engineering practices.

By providing subtle, context-specific insights, a data-driven approach improves NDT techniques. Though usually successful for subsurface defect detection, methods like ultrasonic testing might exhibit different performances depending on material qualities or flaw direction. For their lightweight and great strength, composite materials are preferred, for instance, in the aircraft sector. Their varied character, however, can make typical ultrasonic testing less able to detect delaminations or voids precisely. Engineers can improve detection algorithms, spot fault trends, and increase flaw detection in these complicated materials by digitizing inspection data and using machine learning.

In other words, digitalization and data-driven approaches are changing NDT methods from conventional empirical procedures to dynamic, data-informed processes. Through pattern recognition and bottleneck identification, this change improves fault detection dependability and simplifies operations. By use of predictive models, statistical analysis of digital data aids risk assessment and enhances decision-making.

Pipelines in the oil and gas industry are routinely radiographically tested for corrosion and cracks. Advanced image processing and machine learning analysis made possible by digitizing radiography pictures allows one to forecast corrosion likelihood under various operating conditions. This kind of focused maintenance improves safety and lowers downtime.

Effective data use in system maintenance can be explored to depend on digitization. Turning analog or pictorial data into digital forms lets variables have numerical values assigned, hence facilitating exact analysis and integration into MATLAB and other software tools. In automotive engineering, for example, digital data from dynamometer tests helps improve engine performance models, therefore guiding more effective designs. In power plants, too, eddy current testing data can be digitized to expose patterns in material properties resulting from continuous operation, therefore supporting early defect discovery and maximizing maintenance schedules. By concentrating on pertinent data, improving it through further tests if necessary, and revealing latent linkages and trends, a data-driven strategy helps one to better decision-making.

Although Excel and other conventional tools have been adequate for simple trend analysis, the increasing complexity of data calls for more advanced techniques like machine learning algorithms and related software. Real-time data from acoustic emission sensors on turbine blades can be digitized in the wind energy field to track stress and find microcracks. By means of proactive maintenance, the blade lifetime is extended, and the catastrophic failure risk is lowered.

Recent developments in machine learning algorithms have greatly improved the capacity to analyze vast datasets, find hidden links, and create predictive models, thereby optimizing and guiding decision-making processes. Digital sensor data collecting of stress, strain, and vibration in structural health monitoring of bridges can be handled using machine learning to detect possible structural flaws. By means of prompt interventions, this helps to prevent mishaps and guarantees the integrity and safety of the infrastructure.

7.2 TRADITIONAL CURVE REPRESENTATIONS IN ENGINEERING

Development of any design or manufacturing model normally requires many experiments to ensure correctness and dependability. Still, we may digitize the easily available graphs from textbooks, standards, or other sources and use such data and data-driven methods to provide interesting analysis and create a first model. By lowering the demand for intensive research, this method greatly simplifies the model-building process.

Although data is quite useful in providing insights, it is still important to execute a few tests to confirm theories against pragmatic environments. Instead of thousands of tests, reliable results can come from completing between 50 and 100 experiments and modifying the created initial model—a more sensible approach. Data analysis for trends, anomalies, and bottlenecks helps us optimize operations more effectively. Furthermore, digital curves serve as a supporting tool, enhancing our analytical approach.

To illustrate these concepts, let's consider a common example of the surface finish factor, as shown in Figure 7.1, which is used to adjust the material's endurance limit. As fatigue cracks often begin at the surface, the condition of the surface plays a critical role in a component's fatigue resistance. To reduce the influence of surface finish, test specimens undergo polishing. The level of surface damage is affected by both the processing methods and the material's strength, with higher-strength materials being more susceptible to surface damage. To account for this in the analysis, the surface finish factor is defined as a function of ultimate tensile strength and various manufacturing finishing processes. Using various established graphs (such as shown in Figure 7.1), students select K_{SF}.

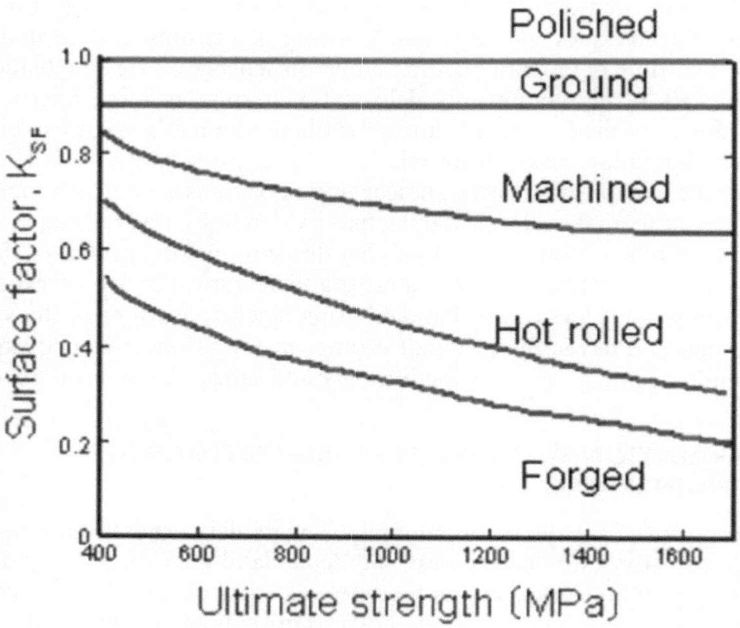

Figure 7.1 Surface finish factor versus UTS.[2]

Table 7.1 Constant and exponent related to surface finish factor

Finishing method	*Constant a*	*Exponent b*
Ground	1.58	–0.085
Machined or cold drawn	4.51	–0.265
Hot rolled	57.7	–0.718
Forged	272	–0.995

In practice, selecting the surface finish factor K_{SF} may serve as a good starting point, but it should be adjusted to reflect the specific in-situ conditions. To facilitate this adjustment, the digital equation $K_{SF} = a\left(S_{ut} \text{ in MPa}\right)^b$ is employed. Initial values for the constant aaa and the exponent bbb are provided in Table 7.1. By conducting experiments that vary the surface finish, K_{SF} can be refined using the collected data, allowing for modifications to the expression $K_{SF} = a\left(S_{ut} \text{ in MPa}\right)^b$.

An important classic example of conventional curve representations is the Wear Map, a necessary instrument in materials science and tribology. The Wear Map helps one to see and understand the wear mechanisms of materials under several operational environments. Through observable criteria like pressure and velocity, this depiction helps operators and engineers project wear behavior. Verifying these approximations through a limited number of trials enables more informed decisions regarding material selection and maintenance techniques.

7.2.1 Key components of the wear map

- *Pressure and relative velocity:* In the context of the present, it can be said that the interaction between components during operation is greatly influenced by pressure and velocity. A wear map can be utilized for estimating the wear behavior of the material pair.
- *Field boundaries:* The wear maps indicate regions where different wear mechanisms yield the same wear rate. The wear maps provide a visual representation of when one wear mechanism may dominate over another. Understanding these boundaries helps identify operational conditions under which specific wear mechanisms are more likely to occur.
- *Wear rate contours:* Within the wear map, contours represent the total wear rate, reflecting the cumulative effect of all contributing wear mechanisms, such as adhesive, abrasive, and corrosive wear. The contours visually depict wear rates across different combinations of pressure and velocity, assisting operators in identifying areas of high wear potential and adjusting operational parameters accordingly.
- *Influencing factors on oxide layer thickness:* The thickness of the protective oxide layer, which is critical for wear resistance, is influenced by three main factors:
    - *Rupture time:* The time required for the oxide layer to break down under mechanical stress. A shorter rupture time suggests a weaker oxide layer that is more susceptible to wear.
    - *Re-oxidation time:* The duration needed for the oxide layer to reform after rupture. A longer re-oxidation time may leave the material vulnerable to wear during this period, as the protective oxide layer is absent.
    - *Formation rate:* The speed at which the oxide layer can form under specific conditions. A higher formation rate enhances wear resistance by rapidly replenishing the protective oxide layer after damage.

The wear map is a crucial tool for understanding wear mechanisms, providing a visual representation of the relationships among pressure, velocity, wear rate, and the dynamics of the oxide layer. By utilizing this map, engineers and operators can make informed decisions that improve material performance, extend component life, and lower maintenance costs in various industrial applications. Its integration with modern data-driven approaches reflects the evolving engineering landscape and underscores the significance of informed decision-making for optimizing performance and efficiency. Additionally, wear maps allow maintenance teams to assess how different factors affect wear rates, enabling proactive planning for maintenance and material replacements before critical failures occur. They also highlight areas of high wear potential, guiding operators to prioritize maintenance efforts on the most vulnerable components, thus optimizing resource allocation. Ultimately, insights from wear maps empower engineers to make strategic choices regarding material selection and operational adjustments, enhancing overall reliability and performance.

Machine learning algorithms have the capability to analyze large volumes of operational and wear-related data collected from various sources (book, manual, and laboratory) and update those using in situ sensors, facilitating a deeper understanding of wear patterns and behaviors beyond what traditional methods offer. By training these algorithms on historical data, engineers can predict future wear rates and maintenance needs with greater accuracy, uncovering hidden relationships and patterns that may not be visible in conventional wear maps. For example, in a manufacturing setting, machine learning can analyze sensor data to detect subtle changes in vibration patterns that correlate with increased wear in machinery components. This integration allows for real-time monitoring, as predictive algorithms continuously assess sensor data, providing timely alerts for maintenance before significant wear occurs. Unlike static wear maps, machine learning can dynamically update predictions based on new data, ensuring that maintenance strategies adapt effectively as operational conditions change. Additionally, machine learning enhances wear maps by identifying anomalies in wear patterns that may indicate underlying issues, enabling proactive maintenance interventions that prevent unexpected failures. In summary, while traditional wear maps are essential for visualizing wear mechanisms and supporting informed decision-making, their effectiveness is significantly improved when integrated with machine learning algorithms, resulting in enhanced predictive maintenance strategies that optimize component lifespan and minimize downtime.

7.3 CHALLENGES ASSOCIATED WITH MANUAL DATA HANDLING AND ANALYSIS

Conventional methods of manually organizing and interpreting data create considerable challenges to modern engineering and maintenance practices. This is especially significant considering the increasing complexity and interconnections among systems. One of the most important of the several challenges manual data handling offers is the inherent risk of human error.

Among the problems that could arise during data entry, analysis, and interpretation are typographical errors, erroneous computations, and inappropriate data input. These errors have the potential to lead to erroneous conclusions, potentially compromising the integrity or safety of the system. The possibilities of mistakes increase with increasing volumes of data.

Manual data collection, organizing, and analysis can considerably slow down the decision-making process. This is particularly true in companies where quick responses are vital. Regarding maintenance and operations, for instance, delays caused by hand data processing could cause longer periods of downtime, higher running costs, and lost opportunities for early interventions. Achieving both accuracy and efficiency in the fast-changing fields of engineering and maintenance depends on the transition from human data processing to automated, data-driven methods. Although hand data processing has significant historical relevance, it presents several challenges that might compromise operational efficiency, decision-making, and system general performance.

Manually driven data handling often results in ineffective human resource utilization. Skilled professionals could find themselves devoting a lot of time to menial data tasks instead of focusing on more significant operations. This misallocation could cause poorer productivity and a slower rate of invention. This is because staff members will be occupied with low-value, repetitive tasks instead of contributing to initiatives with significant impact.

As mentioned in the previous subsections, modern maintenance approaches require to combine data from several sources, including sensors, operational logs, and outside databases. Manual methods find it challenging to smoothly combine these many data sources into a meaningful analysis. The result of this fragmentation could be partial or scattered insights that make it challenging to have complete knowledge and to make wise decisions.

Overcoming these obstacles requires the incorporation of data-driven techniques and digitization, as presented in the previous subsection. Automated data collection, enhanced analytics, machine learning, and the digital twin idea can revolutionize data handling operations. By providing extensive data integration, real-time insights, and predictive maintenance capabilities, these solutions allow organizations to fully use their data.

A recent research trend involves the use of "Digital Twin" models, which update a virtual model of a physical asset in real time using data collected from sensors and computer models. By reflecting the asset's current condition and projecting its future performance, this technology enables preventive maintenance and optimal resource allocation. Machine learning algorithms uncover patterns and abnormalities to enable informed decision-making and early error detection. In simple language, the "Digital Twin" is an electronic copy of a real-life item or system that is structurally and behaviorally identical to its physical counterpart. This digital representation can simulate, analyze, and monitor in real time as it is constantly receiving data from its physical counterpart. A complete digital representation of the physical asset is generated by combining data from multiple sources, such as simulations, sensor readings, and wear debris analysis, to represent the asset's present condition and performance.

In the previous paragraph, the word "asset" has been used, and "Digital Twin" is meant for "Asset Management." Here, "Asset Management" refers to the systematic approach of managing and maintaining a company's physical assets throughout their lifecycle. This involves monitoring, analyzing, and optimizing asset performance to ensure they remain functional, efficient, and cost-effective. In engineering and industrial contexts, asset management often includes predictive maintenance, where data from sensors and Internet of Things (IoT) devices are used to forecast failures and schedule maintenance proactively. The integration of digital twins—virtual replicas of physical assets—enhances asset management by providing real-time insights and enabling data-driven decision-making, which helps improve reliability, reduce costs, and optimize resource allocation.

To illustrate this, we can refer to the research article by Zahmatkesh et al.,[3] which analyzed the wear mechanisms on a material's surface. The study presented "Worn-out Surfaces" identified by the presence of scratches and grooves, signifying material removal due to hard particles or asperities ploughing through the surface. Additionally, evidence of smoother areas indicated material transfer between surfaces, implying localized bonding and subsequent tearing at points of contact.

To enhance understanding, the researchers provided detailed micrographs illustrating the morphology of wear debris. Comprehensive knowledge derived from literature on debris shapes and sizes offers valuable insights into wear mechanisms and the extent of surface degradation over time. Specifically, coarser particles suggest more severe wear conditions, while finer particles point to milder abrasion.

From the perspective of "Digital Twin" or "Cyber–Physical System" applications, wear debris analysis can be digitized using sensors and IoT devices. These data serve as the foundation for developing a digital twin—a dynamic, real-time virtual representation of the material's condition. By integrating wear debris information, the digital twin framework provides continuous insights into the physical asset's health, enabling the prediction of failures and informing data-driven maintenance or component replacement decisions.

To implement an effective Digital Twin for maintenance, it is essential to integrate multiple data sources, such as wear debris analysis, real-time sensor data, and simulation outcomes. This comprehensive integration of Digital Twin technology and cyber–physical systems provides significant advantages in maintenance management, including real-time insights, predictive maintenance capabilities, and enhanced operational efficiency. By leveraging this advanced integration, organizations can optimize resource allocation, minimize unplanned downtimes, and extend the lifespan of physical assets through data-driven strategies.

- *Data integration*
  - *Wear debris analysis:* Data collected from sensors and IoT devices concerning operational wear debris illuminate the mechanisms of wear and material degradation affecting the physical asset. Analyzing this data helps identify wear patterns and potential failure modes.
  - *Sensor data:* Real-time operational data, such as temperature, pressure, and vibration, are continuously gathered by sensors embedded within the machinery. This data is vital for understanding the system's current operating state and identifying anomalies that could signal impending breakdowns.
  - *Simulation data:* Computational simulations model various operational scenarios, predicting how changes in conditions may influence system performance over time. This predictive capability enables engineers to address potential issues before they escalate into major problems.
- *Representation of a time-deteriorated system:* The Digital Twin evolves into a "time-deteriorated system," capturing not only the asset's current state but also its historical performance and degradation patterns. As the digital model updates in real time, operators gain a comprehensive understanding of the asset's lifecycle, allowing for the implementation of more informed and strategic maintenance policies.
- *Integration of cyber and physical systems:* The concept of the Digital Twin is often associated with "cyber–physical systems," which integrate computational algorithms with physical processes. By connecting the digital and physical realms, these systems facilitate advanced automation, control, and monitoring, thereby enhancing operational efficiency. Cyber–physical systems leverage data analytics and machine learning to process extensive datasets, offering deeper insights into asset performance.
- *Predictive maintenance and real-time insights:* One of the primary advantages of employing a Digital Twin is its ability to deliver "real-time insights" into the physical asset's condition. By continuously analyzing data from various sources, the Digital Twin can:
  - *Predict failures:* Identify patterns in the data to forecast potential issues, enabling organizations to proactively schedule maintenance and minimize downtime while maximizing asset longevity.

- *Support maintenance decision-making:* Provide actionable recommendations for repairing or replacing components based on the asset's current state. This data-driven approach results in reduced maintenance costs and optimized resource utilization.

In summary, the process involves collecting and analyzing diverse data streams to create a dynamic, real-time virtual model that informs strategic maintenance decisions and improves overall asset performance.

7.4 DIGITIZATION TECHNIQUES

Digitization, the process of converting analog data into digital formats, facilitates more efficient storage, retrieval, and analysis. Various methodologies are employed across different domains to achieve effective digitization. "Document scanning" uses scanners to convert physical documents, such as text, images, or diagrams, into digital files, often accompanied by advanced software that enhances image quality and simplifies text recognition. "Optical Character Recognition (OCR)" technology identifies text characters within images, transforming scanned documents and PDFs into searchable and editable data, which is particularly useful for digitizing printed materials. "3D scanning" captures the 3D geometry of objects to create digital 3D models, widely applied in manufacturing, and virtual reality. "Audio and video digitization" transfers analog recordings from tapes to digital formats, employing analog-to-digital converters and specialized software to preserve quality. Finally, "Intelligent Document Processing (IDP)" leverages artificial intelligence, including machine learning and natural language processing, to automatically extract and process data from various documents, enhancing efficiency by reducing the need for manual data entry. By adopting these digitization techniques, organizations can streamline workflows, safeguard information, and enhance data analysis and decision-making. In the context of the present book, "our discussion" is restricted to "image/graph" digitization.

7.4.1 Image digitization

Image digitization refers to the process of converting physical images or documents into a digital format that can be analyzed and manipulated using software tools. This technique is crucial for data analysis, especially when working with software like Origin, which is designed for graphing and data manipulation. The steps involved in image digitization are as follows:

- *Import the image:*
    - Begin by opening Origin and creating a new project.
    - Import the image file by navigating to the "File" menu and selecting "Open" or using the keyboard shortcut. Choose the

desired image file (e.g., a scanned graph or chart) and click "Open" to load it into the project.

- *Image calibration:*
  - Calibration is essential to ensure accurate measurements from the digitized image.
  - Identify a known reference point or object within the image (e.g., a scale or a marked feature) and use Origin's measurement tool to measure its dimensions. Input the actual value corresponding to this reference to allow Origin to calculate the scale for the entire image.
- *Image tracing:*
  - After calibration, use Origin's tracing tools to trace features or data points in the image. Tools such as freehand, spline, or line tools are available for precise tracing.
  - It is critical to trace along the edges or significant data points accurately to ensure the integrity of the data extraction process.
- *Data extraction:*
  - Once the features are traced, convert the traced lines or curves into numerical data points.
  - Origin provides tools like "Digitize" or "Data Extraction" that allow users to click on traced curves or lines to obtain x–y coordinates. This step transforms visual data into quantifiable data for analysis.
- *Data analysis and plotting:*
  - Utilize Origin's analytical features to perform various operations on the extracted data. This may include statistical analysis, curve fitting, or application of mathematical functions.
  - Create plots and charts using Origin's graphing capabilities, allowing for a visual representation of the analyzed data.
- *Export and save:*
  - After digitization and analysis, export the data or generated plots for further use.
  - Origin allows saving data in multiple formats, such as CSV or Excel, and exporting plots as images or in different file formats suitable for publications or presentations.

7.4.2 Trendline fitting using Excel 365

Fitting a trendline to data in Excel 365 enables users to analyze relationships between variables:

- *Data entry:* Input your data into an Excel worksheet, organizing it into two columns: one for the independent variable (x-values) and another for the dependent variable (y-values).

- *Select data range:* Highlight the range of data that you want to analyze.
- *Insert scatter plot:* Navigate to the "Insert" tab in the Excel ribbon and select a scatter plot type to visualize the data points.
- *Activate chart design:* After inserting the scatter plot, access the "Chart Design" and "Chart Format" tabs for further customization.
- *Add trendline:* From the "Chart Design" tab, click "Add Chart Element," then select "Trendline" and choose the desired type (e.g., linear, polynomial). Excel provides options for customizing the trendline's appearance and properties.
- *Display equation and R-squared value:* Customize the trendline to display the mathematical equation and the R-squared value on the chart, which indicates the goodness of fit for the trendline.

7.4.3 Plotting using MATLAB® 2020b

MATLAB® is a powerful tool for plotting functions and visualizing data:

- *Open MATLAB®:* Launch MATLAB® and open a new script or command window.
- *Define the function:* Define the mathematical function you want to plot, specifying the range over which the function will be evaluated.
- *Generate the plot:* Use the `fplot` function to create the plot. For example, `fplot(@(x) sin(x), [-pi, pi])` would plot the sine function from $(-)\pi$ to $(+)\pi$ as demonstrated in Figure 7.2.
- *Customize the plot:* Enhance the plot using additional MATLAB® functions, such as title (title('Plot of Sine between -pi and pi')), axis labels (xlabel('-pi < x < pi'), ylabel('Sine Values')), or modifying visual elements to improve clarity and presentation.

To illustrate digitization, we can refer to figures from the research article by Suresh et al.[4] We aim to digitize "Figure 1" and "Figure 2" from this study.

Figure 1 in the article shows a graph illustrating the corrosion rate of Al 7075 alloy and its nanocomposites in H_2SO_4 over time. The y-axis represents the corrosion rate in millimeters per year (mm/year), ranging from 0 to 0.16, while the x-axis indicates time in hours, ranging from 0 to 160 hours. The graph includes four different curves:

- "0%": Represents the pure Al 7075 alloy, showing the highest initial corrosion rate that gradually decreases over time.
- "2%": Represents the Al 7075 alloy with 2% (SiC) nanocomposite reinforcement, exhibiting a lower corrosion resistance (more wear rate) compared to the pure alloy.

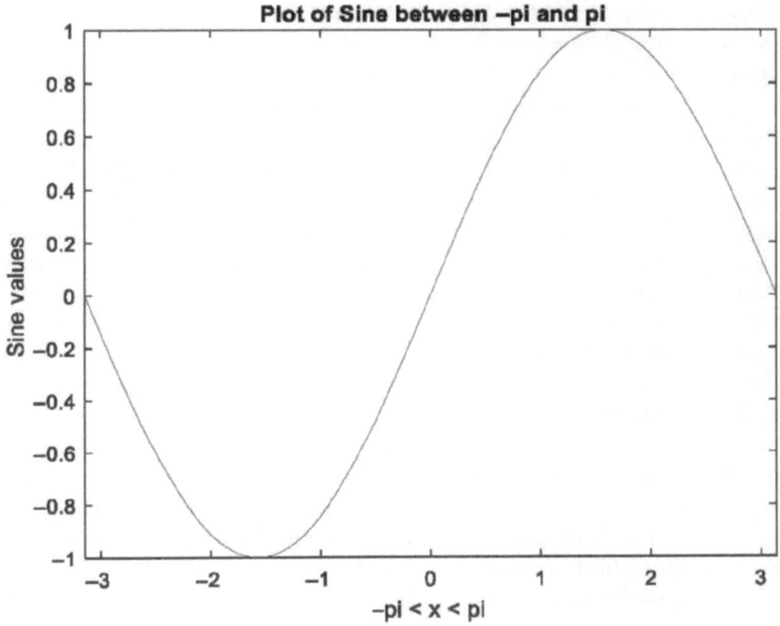

Figure 7.2 Plot of sine between (−)π and (+)π.

- "4%": Represents the Al 7075 alloy with 4% (SiC) nanocomposite reinforcement, showing a further reduction in the corrosion rate relative to the 2% reinforced alloy, but higher wear rate compared to the pure alloy.
- "HYB": Denotes a hybrid nanocomposite, which displays the lowest corrosion rate among the samples, indicating superior corrosion resistance.

Overall, the corrosion rate decreases over time for all samples, with the hybrid nanocomposite (containing an equivalent quantity of nano Al_2O_3 and SiC) demonstrating the most effective corrosion resistance in H_2SO_4.

Figure 2 in the article illustrates a graph of the corrosion rate of Al 7075 alloy and its nanocomposites in HCl over time. The y-axis shows the corrosion rate in millimeters per year (mm/year), ranging from 0 to 0.1, and the x-axis represents time in hours, ranging from 0 to 160 hours. The graph includes four distinct curves, analogous to those explained for the H_2SO_4 case. The overall trend remains consistent: the corrosion rate declines over time for all samples, with the hybrid nanocomposite exhibiting the most significant corrosion resistance in the HCl environment.

Figure 7.3 shows the digitization done in Origin software and plotting in the Excel software. The trend lines have also been incorporated. Figure 7.4 shows the plotting performed in the MATLAB software.

Figure 7.3 shows very strong relationships between independent (time) and dependent variables (corrosion rate). With available data, we can analyze the corrosion rates in H_2SO_4 and compare them to those in HCl. Using the trend line equations, we can further optimize the percentage of nano Al_2O_3 and SiC, a variable that remains unexplored in current literature—through MATLAB® coding. By incorporating digitized data points and using time (variable y_1) and nanocomposite percentage (y_2, y_3) as variables, we can develop a more comprehensive model. This optimizing process may require additional experiments to validate the simulation results, for which the preparation of materials samples is necessary.

The fabrication of Aluminum 7075 matrix hybrid composites involves several intricate steps:

- *Material source:* Aluminum 7075 ingots are melted using a heating system. This alloy, known for its excellent strength-to-weight ratio, is commonly used in aerospace and structural applications.
- *Incorporation of nanoparticles:* A mechanical stirrer creates a vortex in the molten aluminum, facilitating the uniform dispersion of nano-sized particles such as Al_2O_3 (alumina) and SiC (silicon carbide). These nanoparticles enhance the mechanical properties of the aluminum matrix.
- *Hybrid composite formation:* Optimized quantities of nano-Al_2O_3 and nano-SiC powders (as one sample) are introduced into the vortex and thoroughly mixed with molten aluminum. The mixture is then poured into cylindrical metal molds to create the hybrid composite material.
- *Variable ratios:* Additional samples can be prepared based on sensitivity analysis, varying the optimized percentage of nanopowders by ±5%.
- *Specimen preparation:* The composite samples are cast into cylindrical bars (20 mm × 20 mm) and undergo surface preparation using different grades of emery paper for a smooth finish. The samples are then cleaned with acetone to remove contaminants and dried in preparation for testing.
- *Corrosion testing:* The corrosion resistance of hybrid composites is assessed using the weight loss method. Samples are weighed using a digital balance before testing, then immersed in corrosive solutions (0.1 M HCl and 0.1 M H_2SO_4) to simulate real-world conditions. The testing duration is carefully controlled to ensure adequate exposure, and the weight loss is measured afterward to calculate the corrosion rate.

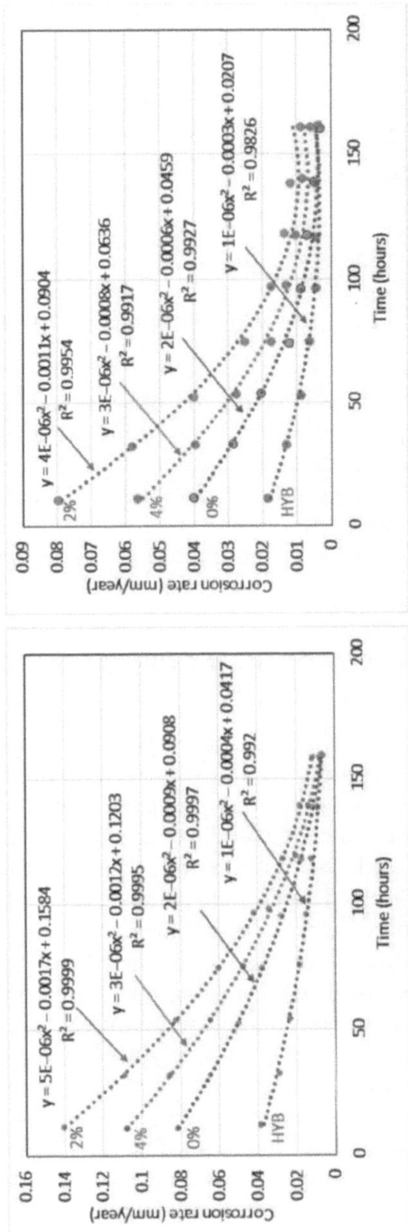

Figure 7.3 Excel plot and fitting (left: H_2SO_4 environment, right: HCl environment).

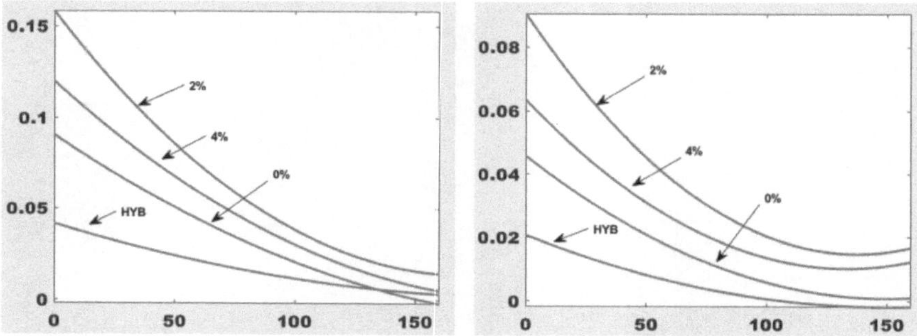

Figure 7.4 MATLAB® plot (left: H_2SO_4 environment, right: HCl environment).

Advantages of the approach: This method offers several advantages: precise control over material properties through optimized nanoparticle dispersion, improved corrosion resistance and mechanical performance, and the ability to simulate real-world conditions accurately. Additionally, integrating digitized data and predictive modeling enhances the efficiency and reliability of the experimental process, leading to a more thorough understanding and better material design.

7.5 LEVERAGING DATA-DRIVEN APPROACHES FOR PREDICTIVE MAINTENANCE

In today's industrial landscapes, the rise of data-driven technologies has transformed how organizations approach equipment maintenance. Traditional methods of reactive and preventive maintenance have given way to predictive maintenance, which leverages data analytics and real-time monitoring to foresee and address potential equipment failures. This transformation has revolutionized asset management, reducing costs, optimizing resource allocation, and enhancing overall reliability.

Predictive maintenance leverages both historical and real-time data collected from sensors installed on machinery. By combining stored information about the system's performance under different operating conditions with a continuous stream of live data, maintenance teams gain valuable insights into the health of the equipment. This enables them to detect anomalies early and take proactive actions to prevent failures.

One of the most significant benefits of a data-driven approach is its ability to uncover patterns and trends that may not be readily apparent through traditional maintenance methods. For instance, advanced analytics and machine learning models can analyze vast amounts of sensor data to detect subtle changes in equipment performance, such as minor increases in vibration or slight deviations in temperature, which might otherwise go unnoticed.

Consider a manufacturing plant that depends on large rotating machinery. With predictive maintenance in place, sensors continuously monitor key parameters such as vibration, load, and temperature. Over time, machine learning models learn the normal operating behavior of the equipment and can detect even subtle anomalies. For example, if the system identifies a gradual increase in vibration levels at the bearing level, this could be an early indicator of bearing wear. Rather than waiting for a catastrophic failure or adhering strictly to a scheduled maintenance plan, the maintenance team can proactively investigate and address the root cause of the vibration to prevent further damage. Additionally, they can replace the bearing at a convenient time, such as during a planned production shutdown. This approach helps prevent unexpected downtime, reduces repair costs, and ensures smoother, uninterrupted operations.

Another example is found in the energy sector, where wind turbines operate in harsh and fluctuating environmental conditions. Predictive maintenance systems collect data from sensors that monitor blade pitch, wind speed, generator temperature, and vibration. When a predictive algorithm detects an unusual pattern, such as a sustained increase in generator temperature, it may indicate an impending failure. In response, the maintenance team can first reduce the load on the affected turbine to prevent further damage, and then strategically schedule maintenance during favorable weather conditions. This approach minimizes disruptions and maximizes energy production, ensuring efficient and reliable operation.

Condition monitoring forms the backbone of predictive maintenance. It involves the continuous observation of machinery through sensors that collect various operational data. The real-time analysis of this data allows maintenance teams to detect any deviations from normal performance, often referred to as anomaly detection. For example, an unexpected increase in vibration levels may signal bearing wear or misalignment, prompting further investigation and immediate intervention. This ability to react swiftly to emerging issues helps prevent minor problems from escalating into major, costly failures.

7.5.1 Analytical techniques and predictive modeling

In today's data-driven world, the application of analytical techniques and predictive modeling has become a critical component of effective decision-making across various industries. By leveraging large datasets and advanced computational methods, organizations can gain deeper insights, forecast future trends, and optimize their operations. This subheading explores the key analytical techniques used in predictive modeling, their applications, and the benefits they offer in various fields.

Analytical techniques form the foundation of predictive modeling and involve the systematic use of data to uncover meaningful patterns, relationships, and trends. These techniques are essential for interpreting large volumes of data and converting raw information into actionable insights. Common analytical techniques include regression analysis, classification, clustering, and dimensionality reduction.

- *Regression analysis* is a widely used method for predictive modeling that establishes a mathematical relationship between a dependent variable and one or more independent variables. Figures 7.3 and 7.4 display trendlines derived from regression analysis. It is crucial to account for the statistical nature of the parameters, which includes factors such as inherent variability, distribution patterns, and the probabilistic characteristics of data collected from observations, literature, or experiments. By incorporating these statistical properties, regression analysis adheres to established principles, effectively modeling the relationships between variables, as demonstrated by the trendline equations in Figure 7.3. The effectiveness and validity of a regression model are heavily influenced by the statistical characteristics of the analyzed data. The coefficient of determination, R^2, shown in Figure 7.3, is an essential metric that measures how well the independent variables explain the variability of the dependent variable, providing a clear indication of the model's goodness of fit.

  It is also important to be aware of overfitting, which occurs when a regression model captures noise or random fluctuations in the dataset instead of the true underlying relationship between variables. An overfitted model fits the training data very well but performs poorly on new, unseen data, as it is overly complex and fails to generalize. To ensure the model's reliability and predictive accuracy, it is essential to use appropriate techniques to prevent overfitting.
- *Classification techniques* are used when the goal is to categorize data into predefined groups. Decision trees, support vector machines, and neural networks are commonly used classification methods. For example, Gupta et al.[5] used classification analytical techniques, particularly focusing on automated methods to detect and categorize wear particles. From their article, the key points about classification techniques are as follows:
  - *Wear particle classification:* Automated classification techniques have been employed to identify different types of wear particles, such as cutting, spherical, fatigue, rubbing, and severe sliding particles. These methods help in diagnosing the condition of mechanical components like gears and predicting potential failures.

- *Use of convolutional neural networks (CNNs):* CNNs are a fundamental tool in modern classification tasks. Gupta et al.[5] customized CNN model, WPCNet, which is used to extract features from images of wear particles. The CNN model applies multiple filters to analyze complex features, enabling efficient and automated classification of particles without manual intervention.
- *Integration with cascade classifiers:* The research integrates CNN models with cascade classifiers to enhance the detection and classification of wear particles. Cascade classifiers use Haar-like features to differentiate between particles and non-particles, ensuring accurate identification.
- *Data augmentation:* To improve model performance and avoid overfitting, data augmentation techniques are used. These include transformations such as rotation, rescaling, and flipping, which increase the size of the training dataset and enhance model robustness.
- *Applications and benefits:* The described techniques are critical for real-time monitoring of industrial machinery, improving predictive maintenance strategies, and reducing manual labor. The automated models significantly increase efficiency, accuracy, and reliability in diagnosing machine conditions.

- *Clustering* is an analytical technique used in data analysis to group similar data points into clusters based on their characteristics. As an *unsupervised machine learning method*, it does not require predefined labels or categories (unlike classification, as explained in previous paragraphs, which assigns data points to existing categories) but instead identifies natural groupings within datasets. To illustrate this concept, consider examples in the fields of mechanical design, corrosion, and surface engineering:
  - *Mechanical design:* In mechanical design, clustering can be used to group different design configurations or components based on performance metrics such as stress distribution, load-bearing capacity, or thermal properties. For example, clustering algorithms can help engineers identify sets of designs that share similar structural characteristics or failure modes, enabling them to analyze and optimize these groups collectively. This approach can lead to more efficient design processes by recognizing patterns and relationships between different mechanical components. Additionally, clustering can assist in classifying mechanical parts for modular design, where components are grouped based on their functionality or similarity in geometric features, making manufacturing and assembly processes more efficient.

- *Corrosion:* In the field of corrosion, clustering can be used to categorize different materials or surface treatments based on their corrosion resistance under various environmental conditions. By analyzing data such as corrosion rate, pH levels, and temperature exposure, clustering algorithms can group materials with similar corrosion behaviors. This helps researchers and engineers identify the most suitable materials for specific applications or environments. For example, clustering can help in understanding the performance of coatings or alloys by grouping them into clusters that exhibit similar degradation patterns, thereby streamlining material selection and corrosion prevention strategies.
- *Surface engineering:* In surface engineering, clustering can be applied to group surface treatments or coatings based on their effectiveness in improving properties like hardness, wear resistance, or friction reduction. By analyzing data from experiments or simulations, such as surface roughness, coating thickness, and microstructural characteristics, clustering algorithms can identify which treatments yield similar performance outcomes. This technique can also be used to optimize surface engineering processes by grouping treatments that show comparable effects on material properties, helping engineers select or modify surface finishes that best meet design requirements. For instance, clustering can identify which coatings provide optimal protection for specific operating conditions, such as high-temperature or high-friction environments.

- *Dimensionality reduction* techniques, such as principal component analysis (PCA), are used to simplify datasets by reducing the number of variables while preserving as much information as possible. These techniques are valuable in making data more manageable and improving the performance of predictive models, especially when dealing with high-dimensional data.

Predictive modeling builds on analytical techniques by using historical and current data to forecast future outcomes. This is accomplished through algorithms and statistical methods that learn from past patterns and relationships within the data. Machine learning models are often employed in predictive modeling to enhance accuracy and reliability.

One key application of predictive modeling is in "predictive maintenance," where it helps forecast equipment failures before they occur. By analyzing sensor data from machinery, predictive models can identify early warning signs of surface degradation, allowing maintenance teams to intervene before a breakdown happens. This proactive approach minimizes downtime, reduces repair costs, and extends the lifespan of assets.

7.5.1.1 *Benefits of analytical techniques and predictive modeling*

Integrating analytical techniques with predictive modeling brings substantial benefits, foremost among them the ability to make "data-driven decisions." Rather than relying on intuition, organizations can base their strategies on solid evidence, which increases the likelihood of positive outcomes. This approach enhances both efficiency and accuracy, enabling effective resource allocation. Another crucial advantage is "risk mitigation"; predictive models help identify potential risks early, allowing organizations to take proactive measures. Estimating the "remaining useful life (RUL)" of machinery is also essential in predictive maintenance. Using historical performance data for degradation analysis, RUL metrics guide maintenance teams on when to perform maintenance or replace parts, which helps avoid unexpected failures and optimizes resource utilization. Additionally, predictive modeling fosters "operational optimization," as insights from these models lead to cost savings and better resource management, ultimately enhancing organizational profitability.

While predictive modeling offers numerous benefits, it also faces challenges, particularly regarding data quality. Predictive models rely heavily on the quality of the data they are trained on, and poor data quality can result in inaccurate predictions. Ensuring data integrity and handling incomplete or inconsistent data pose significant hurdles. To address these challenges, investing in high-quality sensors, advanced data acquisition systems, and reliable processing speeds is essential. However, these resources can be costly, and their expense can be difficult to justify relative to the product's value. Nevertheless, when considering the broader costs associated with sustainability, resource depletion, and environmental impact, alongside system costs, a data-driven approach emerges as not only justifiable but also essential for societal benefit.

7.5.2 Data-driven simulations

Data-driven simulations utilize real-world data to build virtual models of physical systems, enabling industries to predict, analyze, and optimize their operations. This approach provides critical insights that are often difficult to capture using traditional methods. By employing data-driven simulations, organizations can effectively link predictive maintenance with operational streamlining and enhanced product quality.

Unlike conventional simulations that rely on theoretical assumptions, data-driven simulations replicate complex processes by integrating data from sensors, historical records, and environmental variables. Grounded in actual data, these simulations are highly accurate and reflective of real-world conditions, providing a reliable foundation for decision-making and process optimization.

- *Role in process optimization:* Data-driven simulations are instrumental in optimizing industrial processes. In sectors such as chemical manufacturing, simulations model intricate processes, allowing operators to adjust key variables such as temperature, pressure, and chemical concentrations to achieve maximum output. By continually refining these variables with real-time data, companies can increase operational efficiency, minimize waste, and improve product quality. This optimization supports competitive advantage by maximizing resource utilization.
- *Impact on product development:* Data-driven simulations also contribute significantly to product development by enabling virtual testing of prototypes before physical production. Engineers and designers can use simulations to evaluate product performance under a range of conditions, such as stress, heat, and wear. For instance, in the automotive industry, simulations can assess the durability of a new component under extreme driving conditions, allowing for necessary adjustments before physical prototyping. This process accelerates development cycles, reduces prototyping costs, and leads to more reliable products.
- *Environmental management and safety:* In environmental management and safety planning, data-driven simulations play a critical role. For example, in the energy sector, simulations can model the impacts of oil spills or forecast emission levels under different operational scenarios. These simulations enable organizations to assess and mitigate potential environmental risks proactively. In workplace safety, simulations can model emergency scenarios, allowing companies to design safer environments and develop effective response plans.
- *Advantages of data-driven simulations:* Data-driven simulations offer several key advantages, including cost reduction, flexibility, improved decision-making, and risk mitigation. By enabling virtual testing and optimization, simulations reduce the need for trial-and-error approaches in developing and improving the costly physical assets, allowing manufacturing companies to refine processes without halting production, thus reducing downtime and material waste. These simulations are also highly adaptable, easily incorporating new data or scaling to larger systems, which is especially valuable in fast-evolving industries where real-time updates keep simulations relevant and actionable. Additionally, data-driven simulations provide organizations with a deeper understanding of their systems, revealing hidden inefficiencies or bottlenecks that facilitate targeted improvements and boost productivity. Furthermore, simulations create a safe environment to test various scenarios, allowing companies in safety-critical industries, such as aerospace and energy,

to evaluate risks and develop contingency plans, enhancing overall safety and preparedness.

- *Challenges of data-driven simulations:* While data-driven simulations offer numerous advantages, they also present several challenges. A key challenge is "data quality"; the accuracy of simulations depends on the quality of the data they utilize. Inaccurate, incomplete, or outdated data can lead to unreliable predictions, making it essential to invest in robust data acquisition and processing systems to ensure data integrity. Another significant challenge is the "computational cost," which can be unaffordable for many industries. Data-driven simulations, particularly those for complex systems, demand substantial computational power, advanced algorithms, and reliable interpretability, often requiring costly tacit knowledge. For small businesses, these requirements may be prohibitively expensive, although cloud computing and shared data platforms are gradually making such simulations more accessible.
- *Future of data-driven simulations:* The future of data-driven simulations is promising, especially with advancements in artificial intelligence, machine learning, and data analytics. As these technologies evolve, simulations will become even more accurate and sophisticated, allowing for more comprehensive and realistic modeling of complex systems. Emerging trends such as digital twins—real-time, data-driven virtual replicas of physical systems—will further enhance the capabilities of data-driven simulations, offering unprecedented insights into system performance and enabling continuous optimization.

7.6 CASE STUDIES

7.6.1 Case study 1: Thermographic analysis

Thermographic analysis (or thermographic testing, which is an NDT technique and was covered in previous chapter) uses infrared imaging to detect variations in thermal patterns, which indicate potential defects or issues in materials and components. In this context, "digital twin technology" combined with "data-driven methodologies" plays a significant role in improving maintenance and operational efficiency. Thermographic techniques generate thermal images that highlight surface anomalies, which can signal areas of wear or structural weakness. However, "low signal-to-noise ratios" in thermographic data often pose challenges, as noise can obscure subtle thermal variations that are essential for detecting early-stage defects. To mitigate this, "robust filtration systems" are used to enhance image quality, making it easier to interpret the data accurately.

To address this challenge, "principal component analysis" can streamline the diagnostic process by condensing numerous variables into key components, helping analysts focus on factors most relevant to defect detection. By filtering out minimal or random variations, PCA enhances clarity, particularly in complex systems with multiple interacting variables. Research indicates that PCA can reduce data dimensionality by 40% to 50% without sacrificing accuracy, thus optimizing the identification and evaluation of thermal patterns in thermographic analysis.

In a study by Simonov et al.,[6] thermographic analysis results were shown to vary with the material thickness L. For L = 2 mm, the analysis achieved the best results, detecting "16 to 18 defects out of 30" (using the 4th principal component), with signal-to-noise ratio (SNR) values ranging from "5.6 to 23.1," where higher SNR values indicated better defect visibility. At L = 4 mm, "10 defects out of 30" (using the 2nd principal component) were detectable, with SNR values between "4.7 and 11.3." For L = 6 mm, "9 defects out of 30" (using the 2nd principal component) were detected, with SNR values ranging from "1.9 to 5.8." Finally, at L = 8 mm, only "2 defects out of 30" (using 1st principal component) were identified, with SNR values between "1.7 and 4.4." These findings underscore the effect of material thickness on defect detectability, as thicker materials tend to dampen thermal signals, leading to a reduced detection rate compared to thinner materials.

In summary, as the material thickness increases, the number of detectable defects decreases, and the SNR values drop, illustrating that thicker materials dampen thermal signals. PCA helps to identify the most effective principal component for defect detection at each thickness level, with thinner materials yielding clearer results.

7.6.2 Case study 2: Corrosion analysis

Corrosion analysis is essential for assessing material durability, particularly in harsh environments where exposure to chemicals accelerates degradation. A critical part of corrosion analysis involves selecting the appropriate "wear map" for a given material and application, as wear maps guide the interpretation of corrosion rates and patterns under various conditions. Leveraging existing studies to identify research gaps allows for the definition of specific experimental parameters, which in turn informs targeted and effective maintenance strategies.

In the study by Kaewunruen et al.[7] on the application of digital twin technology in bridge infrastructure weakened by atmospheric corrosion, the digital twins were employed to address maintenance challenges and optimize risk management strategies for bridge structures. This approach underscores the value of digitizing data to monitor asset health, predict deterioration, and extend the lifespan of physical structures through proactive maintenance. Here, digital twins served as virtual replicas of physical bridges, allowing researchers to model structural vulnerabilities under various stressors such as extreme weather including corrosion, earthquakes, and load conditions. These virtual models were continuously updated with real-time data from sensors and historical records, enabling precise risk assessments and supporting maintenance planning based on accurate, up-to-date information.

In the context of other types of corrosion, similar digital twin applications can be applied. By digitizing corrosion data, as seen in bridge maintenance, discrepancies due to factors such as sensor inaccuracies, noisy data, and inconsistencies between studies can be identified early. For example, in monitoring chloride-induced corrosion (Kaewunruen et al.)[7] in bridge reinforcement, digital twins can simulate the impact of environmental exposure over time, helping maintenance teams predict when interventions might be required to prevent structural failures. This approach is particularly relevant in ensuring the accurate prediction of RUL, as digital twins integrate both historical data and real-time sensor readings, providing a comprehensive view of asset degradation. The bridge case study illustrates that by using digital twins, corrosion monitoring can be more efficient, extending asset life by enabling targeted and timely maintenance actions.

7.6.3 Case study 3: Gear debris analysis

Gear debris analysis plays a crucial role in assessing "surface degradation" and identifying "failure modes" in gearbox components. By examining the wear debris collected from gear oil samples, engineers can detect early signs of degradation, allowing for timely intervention before significant damage occurs. In a specific setup involving a monitored test gearbox, debris analysis focused on particle sizes ranging from "10 μm to larger flag-like particles," which can reveal information about wear mechanisms. Smaller irregularly shaped particles typically indicate "abrasive wear," while larger flat particles or regularly shaped may suggest "sliding wear" or "fatigue failure."

Scanning electron microscopy (SEM) is widely used to analyze debris particles, offering detailed insights into their morphology and elemental composition. SEM analysis allows differentiation between various types of wear: cutting wear, typically characterized by abrasive particles; sliding wear, often indicated by elongated particles; and fatigue wear, usually represented by spherical particles. To enhance the digital twin model for wear analysis, incorporating multiple cameras to capture high-resolution images of debris is essential. These images can then be processed with advanced image analysis tools to accurately determine the type and size of wear debris, providing greater clarity and precision in monitoring wear mechanisms.

Statistical measures such as "kurtosis" and "skewness" are also employed to evaluate debris patterns and establish degradation trends. Studies have shown that high kurtosis values often correlate with severe wear, while skewness can indicate whether wear distribution is centered or skewed due to specific operating conditions. An ongoing feedback loop for debris analysis ensures that maintenance teams receive continuous updates on surface degradation. By regularly collecting and analyzing oil samples, maintenance teams can adjust their strategies based on the observed wear patterns. For example, the presence of larger particles (over 100 μm) may necessitate immediate action, as they indicate significant surface damage. Conversely, the absence of such particles may suggest stable operating conditions, allowing for extended maintenance intervals. This data-driven approach optimizes maintenance timing, reduces costs, and prolongs the life of gearbox components. The integration of digital twin technology further supports predictive maintenance by simulating potential wear scenarios, providing valuable insights into how varying operational factors affect gear performance over time.[8]

Feng et al.[8] integrated two primary components:

- *High-fidelity digital twin model:* This model bridges virtual and physical realms by accurately simulating the dynamic response of a physical gearbox transmission system. The model includes gear transmission dynamics, wear progression, and specific degradation mechanisms, such as abrasive wear and fatigue pitting. A 21-degree-of-freedom dynamic model simulates the interactions within the gearbox, accounting for components like the motor, shafts, coupling, gears, and casing. Parameters are derived from both empirical formulas and real-time measurements to ensure high coherence between simulation and actual responses. Real-time adjustments in the simulation parameters ensure that the model remains aligned with actual machine responses. By continuously updating these parameters, the digital twin adapts to the evolving state of the physical system, which is particularly useful in tracking wear progression. The digital twin model integrates key "wear and degradation mechanisms," specifically "abrasive wear" and "fatigue pitting," to accurately simulate wear propagation within the gearbox. "Abrasive wear" occurs when particles in the lubricant erode material from gear surfaces. This is modeled using an enhanced Archard wear model that incorporates sliding velocity, Hertzian contact pressure, and wear coefficients, allowing for realistic simulation of wear depth across varied operating conditions. "Fatigue pitting," on the other hand, results from cyclic stress on gear teeth, leading to pits or cracks on the surface. The model tracks the density of these pits, factoring in shear stress, contact pressure, and the stressed volume over multiple cycles. Together, these degradation models enable the digital twin to predict specific wear types under different loads, providing reliable estimates of the RUL of the gears.

- *Transfer learning algorithm:* This algorithm in digital twin-driven gear degradation assessment enables effective knowledge transfer from digital simulations to real-world gear wear analysis. This approach uses a domain-adaptive model that applies insights from simulated data (source domain) to real-world gear wear data (target domain), which is particularly beneficial in situations with limited real-time data. By incorporating a "feature extractor" and "classifier," the algorithm aligns high-level features between simulated and measured data, optimizing prediction accuracy and domain invariance. This process highlights the adaptability and robust predictive power of digital twin technology for industrial applications, significantly advancing predictive maintenance and asset management.
- The "domain adaptation" function of the transfer learning algorithm aligns data from the simulated and real-world environments, ensuring the model interprets information consistently across sources. A feature extractor identifies high-level attributes of wear, such as changes in vibration patterns, temperature fluctuations, and variations in rotational speed, which are crucial for understanding wear progression. Advanced techniques like CNNs strengthen feature extraction, ensuring robustness and enabling the digital twin to identify and correlate wear data from both virtual and physical domains. This alignment allows for the application of insights derived from simulations to actual wear conditions, providing a cohesive understanding of gear health.
- A "classification and prediction" module within the transfer learning algorithm assigns wear levels—normal, early wear, moderate wear, or severe degradation—based on the extracted features. This classifier is trained using labeled data from both simulations and actual operations, making it adept at distinguishing between different stages of gear degradation.

These case studies illustrate how data-driven approaches, combined with digital twins, optimize maintenance strategies across diverse applications, from thermographic defect detection and corrosion monitoring to gear wear analysis, significantly enhancing operational efficiency and equipment reliability.

7.7 CONCLUSIONS

The shift toward digitization and data-driven methodologies in engineering is transforming maintenance strategies and improving safety with enhanced performance. This chapter underscores how digitized data enhances analysis, enabling real-time, data-driven decisions that optimize asset reliability and lifespan. Data-driven approaches, including predictive maintenance, incorporate machine learning algorithms to detect patterns and predict equipment health, thereby reducing downtime and costs. Case studies, such as thermographic testing, corrosion analysis, and gear debris analysis, demonstrate the practical benefits of digitization across various applications. The integration of digital twins further strengthens predictive capabilities, allowing virtual modeling and real-time insights for proactive maintenance strategies. As the complexity of engineering systems grows, embracing these methods offers a sustainable path toward efficiency, safety, and resilience in industrial operations.

REFERENCES

1. Software for Graphing and Analysis, www.originlab.com/
2. Software for state-of-the-art fatigue analysis, www.efatigue.com/constantamplitude/background/stresslife.html
3. Zahmatkesh, B., Enayati, M. H., & Karimzadeh, F. (2010). Tribological and microstructural evaluation of friction stir processed Al2024 alloy. *Materials & Design*, *31*(10), 4891–4896.
4. Suresh, S., Gowd, G. H., & Devakumar, M. L. S. (2018). Corrosion behaviour of Al 7075/Al2O3/SiC MMNCs by weight loss method. *Journal of Bio- and Tribo-Corrosion*, *4*, 1–6.
5. Gupta, K.K., Hirani, H., & Muzakkir, S. M. (2023). Online gear wear particle detection and categorization using a convolutional neural network algorithm integrated with cascade classifier. *Tribology in Industry*, *44*(2), 212.
6. Simonov, D., Vavilov, V., & Chulkov, A. (2020). Infrared thermographic detector of hidden corrosion. *Sensor Review*, *40*(3), 283–289.
7. Kaewunruen, S., Sresakoolchai, J., Ma, W., & Phil-Ebosie, O. (2021). Digital twin aided vulnerability assessment and risk-based maintenance planning of bridge infrastructures exposed to extreme conditions. *Sustainability*, *13*, 2051. https://doi.org/10.3390/su13042051.
8. Feng, K., Ji, J. C., Zhang, Y., Ni, Q., Liu, Z., & Beer, M. (2023). Digital twin-driven intelligent assessment of gear surface degradation. *Mechanical Systems and Signal Processing*, *186*, 109896.

Chapter 8

Failure analysis and life-cycle assessment approaches

8.1 INTRODUCTION

To design more reliable and durable systems, engineers need a clear understanding of how and why failures occur. This understanding is achieved through failure analysis and life-cycle assessment (LCA), which are essential tools for evaluating system limitations, environmental impacts, and opportunities for optimization. LCA involves assessing a component's remaining useful life and evaluating the economic implications. When maintenance is required, it is crucial to determine whether repairing the system or replacing it entirely is the most cost-effective approach. These considerations are essential for informed decision-making.

Together, failure analysis and LCA help ensure system and component safety, efficacy, and environmental sustainability. By combining these approaches, engineers can better assess technical and environmental factors affecting material durability, especially under corrosive conditions. Understanding material failure modes and their environmental impacts is crucial for creating durable, sustainable products.

Failure analysis is crucial for identifying degradation mechanisms in materials exposed to harsh environments, such as those that accelerate corrosion and wear. For instance, pipelines in the oil and gas industry, which are frequently exposed to saltwater, often experience corrosion forms like pitting and stress corrosion cracking. Failure analysis allows engineers to pinpoint these mechanisms and implement preventive solutions, such as protective coatings or corrosion-resistant materials. In aerospace, this analysis aids in understanding alloy degradation, where aluminum components exposed to salt spray and temperature fluctuations may suffer intergranular corrosion. Such insights enable manufacturers to refine materials and processes to enhance safety and durability. An example of this is corrosion in a brake line, where environmental exposure can create a hidden fault that gradually weakens the line. Over time, as internal pressure increases, this weakening can lead to a sudden rupture and catastrophic brake failure. This

DOI: 10.1201/9781003545019-8

illustrates the importance of early fault detection and LCA to avoid severe consequences.

This chapter emphasizes the importance of a solid understanding of failure analysis techniques, enabling engineers to anticipate potential issues and implement preventive measures that extend component lifespans. Additionally, LCA plays a key role by examining the environmental impacts of a product's entire life, from material extraction to disposal. In corrosion and surface engineering, LCA helps engineers select materials and processes that minimize environmental harm while enhancing durability. For instance, in coastal infrastructure, material choices must balance corrosion resistance with environmental footprint. While stainless steel alloys offer exceptional durability, their high-energy manufacturing process has significant environmental costs. However, their long lifespan and low maintenance requirements make them a sustainable choice over time. Similarly, surface treatments like galvanization and anodization enhance corrosion resistance but often involve hazardous chemicals. LCA can help identify alternative coatings, such as organic options, which provide similar protection with lower environmental impact.

By incorporating LCA, this chapter encourages readers to consider both sustainability and durability when making material choices. This approach aligns with contemporary engineering's focus on sustainable practices, driven by industry goals to reduce environmental impact. By integrating failure analysis with LCA, engineers can better balance product longevity with ecological responsibility. For instance, in the automotive industry, failure analysis can reveal why components like undercarriages and brake parts corrode when exposed to road salt. LCA can then assess the environmental impact of various materials and coatings. Together, these assessments guide manufacturers toward materials that resist corrosion while minimizing environmental impact, resulting in vehicles that are both resilient and eco-friendly. Similarly, in power plants, high temperatures and corrosive gases accelerate the degradation of boilers and turbines. Failure analysis can identify specific types of corrosion, like high-temperature sulfidation, and LCA assesses the long-term impacts of different material options. This combined approach supports material choices that reduce maintenance needs and lower environmental costs.

In addition to identifying root causes of failures, failure analysis supports improved design and reliability through systematic investigation of component, system, or process vulnerabilities. Techniques such as fault tree analysis (FTA) and failure mode and effect analysis (FMEA) allow engineers to proactively address high-risk areas. Aligning failure analysis with LCA extends product life and conserves resources. LCA provides a comprehensive view of a product's environmental impact across its life cycle, revealing hidden costs like water use and greenhouse gas emissions. While failure analysis focuses on reliability, LCA emphasizes environmental effects, facilitating well-rounded decision-making. By following LCA's four stages—goal setting, inventory analysis, impact assessment, and interpretation—manufacturers can identify high-impact phases to inform sustainable design. For instance, in vehicle design, using lightweight materials can reduce fuel use during operation, but LCA also considers the environmental impact of manufacturing these materials.

Although distinct in focus, both LCA and failure analysis aim to enhance product efficiency and lifespan. LCA uses reliability data from failure analysis to refine product lifespan assumptions, and insights from failure analysis can inform design adjustments to reduce environmental impact. For example, failure analysis of magnetic bearings may reveal common wear mechanisms, while LCA evaluates the environmental implications of using different magnetic materials. Together, these analyses ensure that components are both functional and eco-friendly.

While each method has challenges—failure analysis often relies on historical data, and LCA requires sophisticated data collection—advances in data analytics, the Internet of Things (IoT), and digital twin technology hold promise for overcoming these limitations. These technologies improve the accuracy of both proactive failure analysis and real-time environmental impact assessments, allowing LCA to adapt to actual usage conditions. Failure analysis and LCA are essential for developing sustainable, high-performance systems. Failure analysis enhances reliability, while LCA addresses environmental implications, guiding engineers toward resilient and sustainable solutions. By integrating these approaches, engineers can strike a balance, creating products that are both durable and environmentally responsible. In a world focused on sustainability, combining these methods will be crucial for designing systems that are both resilient and eco-friendly.

8.2 FAILURE AND FAULT ANALYSIS

A system is considered to have failed when it no longer serves its intended function, either by completely ceasing to operate or by losing its effectiveness. For instance, a 15-year-old laptop may be deemed obsolete compared to current technology, even if it still powers on, as advancements such as transitioning from an i3 to an i9 processor make it functionally inadequate. In cases of obsolescence, failure analysis is often unnecessary.

Failure detection is generally straightforward, as malfunctions or sharp drops in performance are often clear indicators. However, pinpointing the root cause of a failure requires a more detailed approach. Root cause analysis (RCA) involves tracing the origin of a defect and observing its progression or interactions with other issues, which can evolve into complex failures. Rather than solely examining failures, engineers aim to detect faults—early-stage anomalies that diminish performance before a full breakdown occurs.

Fault analysis is therefore critical and challenging, as it reveals underlying issues that are essential for failure prevention. While reactive measures, such as failure analysis, seek to avoid the recurrence of similar failures, fault detection represents a proactive approach. This book emphasizes proactive strategies to help reduce failures to a minimal level, fostering a more reliable system.

Mechanical failure includes situations where components underperform without obvious damage. For instance, if a system expected to operate at 98% efficiency delivers only 95%, it can be classified as a failure. In engineering, "failure" and "fault" are related but distinct concepts: a failure is the inability to meet specified parameters due to factors like design flaws or material degradation, while a fault is an anomaly that precedes failure but does not immediately stop functionality. Identifying faults early can maintain reliability and extend operational life. Micro-level fault analysis and macro-level failure analysis are essential for understanding causes and preventing recurrence.

Determining failure probability requires a statistical approach. For instance, with historical data showing failures at 308, 500, and 1000 hours, statistical modeling can predict the likelihood of failure in new systems. Depending on the analysis, failure probabilities might be estimated at 5%, 1%, or even 10%, highlighting the need for nuanced approaches. A thorough analysis involves defining key performance indicators (KPIs) and setting up effective inspection protocols. These protocols depend on proper sensor selection and configuration for continuous monitoring and precise assessment.

Our discussion has progressed from understanding failures to examining faults and fault detection, moving from straightforward failure identification to the more complex task of fault detection. Historical data on failure probabilities and initial models can serve as a foundation, helping us to pinpoint KPIs for monitoring. When critical KPIs are unclear, we can start by measuring multiple parameters and refining the selection iteratively. Ultimately, our goal is to design a reliable, safe, and efficient system that benefits society while minimizing environmental impact—a true win–win objective. Key elements in achieving this goal include modeling, signal capturing and processing, and computational intelligence, which together form the backbone of a data-driven approach. Modeling provides a valuable reference point by comparing new data with previous knowledge. Meanwhile, accurate processing of signals from sensors is crucial for precise analysis. Emphasizing a data-driven approach allows for model refinement based on newly available information and evolving conditions. While analytical models establish a foundation during the early development of products, processes, or systems, real-world applications benefit from sensors and continuous fault-related data collection. This proactive approach enables preventive measures to avoid failures, issue early warnings, and support the development of effective maintenance strategies.

Failures are often difficult to analyze, as the root cause or initial failure point may not be immediately evident. Accurately tracing the origin typically requires a multi-faceted approach. The best strategy for controlling failure is early detection of potential failure points, allowing for more manageable diagnostics and solutions as the system operates or experiences load. This proactive approach relies on continuous monitoring of key features to maintain reliability. An effective control strategy involves gathering information and acting before a failure occurs, promoting a shift from traditional corrective maintenance to predictive maintenance, which is condition-based and responds to real-time system status. This approach, along with predictive maintenance strategies and relevant case studies, will be explored further in the next chapter.

A mechanical failure can be analyzed using a fishbone diagram, also known as a cause-and-effect or Ishikawa diagram. This visual tool helps identify and examine the root causes of failures systematically. This brainstorming tool allows us to consider possibilities and pinpoint plausible causes contributing to a failure. Named after its resemblance to a fish skeleton, it places the failure or effect at the head and the causes along branching fishbones. These branches often represent categories such as Manpower, Method, Machine, Material, Measurement, and Mother Nature (6 Ms), helping us think systematically about contributing factors/branches/categories.

When evaluating manpower, we consider the resources available, including skills, training, and expertise. For methods, we analyze the processes within the system, and for machines, we focus on the tools and technology involved. Materials refer to the resources utilized, while measurement involves methods of quantification—such as assessing surface roughness and detecting surface or subsurface cracks—often applying statistical techniques to gauge performance and quality. A critical aspect is crack formation; I specifically use "crack formation" rather than "crack initiation" because initial cracks are typically present in most materials. These may result from grain boundary issues or minor manufacturing flaws. Although these initial cracks or lines may not be immediately visible, they exist and can progress to crack formation, where cracks begin to propagate. Distinguishing between initial cracks and those that progress is essential for understanding material behavior.

Mother Nature covers environmental factors, like humidity and temperature, which may impact results (e.g., corrosion). Failure analysis is essential for ensuring the reliability and safety of complex systems, from manufacturing equipment and automotive parts to critical infrastructure and advanced electronics. By uncovering the causes of failures, engineers can enhance designs, implement preventive strategies, and lower operational risks and maintenance costs.

Various systems and materials are susceptible to specific surface degradation mechanisms, as outlined in Chapters 2 to 4 of this book. These mechanisms are influenced by factors such as design, environmental exposure, and operational conditions. Common degradation types include fatigue, which is a critical issue for automotive and aerospace components subjected to repeated cyclic loads; corrosion, particularly in metals exposed to moisture, salt, or chemicals; creep, affecting high-temperature applications like turbine blades and power plant components in materials such as metals and polymers; wear, which impacts moving parts such as bearings, gears, and pistons; thermal shock, causing stress and cracking in materials such as ceramics and glass due to rapid temperature changes; and the synergistic effects of these degradation mechanisms combined.

Failure and fault analysis employs systematic techniques to identify, diagnose, and evaluate issues in systems, with several key methodologies commonly applied across industries. A few of these methodologies include:

- *Failure Mode and Effect Analysis:* FMEA is a proactive approach used to identify potential failure modes, assess their impact, and prioritize them based on severity, likelihood, and detection capability. By concentrating on high-risk failure modes, engineers can implement preventive measures early in the design phase, making FMEA especially valuable at the initial stages of a project. This methodology is particularly useful when modifying materials, changing lubricants, introducing new surface engineering techniques, or applying innovative approaches. In these cases, FMEA provides a systematic framework for evaluating potential risks, helping to anticipate and address possible issues before they arise. Widely applied in manufacturing and product development, FMEA enables teams to proactively mitigate risks. The following sections of this chapter will delve deeper into this essential methodology.
- *Fault tree analysis:* FTA uses a deductive, tree-like structure (as shown in Figure 8.1) to illustrate the causal relationships among faults and failures. Starting with a top-level failure, the tree maps out contributing factors and root causes, making it an effective visual tool for analyzing complex systems with multiple failure modes.

To understand FTA, let us examine the steps using fatigue-induced fracture as an example. Fatigue failure results from repeated cyclic or fluctuating loads on a material, which gradually causes cracks to develop over time. Most materials contain voids or micro-cracks that can initiate and propagate under cyclic loading. This happens when the material experiences alternating strain, shifting from tension to zero or even to compression. Similarly, stress fluctuations between positive, zero, and negative levels, along with repeated cycles of stress, strain, or thermal changes, can lead to fatigue and eventual failure. What may start as a minor discontinuity in the

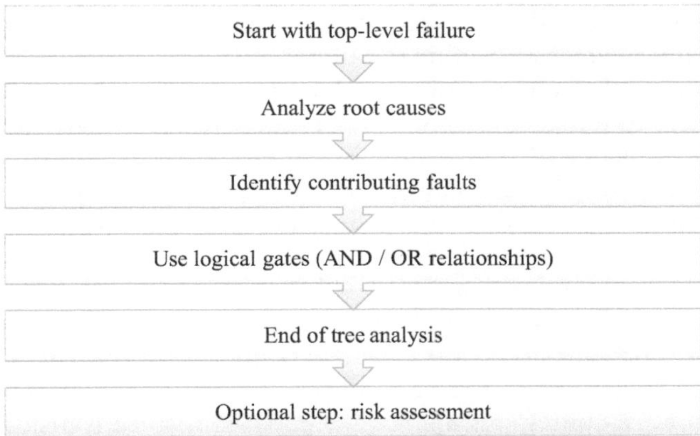

Figure 8.1 Fault tree analysis.

material can evolve into a micro-crack that propagates, ultimately resulting in fracture.

Crack initiation and crack formation are distinct stages: crack formation occurs when an initial crack gains sufficient energy to develop into a detectable crack, which can be identified using in situ non-destructive testing (NDT) techniques. Once formed, this crack can lead to failure if it receives additional energy. Consequently, we examine crack formation, crack propagation, and eventual failure. Crack initiation commonly occurs at stress concentration points, and designing components to distribute stress evenly can help reduce failure risk. However, initial cracks or material discontinuities may still be present, acting as potential sites for crack initiation. This failure path typically starts with microscopic cracks forming at stress concentrators—such as notches, holes, or inherent material flaws. These minute cracks, often undetectable by visual inspection or NDT, grow under cyclic loading due to localized stress concentrations at the crack tips. As these cracks propagate, usually perpendicular to the applied stress, they may exhibit brittle or ductile behavior depending on the material properties and loading conditions.

In FTA, each step contributing to fatigue failure is represented as a contributory event, which builds toward the primary failure event. The initial events include factors such as material properties, stress concentrators, load frequency, and environmental conditions—all of which contribute to crack initiation. This is followed by propagation events, where established theoretical models help relate crack growth rate, localized stress, and the critical crack size beyond which the material's remaining cross-section fails to support applied loads. Critical crack size, a key threshold, depends on material properties, load intensity, and crack geometry. These factors and events are systematically analyzed in fault tree steps to identify root causes and contributing factors leading to failure.

By pinpointing these contributory factors, designers can target steps in the fault tree to mitigate or slow crack propagation, thereby enhancing the component's service life under fatigue loading. Effective control over these factors enables the design to align with sustainability criteria by extending the component's life and reducing the likelihood of sudden, catastrophic failure.

Consider a second example related to creep-induced failure. Unlike fatigue, which arises from cyclic loading, creep failure results from prolonged exposure to constant stress over extended periods. This gradual deformation under sustained stress is a critical concern for materials subjected to high temperatures or continuous loads. Like corrosion, creep is a slow and persistent process, occurring progressively over time rather than immediately. It is characterized by continuous strain and deformation, even when the applied load remains constant. In contrast to fatigue, where varying loads, strains, or temperature fluctuations are typical, creep progresses steadily without the need for changes in external conditions.

Puthoff et al.[1] demonstrated how hardness testing can be effectively used to explore creep-related behavior in materials. In their method, a controlled load was applied with a hard indenter held on the material's surface, causing indentation depth to increase due to plastic deformation, with hardness then measured. Extending the hold time from 0.01 to 50 seconds increased the indentation mark by 2.6 times (though this can vary under different operating conditions), highlighting significant creep behavior. The authors employed broadband nanoindentation creep (BNC) to gather detailed data, monitoring the indenter's gradual penetration under a constant load and capturing instantaneous hardness and strain rate over time. High-speed nanoindenters enabled data collection across multiple strain rates, facilitating hardness versus strain rate curve construction and insights into deformation mechanisms. Analysis of this data provided activation energy and volume, key parameters for understanding the material's stress response. This approach shows that creep characteristics can be assessed using existing tools and sensors rather than new ones, with a data-driven methodology enhancing accuracy by refining sensor relationships. In their tests, applying a 10-mN load produced an indentation with distinct contact edges, marked by a three-point star shape. As creep time increased, the contact edges expanded, yielding a larger indentation that effectively captured the material's creep response.

FTA effectively models this time-dependent failure mechanism, illustrating how cumulative events contribute to the primary failure event. In that case, the root node (primary failure event) is driven by constant or high stress over time. Several intrinsic and extrinsic factors impact the creep process, affecting the rate of deformation, ultimate strength, and service life of the material. These include:

Stress levels: Higher stress intensifies deformation and accelerates creep.

Temperature: Elevated temperatures significantly speed up creep, reducing material lifespan.

Material characteristics: Properties like grain size, phase composition, and resistance to grain boundary sliding impact creep resistance.

- *Ductility:* Materials with low ductility may experience brittle fracture, while ductile materials may deform before failing.
- *Material strength and rupture time:* Material strength and rupture time play critical roles in creep resistance, with stronger materials enduring creep for longer periods, while rupture time marks the point of ultimate failure under sustained stress. Creep progresses through three stages: in the primary stage, deformation starts rapidly but gradually slows as the material adjusts to the load; in the secondary or steady-state stage, deformation proceeds at a constant rate due to a balance between strain-hardening and recovery; and in the tertiary stage, deformation accelerates toward failure, driven by microstructural changes such as void formation and grain boundary separation. This progression resembles a "bathtub curve, where the initial high deformation rate declines to a steady state before accelerating in the final stage.

Contributory events: Within the fault tree, these stages are further detailed by contributory events that amplify the risk and timing of failure:

- *No preventive maintenance:* A lack of inspection or repair increases the risk of undetected creep progression. Preventive maintenance includes scheduled activities like regular inspections, cleaning, and lubrication, which may be carried out every 24 or 48 hours, weekly, or bi-weekly, depending on specific requirements.
- *Unexpected stress events:* Sudden loads can expedite creep deformation and bring forward failure.
- *High-temperature exposure:* When materials experience temperatures beyond their design limits, creep rates surge, leading to early onset of tertiary creep.

Mapping creep in FTA: In FTA, each stage of creep is depicted as a branch, detailing primary and contributing factors that drive failure. This structured approach allows for a focused examination of critical intervention points to help reduce creep risks. For example, primary creep can be managed by selecting appropriate materials and optimizing design, secondary creep risks by regulating environmental factors like temperature, and tertiary creep through preventive maintenance to detect early signs of advanced creep, enabling timely repairs or replacements. A fine-grained material structure is particularly beneficial in minimizing creep susceptibility, as fine grains tend to limit crack propagation by halting its progression at grain boundaries, assuming stress levels are controlled.

This underscores the role of fine grains in limiting crack expansion at interfaces, especially around voids, which encourages us to go beyond traditional failure analysis and focus on fault analysis. This shift enables the early detection of potential failures, helping to prevent breakdowns, excessive wear, financial loss, and environmental impact, thereby supporting sustainability goals. Achieving this requires a well-defined sensing strategy—whether time-based or frequency-based—and an understanding of the unique signatures and behaviors associated with each component.

Sensing trends or frequencies unique to specific components are influenced by their function, design, and stage of wear. For example, ultra-mild wear exhibits distinct characteristics compared to medium wear. Through time-based or frequency-based analysis, we can identify key indicators for ongoing monitoring, recording, and analysis. This approach underpins our data-driven methodology, where signals can be in the time domain, frequency domain, or a combination of both, depending on which is most informative for the context. The data-driven approach begins with foundational signatures, models, and insights from the literature, integrating these into our sensing and computational frameworks. As new data is collected, these models are continuously refined and updated, enhancing both accuracy and adaptability.

Detailed mechanisms: Creep-induced failure often arises from complex mechanisms involving grain-boundary cavities and micro-voids. Grain-boundary cavities are diffusion-driven and form at grain boundaries under prolonged stress, often initiating at inclusions or impurities. Their growth follows either a diffusion-based or power-law model, and these cavities can coalesce, even at low strains, leading to intergranular fracture. In contrast, micro-voids are formed through slip mechanisms at the microscale and are typically associated with dimple rupture. Understanding the distinctions between these mechanisms is crucial for accurately diagnosing the origins of failure.

By breaking down these mechanisms within a fault tree, FTA allows analysts to track how contributory events incrementally increase the likelihood of creep failure. This approach supports the development of maintenance schedules, operational guidelines, and design enhancements to extend the component's lifespan. With FTA, engineers can pinpoint weak links, determine probable failure points, and implement preventive measures that preserve material integrity under sustained load conditions.

- *Root cause analysis:* A reactive method used after a failure, RCA aims to identify primary causes of failure, distinguishing root causes from symptomatic issues. By addressing the fundamental cause, RCA supports corrective actions[2] that prevent recurrence, rather than just resolving immediate symptoms. In the era of digit twin, there is a pressure to move from offline RCA to online RCA, a transition that is currently in progress. Traditionally, failures are detected using off-line methods, such as scanning electron microscopy (SEM) or other advanced microscopy techniques, only after a failure has occurred. These analyses are performed post-failure in an offline mode. The current objective, however, is to develop an online system capable of making real-time decisions and judgments autonomously. This is the direction we need to pursue for more proactive fault detection.
- *Condition monitoring and predictive maintenance*: Through real-time data analysis, condition monitoring uses techniques such as vibration analysis, thermal imaging, oil analysis, acoustic analysis, etc. to assess system health continuously. Predictive maintenance, often enhanced by machine learning algorithms, forecasts maintenance needs, thus reducing unplanned downtime and enhancing system reliability.
- *Non-destructive testing:* NDT methods (covered in Chapter 6 of this book) allow engineers to inspect materials and components without damaging them. This is especially crucial in fields like aerospace, where even small defects can lead to catastrophic failure.
- *Fault detection and diagnostics (FDD):* This method, often termed fault diagnosis and health monitoring, believes that failure analysis will naturally follow if we have a reliable FDD system in place. The FDD approach relies on modeling, signal processing, and computational intelligence. The FDD approach can be model-based, data-driven, or knowledge-based. If historical data is available, we can begin with a knowledge-based model, which explains the need for accurate modeling. If we are dealing with a completely new system, initial samples may be created to generate data, allowing us to build an approximate model that can be continuously refined as more data is collected. Fault monitoring involves analyzing deviations from expected performance levels—detecting issues before they escalate to failures. A data-driven approach simplifies this process. Including past data allows us to draw meaningful conclusions by examining relationships and trends, making it easier to understand and manage faults within a system.

Combining methodologies such as FMEA, FTA, RCA, condition monitoring and predictive maintenance, NDT, and FDD creates a robust framework for understanding and mitigating degradation, thereby enhancing reliability and safety across various industries. For example, in aerospace manufacturing, FMEA allows engineers to proactively identify and prioritize high-risk failure modes during the design phase, anticipating issues before they arise. FTA, shown in Figure 8.1, offers a tree-like map to visually analyze complex systems with multiple failure pathways, tracing a critical fault in a component back to root causes. In cases where failures occur, RCA identifies fundamental causes, as in the study by Hirani[2] on the fracture of an outer ring in a four-row cylindrical roller bearing, where RCA distinguished underlying causes from symptoms to guide effective corrective actions. In parallel, condition monitoring and predictive maintenance leverage real-time data, with tools like vibration and thermal analysis detecting early signs of wear and predicting maintenance needs, helping industries like power generation prevent unplanned downtime. Finally, NDT methods, discussed in Chapter 6, enable safe, non-intrusive inspection of components, critical for high-stakes applications such as detecting material defects in aerospace components, where even minor flaws can have severe consequences. Together, these methods build a comprehensive approach to improving the durability and operational safety of systems in demanding applications.

Advances in sensor technology, data analytics, and machine learning have made data an invaluable resource in modern failure and fault analysis. Fault analysis provides a structured approach to diagnosing malfunctions within systems, with data-driven techniques significantly enhancing accuracy and depth. Access to reliable data enables more precise decision-making, while relying solely on models may limit the adaptability and predictive reliability needed for long-term planning. Data-driven approaches begin with gathering historical data to identify relationships between various inputs and failure events, forming the foundation for comprehensive analysis. Integrating traditional statistical methods, such as regression and correlation, with advanced clustering and classification algorithms enables more nuanced insights into system performance.

To organize and interpret data effectively, classification methods can be applied, categorizing failures into distinct types or "buckets." For example, different types of failures—such as cracks, pits, or dimples—can each be grouped into their respective categories, allowing targeted responses. KPIs are essential in this process, often requiring machine learning algorithms for effective clustering and classification. While we will not go deeply into machine learning methods in this course, we will highlight important features and examine a case study to illustrate practical application.

For a thorough fault analysis, multiple data sources must be considered. Sensor data alone is insufficient; operational parameters should be reviewed to link sensor readings to real operating conditions. Maintenance history, including timing and duration of activities, provides context, and environmental factors like humidity and temperature further enhance the analysis. Reviewing these sources enables a comprehensive understanding of system performance, offering a robust foundation for diagnosing and mitigating faults effectively.

In data preprocessing, it is crucial to clean and organize the dataset by addressing missing values, removing outliers, standardizing variables, and engineering relevant features. Once a model is trained and validated, predictive analysis helps anticipate failures or identify underlying issues. This predictive capability enables continuous improvement, as new data refines the model's accuracy in forecasting and explaining failures.

Real-time monitoring and continuous data collection empower engineers to detect subtle anomalies, recognize patterns, and predict potential failures before they occur. For example, vibration data can reveal early signs of wear or imbalance in rotating machinery, enabling timely intervention to prevent complete breakdowns. Machine learning algorithms further enrich fault analysis by identifying complex, non-linear relationships in large datasets that may be overlooked by traditional methods. Predictive models trained in historical data offer insights into how specific faults may develop under certain conditions, enabling more accurate predictions of failure likelihood and timing. Particularly in critical systems, like power grids, where failures have significant impacts, big data analytics combined with real-time condition monitoring supports predictive maintenance, optimized repair scheduling, and minimized downtime. These data-driven techniques complement traditional methods, strengthening efforts to enhance reliability and safety across complex systems.

Benefits of failure and fault analysis: Failure and fault analysis provides substantial advantages to organizations and industries, improving reliability by identifying and mitigating potential failure modes to ensure products and systems perform consistently over time. This proactive approach enables cost savings by reducing the need for repairs, replacements, and associated expenses in materials and labor, while also helping to prevent revenue losses. In safety-critical fields like aviation, healthcare, and energy, failure analysis is especially vital, as it prevents accidents, ensures compliance with regulatory standards, and enhances the safety of personnel and the public.

Challenges and future directions in failure and fault analysis: Even while it has its uses, failure and fault analysis is not always easy, especially when dealing with complicated modern systems that have a lot of moving parts and variables that can cause unexpected outcomes. To properly understand the data, sophisticated analytical techniques and professional knowledge are required due to the intricacy of the problem. Incomplete or noisy data makes it harder to conduct exact analyses and make educated decisions; therefore, data availability and quality are additional potential obstacles. Future developments in failure and defect analysis are anticipated to be facilitated by the likes of the IoT, digital twins, and artificial intelligence (AI). Engineers can improve prediction accuracy and fault detection with the use of digital twins, which are virtual representations of actual systems. These allow engineers to test and model different failure scenarios. IoT devices will increase the breadth of data collection, allowing for real-time insights into the performance and health of equipment.

8.3 FAILURE MODE AND EFFECT ANALYSIS

FMEA[3] is a structured methodology designed to identify potential failure modes in a product, process, or system, assess their effects, and prioritize actions to reduce or eliminate associated risks. While FMEA can be a reactive tool for analyzing existing products, it serves as a proactive approach when insights gained are applied to improve future products or services.

Widely adopted across industries including automotive, aerospace, power sector[4] and manufacturing, FMEA plays a critical role in risk management, quality assurance, and product development. Addressing potential failures during the design phase helps organizations enhance reliability, safety, and customer satisfaction while reducing unexpected failure costs. The primary objective of FMEA is to lessen both the severity and likelihood of failures, supporting improved product quality, reliability, and safety—goals that align with this course's focus on maintaining product integrity and preventing environmental or material degradation.

In guiding maintenance strategies and surface engineering requirements, FMEA links failure types to their underlying mechanisms and modes, allowing for structured prioritization using the Risk Priority Number (RPN). High-RPN failures, which represent significant risks, are addressed promptly, while lower-RPN issues, especially those with minimal risks and impacts, may be deferred to optimize resources. FMEA's proactive approach enables the early detection and management of potential failures throughout the product life cycle, with consequences ranging from minor disruptions to severe safety concerns.

The prioritization process in FMEA resembles principal component analysis (PCA), where the focus is on the largest eigenvalues that capture the most variance, while smaller values are often disregarded to streamline the analysis. In FMEA, concentrating on failure modes with the highest RPN values allocates resources effectively to the most critical issues. Lower-RPN modes can be managed with flexibility, either delayed or omitted if their risk contribution is minimal. For instance, if 40 failure modes are identified, and the top 5 account for 85% of the overall impact, it may be efficient to deprioritize the remaining 35 modes, as they collectively contribute only 15% of the total risk.

8.3.1 FMEA processes key steps

- *Define the scope:* The first step in FMEA is defining the scope of analysis. Teams identify the system, product, or process under evaluation, and break it down into subsystems or components for a detailed examination. This step is often referred as "Product Decomposition: Breaking down a system into its components."
- *Identify failure modes:* For each component or subsystem, teams brainstorm and list all possible failure modes. These could include wear, fatigue, corrosion, misalignment, or any other condition that could impair the component's function.
- *Determine effects of each failure mode:* Next, teams analyze and document the potential effects of each failure mode on the system, operator, or end-user. Effects can range from minor quality issues to severe impacts on safety and functionality. Understanding these effects helps assess the significance of each failure mode. This step is often referred to as "Effect evaluation: Assessing potential consequences of each failure."
- *Identify causes and current controls:* For each failure mode, teams explore possible causes, such as material defects, design flaws, improper assembly, or environmental factors. They also review any existing controls, such as testing procedures or quality checks, aimed at preventing or detecting the failure. This step is often divided into two parts: "Root cause analysis: Investigating causes of each failure mode" and "Current controls assessment: Reviewing existing preventive measures."
- *Assign severity, occurrence, and detection ratings:* FMEA uses a rating system to evaluate each failure mode based on three factors:
    - *Severity:* This metric assesses the severity of the failure's impact, with higher severity ratings assigned to failure modes that have more serious consequences. Understanding failure probabilities and categorizing failures as low, medium, or high offers a structured approach to risk assessment. A medium failure is identified when a material or system performs approximately as expected; for example, if a system designed for a 3-year lifespan fails around that time, it would be classified as a medium failure. Conversely, a failure occurring significantly earlier—such as within 1 month for a system expected to last three years—would be considered highly critical.

  - *Occurrence:* This evaluates the likelihood of the failure mode happening, based on historical data, statistical analysis, or expert judgment.
  - *Detection:* This assesses the probability that the failure will be detected before it impacts the end-user. Lower detection ratings indicate a higher chance of the failure going undetected.
- *Calculate the RPN:* The RPN is calculated by multiplying the severity, occurrence, and detection ratings, yielding a numerical value that reflects the overall risk level of each failure mode. This helps teams prioritize issues based on their potential impact. Assigning these scores, typically on a scale from 1 to 10, can involve some subjectivity; for instance, one person might rate a scenario as 1 while another might score it as 2 or 3. This subjectivity can be minimized through practical experience or by consulting established RPN guidelines from reliable sources like the ASM handbook, which offers standardized ratings and comparisons to promote consistency in scoring.

  Both literature and field experience play crucial roles in standardizing RPN assignments, making FMEA a powerful tool for aligning current conditions with desired outcomes. It provides a structured approach, where higher RPN scores indicate a need for improvement. By focusing on high-RPN items, teams can drive more efficient progress and achieve a higher return on investment.
- *Develop and implement mitigation actions:* Based on the RPN, teams focus on high-priority failure modes and develop action plans to mitigate or eliminate them. These actions may include design modifications, process improvements, additional testing, or enhanced quality control measures.
- *Review and update the FMEA:* FMEA is an iterative process, and as new data or insights become available, the analysis should be updated. Continuous improvement and regular reviews ensure that risk management remains effective throughout the product's life cycle.

8.3.2 Types of FMEA

FMEA can be tailored to various applications, depending on the context and purpose of the analysis. The main types of FMEA include:

- *Design FMEA (DFMEA):* DFMEA is applied during the design phase of a product or system to identify potential failures resulting from design flaws. By addressing these issues early, DFMEA helps prevent costly redesigns and ensures that the product meets reliability and safety standards. For example, in a journal bearing system, potential failure modes might include material wear, improper geometry, incorrect lubricant selection, and lack of filtration to minimize lubricant contamination. Identifying and resolving these issues reduces the likelihood of failure and extends the system's life cycle.

- *Process FMEA (PFMEA):* PFMEA is used in manufacturing and production to evaluate potential failure modes within processes, such as assembly, machining, or packaging. By analyzing process-related risks, PFMEA enables companies to optimize manufacturing quality and minimize production defects.
- *System FMEA:* This type of FMEA assesses risks within an entire system, considering interactions between subsystems and components. System FMEA is valuable for complex systems where failures in one part could have cascading effects on other areas.

8.3.3 Benefits of FMEA

FMEA's primary advantage is its proactive approach, allowing organizations to identify and address potential failure modes early, thereby reducing costly and time-consuming corrections later. Its benefits include enhanced product reliability, as it enables the identification and elimination of potential failure points to create more robust designs, which increases customer satisfaction. In industries where safety is paramount, like automotive, aerospace, and healthcare, FMEA ensures safer designs by systematically assessing risks, thereby protecting users and minimizing liability. It also yields significant cost savings by identifying issues early, reducing the need for repairs, recalls, and downtime, ultimately lowering maintenance costs and boosting operational efficiency. Furthermore, FMEA's structured framework enhances problem-solving by facilitating clear communication about risks and helping teams develop effective mitigation strategies, supporting continuous improvement.

8.3.4 Challenges in implementing FMEA

While FMEA offers significant advantages, it also faces challenges, such as accurately estimating occurrence and detection ratings, which often rely on limited historical data and expert judgment; without robust data, teams may misjudge risks, leading to inaccurate scores. The effectiveness of FMEA hinges on the expertise and rigor of the team, as overlooked failure modes or superficial analysis can leave critical risks unaddressed, resulting in incomplete risk management. Additionally, the process can be time-consuming, especially in complex systems with numerous components, demanding considerable resources. Integrating FMEA into the organizational culture poses another challenge, as it requires regular updates and a commitment to continuous improvement, which can meet resistance. Ensuring FMEA becomes a core aspect of design and manufacturing processes is essential to fully realize its benefits.

8.3.5 Future directions in FMEA

Technological advancements are enhancing FMEA's capabilities by enabling more accurate and efficient risk assessments. Digital twins—virtual models of physical systems—allow teams to simulate various failure scenarios and analyze potential impacts in a controlled environment, improving FMEA accuracy through real-time data and refined occurrence and detection ratings. The integration of machine learning and data analytics further advances FMEA by analyzing large datasets from similar systems, helping to identify potential failure modes and predict occurrence rates with greater precision. This data-driven approach is particularly beneficial in industries with abundant historical data. Additionally, the IoT enables real-time monitoring and condition-based maintenance by using sensors to track component health continuously, providing essential data for timely FMEA updates. This dynamic, data-rich framework enhances risk assessments, bolstering reliability throughout the life cycle of products and systems.

8.3.6 Case study: Journal bearing system

Figure 8.2 illustrates a structured approach to performing FMEA with a focus on calculating the RPN and determining if corrective action is necessary. The following are the important steps:

- *Define scale for severity, occurrence, and detection:* Establish a rating scale for evaluating the severity, likelihood of occurrence, and detectability of each failure mode. The question of how to rate severity depends on familiarity with the system, with scales like 1 to 5, 1 to 10, or even 1 to 100 being options. For systems with limited data, a 1 to 5 scale may be appropriate, while a 1 to 10 scale is common for more familiar systems. Highly familiar systems might use a 1 to 100 scale, although this course will primarily use the 1 to 10 scale. This scale allows us to assess the impact and seriousness of a failure mode, where a score of 1 represents minimal impact, and higher scores like 5 or 10 indicate more severe consequences, potentially involving safety risks, catastrophic failures, or significant financial losses. Some subjectivity may occur in assigning scores—different evaluators might rate the same scenario as a 1 or a 2—but each rating should be justified with a clear rationale. After determining the severity of a failure mode, the next step is to identify its root causes. When data from literature or past experiments is

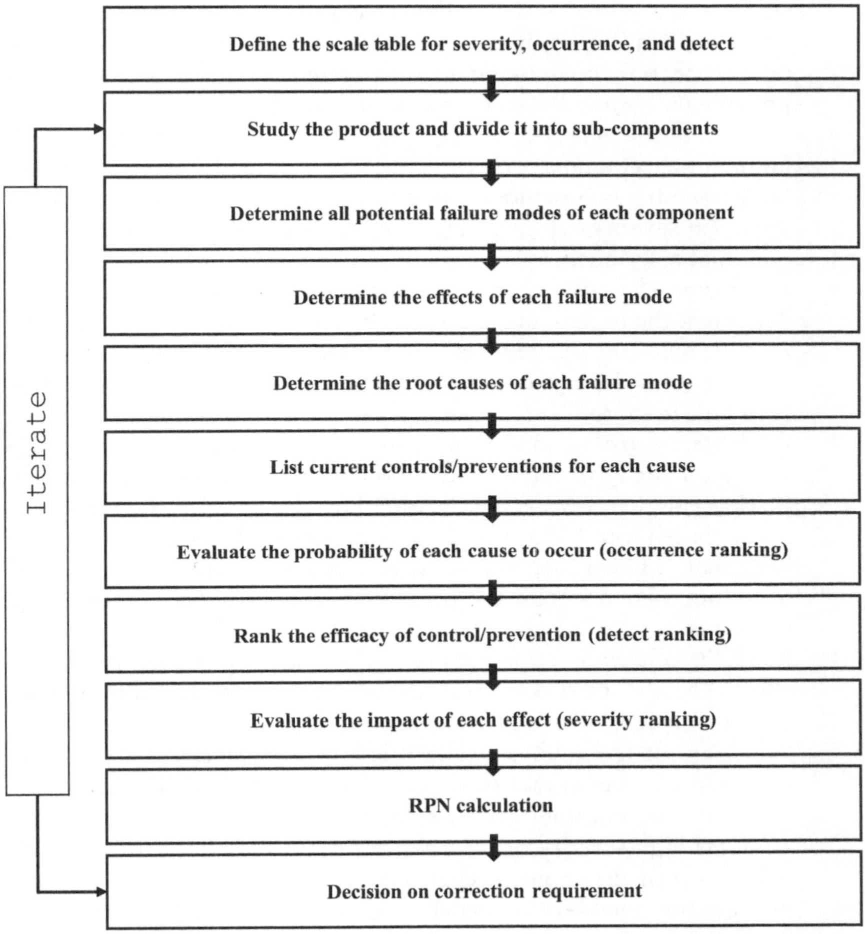

Figure 8.2 FMEA process flowchart.

available (e.g., 5 out of 10 tests showing failure), it helps in assigning an occurrence ranking or estimating the probability of failure (POF).

- *Study product and divide it into sub-components:* Understand the product and break it down into manageable sub-components for detailed analysis. To start, identify the part or system requiring analysis, which should be clear to all involved. Begin by pinpointing the exact item for examination and proceed by analyzing its components. For large assemblies, subdivide into subassemblies, then individual components, and finally specific features. This top-down approach helps ensure comprehensive coverage.

- *Determine all potential failure modes of each component:* Failure here refers to any instance where the system does not fulfill its intended purpose. Identify possible ways each component might fail. Once the relevant feature, component, or subsystem is identified, determine its possible failure modes, as each element could fail in unique ways.
- *Determine the effects of each failure mode:* Assess the impact each failure mode could have on the system and the end-user. After identifying potential failure modes, assess the impact of each mode to determine the severity ranking. This step is essential for understanding potential consequences.
- *Determine the root causes of each failure mode:* Investigate and document the underlying causes of each identified failure mode.
- *List current controls/preventions for each cause:* Identify and list existing controls or preventive measures currently in place to address each failure cause.
- *Evaluate the probability of each cause to occur (occurrence ranking):* Assign a score based on the likelihood of each failure mode occurring (POF). This step involves estimating how frequently a failure might occur based on historical data, experimental results, or expert judgment. Higher occurrence scores suggest a greater likelihood of failure, which signals the need for increased focus on preventive measures
- *Rank the efficiency of control/prevention (detect ranking):* Assess the effectiveness of each control measure in detecting the failure mode. This involves evaluating the detection capability—specifically, the likelihood of identifying a failure before it escalates. For example, assessing whether cracks or other defects can be reliably detected is essential. Detection effectiveness plays a critical role in risk evaluation, as high severity or occurrence ratings may not pose as much risk if robust detection mechanisms are in place. Detection scores indicate the reliability of monitoring and diagnostic tools in identifying faults early, enabling timely intervention. A low detection score suggests that faults are difficult to detect, thereby increasing the risk associated with that failure mode.
- *Evaluate the impact of each effect (severity ranking):* To evaluate the impact of each failure mode, we assign a severity score based on the potential consequences or seriousness of its impact on the system. This process involves examining how each failure mode could affect performance, safety, reliability, or customer satisfaction. A higher severity score reflects a more critical impact, such as a safety hazard, catastrophic failure, or substantial financial loss, whereas a lower score indicates minimal disruption or inconvenience. The severity ranking thus helps prioritize which failure modes require urgent attention.

- *RPN calculation:* The three factors—severity, occurrence, and detection—together form a structured approach for analyzing failure modes. These factors enable teams to calculate the RPN, which helps in prioritizing actions according to the overall risk associated with each failure mode. By assessing each mode's impact, likelihood, and detection effectiveness, organizations can allocate resources strategically, focusing on the most critical issues. The RPN is determined by multiplying the assigned scores for occurrence (O), detection (D), and severity (S): RPN = O × D × S.
- *Decision on correction requirement:* If the RPN exceeds a certain threshold, corrective action is required. If it is within acceptable limits, no correction is needed.
- *End or iteration:* If corrective action is needed, the process cycles back to the necessary steps until the RPN is acceptable. If no further action is needed, the analysis ends.

The FMEA begins with examining engineering drawings to identify possible assembly issues, such as interference, clearance, and overall impact on the system. Using three-dimensional (3D) models, we can further understand stress distribution, pinpoint areas of high-stress intensity, and detect potential crevice formations. Both 2D and 3D models, along with physical prototypes, if available, provide a comprehensive view of how different parts of the system interact, enhancing the accuracy of our analysis.

To illustrate the FMEA process, consider a case study involving a journal bearing. In the simplified model of a journal bearing system, there are three main components:

- *Journal:* A cylindrical shaft that rotates within the bearing. The journal is made of metals.
- *Bearing:* A surrounding cylindrical housing in which the journal is rotated. The bearing allows smooth rotation and supports the journal, ensuring minimal friction between the journal and itself. Generally, clearance fit is used between the journal and the shaft.
- *Lubricant supply system:* It can be gravity-fed or pressurized by an external pump. Connected with the bearing mostly through an oil hole positioned into the bearing, this system provides lubricant into the bearing, reducing friction and wear between the journal and bearing surfaces.

Figure 8.3 presents a 3D view of a journal bearing along with a schematic of the testing setup for a journal bearing system. The setup includes various components labeled by numbers, which are detailed in Figure 8.3(b):

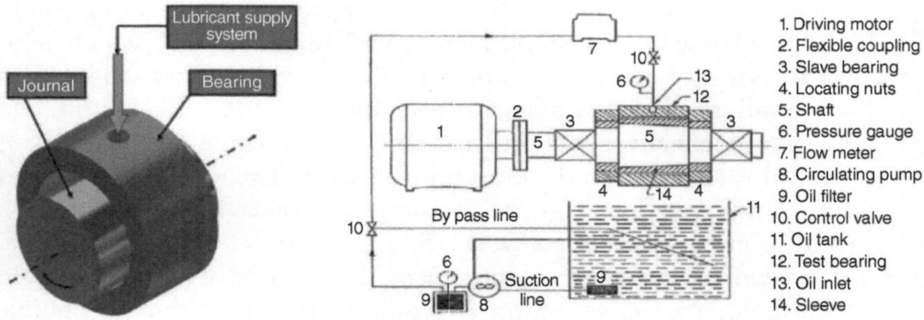

(a) Gravity-fed journal bearing (b) Circulating-pump-fed journal bearing

Figure 8.3 Simplified journal bearing system overview.

- *Driving motor:* The primary power source that drives the shaft.
- *Flexible coupling:* Connects the motor to the shaft, allowing for slight misalignments.
- *Slave (supporting) bearings:* Support the shaft on the opposite end.
- *Locating nuts:* Secure the bearing assembly in place.
- *Shaft:* The central rotating element within the bearing assembly.
- *Pressure gauge:* Measures the pressure of lubricant to be supplied to the test bearing.
- *Flow meter:* Monitors the lubricant flow rate.
- *Circulating pump:* Ensures the continuous flow of the lubricant through the system.
- *Oil filter:* Removes contaminants from the lubricant.
- *Control valve:* Regulates the lubricant flow within the system.
- *Oil tank:* Reservoir for storing the lubricant.
- *Oil inlet:* The point where oil enters the test bearing.
- *Test bearing:* The bearing being tested within this setup.
- *Sleeve:* Typically, the section of shaft length within the length of the testing bearing is termed a "Journal." However, in the setup (shown in Figure 8.3(b)), a sleeve is mounted on the shaft with an interference fit and can be considered as the "Journal."
- The system features a bypass line that redirects lubricant flow as needed, and a suction line that pulls oil from the tank through the filter and pump, circulating it to the test bearing.

For this journal bearing system, the study focuses on three primary components: the journal, the bearing, and the lubrication supply. Each component is assigned its own RPN, allowing us to prioritize and address the highest RPN issues, recalculating iteratively until RPN values reach acceptable levels. This iterative approach, supported by data from literature,

Figure 8.4 Abrasive wear, corrosion wear, and fatigue wear of journal bearing.

experiments, simulations, and theoretical models, is essential for refining the product design.

The journal bearing system and its subcomponents are susceptible to various failure modes (as shown in Figure 8.4), including abrasive wear, fatigue, and corrosion.

Figure 8.4 shows three different images related to the failure of journal bearings:

1. *Left image:* A close-up of a journal bearing housing with abrasive wear of copper alloy.
2. *Center image:* A photograph of a bearing surface with a visible area of localized damage. The image shows wear patterns and a central region with significant scuffing or scoring, likely due to operational issues such as inadequate lubrication or high loads.
3. *Right image:* An image of a bearing surface showing two symmetrical areas of pronounced wear or erosion near the central lubrication groove. These wear patterns suggest potential concentrated stress points, leading to material degradation over time. Additionally, small pits are visible, likely resulting from fatigue or cavitation effects.

Each failure mode is classified and assessed for severity and likelihood, considering factors such as external loading, material properties, and environmental conditions. For example, abrasive and adhesive wear have unique severity indices based on wear coefficients, while fatigue—whether on the surface or subsurface—requires specific detection methods. Environmental factors like humidity can induce corrosion, sometimes necessitating corrosion-resistant materials, which may increase costs. Failures in the lubrication system may result from issues such as oil leakage, partial supply, clogged filters, or inadequate temperature control, potentially aggravated by

Table 8.1 Severity guidelines for journal bearing FMEA (1–10 qualitative scale)

Effect	*Rank*	*Criteria*
No	1	No effect on bearing performance.
Very slight	2	Very slight increase in friction only.
Slight	3	Slight increase in friction only.
Minor	4	Minor increase in friction only.
Moderate	5	Moderate increase in friction only.
Significant	6	Increase in friction and slight wear occurrence.
Major	7	Significant increase in friction and moderate wear occurrence.
Extreme	8	Significant increase in friction and wear.
Serious	9	Complete failure of lubrication with maximum increase in friction and wear.
Hazardous	10	Catastrophic failure.

Table 8.2 Occurrence guidelines for journal bearing FMEA (1–10 qualitative scale)

Effect	*Rank*	*Criteria*
Almost never	1	Failure unlikely. History shows no failure
Remote	2	Rare number of failures likely
Very slight	3	Very few failures likely
Slight	4	Few failures likely
Low	5	Occasional number of failures likely
Medium	6	Medium number of failures likely
Moderately high	7	Moderately high number of failures likely
High	8	High number of failures likely
Very high	9	Very high number of failures likely
Almost certain	10	Failure almost certain

debris or contamination, which can introduce moisture or accelerate lubricant oxidation. Table 8.1 provides severity guidelines for journal bearing FMEA (1–10 qualitative scale). It presents the severity rankings and criteria used in FMEA for assessing the impact of failure modes on a journal bearing system. Each severity level is assigned a rank from 1 to 10, accompanied by a description of the effect on the bearing system. Table 8.2 tabulates the occurrence guidelines for journal bearing FMEA (1–10 qualitative scale). It categorizes the likelihood of failure occurrences in a journal bearing system. The table provides ranks from 1 to 10, each with associated criteria, to help assess the POF modes based on historical and anticipated data. Table 8.3 lists the detectability guidelines for journal bearing FMEA (1–10 qualitative scale). It ranks the ease of detecting failure modes within a journal bearing system, based on the stage and reliability of the detection methods. The table assigns ranks from 1 to 10, each with specific criteria, indicating the likelihood of identifying potential issues effectively.

Table 8.3 Detectability guidelines for journal bearing FMEA (1–10 qualitative scale)

Effect	*Rank*	*Criteria*
Almost certain	1	Proven detection methods available in concept stage.
Very high	2	Proven computer analysis available in early design stage.
High	3	Simulation and/or modelling in early stage.
Moderately high	4	Tests on early prototype system elements.
Medium	5	Tests on preproduction system components.
Low	6	Tests on similar system component.
Slight	7	Tests on product with prototype and system components installed.
Very slight	8	Proving durability tests on product with system components installed.
Remote	9	Only unproven or unreliable techniques available.
Almost impossible	10	Detection of failure is almost impossible with current methods.

For each identified failure, we evaluate severity, occurrence, and detection to calculate the RPN, enabling risk-based prioritization. Severity ratings reflect the failure's impact, from minor issues to catastrophic risks, while occurrence ratings, informed by research or experimental data, indicate failure probability. The detection rating measures our ability to identify faults early.

Multiplying these ratings yields the RPN for each failure mode (as shown in Table 8.4), guiding priority in mitigation efforts. For example, if abrasive wear in a journal bearing has a severity of 4, occurrence of 3, and detection rated at 1, the RPN is calculated as $4 \times 3 \times 1$, resulting in an RPN of 12. In contrast, subsurface fatigue might have a higher detection difficulty, resulting in a higher RPN. Table 8.4 shows various potential failure modes for a journal in a bearing system. Each failure mode is listed alongside its potential effects, as well as ratings for severity, occurrence, detection, and the resulting RPN.

Detection of lubrication-related issues can be relatively straightforward, as visible signs like a nonfunctioning pump or lack of oil flow are easily observed. However, for more complex assemblies, advanced monitoring tools may be necessary, especially when components are not easily accessible.

The final step involves ranking failure modes hierarchically, starting with the highest RPN. High-RPN issues are addressed first, while lower-RPN issues may be deferred or excluded based on available resources. This structured approach enables targeted improvements, making FMEA an invaluable tool for managing the reliability and safety of journal bearing systems and other complex mechanical assemblies.

Let's delve into the journal-related data analysis. The highest RPN values observed for the journal are 140, 128, 126, and 126. Depending on our

Table 8.4 FMEA worksheet for journal

Potential failure mode	*Potential effect(s) of failure*	*Severity*	*Occurrence*	*Detection*	*RPN*
Abrasive wear	Temperature rise of lubricant, decrease in lubricant viscosity, increase in friction and wear	4	3	1	12
Adhesive wear	Temperature rise of lubricant, decrease in lubricant viscosity, increase in friction and wear	4	4	1	16
Subsurface fatigue	Formation of subsurface cracks, increase in stresses due to stress concentration. reduction in load-carrying capacity	6	3	7	126
Surface-initiated fatigue	Formation of pits on the surface, increase in friction and wear	6	3	7	126
Moisture corrosion	Lubricant contamination, increase in friction	7	3	1	21
Frictional corrosion	Surface damage in the form of increased surface roughness	8	4	3	96
Plastic deformation	Change in journal surface profile, increases friction and wear	6	3	3	54
Indentation	Formation of uneven valleys, deteriorating surface smoothness, and increasing friction	8	4	4	128
Fatigue fracture	Journal failure, catastrophic failure	10	2	7	140
Thermal cracking	Reduction in load-carrying capacity	9	2	4	72

specific requirements, we can prioritize these high-RPN failures or opt to address only those with an RPN above a certain threshold, such as 50 or 100.

Table 8.5 presents an FMEA worksheet focused on the lubrication system component, detailing potential failure modes, their effects, severity, occurrence, detection rankings, and the resulting RPN. The RPN values in this table indicate that the lubrication system does not present significant risks or challenges.

Shifting the focus to the bearing, we see even higher RPN values (Table 8.6), such as 324, 294, 288, and 280. These elevated values primarily result from detection challenges; bearing failures are harder to detect, which leads to higher detection scores. Improved diagnostic tools could lower these numbers. This analysis identifies the bearing as a critical component within the system, which comprises the journal, bearing, and lubrication (and sometimes additional subsystems). Due to wear and dimensional

Table 8.5 FMEA worksheet for lubrication system component

Potential failure mode	*Potential effect(s) of failure*	*Severity*	*Occurrence*	*Detection*	*RPN*
Fails to supply any lubricant to the bearing	Bearing seizure	10	5	1	
Supply the lubricant only partially	Contact between journal and bearing, increased friction and wear	8	5	1	40
Supply lubricant without filtration of debris	Surface scratching, increase in friction and wear	8	5	1	40
Supply lubricant at an undesirable temperature	Undesirable lubricant viscosity, variation in frictional losses	7	5	1	35
Supply lubricant with other contaminants	Surface scratching, increase in friction and wear	7	5	1	35

reduction that increases clearance, bearings often need replacement every 6 months to 1 year, serving as sacrificial components.

From the analysis, we conclude that failure modes with high RPN values (as summarized in Table 8.7) should be prioritized, as they present the most significant risk. Step 9 of the FMEA process focuses on actions to mitigate or eliminate these high-risk failures. For this case, failure modes like indentation, corrosion, plastic deformation, frictional and abrasive wear, and improper geometry, with RPN values as high as 224, are flagged for immediate corrective action. Even issues like geometry variations require careful attention to maintain performance within acceptable tolerances. With precise manufacturing, better measurement, and effective detection methods, we can significantly improve component life and system reliability.

The journal and lubrication systems show relatively low RPNs, likely due to the presence of a lubricant sensor, which aids in detection. Conversely, if there were no wear sensor to detect debris, the RPN would increase. Hence, special attention must be paid to bearing characteristics, such as surface roughness and geometry. For instance, incorporating micro-grooves in the bearing surface can manage dust and debris, while selecting appropriate material hardness or surface coatings (bi-metallic or tri-metallic bearings) can enhance wear resistance based on specific requirements.

This case study applied FMEA to journal bearings, identifying 40 potential failure modes and calculating their respective RPNs to prioritize high-risk issues. Key failure modes, including indentation, corrosion, plastic deformation, and wear, were analyzed with recommendations for addressing these issues through design optimization, material selection, and manufacturing

Table 8.6 FMEA worksheet for bearing

Potential failure mode	*Potential effect(s) of failure*	*Severity*	*Occurrence*	*Detection*	*RPN*
Non-uniform wear causing geometrical variations	Change in bearing surface profile, Increase in friction and wear	7	8	4	224
Frictional corrosion	Surface damage in the form of increased surface roughness	8	7	5	280
Adhesive wear	Temperature rise of lubricant, decrease in lubricant viscosity, increase in friction and wear	8	8	4	256
Improper geometry of the bearing due to manufacturing inaccuracies	Non-uniform bearing surface profile, increase in friction and wear	8	8	4	256
Spalling	Abrasive wear of the surface, increase in friction and wear	8	6	4	192
Abrasive wear	Temperature rise of lubricant, decrease in lubricant viscosity, increase in friction and wear	8	8	4	256
Subsurface fatigue	Formation of subsurface cracks, increase in stresses due to stress concentration, reduction in load-carrying capacity	8	3	7	168
Moisture corrosion	Lubricant contamination, increase in friction	7	7	6	294
Plastic deformation	Change in bearing surface profile, increases friction and wear	8	6	6	288
Indentation	Formation of uneven valleys, deteriorating surface smoothness, and increasing friction	9	6	6	324
Fatigue fracture	Bearing fracture, catastrophic failure	8	3	6	144
Thermal cracking	Reduction in load-carrying capacity	7	2	6	84
Fretting	Increased vibrations and noise, increased friction and wear	7	2	6	84
True brinelling	Increased vibrations and noise, increased friction and wear	5	2	6	60

Table 8.7 Failure modes with high RPN values

Indentation	Formation of valleys deteriorating surface smoothness and increasing friction	324
Moisture corrosion	Lubricant contamination, increase in friction	294
Plastic deformation	Change in bearing surface profile, increases friction and wear	288
Frictional corrosion	Surface damage in the form of increased surface roughness	280
Adhesive wear	Temperature rise of lubricant and decrease in viscosity, increase in friction and wear	256
Abrasive wear	Temperature rise of lubricant and decrease in viscosity, increase in friction and wear	256
Improper geometry of the bearing due to manufacturing inaccuracies	Non-uniform bearing surface profile, increase in friction and wear	256
Non-uniform wear causing geometrical variations that deteriorates performance	Change in bearing surface profile, increase in friction and wear	224

processes. Suggested solutions include the use of lubricants with anti-wear and anti-corrosion additives, the application of soft coatings, and the refinement of surface roughness to minimize wear. Additional enhancements may involve grooving the bearing surface to manage debris accumulation and employing precise manufacturing techniques focused on critical contact zones, such as a 90-degree bearing segment, to improve durability and performance.

8.3.7 Case study 2: Permanent magnetic bearing

This case study examines the FMEA of permanent magnetic bearing (PMBs). PMBs are a relatively novel technology, and there is limited literature addressing their failure modes. Most existing research focuses on the mathematical modeling of PMBs rather than identifying potential failures and assessing their operational impact. However, understanding these failure modes is essential for detecting and addressing critical issues. This case study draws on the work of Lijesh et al.[5] Here, we aim to focus solely on FMEA without delving into complex mathematical models or finite element analyses typically used in magnetic calculations. For detailed mathematical calculations, readers are encouraged to review Lijesh et al.[5]

PMBs provide contactless support using magnetic forces, which reduces friction and wear, enables high-speed operation, minimizes vibration, and enhances reliability. Although PMBs are ideal for applications in fields like aerospace and high-speed machinery due to their low friction and extended lifespan, they may require a continuous power source when combined with active magnetic bearings (AMBs) for versatility and have limitations in load capacity. This case study explores various failure modes observed in PMBs, such as wear, cracking, demagnetization, and fracture, which are often triggered by excessive load or environmental conditions. By conducting experiments, we assessed the severity of these potential failures and developed methods for early detection, establishing an essential foundation for failure prevention. We used the FMEA framework to calculate the RPN, which helps prioritize the most critical issues. The iterative process involves focusing on high-RPN issues, implementing corrective actions, and re-evaluating through further experiments. Each modification is documented in the FMEA sheet, and subsequent testing validates the effectiveness of the solutions, ultimately enhancing PMB performance and durability.

The initial step in FMEA is a thorough product understanding—whether through studying an existing product or, for a new product, conducting experiments. This involves developing a mathematical model, fabricating a prototype, performing tests, and then initiating FMEA. For PMBs, this approach parallels that of traditional journal bearings. In journal bearings, a shaft (journal) rotates within a bearing housing, while in magnetic bearings, the equivalent part is called the rotor, though the function is similar.

In a PMB, the rotor or journal is suspended in air, eliminating direct contact with the bearing sleeve and reducing friction to almost zero. Friction, if any, arises only from air resistance, as the system relies on a strong magnetic repulsion between surfaces. Magnetic rings or elements are typically mounted on the rotor and bearing housing, creating a continuous repulsive force that maintains separation and prevents mechanical wear. The absence of surface contact means that additional lubricants are unnecessary, resulting in zero wear and minimal friction. As air has low density and viscosity, it offers minimal resistance to rotation—except at extremely high speeds, such as 100,000 rpm. Consequently, friction in such setups remains almost negligible.

Another key point is the availability and affordability of neodymium iron boron magnets, which have become widely accessible at reasonable prices. These high-strength magnets offer numerous advantages, including near-zero friction, zero wear, and no need for lubrication, making them ideal for applications like flywheels, artificial hearts, and molecular pumps. However, a challenge arises in their application and control, as current documentation on magnetic bearings is often incomplete and acquiring the technology from the market can be very costly.

One of the primary challenges with neodymium iron boron magnets, especially in larger sizes, is the non-uniformity of the magnetic field. For instance, a half-sector magnet spanning 180 degrees rarely maintains uniformity across its surface. Over time, this uneven field can deteriorate further due to demagnetization, a process that can begin within a few years and manifests in various forms. It is important to note that PMBs should be customized to specific load, rotational speed, or tangential speed requirements based on operating conditions. Minor magnetic field variations within a 5% tolerance are often manageable without significantly affecting performance, but deviations above 10% can introduce unwanted vibrations and cause system imbalance. In such cases, AMBs can supplement PMBs, offering enhanced control by adjusting shaft position—something PMBs alone cannot achieve. However, the inherent field variations in PMBs may still lead to unsteady rotor motion, potentially causing damaging vibrations and, in extreme cases, rotor–stator contact. This contact risks catastrophic failure, as the brittle nature of permanent magnets makes them susceptible to breakage on impact. Therefore, while PMBs offer notable advantages, they also present risks that must be carefully managed. Calculating the RPN is crucial for proactively identifying and mitigating these risks.

Lijesh et al.[5] examined three distinct configurations of PMBs with varied magnet arrangements to explore the impact of magnet shape and setup on the performance of PMBs. Figures 2–4 of their article illustrate these configurations.

In "Configuration 1" (Figure 2), a full-ring stator magnet with an axial length of 50 mm was paired with a full-ring rotor magnet of 48 mm, assembled from six smaller magnets. This setup achieved a high-stiffness design with limited load-carrying capacity. The PMB system allowed the rotor to levitate via repulsive forces between the rotor and stator magnets, preventing contact and enabling frictionless rotation.

"Configuration 2" (Figure 3) incorporated a 180-degree half-ring stator magnet, 30 mm in axial length, paired with a 32 mm full-ring rotor magnet formed from four smaller magnets. This configuration increased load capacity compared to Configuration 1 but offered slightly lower stiffness.

In "Configuration 3" (Figure 4), a 90-degree sector stator magnet was positioned at the bottom, supplemented by an additional 45-degree arc at the top for enhanced stability, yielding a stator and rotor length of 24 mm. Each configuration—full-ring, half-ring, and quarter-segment—was used to study how variations in cylindrical segment shapes and magnet arrangements influenced the bearing's performance, providing insights into optimal designs for various applications.

In each case, the rotor's outer diameter (OD) remained consistent at 48 mm, with an inner diameter (ID) of 5 mm to fit the 5 mm shaft. The stator magnet length varied slightly: in one configuration, it was longer than the rotor, in another shorter, and in the third reduced in strength. The relevant parameters for PMBs used in this case study are summarized in Table 8.8.

Each configuration had unique strengths and limitations: Configuration 1 offered high stiffness but limited load capacity; Configuration 2 increased load capacity by allowing downward repulsion without upper resistance,

Table 8.8 Parameters of different configurations of magnetic bearings

Parameters	*Configuration 1*	*Configuration 2*	*Configuration 3*
Rotor magnet			
Axial length	48 mm	32 mm	24 mm
Inner radius	5 mm	5 mm	5 mm
Outer radius	48 mm	48 mm	48 mm
Remanence	1.04T	1.04T	1.04T
Stator magnet			
Axial length	50 mm	30 mm	25 mm
Inner radius	25 mm	25 mm	25 mm
Outer radius	35 mm	50 mm	30 mm
Remanence	1.4T	1.04T	0.7T
Experimental condition			
Rotor weight	0.6 kg	0.4 kg	0.3 kg
Static load	2 kg	373 kg	2 kg

though with reduced stiffness; and Configuration 3 achieved a balanced performance with moderate load capacity and stiffness, adaptable to applications requiring a middle ground. This analysis clarified the trade-offs between load-carrying capacity and stiffness across the different magnetic bearing configurations.

An experimental test setup was assembled to evaluate these configurations. The setup included a disc with multiple mounting holes, allowing adjustable bolt placements to create imbalances or eccentric loads, simulating controlled conditions. The housing, made of aluminum or stainless steel, was mounted on a T-channel base for structural rigidity, while the rotor magnets were mounted on a stainless-steel shaft to prevent magnetic field dissipation that would occur with ferromagnetic materials. In this configuration, one end of the shaft remained free, while the other end was coupled to an induction motor. The setup included a spiral coupling and an Asea Brown Boveri (ABB) frequency drive to vary rotational speed, enabling observation of the effects of speed changes on the magnetic bearing. During continuous operation, the bearing performed well; however, significant instability was observed during acceleration and deceleration phases, especially when ramping up or down to certain speeds. This highlighted a key limitation of PMBs under transient conditions, suggesting that additional support—such as hydrodynamic or AMBs—may be required to maintain stability during speed fluctuations.

To apply FMEA, the initial step involved a thorough analysis of the product requirements, including dimensions, manufacturing processes, and assembly techniques. Failure-prone components were then identified, beginning with the rotor magnet mounted on a stainless-steel shaft, which appeared most susceptible to failure. Other components at risk included the stator magnet, the casing, and the coupling that connects the assembly to the motor. Focusing on rotor-related failures, five primary failure types were identified through experimentation: wear, cracking, demagnetization, new pole formation, and fracture.

In principle, wear should not occur in PMBs, as they are designed for wear-free operation. However, wear was observed in this case, suggesting potential issues such as limited knowledge of application specifics, manufacturing defects, or assembly errors. Notably, the rotor magnet's surface coating began peeling under operational conditions, with initially shiny magnets showing large worn patches after extended use. This wear was especially pronounced in Configuration 2, where visibly worn surfaces were observed.

New pole formation was also observed, a phenomenon commonly identified through magnetic particle NDT. When subsurface cracks exist, additional magnetic poles can emerge, significantly altering the magnetic field and raising concerns for PMBs. This effect, along with other identified failure modes, underscores the need for comprehensive analysis and focused corrective measures to improve the durability and reliability of PMBs.

Improper design of permanent magnets can contribute to rotor wear, as evidenced in this study. Another significant issue encountered was cracking, which appeared on the surface of the permanent magnets, particularly in Configurations 1 and 3. Due to the brittle nature of permanent magnets, even slight impacts may lead to cracks or, in severe instances, fragment the magnet. This occurrence was carefully documented in the study. Rotor wear was evident across all configurations, as depicted in the figures, along with a notable temperature increase that further aggravated the wear.

Ideally, a PMB should exhibit no temperature rise due to the absence of mechanical contact. However, frictional contact between surfaces, coating wear, and scrapping caused friction to increase, leading to higher temperatures. Consequently, the expected low-wear, low-friction, and minimal-temperature design did not hold in this case. Cracks were largely attributed to unsteady rotor motion, where the shaft failed to rotate concentrically within the bearing. This erratic motion, coupled with insufficient stiffness, occasionally resulted in collisions between the stator and rotor magnets, initiating cracks. This issue was especially pronounced during the acceleration and deceleration phases. For instance, as the motor accelerated from 0 to 1500 rpm or slowed to a stop, rotor misalignment often led to repeated stator contacts. These findings suggest that an additional support mechanism could be beneficial to stabilize the rotor during speed transitions, a key takeaway from these experiments.

The study also uncovered issues with demagnetization. When there was an imbalance in magnetic strength between the rotor and stator magnets, the stronger magnet could demagnetize the weaker one. In this case, the stator had a magnetic strength of 1.4 T, while the rotor measured 1.04 T, leading to rotor demagnetization while the stator remained unaffected. This finding underscored the importance of matching magnetic strengths to prevent demagnetization. Initially, obtaining stator and rotor magnets with equal strengths proved challenging, necessitating adjustments to the configuration, which will be detailed in subsequent slides.

In summary, the rotor experienced demagnetization due to its lower magnetic strength relative to the stator. Using magnets of equal strength could reduce this risk, although mechanical contact between magnets could still lead to demagnetization. If the rotor collided with the stator, an offset could occur, preventing the rotor from remaining centered within the stator. This offset disrupted the self-centering property of the PMB, causing further operational challenges.

PMBs are typically self-centering, but this characteristic was not observed in Configuration 1, where signs of demagnetization emerged. Upon examining an unused rotor, consistent magnetic field strength was noted. However, the used rotor showed diminished magnetic strength in several areas, while one side retained similar strength. This suggested that contact may have occurred between the rotor and the housing or stator during operation. Given these findings, it is crucial to apply FMEA to identify corrective actions and solutions for moving forward.

To evaluate these failures, each was ranked by severity, occurrence, and detection on a scale from 1 to 5, given the limited data available for this relatively new system. All the values have been listed in Tables 8.9 to 8.11.

Severity rankings were established through a combination of theoretical modeling, experimental testing, and software analysis. For instance, a pole finder sheet helped identify multiple north and south pole formations, which revealed surface cracks and contributed to understanding the severity of each failure mode. In summary, the FMEA of PMBs identified 13 critical failure modes, with decreased magnetic field strength and non-uniformity highlighted as primary concerns. Table 8.12 provides an FMEA worksheet for a PMB, detailing components, potential failure modes, and their severity,

Table 8.9 Severity guidelines for passive magnetic bearing FMEA (1–5 qualitative scale)

Effect	*Rank*	*Criteria*
No	1	No effect
Slight	2	Wear of stator/rotor or rubbing on casing, or crack in the rotor/stator
Moderate	3	Demagnetization in the rotor and stator
Major	4	Crack in rotor
Serious	5	Pole formation and fracture in the stator/rotor

Table 8.10 Occurrence guidelines for passive magnetic bearing FMEA (1–5 qualitative scale)

Failure mode	*Rank*
Wear of stator/rotor	3
Crack of stator/rotor	4
Demagnetization of stator/rotor	2
Coupling failure	1
Pole formation	1
Rubbing on casing	3

Table 8.11 Detection guidelines for passive magnetic bearing FMEA (1–5 qualitative scale)

Failure mode	*Rank*
Wear of stator/rotor	2
Crack of rotor	2
Crack of stator	3
Demagnetization of stator/rotor	3
Pole formation of stator/rotor	1
Fracture of stator/rotor	1
Rubbing on casing	2
Coupling failure	1

Table 8.12 FMEA worksheet for a PMB

Component	*Potential failure mode*	*Severity*	*Occurrence*	*Detection*	*RPN*
A—Housing	Wear	2	3	2	12
B—Rotor	Wear	2	3	2	12
	Crack formation	4	4	2	32
	Demagnetization	3	2	3	18
	Pole formation	5	1	1	5
	Fracture	5	1	1	5
C—Stator	Wear	2	3	2	12
	Crack formation	2	4	3	24
	Demagnetization	3	2	3	18
	Pole formation	5	4	1	20
	Fracture	5	1	1	5
D—Coupling	Breakage	5	1	1	5

occurrence, and detection scores, along with the calculated RPN for each mode. Higher RPN values point to more critical risks requiring prioritized attention.

By utilizing square magnets, rubber isolators, and aluminum frameworks, effective damping was achieved, reducing the likelihood of catastrophic failure. This iterative process, which culminated in a design registration[6], illustrates how systematically addressing each failure mode with innovative solutions can enhance system durability and performance. This approach may serve as a model for similar projects.

8.4 LIFE-CYCLE ASSESSMENT

LCA is a systematic approach for evaluating the environmental impacts of a product, process, or service throughout its entire life cycle, from raw material extraction to disposal or recycling. As concerns about environmental sustainability grow, LCA has emerged as a critical tool for companies, governments, and researchers seeking to understand and minimize the ecological footprint of their products and processes. By providing insights into resource use, emissions, and waste generation, LCA helps guide sustainable decision-making across industries.

8.4.1 Understanding LCA

LCA aims to capture the full environmental impact of a product or process by assessing it across four main phases: raw material extraction, manufacturing, use, and end of life. Often referred to as a "cradle-to-grave" approach, LCA goes beyond immediate usage to consider upstream and downstream impacts, providing a complete picture of a product's environmental footprint.

The LCA process is governed by a set of standardized steps, typically following guidelines from the International Organization for Standardization (ISO 14040 and ISO 14044). These steps include the following:

- *Goal and scope definition:* In this phase, the objectives of the LCA are established, including what the assessment aims to accomplish and the boundaries of the study. Boundaries determine which stages or components of the product's life cycle will be assessed, such as excluding transportation or focusing only on manufacturing.
- *Inventory analysis:* This step involves collecting data on all inputs and outputs associated with each stage of the product's life cycle. Inputs include raw materials, energy, and water, while outputs include emissions to air, water, and land, as well as waste byproducts. Inventory analysis creates a detailed account of the resources consumed and emissions generated at each life-cycle stage.

- *Impact assessment:* In this phase, the data collected during inventory analysis is translated into potential environmental impacts. These impacts are categorized into areas like global warming potential, ozone depletion, acidification, and resource depletion. Each impact category reflects a specific type of environmental burden, allowing for a nuanced understanding of a product's ecological footprint.
- *Interpretation:* The final step involves analyzing the results of the impact assessment to identify significant findings, draw conclusions, and make recommendations. This step highlights areas for improvement, guiding design and production decisions that can reduce environmental impacts.

8.4.2 Types of LCA

LCA can be tailored to meet various objectives, resulting in different types of assessments:

- *Cradle-to-grave:* This is the traditional form of LCA, covering the entire life cycle of a product from raw material extraction to disposal. It provides a complete overview of environmental impacts, making it the most comprehensive LCA type.
- *Cradle-to-gate:* This type of LCA assesses environmental impacts from raw material extraction through to the point at which the product leaves the manufacturing facility, excluding distribution, use, and disposal. It is commonly used in manufacturing to improve production processes.
- *Gate-to-gate:* A gate-to-gate assessment focuses on a specific part of the production process. For example, an LCA might analyze the environmental impact of a particular manufacturing stage, such as assembly or packaging, rather than the entire product life cycle.
- *Cradle-to-cradle:* This type of LCA follows the product from raw material extraction through disposal, with a focus on recycling or reusing materials at the end of life. The goal is to create a closed-loop system where resources are continually reused, minimizing waste.

8.4.3 Benefits of LCA

The comprehensive nature of LCA offers valuable benefits for companies, governments, and other stakeholders. It supports informed decision-making by providing data-driven insights into environmental impacts, helping organizations choose more sustainable materials, energy sources, and processes by revealing "hidden" environmental costs. LCA enhances resource efficiency by identifying areas for conserving resources throughout the life cycle, leading to improved energy and material efficiency while reducing costs and minimizing environmental impacts. It also reduces emissions and waste by highlighting high-output areas, enabling targeted actions that not only lessen environmental impacts but also ensure regulatory compliance. Additionally, LCA supports eco-friendly product design, guiding designers in selecting materials and processes that reduce ecological footprints, which appeals to environmentally conscious consumers and creates a competitive edge. Finally, LCA aids regulatory compliance and sustainability reporting by supplying essential data for sustainability reports, demonstrating environmental responsibility, enhancing reputations, and aligning with sustainability standards.

8.4.4 Applications of LCA in industry

LCA has applications across a wide range of industries, each of which benefits from insights into environmental impacts and resource use:

- *Manufacturing:* Manufacturers utilize LCA to evaluate the environmental impact of raw material sourcing, production processes, and waste management. By identifying energy-intensive stages, material waste points, and areas of quality loss, companies can optimize operations, reduce costs, and increase efficiency.
- *Automotive:* The automotive industry uses LCA to evaluate the environmental impacts of materials, such as steel and aluminum, and of alternative fuels like electric and hybrid power sources. LCA is critical for understanding the trade-offs between fuel efficiency, emissions, and material choices.
- *Energy:* In the energy sector, LCA is used to compare the environmental impacts of different energy sources, such as coal, wind, and solar. For example, a solar panel might have a higher environmental impact during production, but significantly lower impacts during its use phase compared to fossil fuels.
- *Construction:* The construction industry uses LCA to assess building materials, energy consumption, and waste management. By evaluating options such as concrete, steel, and sustainable building materials, LCA helps reduce the carbon footprint of construction projects and promotes green building practices.

8.4.5 Challenges in implementing LCA

Although LCA provides valuable insights, it also faces several challenges. One major issue is data availability and accuracy, as LCA depends on comprehensive data about material use, energy consumption, emissions, and waste, which may not always be accessible or reliable, especially for impacts like biodiversity and human health. The scope and boundaries of an LCA further influence its outcomes; a narrow scope can miss essential life-cycle stages, while a broad one complicates data collection and analysis, often making the process more resource-intensive. To manage these challenges, organizations may rely on industry databases, modeling tools, or external consultants, although these solutions can raise costs. Another challenge lies in interpreting results, as different impact categories may conflict; for example, a material that lowers greenhouse gas emissions might increase water consumption, requiring careful trade-off analysis and sometimes subjective decision-making. Additionally, LCA is inherently time-consuming and demands specialized expertise, which can make it difficult for smaller businesses or organizations with limited resources to conduct effectively.

8.4.6 Future directions in LCA

As technology and environmental awareness progress, LCA is becoming more precise and accessible. Advances in data analytics, machine learning, and digital twins enable more detailed and dynamic assessments, with real-time data collection from IoT devices allowing continuous monitoring of environmental impacts for greater accuracy. The development of industry-specific LCA databases and software also enhances accessibility, making it feasible for organizations of all sizes to conduct assessments. Additionally, with the rise of the circular economy, LCA is increasingly used to evaluate not only end-of-life disposal but also opportunities for recycling, remanufacturing, and material reuse. Regulatory agencies are contributing as well, introducing policies that encourage or mandate LCA in product design and reporting, positioning LCA as a key tool in sustainable development efforts.

8.5 PRINCIPAL COMPONENT ANALYSIS

In discussing failure modes, our goal is to identify different types of failures early to enable timely intervention. This requires high-quality sensing devices capable of delivering a variety of signals in the frequency domain, time domain, or both. A data-driven, algorithmic approach is essential for informed decision-making, relying on accurate data that is noise-free and minimally correlated across variables. Techniques such as PCA are valuable here, helping us select the most critical features from large datasets and reduce dimensionality. For example, PCA can isolate three or four key features out of a larger set of ten that explain most of the variance (85%–90%), allowing us to focus on the most impactful data points. This dimensionality reduction is crucial for effective failure mode analysis, early detection, and corrective actions.

PCA is a statistical method designed to simplify complex datasets while retaining essential information, widely used in data science, machine learning, and engineering. Applications include feature extraction, image compression, pattern recognition, data visualization, and noise reduction.

By reducing high-dimensional data to a few uncorrelated components, PCA enables the creation of 2D or 3D visualizations, uncovering hidden trends, clusters, and patterns in the data.

Data quality and signal-to-noise ratio are both improved using PCA in noisy datasets, which retains components with the biggest variance while filtering out less significant ones. Feature selection using PCA is a common machine learning technique that can improve model accuracy and speed up computation. By using principal components instead of pixel values to represent images, PCA in image processing minimizes storage requirements while keeping critical features. Anomaly identification and pattern recognition are two further areas where PCA shines, particularly in industrial quality control. PCA has become an essential tool for data analysis and modeling as it breaks down original variables into a new collection of uncorrelated components, which simplifies data processing, visualization, and interpretation.

8.5.1 Understanding PCA

In the age of big data, it is common to encounter datasets with many variables (or features), which can lead to challenges in analysis and computation. High-dimensional data can be difficult to visualize and often contains redundant or correlated features that add complexity without contributing additional information. PCA addresses these challenges by reducing the number of variables in a dataset, effectively capturing the most important patterns while filtering out noise.

Figure 8.5 illustrates the effect of PCA on a 2D dataset. On the left side, there is a scatter plot with points distributed in an elliptical shape along the plane defined by the x- and y-axes. An arrow labeled "PCA" points to the right side of the plot, showing the result after applying PCA. On the right side, the data has been transformed so that the points lie along a single axis, showing a reduced dimensionality representation. The primary direction x' captures the main variance in the dataset, and the secondary direction y' captures minimal variance.

PCA works by transforming the original set of correlated variables into a smaller set of uncorrelated variables called principal components. Each principal component is a linear combination of the original variables and is oriented to capture the maximum variance in the data. By focusing on the directions in which data variability is greatest, PCA simplifies the dataset and reveals its underlying structure. The PCA process can be broken down into the following steps:

- *Standardization:* As PCA is sensitive to the scale of data, the first step involves standardizing the dataset so that each variable has a mean of zero and a standard deviation of one. This ensures that variables with larger ranges do not dominate the principal components.

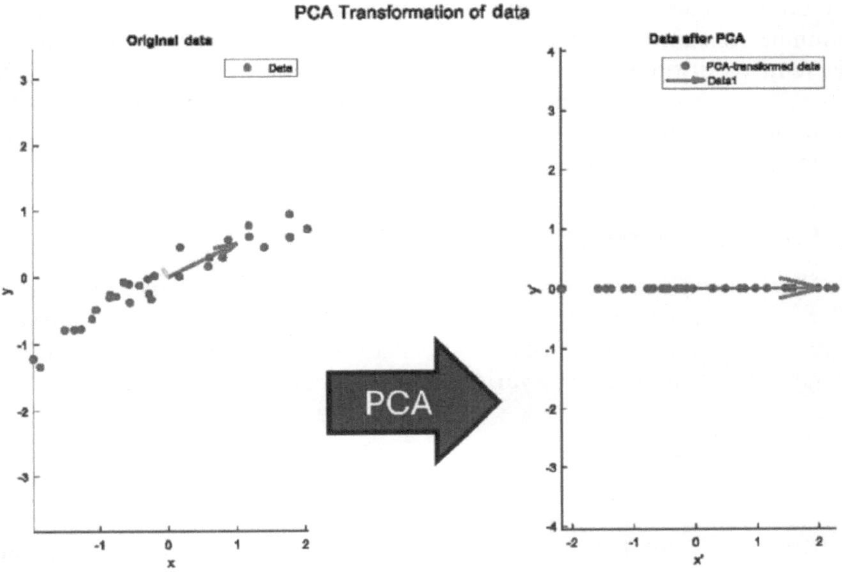

Figure 8.5 Effect of PCA on a 2D dataset.

- *Covariance matrix computation:* After standardization, a covariance matrix is calculated to quantify the relationships between variables. This matrix shows how each pair of variables varies together, highlighting any correlations within the dataset.
- *Eigendecomposition:* The covariance matrix undergoes eigendecomposition to extract its eigenvalues and eigenvectors. Eigenvalues indicate the amount of variance explained by each principal component, while eigenvectors define the directions of the principal components.
- *Principal component selection:* The eigenvalues are sorted in descending order, and the corresponding eigenvectors are arranged accordingly. Typically, only the principal components with the highest eigenvalues are retained, as they capture the most significant variance in the data. The choice of how many components to retain depends on the desired balance between dimensionality reduction and information retention.
- *Projection:* The final step involves projecting the original data onto the selected principal components. This creates a transformed dataset with reduced dimensions, where each data point is represented by its coordinates along the principal components rather than the original variables.

By reducing a dataset's dimensions, PCA enhances efficiency and interpretability, particularly in tasks such as data visualization, clustering, and training machine learning models. For instance, in gear failure analysis, PCA helps distinguish normal operating conditions from faults by identifying key features critical for accurate fault classification. By simplifying data, PCA streamlines the analysis process, improving processing speed and reliability in fault detection.

8.5.2 Mathematical foundations of PCA

PCA relies on concepts from linear algebra, particularly eigendecomposition. For a dataset with multiple variables, PCA uses the covariance matrix to capture the variances and covariances between these variables. This process begins by calculating the covariance matrix **C** of the dataset **X**:

$$C = \frac{1}{n-1} \mathbf{X}^T \mathbf{X}$$

where n is the number of observations and **X** is the matrix of centered data (each variable has a mean of zero).

8.5.2.1 Eigendecomposition of the covariance matrix

To identify the principal components, PCA decomposes the covariance matrix **C** into eigenvalues and eigenvectors:

$$Cv = \lambda v$$

- λ (eigenvalues) represent the amount of variance in the data along each direction defined by the principal components.
- v (eigenvectors) defines the direction of each principal component.

The eigenvalues are sorted in descending order, and the eigenvectors associated with the largest eigenvalues are selected to form the principal components. The first principal component captures the highest variance, followed by the second, and so forth. By maximizing variance along these new, orthogonal directions, PCA ensures that each principal component is uncorrelated with the others, capturing unique information about the dataset.

8.5.3 Benefits of PCA

PCA provides significant advantages in data analysis by transforming high-dimensional data into a smaller set of principal components, which enhances processing efficiency and enables clear data visualization for pattern recognition. It also acts as a noise filter, preserving only the components with the most variance, thus improving data clarity and reducing the influence of outliers. Additionally, by lowering the number of features, PCA reduces computational demands, facilitating faster model training and more efficient data storage. In machine learning, this dimensionality reduction can enhance model performance, mitigating overfitting and aiding algorithms sensitive to high-dimensional data.

8.5.4 Challenges in implementing PCA

While PCA offers numerous advantages, it also has notable limitations. One challenge is the interpretability of principal components, as they are linear combinations of original variables and may be difficult for non-technical stakeholders to understand. PCA is also sensitive to data scaling and outliers, requiring standardization and careful handling to avoid skewed results. Additionally, PCA involves discarding some information by focusing on high-variance components, which can lead to a loss of subtle but important details in certain applications. Finally, PCA assumes linear relationships in the data, which may limit its effectiveness on datasets with complex, non-linear patterns, where alternative methods like kernel PCA or t-stochastic neighbor embedding (SNE) might be more suitable.

8.5.5 Future directions for PCA

As data complexity grows, PCA is advancing with innovations in data science. Variants like kernel PCA, which apply non-linear kernels, enable the capture of complex, non-linear relationships, making PCA more applicable to fields such as image recognition and genomics. Its integration with machine learning is enhancing PCA's role in feature selection and dimensionality reduction, boosting model performance across domains like natural language processing and computer vision. Moreover, combining PCA with real-time data processing via IoT and big data analytics offers new possibilities for dynamic data reduction and anomaly detection. The development of sophisticated algorithms, including deep learning methods, expands PCA's utility in applications with non-linear patterns and vast data volumes.

8.5.6 Case study: Fault monitoring using PCA

Fault monitoring in industrial machinery is essential for the early detection of potential failures, ensuring operational safety, efficiency, and cost savings. In the present subheading, we will discuss a case study covered by Li et al.[7] This case study examines fault monitoring in gearboxes using PCA, a statistical tool used to simplify complex datasets by identifying key components that capture the most significant variation within the data. The study explores how PCA can differentiate between various gear conditions, focusing on both healthy and faulty gears. Specifically, it examines vibration signal analysis from gearboxes to classify fault types, using temporal and frequency features as diagnostic indicators. They categorize gear conditions in the following three main types:

- *Normal (healthy) gear:* Operating in expected condition without faults.
- *Tooth cracked:* Faulty gear with subsurface cracks, invisible to the naked eye.
- *Tooth broken:* Faulty gear with visible damage detectable through visual inspection.

The aim of this analysis is to distinguish between gear faults and complete failures, providing valuable insights for fault diagnosis. Faulty gears can eventually lead to failure. To perform the study, they followed the following steps:

- *Data collection and feature extraction:* Vibration signals were collected from an industrial gearbox using an accelerometer to understand how the gears behave under different conditions. These signals contain valuable information about the gears' health but are often complex and affected by random noise, which makes it harder to get clear information.

  To identify any faults in the gears, a feature extraction algorithm was used with anticipation that derived features would encapsulate discriminative information pertinent to gear faults, facilitating accurate classification of various fault types. This algorithm picks out important information from the signals using techniques that help highlight signs of damage more accurately. Effective feature extraction not only improves accuracy but also keeps the data processing efficient by simplifying complex data into clear, useful indicators. By focusing on the most relevant parts of the signals, this method reduces unnecessary information and nsarrows down the data to what is essential. Two primary types of features were extracted from the signals:

  - *Temporal features:* These properties describe a signal's behavior over time and are represented by statistical measures such as mean, standard deviation, skewness, and kurtosis. Time-domain features capture information about variations and patterns in the vibration signal, which can reveal specific faults. By examining these characteristics, we can gain insights into the temporal behavior of the signal, making these measures valuable for identifying potential issues in machinery.

 - *Frequency features:* Frequency features capture the spectral characteristics of a signal, providing information about dominant frequencies, harmonics, and other key aspects of the spectrum. These features reveal how various faults impact the vibration patterns of gears, helping to identify abnormal conditions. By analyzing the signal in the frequency domain—using methods like Fourier transform or wavelet transform—frequency features offer valuable insights, including peak frequency, spectral centroid, spectral bandwidth, and spectral entropy, all of which contribute to a comprehensive understanding of the signal's behavior.

- *Use of PCA in fault monitoring:* PCA helps reduce the dimensionality of the vibration signal data, isolating the most significant components that capture the majority of the variation within the data. The following steps were followed to implement PCA:
  - *Data standardization:* To ensure comparability across features, the vibration data were standardized, adjusting each variable to have a mean of zero and a standard deviation of one.
  - *PCA application:* PCA was applied to the standardized dataset to transform the data into a set of uncorrelated principal components.
  - *Fault classification:* The transformed data were then analyzed to classify gear conditions into predefined categories. Faults were categorized by PCA into different classes: "Case 1: Healthy gear," "Case 2: Gear with cracked tooth," "Case 3: Gear with broken tooth."
  - *Result analysis:* By isolating critical features, PCA allowed the system to identify cracked and broken teeth based on variations in the vibration signals. In the study of Li et al.,[7] they identified the following ten (10) features to be extracted from the collected vibration raw data:
    1. *Mean value:* The arithmetic average of a dataset, computed by summing all values and dividing by the total number of data points ($M = \frac{\text{Sum of terms}}{\text{Total number of terms}}$). Li et al.[7] considered absolute means $M = \frac{\sum_{1}^{n}|x_i|}{\text{Total number of terms}}$) as "feature 5."
    2. *Standard deviation (feature 1):* A measure of the dispersion or spread of data points around the mean, indicating the average distance between each data point and the mean ($\sigma = \sqrt{\frac{\Sigma(x - M)^2}{n - 1}}$).

3. *Skewness (feature 2):* A measure of the asymmetry of the data distribution, determining whether the data skews left (negative skewness), right (positive skewness), or is symmetric (zero skewness). Skew $= \sum_{i=1}^{i=n} \frac{(x_i - M)^3}{n\sigma^3}$.
4. *Kurtosis (feature 3):* A measure of the peakedness or flatness of the data distribution, describing tail heaviness and the presence of outliers compared to a normal distribution (Kurtosis $= \sum_{i=1}^{i=n} \frac{(x_i - M)^4}{n\sigma^4}$).
5. *Maximum peak value (feature 4):* The highest observed value in a dataset, representing the peak or maximum intensity ($\max(x_i)$).
6. *Root mean square (RMS) value (feature 6):* The square root of the average of squared values in a dataset, offering a measure of overall magnitude or intensity ($x_{rms} = \sqrt{\frac{\Sigma (x_i)^2}{n}}$).
7. *Crest factor (feature 7):* The ratio of peak value to RMS value of a signal, indicating the relative magnitude of peaks compared to the overall signal ($\max(x_i)/x_{rms}$).
8. *Shape factor (feature 8):* A dimensionless parameter characterizing waveform shape, describing the relationship between RMS value and waveform average (x_{rms}/M).
9. *Impulse factor (feature 9):* The ratio of peak value to the area under the waveform, representing energy concentration in a short time duration ($\max(x_i)/M$).
10. *Clearance factor (feature 10):* A parameter in vibration analysis assessing clearance between components, defined as the ratio of maximum peak value to absolute mean value in the frequency domain $\frac{\max(x_i)}{\frac{\Sigma |x_i|}{n}}$.

Figure 4 of the Li et al.[7] research article presents a series of graphs that illustrate the variation trends of ten features extracted from vibration data under three gear conditions: normal, cracked tooth, and broken tooth. They analyzed 23 datasets (sampling rate of 12.5 kHz) for each condition. Key observations from the original features, as noted in Figure 4 of Li et al.[7] include:

- Features 1 and 2 show relatively subtle differences across conditions, suggesting they may be less effective in distinguishing fault types.
- Feature 3 exhibits distinct patterns for normal tooth conditions but shows considerable overlap in the responses for cracked and broken tooth conditions.
- Features 5, 7, 8, 9, and 10 display distinct patterns for broken tooth conditions, although there is significant overlap between cracked and broken tooth responses.
- The difficulty in distinguishing signals between normal and cracked tooth conditions, aside from Feature 3, suggests potential feature overlap or similar vibration patterns. This highlights the need for a robust and discriminative analysis technique like PCA to extract meaningful features, facilitating accurate classification of different gear conditions.

As the raw data of Li et al.[7] are not available, for academic understanding we took an alternative approach and digitized the input graph provided by Li et al.[7] The collected data have been plotted in Figure 8.6.

In Figure 8.6, the green line with square markers represents broken teeth, the red line with triangle markers indicates cracked teeth, and the blue line with circle markers corresponds to normal gear conditions, as labeled in the plots. The plots in Figure 8.6 were generated using the following MATLAB code:

- The code loads data from an Excel file specified by the filename variable using the `read matrix` function, storing the data in the `data` variable.
- The data is separated into three variables—`features_normal`, `features_cracked`, and `features_broken`—each containing specific columns that correspond to the features of normal gear, cracked gear, and broken gear, respectively.
- A new figure is created using the `figure` function to visualize all the features.
- The `feature_names` variable is defined as a cell array of strings, listing the names of each feature.
- A loop iterates over each feature to create individual subplots:
- Inside the loop, a subplot is created with the `subplot` function. The subplot layout is specified as two rows and five columns, with the current feature index `i` determining the subplot position.
- A scatter plot is generated for the normal gear data using the `scatter` function, with the x-axis representing the sample index and the y-axis showing the corresponding feature values. The markers are outlined in blue (`'b'`).

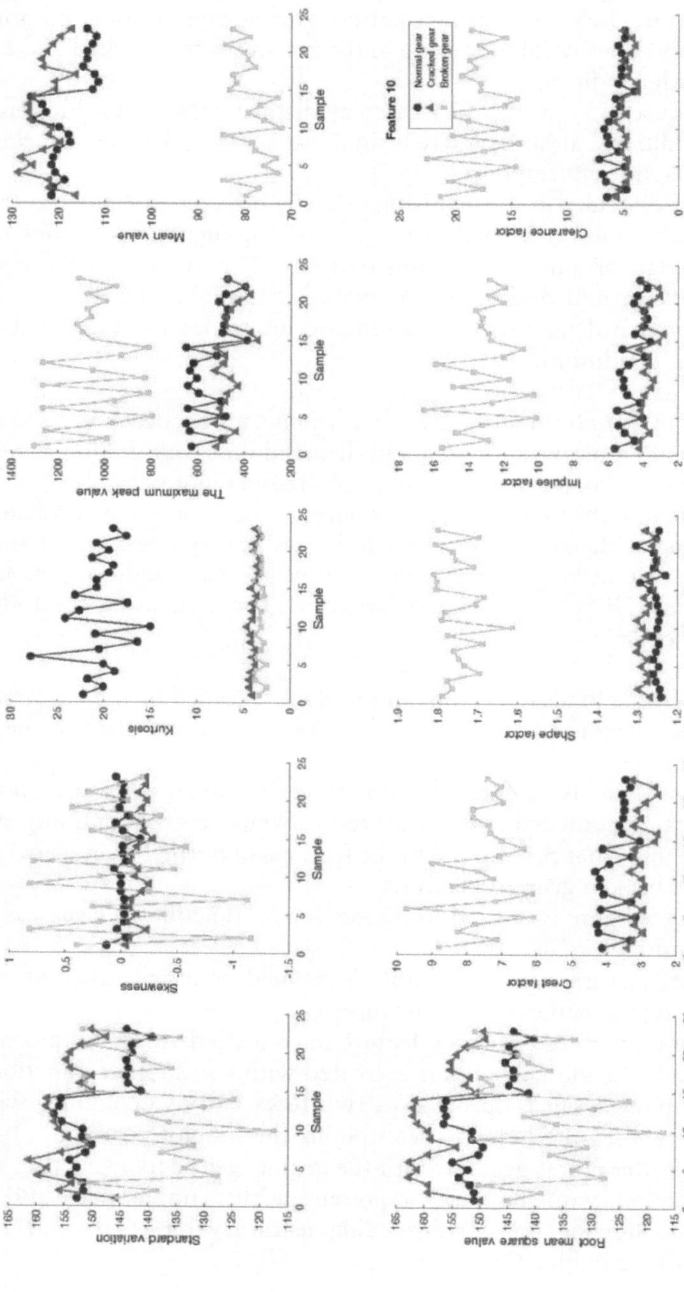

Figure 8.6 Variation trends of originally extracted features.

- Similarly, scatter plots are created for the cracked gear (red markers) and broken gear (green markers).
- The x-axis is labeled as "Sample," and the y-axis label displays the name of the current feature (`feature_names{i}`).
- Each subplot title is set to "Feature" followed by the feature index `i`.

Next, we apply standardization as discussed in Section 8.5.1 of this chapter. Standardization refers to the process of transforming the initial variables to have comparable scales. The main reason for performing standardization prior to PCA is to ensure that each variable contributes equally to the analysis. In Figure 8.6, each feature is plotted within a different range: feature 1 spans 115 to 165; feature 2, –1.5 to 1; feature 3, 0 to 30; feature 4, 200 to 1400; feature 5, 70 to 130; feature 6, 115 to 165; feature 7, 2 to 10; feature 8, 1.2 to 1.9; and feature 10, 0 to 25. Standardizing these values to an approximate range of –3 to 3 ensures uniformity across features. Although this process can make many curves appear nearly identical with significant overlap—complicating feature differentiation—it is crucial to achieve a mean of 0 and a variance of 1, thereby maintaining consistency in the data for accurate analysis. Standardization removes the biases caused by differences in scales, enabling a fair comparison and interpretation of the principal components.

Explanation of MATLAB code for Standardization:

- *Standardizing the features:*
  - features_normal_std = zscore(features_normal); This line applies the `zscore` function to standardize the data for the normal gear (`features_normal`). The `zscore` function calculates the z-scores of each column, indicating how many standard deviations each data point is from the mean. This standardization process transforms the data to have a mean of zero and a standard deviation of one.
  - Similarly, `features_cracked_std = zscore(features_cracked);` and `features_broken_std = zscore(features_broken);` standardize the data for cracked and broken gears, respectively.
- *Plotting the standardized features:*
  - The following code block contains a loop that iterates over each standardized feature for normal, cracked, and broken gears (`features_normal_std`, `features_cracked_std`, and `features_broken_std`).
  - `subplot (2, 5, i);`: Sets up a 2x5 grid of subplots, selecting the ith subplot to plot the current feature.
  - `scatter` functions: Used to create scatter plots of the standardized feature values for each gear condition, with markers and colors

to differentiate the conditions (circle for normal, triangle for cracked, and square for broken).

- `plot` functions: Connect the scatter points with lines for each gear condition, creating line plots (Figure 8.7) that visually represent variations in the standardized feature values.
- The x-axis represents the sample index (`1:size(features_normal_std, 1)`), while the y-axis shows the standardized feature values (`features_normal_std(:, i)`, `features_cracked_std(:, i)`, `features_broken_std(:, i)`).
- `xlabel`, `ylabel`, and `title` functions are used to set the labels and title for each subplot.

The next step in PCA is to calculate the covariance matrix (as shown in Figure 8.8), which displays the relationships between features. This square matrix has dimensions n*n, where n is the number of dimensions or variables in the dataset. Each entry in the covariance matrix represents the covariance between two variables. For instance, the covariance between the first and second features might be 0.19, between the first and third features 0.31, and between the first and fifth features 0.95.

Importantly, the diagonal elements of the covariance matrix represent the variance of each feature, which should be 1 after standardization. The off-diagonal entries denote covariances between pairs of variables, measuring how they vary together. A positive covariance indicates that as one variable increases, the other does as well, while a negative covariance implies an inverse relationship. The covariance matrix is symmetric, meaning that the values in the upper triangular portion mirror those in the lower portion, as the covariance between variables a and b is equal to that between b and a.

As noted previously, standardization adjusts the mean of each feature to 0 and the variance to 1. Consequently, each diagonal element in the covariance matrix should also be 1, which has been confirmed through MATLAB code validation.

Analyzing the covariance matrix helps us understand how variables are related to each other, revealing any strong correlations or redundancies. In PCA, this is important because we can either combine or remove these highly correlated variables, reducing the data's complexity while keeping the essential information. For example, one can estimate eigenvalues and eigen vectors for the covariance matrix (Figure 8.8) using:

Using a percentage-based approach in PCA helps identify the principal components that have the greatest impact on data analysis. In Figure 8.9, the first eigenvector captures 54.6% of the variance, the second 18.02%, the third 10.31%, the fourth 8.69%, and the fifth 5.62%, collectively accounting for 96.88% of the total variance—exceeding the commonly used threshold of 95%. Depending on the specific application, thresholds of 85% or 90% may also be used. For instance, Li et al.[7] utilized the top four

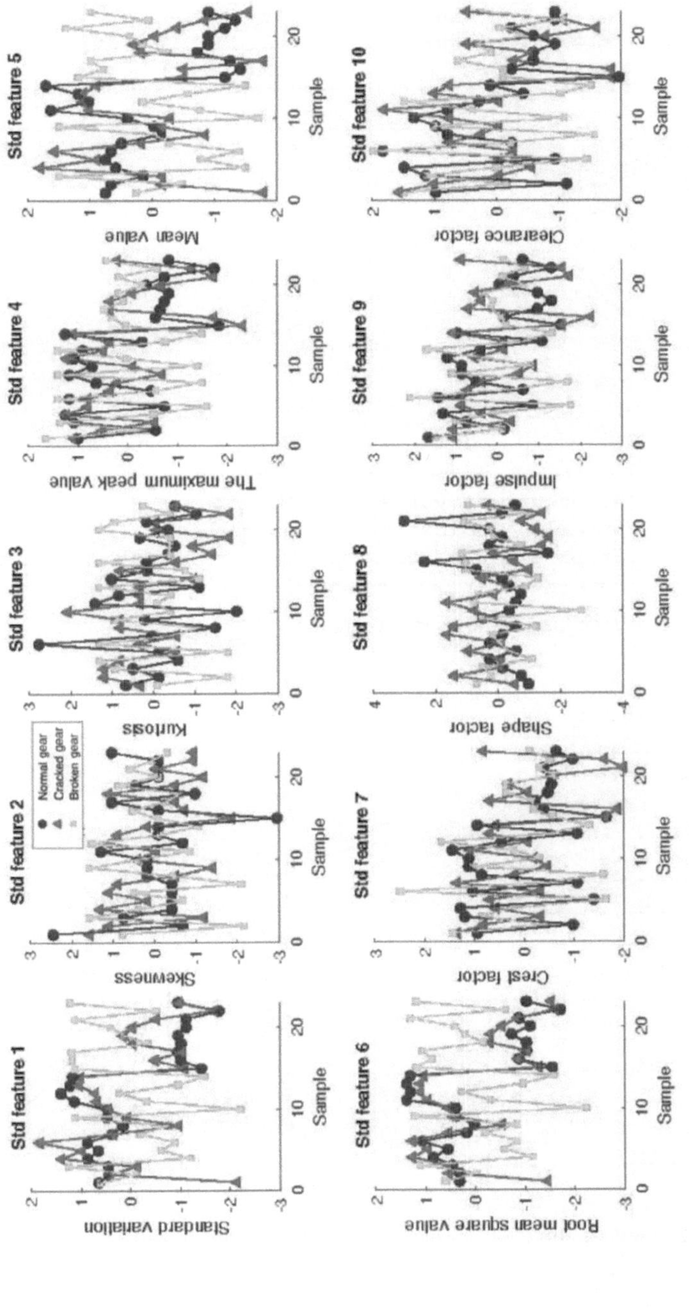

Figure 8.7 Variation trends of originally extracted features (shown in Figure 8.6) after standardization.

Represent the covariances between variables, indicating their relationship and how they vary together Covariance Matrix

1	0.19	0.31	0.8	0.95	0.99	0.55	−0.37	0.66	0.62
0.19	1	0.06	0.4	0.17	0.19	0.48	−0.33	0.45	0.42
0.31	0.06	1	0.36	0.32	0.35	0.28	0.1	0.4	0.25
0.8	0.4	0.36	1	0.7	0.8	0.92	−0.19	0.93	0.9
0.95	0.17	0.32	0.7	1	0.95	0.46	−0.42	0.59	0.48
0.99	0.19	0.35	0.8	0.95	1	0.57	−0.3	0.66	0.62
0.55	0.48	0.28	0.92	0.46	0.57	1	−0.12	0.9	0.91
−0.37	−0.33	0.1	−0.19	−0.42	−0.3	−0.12	1	−0.09	−0.05
0.66	0.45	0.4	0.93	0.59	0.66	0.9	−0.09	1	0.88
0.62	0.42	0.25	0.9	0.48	0.62	0.91	−0.05	0.88	1

Represent the variances of each variable, indicating how each variable varies from its mean value.

Figure 8.8 Covariance matrix and its interpretation.

Eigenvalues of covariance matrix in descending order
5.30
1.75
1.00
0.84
0.51
0.14
0.05
0.05
0.04
0.02

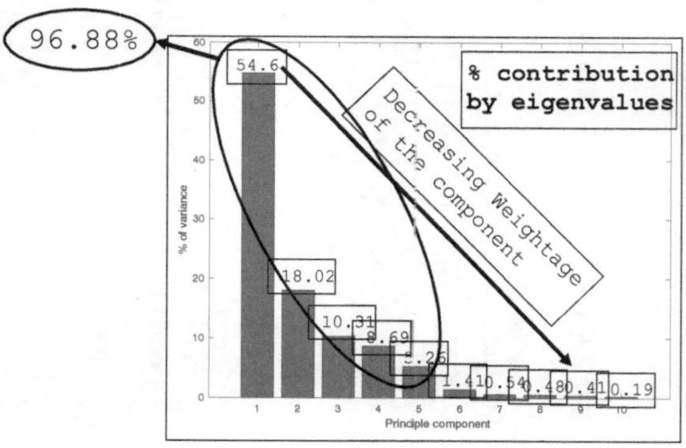

Figure 8.9 Eigenvalues and their percentage contribution.

eigenvectors and demonstrated that the first three, termed "New Feature 1," "New Feature 2," and "New Feature 3," effectively distinguished between gear conditions: normal, cracked, and broken. These features provided clearer differentiation, with "New Feature 1" ranging from approximately -2000 to 1000, "New Feature 2" from -200 to 400, and "New Feature 3" from -20 to 40. Normal conditions occupied lower values, while cracked and broken conditions appeared at higher values.

PCA derives these principal components, or new axes, by examining the covariance matrix's eigenvalues and eigenvectors. Geometrically, these components represent directions of the highest variance within the data: the first component captures the most variance, and each subsequent component, being orthogonal to the previous one, captures as much of the remaining variance as possible. This transformation creates a new coordinate system that aligns with the data's main patterns.

This approach underscores the importance of establishing a standard "normal" signal to detect deviations effectively. Maintenance decisions are informed by comparing and ranking signals, emphasizing fault detection to enable proactive measures.

8.6 CONCLUSIONS

Failure analysis and LCA are indispensable for designing resilient, sustainable systems. Failure analysis identifies potential degradation mechanisms and enables preventive solutions, while LCA assesses environmental impacts across a product's life cycle. Together, they ensure products are not only durable but also environmentally responsible. Techniques such as FTA, FMEA, and RCA, play crucial roles in systematically identifying and addressing high-risk failure modes, facilitating proactive risk management. Case studies on journal bearings and PMBs underscore the practical application of these frameworks, showing how prioritizing critical failure modes can lead to improved durability and functionality. Advancements in data analytics, IoT, and digital twin technology further strengthen failure analysis and LCA, allowing for real-time monitoring and adaptive maintenance strategies. This integration supports a balanced approach to product longevity and ecological responsibility, aligning with modern engineering goals for sustainable development.

REFERENCES

1. Puthoff, J. B., Jakes, J. E., Cao, H., & Stone, D. S. (2009). Investigation of thermally activated deformation in amorphous PMMA and Zr–Cu–Al bulk metallic glasses with broadband nanoindentation creep. *Journal of Materials Research*, *24*, 1279–1290.

2. Hirani, H. (2009). Root cause failure analysis of outer ring fracture of four-row cylindrical roller bearing. *Tribology Transactions*, *52*(2), 180–190. https://doi.org/10.1080/10402000802180151
3. Stamatis, D. H. (2003). *Failure Mode and Effect Analysis*. Quality Press.
4. Arabian-Hoseynabadi, H., Oraee, H., & Tavner, P. J. (2010). Failure modes and effects analysis (FMEA) for wind turbines. *International Journal of Electrical Power & Energy Systems*, *32*(7), 817–824.
5. Lijesh, K. P., Muzakkir, S. M., & Hirani, H. (2016). Failure mode and effect analysis of passive magnetic bearing. *Engineering Failure Analysis*, *62*, 1–20.
6. Hirani, H., & Lijesh, K. P. (2005). "Stator," Application number: 259883, Design Accepted and Published, Journal No. 17/2015, Journal Date: 24/4/2015.
7. Li, W., Shi, T., Liao, G., & Yang, S. (2003). Feature extraction and classification of gear faults using principal component analysis. *Journal of Quality in Maintenance Engineering*, *9*(2), 132–143.

Chapter 9

Mitigation strategies for surface degradation

9.1 INTRODUCTION TO MAINTENANCE STRATEGIES

Maintenance, as defined in the dictionary, is the act of keeping something in good condition. This concept applies broadly to all types of machinery and equipment, from everyday tools like bicycles and motorbikes to complex industrial systems. In this chapter, we will focus specifically on maintenance as it relates to machines.

In the previous chapter, I highlighted the importance of detecting and addressing issues before they escalate into full-blown failures. Once a failure has occurred, it is often too late to implement an immediate fix; at best, design modifications can be made for the future. This is why we refer to "potential failure"—a fault that, if left unattended, could disrupt operations. Identifying these potential issues allows for proactive management and minimizes operational risks.

Maintenance is the systematic monitoring of machinery to identify and address potential issues before they escalate into full-scale failures. The two primary maintenance approaches are preventive and corrective maintenance. Preventive maintenance involves routine inspections—such as cleaning, lubrication, and identifying potential issues—conducted at scheduled intervals, like weekly or biweekly, based on equipment requirements. Corrective maintenance, in contrast, relies on keeping spare parts on hand to quickly restore functionality after a failure. These methods are commonly used in small-scale industries where the cost of continuous monitoring may be challenging to justify.

In many such companies, maintenance is viewed as an unavoidable but unwelcome overhead, placing strain on operations. Consequently, maintenance budgets are often limited to around 5% of total expenditure, encouraging cost-cutting that may inadvertently compromise equipment reliability. These industries often overlook hidden costs associated with neglecting maintenance, including the risks of parts becoming obsolete, downtime costs from unavailable parts, and carrying costs for excess inventory. This

DOI: 10.1201/9781003545019-9

perception undervalues maintenance as a strategic investment in equipment longevity and reliability.

The 1984 Bhopal gas tragedy in India is a stark example of the devastating consequences of neglected maintenance and industrial safety failures. On the night of December 2–3, 1984, a massive leak of methyl isocyanate (MIC) gas from the Union Carbide India Limited (UCIL) pesticide plant in Bhopal resulted in approximately 2500 immediate fatalities, with over 200,000 individuals suffering long-term health impacts. This tragic incident, recognized as one of the worst industrial disasters in history, underscored the severe repercussions of industrial negligence and highlighted significant gaps in regulatory oversight, corporate accountability, and environmental safety standards. The societal and economic toll was staggering, and the company responsible ultimately faced severe financial repercussions, losing far more than it saved through cost-cutting.

A key factor in the Bhopal disaster was the shutdown of a refrigeration system that was meant to keep the highly toxic and volatile compound MIC at a safe temperature of around 0°C. As a reactive chemical, MIC required strict handling and storage protocols, yet safety measures at the plant were systematically weakened over time due to a combination of technical and managerial shortcomings. Maintenance protocols and critical safety equipment were routinely neglected; essential systems like refrigeration for MIC storage tanks and gas scrubbers to neutralize potential leaks were either nonfunctional or deliberately shut down to reduce costs. Months before the incident, the refrigeration system had been deactivated, allowing the temperature to rise to 20°C to 40°C—normal in Indian climates but dangerously high for MIC storage. Compounding this risk, the gas scrubber designed to neutralize leaked MIC was also out of operation, and a corroded exhaust pipe remained unrepaired, preventing safe venting of the gas. This cascade of failures in essential safety and maintenance systems ultimately culminated in one of the world's most devastating industrial disasters.

This tragedy underscores the risks of minimizing maintenance expenses at the cost of safety, often resulting in far-reaching damage to both the company and society. It highlights maintenance not as an optional expense but as a crucial investment in operational safety. Although it is important to optimize maintenance costs, neglecting it entirely can lead to catastrophic outcomes.

Okoh and Haugen[1] conducted a comprehensive study of 183 major industrial accidents involving hydrocarbons and hazardous chemicals between 2000 and 2011. They found that 44% of these incidents were linked to maintenance failures. The most common active cause was the "lack of barrier maintenance," accounting for 50% of these incidents. The study primarily focused on accidents in chemical manufacturing facilities, where toxic substances were frequently involved, highlighting the widespread implications of maintenance deficiencies in high-risk industries. The author's analysis revealed distinct regional trends in maintenance-related accidents and emphasized the importance of structured maintenance practices to mitigate risks. The findings emphasize the critical role of proper maintenance practices and organizational oversight in reducing major industrial accidents.

A comprehensive risk analysis is essential for selecting the appropriate maintenance approach, as it helps determine whether routine maintenance is necessary or if corrective actions alone are sufficient. If the analysis reveals that corrective maintenance may not adequately mitigate risks, a proactive strategy—such as condition-based maintenance (CBM)—becomes crucial. In this chapter, we will focus on condition-based monitoring, an efficient and effective maintenance approach that adapts to real-time equipment needs.

As discussed in previous chapters, nearly all physical systems in motion, along with their subsystems and components, deteriorate over time. This natural wear makes maintenance procedures indispensable for ensuring ongoing safety and operational effectiveness. Maintenance plays a particularly vital role in high-risk systems, as it helps manage and minimize potential hazards to people, the environment, and the equipment itself. To safeguard individuals, protect the environment, and maintain company standards, it is essential to prioritize maintenance from the beginning of operations and to keep risks well within acceptable limits.

9.1.1 Maintenance strategies along P–F curve

In the context of maintenance strategy, two key terms are "potential failure" (P) and "functional failure" (F). Potential failure represents the initial stage where a fault has emerged, but the system continues to operate within acceptable limits. In the previous chapter, this was referred to as a "fault," detectable through "fault analysis" using nondestructive testing (NDT) techniques, data capture, and analysis. At this stage, early signs of deterioration—such as unusual vibrations, elevated temperatures, strange noises, or other deviations from normal conditions—suggest that a failure could be imminent. CBM can be applied here to monitor and address these issues proactively, enabling preventive actions to prevent further damage and maintain system integrity.

Functional failure, on the other hand, occurs when the fault progresses to the point where the system can no longer perform its intended function, resulting in downtime, safety risks, or other adverse consequences. At this stage, corrective maintenance becomes the only viable option to restore functionality. It is important to note that functional failure does not necessarily imply a complete breakdown but rather a loss of performance based on the user's criteria. The interval between potential failure and functional failure, known as the "remaining useful life" (RUL), is critical for planning

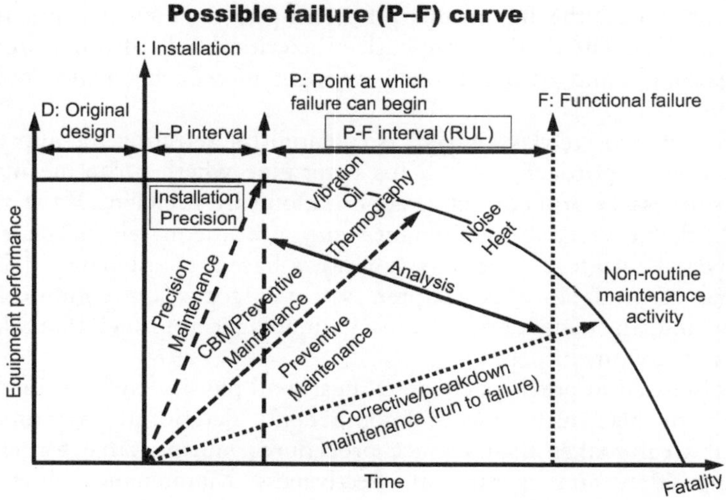

Figure 9.1 Schematic illustrating various maintenance strategies. (A modified figure with information adapted from www.aladon.com/why-the-original-model-of-the-p-f-curve-is-the-correct-model/.)

maintenance actions to ensure timely intervention and avoid unplanned outages.

For clarity, Figure 9.1 presents an expanded version of the P–F curve, known as the "D–I–P–F" curve. This diagram illustrates the relationship between equipment performance and time, showing the progression from initial design and installation through potential failure to functional failure. The key stages on this curve are as follows:

- *D: Original design*—This point represents the initial design of the equipment with its highest performance capability.
- *I: Installation*—Following installation, precision maintenance can be performed to ensure optimal equipment performance and reduce the risk of premature failure. Typically, the equipment supplier is responsible for conducting precision maintenance, a practice widely adopted across industries as it incurs no additional cost for the purchaser of the equipment or machine.
- *P: Potential failure*—The point at which early signs of deterioration (e.g., vibration, oil contamination, abnormal thermography readings, noise, heat) can be detected. This is the beginning of the P–F interval, also known as the "remaining useful life" (RUL).

- *F: Functional failure*—The stage where the equipment can no longer perform its intended function, leading to downtime or other negative outcomes. At this stage, only corrective maintenance or breakdown maintenance (run-to-failure) is viable.

The section of the curve between point "P" (potential failure) and point "F" (functional failure) is known as the "P–F curve." It represents failures that provide warning signs before a complete functional loss, serving as a foundation for CBM strategies focused on early detection and corrective action. Nonroutine maintenance activities are highlighted at the functional failure point, underscoring the critical need for timely intervention to prevent severe consequences, such as total equipment failure.

In equipment maintenance, a range of strategies is applied to address potential failures and prevent functional loss. Each approach along the D–I–P–F curve (shown in Figure 9.1) is designed to optimize equipment performance, longevity, and reliability. Key maintenance strategies along the curve include:

- *Precision maintenance:* This is implemented during the installation phase to ensure precise setup and alignment of equipment. By performing precision maintenance, companies can reduce the likelihood of premature degradation and delay the onset of potential failures.
- *CBM/predictive maintenance:* As soon as early signs of potential failure are detected, CBM is applied. This strategy utilizes diagnostic tools and monitoring techniques—such as vibration analysis, oil analysis, thermography, and noise monitoring—to detect deterioration and manage the interval between potential and functional failure (known as the P–F interval). This predictive approach allows for targeted intervention, potentially extending the equipment's useful life.
- *Preventive maintenance:* Conducted within the P–F interval, preventive maintenance addresses predictable issues—based on existing practical and theoretical knowledge but without real-time data collection—before they worsen. This strategy (also known as "time-based maintenance") involves routine maintenance activities scheduled at regular intervals to manage foreseeable risks and ensure continued functionality, even in the absence of immediate failure indicators.
- *Corrective/breakdown maintenance (run-to-failure):* This approach is applied when functional failure is imminent, and repair or replacement is necessary to restore the equipment's functionality. Corrective maintenance is typically chosen when condition-based or preventive

maintenance is not technically feasible or is not cost-effective—for instance, when the cost of maintenance exceeds the failure costs it is meant to avoid.

Before delving into maintenance strategies from a comprehensive perspective, let us revisit the "bathtub curve" concept (illustrated in Figure 9.2), which was introduced in our discussions on wear mechanisms. This curve, widely used in reliability engineering, represents failure rates over the life cycle of equipment or systems and features three distinct phases. The first phase, known as the decreasing failure rate or "infant mortality" phase, occurs early in the system's life when failures often result from manufacturing defects or installation errors, and the failure rate decreases over time (shown by the dotted line). Next, the constant failure rate phase represents a period of stable, random failures that are relatively infrequent, commonly considered the equipment's "useful life" (illustrated by the green line). Finally, the increasing failure rate or wear-out phase arises as the equipment ages, with failures primarily due to wear and tear as parts reach the end of their operational life (indicated by the orange dotted line). The observed failure rate, shown by the blue curve, combines these three phases to create the characteristic "bathtub" shape, where failure rates start high in the early phase, stabilize during the useful life, and increase again as wear-out failures become more prevalent.

Although the shape of a bathtub curve may vary slightly across different systems, the underlying principles remain the same. This curve, or a similar representation, can also be applied to illustrate potential failure rates over a system's life cycle, offering insights into maintenance planning. While the "bathtub curve" has limitations and drawbacks, as noted in the literature, it is a useful tool to introduce maintenance concepts. The curve suggests that maintenance requirements are minimal during the design phase and begin to emerge during installation.

Companies usually manage the installation process with high precision and often provide complimentary maintenance for a limited period (e.g., 3, 6, or 12 months). During this time, issues are minimal due to the company's direct involvement. However, as this maintenance period ends, the risk of failures may increase, marking the beginning of potential failures at point "P" on the curve. The interval between installation and point "P" typically sees few failures, although this can vary by application. Generally, the time span from "P" (potential failure) to "F" (functional failure) is relatively long, representing the system's residual useful life (RUL), as shown on the right side of the bathtub curve in Figure 9.2. This RUL gradually decreases as the system continues to operate.

For example, if a component's expected life is 3 years and 1 year has already passed, the RUL would be 2 years. When the system reaches functional failure at point "F," it does not necessarily mean the component is

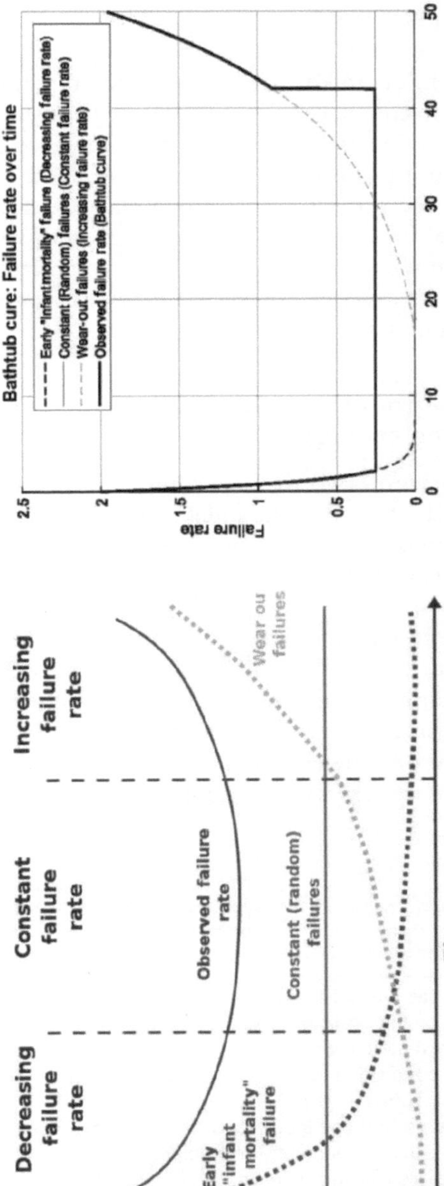

Figure 9.2 Bathtub curve.

completely inoperative but rather that it can no longer perform its intended function effectively. If failure occurs at the system level, immediate repair options may be limited. However, at the component level, breakdown or corrective maintenance can be applied by replacing the failed part, such as a bearing, gear, clutch, or brake. To minimize downtime, many companies maintain a stocked inventory of essential parts. For instance, bearings are often kept on hand so that if one fails, it can be replaced promptly, sometimes restoring system functionality within half an hour. This maintenance approach, known as corrective or breakdown maintenance, is also referred to as "run-to-failure" maintenance, where no preventive action is taken until failure occurs, at which point the component is replaced. This "run-to-failure" strategy can be a viable approach for low-risk products where the impact of failure is minimal.

9.1.2 "Time-dependent" versus "time-independent" maintenance strategies

In reviewing the literature, two distinct maintenance strategies frequently emerge: "time-dependent" and "time-independent" maintenance strategies. While various authors offer differing interpretations, this book defines and outlines the goals of each type (tabulated in Table 9.1) as follows:

- *Time-dependent maintenance:* Time-dependent maintenance strategies are proactive and planned. They aim to prevent failures by conducting maintenance based on a predetermined schedule or continuous monitoring of equipment health. In these approaches, the passage of time and the expected patterns of equipment wear or degradation influence when maintenance activities occur. This category includes:

Table 9.1 "Time-dependent" versus "time-independent" maintenance

Aspect	*Time-dependent maintenance*	*Time-independent maintenance*
Type of approach	Proactive (scheduled or condition-monitored)	Reactive (only after a failure occurs)
Common methods	Preventive maintenance, condition-based maintenance (CBM)	Breakdown maintenance (run-to-failure)
Based on schedule	Yes (scheduled intervals or continuous condition tracking)	No (intervention only after failure)
Timing	Before failure based on anticipated wear or condition data	After failure; no proactive intervention
Goal	To prevent failure by routine checks or condition data	To fix or replace only when a failure happens

- *Preventive maintenance:* This approach relies on a fixed schedule or set intervals (e.g., monthly, quarterly, or after a specific usage period) and is guided by existing theoretical and practical knowledge or manufacturer recommendations. Preventive maintenance is effective when equipment exhibits predictable wear patterns, even without real-time in situ data on its condition. Maintenance is performed at these intervals regardless of the equipment's current health, with the goal of preventing failures before they occur. Preventive maintenance is considered "time-dependent" because it follows a consistent, time-based schedule.
- *CBM:* While CBM is also proactive, it relies on real-time or continuous monitoring of equipment conditions (in-situ data collection), such as temperature, vibration, oil levels, or other health indicators, to decide when to perform maintenance. Unlike preventive maintenance, CBM is more adaptive; maintenance is only performed when specific indicators show that a potential failure may be imminent. However, CBM is still time-dependent in the sense that it monitors changes in condition over time to predict when maintenance is needed. It does not wait for a failure but rather uses trends in the data to anticipate it.

- *Time-independent maintenance:* Time-independent maintenance strategies are reactive. They are applied only after an actual failure has occurred. In this approach, there is no reliance on scheduled or predictive interventions based on time or condition monitoring. Instead, maintenance is carried out as a response to a breakdown. This category includes:
- *Breakdown maintenance (run-to-failure):* This approach involves allowing the equipment to run until it fails completely, at which point corrective action is taken to repair or replace the failed component. There is no routine monitoring or scheduled maintenance, and interventions are only performed in response to a functional failure. This strategy is time-independent because maintenance is not influenced by any schedule or monitoring over time—it is purely reactive.

9.1.3 Overview of condition-based maintenance

CBM serves as a bridge between precision, preventive, and corrective maintenance by offering real-time insights into equipment health. This continuous assessment, known as condition monitoring, allows maintenance actions to be based on the actual status of the equipment rather than following fixed schedules. As such, CBM is especially valuable for systems where continuous monitoring is feasible, making it a more reliable choice for maintaining optimal performance and preventing unplanned downtime.

By categorizing maintenance strategies (refer to Section 9.1.2), we can distinguish between time-dependent and time-independent approaches. Time-dependent methods, including preventive maintenance and CBM, aim to reduce unexpected failures and extend equipment life through proactive intervention. Preventive maintenance, however, follows a fixed schedule regardless of equipment condition, which can sometimes be less cost-effective as demands for reliability, availability, and quality increase. Conversely, CBM uses real-time data to make informed maintenance decisions, enhancing efficiency and reducing unnecessary maintenance activities. Meanwhile, time-independent maintenance, such as breakdown maintenance, waits for equipment to reach functional failure before intervening, which is often suitable when failure has minimal operational impact or low replacement costs.

CBM is built upon two essential components "diagnostics" and "prognostics":

- *Diagnostics:* This process includes the detection, isolation, and identification of faults. Detection identifies anomalies within the system, isolation pinpoints the specific component that is failing, and identification determines the nature of the fault. Diagnostics provide insights into issues after they occur, helping technicians to understand current problems in the equipment.
- *Prognostics:* Prognostics go a step further by predicting the RUL of a component under specific operating conditions. This prediction is based on the historical and real-time data of the system and uses probability models to estimate when a component is likely to fail. Prognostics help in planning maintenance actions ahead of time, thus enabling better resource allocation and minimizing unplanned downtime. While diagnostics focus on analyzing issues after they happen, prognostics focus on anticipating future failures.

When prognostic predictions are insufficient, diagnostics become essential to identify and address existing issues.

9.2 CONDITION MONITORING AND INFORMED DECISION-MAKING

To achieve high levels of availability, quality, and dependability, condition monitoring is an important tool in engineering management for managing data efficiently and making informed decisions. The diagnostic and prognostic capabilities of CBM enable the discovery of faults and the prediction of life expectancy, respectively, allowing for the continuous monitoring of operating equipment. This allows for optimal system performance by handling problems as soon as they arise and fixing them quickly. Condition monitoring is critical for understanding the behavior of a system in real time, which allows for intelligent maintenance decisions to be made using reliable data. Insightful maintenance managers are also required to know how to strategically use this data. Preventing issues from becoming catastrophic failures requires identifying probable or prospective problems early on.

Research highlights the severe consequences of inadequate maintenance, especially in high-risk industries. Okoh and Haugen[1] highlighted 183 major accidents in the chemical industry from 2000 to 2011, even when some maintenance measures were in place. It is anticipated that more comprehensive maintenance could have mitigated or even prevented many of these severe incidents. Around 69% of incidents were attributed to shortcomings in planning, scheduling, and fault diagnosis, underscoring the need for proactive diagnostics rather than reactive measures. With a more adaptive, CBM approach, organizations could use real-time data to address faults early, reducing the chance of escalation into major failures. As equipment enters operational use, regular data collection and analysis are vital for an evolving maintenance strategy. Rather than relying on static, traditional approaches, maintenance should adapt over time, integrating advancements in sensors and technology to meet modern standards and mitigate risk more effectively.

9.2.1 Case study: Condition monitoring of spur gears

In Chapter 6, various NDT techniques essential for effective condition monitoring were discussed. These methods, including eddy current sensors, acoustic emission sensors, digital image correlation (DIC) for strain, thermographic imaging, and ultrasonic sensors, are well-documented in the literature and play an important role in applications like monitoring spur gears. Effective condition monitoring requires not only the selection of appropriate sensors but also a clear understanding of data acquisition and processing techniques to obtain and utilize essential data from these sensors.

Data acquisition from sensors is typically conducted using various methods, often employing a data acquisition system to capture and store information. Once collected, the data undergoes processing to ensure it is ready for analysis, with key steps including denoising, filling missing values, and bridging data gaps to maintain continuity. Proper data processing is crucial for accurately diagnosing system conditions, allowing fault-related data to be isolated from irrelevant noise—a process known as fault isolation.

After isolating faults, the next step is fault identification. In the case of spur gears, current sensor data is compared to baseline data to identify specific issues such as wear, misalignment, thermal degradation, or other known faults. Once a fault is identified, a prognostic approach can estimate the component's RUL. For example, if a 5-mm crack is detected beneath the surface of a spur gear, it may be projected to fail within 2 years under current conditions. This information enables targeted maintenance actions based on RUL estimates, optimizing repair or replacement decisions to avoid unexpected breakdowns.

With all known factors, a major overhaul or targeted adjustments can be implemented to extend the component's life—central to the prognostic approach. By examining the historical performance of the system, prognostics can estimate how long a part will continue to function effectively under certain conditions. For instance, a detected crack in a spur gear may prompt reinforcement, load adjustment, or other modifications to prolong its operational life. The chosen maintenance strategy depends on functional requirements, such as efficiency, temperature tolerance, wear limits, and available resources, which vary according to application and environment.

These considerations determine whether the maintenance approach will be automated or managed by a maintenance professional. In automated systems, continuous data processing and real-time updates are essential for accurate decision-making, while an experienced maintenance manager may interpret data and make effective decisions based on the component's condition.

In CBM, the goal is to ensure quality, dependability, and availability. Whether the system is automated or manually managed, decision-makers must be informed and able to make sound choices. Documenting maintenance strategies and findings is essential to ensure continuity, allowing future personnel to build on existing knowledge rather than starting over due to staff changes. This documentation supports consistency in maintenance practices over time.

As automation in condition monitoring advances, data analytics becomes an invaluable tool for refining maintenance strategies. Analytics assess the impact of previous maintenance actions, helping to evaluate whether these actions improved or degraded performance. This feedback loop enhances maintenance management for future cycles, enabling maintenance personnel or managers to make more informed, data-driven decisions for the optimal performance and longevity of spur gears and other critical components.

At IIT Delhi's online condition monitoring setup for spur gears, vibration analysis, oil analysis, and infrared thermography serve as core diagnostic tools. Wear debris analysis is conducted as part of the oil analysis to gain insights into component wear. For vibration analysis, unidirectional or three-axis accelerometers are used to capture detailed measurements, including temperature changes over time. The data collected is converted from time to frequency domain to diagnose issues such as misalignment, imbalance, bearing wear, or mechanical looseness—making vibration analysis especially effective for monitoring motors, pumps, and other rotating equipment.

Vibration analysis is widely applied in condition monitoring, estimated to be used in 60% to 65% of cases involving rotating machinery. Oil analysis is particularly useful in systems where lubrication is integral, as it allows for measurements of temperature, wear debris, moisture content, and potential corrosion risks. Oil analysis reveals issues such as dust contamination, soot formation, metal fatigue, and oxidation. Changes in oil properties, such as viscosity and additive depletion, indicate oil degradation and the need for replacement, making this analysis common for engines, gearboxes, hydraulic systems, and other lubricated machinery. For systems using oil-based lubrication, oil analysis is preferred, while for systems using only grease without oil circulation, vibration analysis is typically employed.

In situations where vibration analysis is inconclusive, especially with high friction, impact, or localized temperature increases, infrared thermography can be beneficial. This NDT method uses cameras to detect infrared radiation emitted by components, helping identify localized overheating, which may signal electrical failures, insulation issues, or excessive friction. Unlike oil analysis, which provides a general overview of system heating, infrared thermography pinpoints specific overheating areas or potential failures.

In IIT Delhi's lab, these techniques are used in a comprehensive test setup (as shown in Figure 9.3) to monitor the condition of spur gears. This setup includes a mounted spur gear unit, couplings, and a debris sensor to track wear particles in real time. The debris sensor provides feedback on particle levels, offering insights into component wear. Additionally, moisture content and viscosity changes in the oil are measured to assess degradation, creating a robust monitoring system for timely and accurate maintenance decisions.

Figure 9.3 is divided into four labeled sections (a, b, c, and d), each displaying different components of the spur gear condition monitoring setup.

- *Part (a):* This shows the complete setup, including an "eddy current dynamometer," "shaft and couplings," "DC shunt motor," and "spur gearbox." This section also highlights the placement of "proximity sensors" and "wear debris sensors" essential for monitoring system performance.
- *Part (b):* This provides a close-up view of the spur gearbox with various sensor placements. The "proximity sensor" and "accelerometer" positions are clearly marked, which are used for vibration monitoring and detecting positional changes.
- *Part (c):* This displays the oil circulation system and additional sensors, including an "oil pump" for circulating oil and a "magnetic particle filter" for removing metallic debris from the oil to maintain its purity.
- *Part (d):* This focuses on the suite of sensors dedicated to oil quality and contamination detection. It includes a "moisture sensor" to detect

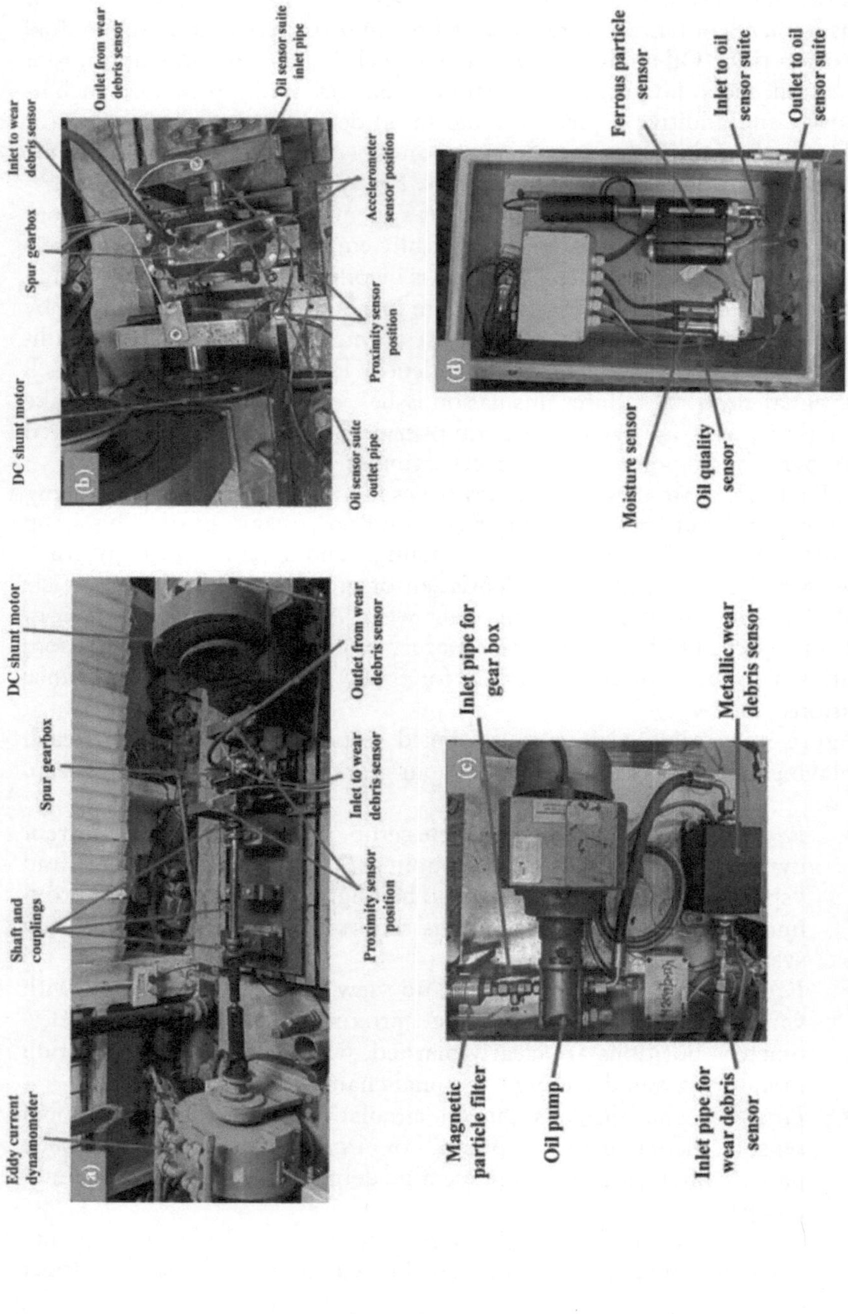

Figure 9.3 Online condition monitoring setup for gears at IIT Delhi.

water content, an "oil quality sensor" to monitor oil degradation, and a "ferrous particle sensor" to identify iron-based debris in the oil.

This integrated setup utilizes multiple sensors to monitor a range of parameters, including vibration, oil quality, wear debris, and moisture levels, allowing for a thorough condition assessment of the spur gear system.

The experimental setup (illustrated in Figure 9.3) consists of a single-stage spur gearbox, driven by a DC shunt motor and loaded using a water-cooled eddy current dynamometer. The dynamometer used in this setup is from the E-50 "LC" series, with a torque capacity ranging from 0 to 75 Nm, and includes a controller that supplies a current output between 0 and 3 A. The gearbox has a reduction ratio of 1.96 and is powered by a 30 kW DC shunt motor, which is controlled by a three-phase 440V/75A motor controller. This controller utilizes pulse width modulation (PWM) to rectify the AC input voltage and control the DC voltage supplied to the motor, allowing precise regulation of speed and torque to maintain target operating conditions.

The gearbox in this experiment is equipped with 20 degrees involute spur gears, featuring a 2-mm module and a face width of 33 mm. These gears are made from EN24 steel with a hardness level of 30 ± 2 HRC, a relatively low hardness chosen intentionally to accelerate wear for testing purposes. Detailed specifications for the pinion and gear, including material, hardness, dimensions, operating parameters, surface roughness, Young's modulus, and lubricant details, are provided in Table 9.2.

Table 9.2 Various parameters related to gearbox

Parameter	*Pinion*	*Gear*
Material	EN24	EN24
Hardness (HRC)	30 ± 2	30 ± 2
Module (mm)	2	2
Pitch circle diameter (mm)	54	106
Base circle diameter (mm)	50.74	99.6
Tip diameter (mm)	58	110
Pressure angle (degrees)	20	20
Face width (mm)	33	33
Number of teeth	27	53
Speed (rpm)	1,200, 500	—
Torque (nm)	40, 50	—
Contact ratio	1.6	1.6
Roughness (Ra) (μm)	0.363 ± 0.027	0.357 ± 0.037
Roughness (Rz) (μm)	0.579 ± 0.066	0.556 ± 0.095
Young's modulus (GPa)	207	207
Lubricant	O1:API GL-4 SAE 90 O2:API GL-4 EP 90 (March 2022) O3:API GL-4 EP 90 (January 2017)	Same as Pinion

The setup in place provides a comprehensive system for diagnosing issues within a gear mechanism. In this experiment, spur gears are used, though other types of gears, including helical gears, could also be tested, as long as lubrication is applied. The gearbox is mounted with an accelerometer to capture signals, enabling condition monitoring through time-critical analysis (TCA) and other analysis tools. By comparing matching frequencies, potential issues such as surface damage or misalignment can be detected.

The setup for condition monitoring includes a variety of sensors and instruments designed to capture detailed data on gear and lubricant conditions. The "Metallic Wear Debris Sensor" (AS-19144-KW) detects both ferrous (40 to 400 µm) and non-ferrous (135 to 450 µm) particles in the gearbox, sampling 12 times per minute. The "Oil Sensor Suite" (ANALEXrs) measures total ferrous debris, oil degradation, and temperature, calculating an oil quality index (OQI) from 1 to 100. This suite also measures moisture content, with one reading per minute. For "pH measurements," the Metrohm 916 Ti-touch instrument, with a range of –13 to +20 pH, resolution of 0.001 pH, and accuracy of 0.003 pH, was used. Lastly, oil viscosity was assessed with an "MCR 102 Rheometer," measuring at a shear rate of 100 per second within a temperature range of –5°C to 200°C, ensuring comprehensive monitoring of lubricant properties.

9.2.1.1 Experimental procedure

Three independent tests were performed, labeled as follows: "Test 1: Lubricant without aqueous HCl," "Test 2: Lubricant with 0.0025% v/v aqueous HCl," and "Test 3: Lubricant with 0.0025% v/v aqueous HCl + Nanoadditive." In these tests, HCl was introduced to simulate corrosive conditions, allowing observation of gear material behavior under realistic operating conditions. Additionally, to assess the effectiveness of mitigating mild acidic environmental effects on gear performance, nanoadditives were incorporated in Test 3 to evaluate their potential to reduce the impact of acidic conditions.

In all the tests, 30 mL oil samples were extracted every 10 minutes for offline testing. To maintain oil levels, a fresh 30 mL sample with matching properties was preheated to the sump temperature and added back. Given the sump capacity of 1500 mL, this 30 mL sample constitutes only 2% of the total oil volume, ensuring minimal impact on the overall oil composition. Offline samples were analyzed for pH and viscosity within 24 hours of extraction.

9.2.1.2 Results and observations

The results are shown in Figures 9.4 and 9.5. Figure 9.4 depicts the wear mass and iron particle concentration (Fe ppm) under three conditions of lubricant: without HCl, with HCl, and with HCl plus nanoadditives.

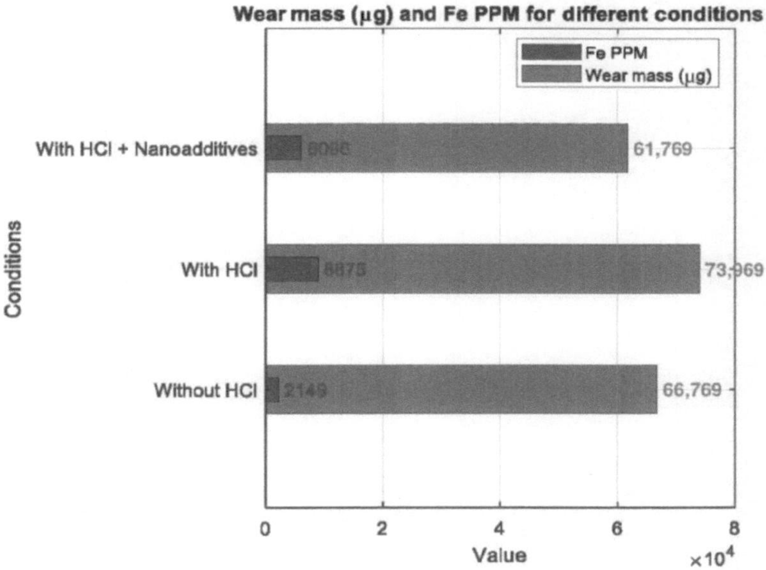

Figure 9.4 Distribution of Fe ppm and wear mass for "Test 1," "Test 2," and "Test 3."

- *Without HCl:* The wear mass is 66,769 μg, and the Fe ppm is 2149.
- *With HCl:* Introducing HCl increases the wear mass to 73,969 μg and the Fe ppm to 8875.
- *With HCl + nanoadditives:* Adding nanoadditives reduces the wear mass to 61,769 μg and the Fe ppm to 6096, indicating that nanoadditives effectively lower wear even in corrosive conditions.

The data in Figure 9.4 shows the effects of lubricant degradation in the presence of aqueous HCl on wear mass and Fe ppm levels. Wear mass increased from 66,769 to 73,969 μg and Fe ppm increased from 2149 to 8875, indicating higher wear rates. However, the addition of nanoadditives to the degraded oil effectively reduced wear, with wear mass dropping to 61,769 μg and Fe ppm decreasing to 6096, demonstrating the wear-reducing benefits of nanoadditives.

Figure 9.5 shows the distribution of wear particles in different size ranges (>40, >60, and >100 μm) under three conditions: without HCl, with HCl, and with HCl plus nanoadditives. The number of wear particles with particle size distribution is as follows:

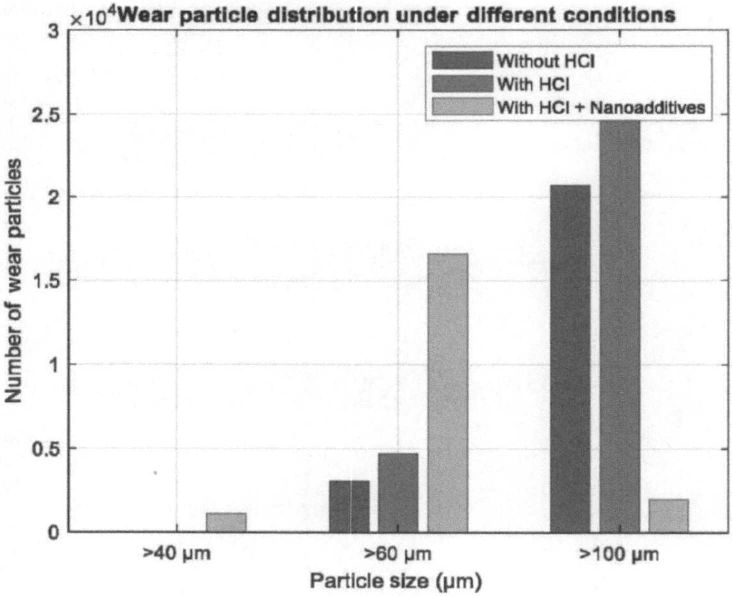

Figure 9.5 Ferrous particle size distribution for "Test 1," "Test 2," and "Test 3."

- *For particles >40 μm:*
  - Without HCl: 57 particles
  - With HCl: 26 particles
  - With HCl + nanoadditives: 1126 particles
- *For particles >60 μm:*
  - Without HCl: 3070 particles
  - With HCl: 4726 particles
  - With HCl + nanoadditives: 16,596 particles
- *For particles >100 μm:*
  - Without HCl: 20,723 particles
  - With HCl: 27,960 particles
  - With HCl + nanoadditives: 1962 particles

The addition of HCl in the experimental setup led to a significant increase in the generation of wear particles, particularly among larger sizes. Conversely, incorporating nanoadditives effectively reduced particle count, indicating a decrease in wear. In Tests 1 and 2, particles over 100 μm were the primary contributors to wear, while in Test 3, wear was mainly attributed to particles exceeding 60 μm. The presence of larger particles correlated with more severe surface damage, mirroring the increase in particle size and volume observed with HCl introduction. Nanoadditives, however, reduced the average particle size and minimized wear and surface damage. This effect suggests that, even at low concentrations, nanoadditives absorb moisture and form a protective layer on contact surfaces, thereby mitigating wear severity.

These findings demonstrate the potential of nanoadditives to reduce wear rates, highlighting the advantages of nanotechnology for managing surface degradation. The results further emphasize how a combined approach of condition monitoring, advanced materials, and nanoadditives can improve gear performance and decrease wear in corrosive conditions. Overall, this research supports nanoadditives as a practical solution for extending the service life of critical components, showcasing modern materials science's role in enhancing equipment durability and reliability in challenging environments.

9.2.2 Case study: Data-driven failure mode and effect analysis to enhance maintenance planning

Chapter 8 has discussed the FMEA approach in detail. In the current section, we will examine a study by Filz et al.,[2] where a data-driven FMEA approach was implemented to improve maintenance planning. This method prompts the question: how can maintenance be further optimized?

The need for a data-driven FMEA framework arises from the increasing availability of operational data and the growing focus on predictive maintenance for industrial equipment, such as aircraft. This data-driven approach reduces the subjectivity inherent in traditional FMEA, leading to more accurate failure predictions and improved maintenance strategies. In traditional FMEA, subjectivity mainly affects the ranking of the risk priority number (RPN). This is due to the qualitative assessment (often on a 1 to 5 or 1 to 10 scale, as discussed in two case studies in Chapter 8) of three key factors: severity, occurrence, and detection. Each factor is typically assigned a numerical value based on the evaluator's judgment, introducing variability across assessments.

It aims to deliver precise risk assessments and enable effective resource allocation, providing valuable support even to less experienced employees. To implement this, the following procedure can be opted:

- *Data collection and preparation:* Relevant operational data, such as environmental conditions, usage patterns, and historical maintenance records, are gathered. The data is preprocessed by handling missing values, normalizing them, and transforming them into a suitable format for prediction models.
- *Model selection and optimization:* It is possible to choose the best (i.e., deep learning) machine learning model for defect prediction

after testing several models. To help choose the right data analytics models, some modeling methodologies are provided. An example would be the use of multivariate regressions or artificial neural networks (ANNs) when dealing with many independent variables. As every dataset has its own set of variable correlations, training an algorithm to provide accurate predictions requires fine-tuning of certain hyperparameters. Unfortunately, no one collection of hyperparameters will provide a perfect match for every piece of data. Predicting the occurrence of specific failure modes is made more accurate and reliable by optimizing the model parameters.

- *Integration into FMEA:* The predicted fault probabilities are converted into occurrence values within the FMEA. Severity and detection values are assigned by experts, and the RPN is calculated for each failure mode.
- *Decision support:* The FMEA framework calculates RPN values, which are displayed in a graphical user interface (GUI) to support maintenance planning. The GUI uses color coding for easy interpretation: high risk (red) for $100 \leq \text{RPN} \leq 1000$, medium risk (orange) for $50 \leq \text{RPN} < 100$, low risk (yellow) for $2 \leq \text{RPN} < 50$, and minimal risk (green) for $\text{RPN} = 1$. These threshold values can be customized based on the specific application. By highlighting high-risk failure modes and recommended maintenance actions, the user-friendly interface enables quick identification of critical risks, facilitating data-driven maintenance decisions.

Filz et al.[2] proposed a framework based on a concept known as the cyber–physical production system (CPPS) to support a data-driven FMEA methodology. This framework is divided into four main domains: the physical realm (I), data acquisition (II), cyber domain (III), and decision support (IV). Through an iterative process, it continuously utilizes the latest operational data to enhance the fault prediction algorithm and the data-driven FMEA.

- The first module (Phase I) represents the "physical world," which refers to the actual system being analyzed. In traditional FMEA, this phase involves examining the system to identify its various failure modes and the maintenance strategies applied to address them. This process was followed here to gain a comprehensive understanding of the system and its associated failure modes. The physical realm (I) encompasses actual maintenance activities and key performance indicators (KPIs) relevant to the technical equipment, such as machinery. This phase is completely similar to what was covered in the FMEA section of Chapter 8.
- The "data acquisition" phase (II) is where operational data is gathered and prepared for analysis in the cyber world. This data flows from

the physical world. Similar to permanent magnetic bearings (PMBs), sensors allow for constant monitoring of critical features and failure modes. Additional data allows for a more thorough comprehension of the system, which in turn permits ongoing enhancements to the system or data processing. The cyber world then incorporates these upgrades. Identifying the likelihood of errors is a critical component of this paradigm. To achieve this, operational data from the technical equipment and components that have been observed is used, which has already been preprocessed. The features and kinds of variables in the dataset are crucial in deciding the data analytics model to use. Collecting operational data from the monitored components or subsystems is part of data acquisition (II), which is then followed by data preparation. The framework starts with operational data collection, capturing environmental conditions, usage patterns, and historical maintenance records. Preprocessing includes handling missing values, normalizing, and transforming data for predictive model use.

- *"Cyber World":* This virtual environment introduces a three-dimensional (3D) model of the real system, often referred to as a "cyber system." The goal of the modeling phase in the data-driven FMEA methodology is to improve model accuracy by tailoring it to specific use cases and datasets. Recognizing that different use cases may require varying parameter values, a generalized process is recommended to optimize model performance. The severity and detection levels, set by experts, remain constant; however, the occurrence rating can be adjusted based on real-time data, allowing for dynamic changes to the RPN. Estimating the likelihood of occurrence can be challenging, particularly when historical data is limited or inconsistent, as seen in the FMEA case study on PMBs. Subjectivity often plays a role, with individuals relying on personal experience or assumptions about failure rates that may vary based on their familiarity with the equipment. For instance, a physical model may initially assign a score of 6 on a 1 to10 scale for failure occurrence, but real-world data could suggest adjusting this to 7 or 5. Predicted fault probabilities for each failure mode are mapped onto a scale of 1 to 10 (e.g., 84% = 8, 27% = 3) to establish the occurrence value for FMEA. Incorporating both real-time and historical data through this iterative process improves the accuracy of the RPN, enabling more adaptive and responsive maintenance planning.
- *Decision support:* In the "decision support" phase (IV), evaluation and deployment occur as maintenance recommendations are formulated based on the RPN findings. Here, the calculated RPN supports maintenance decisions and strategy development, enabling prioritization of failure modes for effective resource allocation.

High-risk failure modes are prioritized, and maintenance actions are fed back into the physical world through this feedback loop. The data-driven FMEA results provide actionable suggestions to guide maintenance efforts and mitigate potential failures.

9.2.2.1 Benefits

This data-driven FMEA approach integrates historical data with real-time insights, creating a dynamic alternative to traditional, predefined maintenance strategies. By leveraging continuous RPN calculations, it allows maintenance plans to adapt in real time. For example, in the case of PMBs discussed in Chapter 8, design changes—such as substituting stainless steel with rubber between components—led to enhanced performance. Each updated RPN informs subsequent maintenance actions, supporting the development of optimized strategies through an iterative and adaptive process. Real-time data also enables ongoing adjustments to the occurrence ranking, making this approach invaluable for accommodating environmental shifts and operational variations, like different flight techniques among pilots, which call for a flexible maintenance strategy.

9.2.2.2 Findings

The case study showed that the data-driven FMEA framework could predict failure occurrences with an average accuracy of 95%. This approach supports more consistent and objective maintenance planning, reduces the need for subjective judgments, and offers actionable insights to improve resource allocation. By scheduling maintenance based on actual condition predictions, the framework helps extend the lifespan of critical components, optimize costs, and reduce unnecessary replacements.

This research aims to combine historical operating data from the usage phase with a structured fault prediction process, supported by FMEA. This integration provides targeted, clear, and practical data-driven support for dynamic maintenance planning, especially for high-value assets like production equipment and aircraft.

9.3 LIFE-CYCLE IMPACT ASSESSMENT

The life-cycle impact assessment (LCIA) approach is a standardized framework developed to evaluate a product's entire life cycle—from raw material extraction to final disposal—and its interactions with the environment. This approach is essential to minimize potential environmental impacts associated with products by assessing each stage comprehensively. LCIA quantifies the materials, energy, and resources used throughout the production, distribution, consumption, and disposal stages. It also monitors emissions, waste, and byproducts generated at each step of the life cycle, offering a holistic view of a product's environmental impact. In an LCIA, the following factors must be accounted for:

- *Waste production:* Quantifying the waste generated during each phase, from manufacturing scraps to end-of-life disposal.
- *Emissions:* Tracking emissions released into the air, water, and soil, including pollutants and greenhouse gases, to understand their potential impact on ecosystems and human health.
- *Energy consumption:* Evaluating the energy consumed throughout production, transportation, usage, and disposal to assess the energy footprint and opportunities for energy conservation.
- *Material use:* Monitoring raw material inputs and evaluating sustainable alternatives or methods for reducing material intensity.

By incorporating these factors, LCIA plays a critical role in sustainable development, especially in mitigating corrosion and environmental degradation. It enables companies to understand and quantify how their operations and material choices impact the environment, helping them identify improvement areas. Through LCIA, organizations can assess potential environmental effects, thereby evaluating the overall sustainability and ecological footprint of their systems. This makes LCIA a valuable tool for informed decision-making and responsible environmental management. For companies focusing on corrosion and surface engineering, LCIA provides insights to adopt strategies that minimize environmental impact, enhance resource efficiency, and contribute to a more sustainable future.

9.3.1 Life-cycle assessment

In this section, we will look at how a product or system affects the environment, society, and economy. These three aspects—environment, society, and economy—are crucial for a complete evaluation. We will discuss ways to measure a product's environmental impact, including factors like CO_2 emissions, energy use, and particle release into the atmosphere. These data points help us make comparisons and establish environmental impact equivalencies, often with the aid of LCIA software tools. Luckily, the International Organization for Standardization (ISO) standards like ISO 14040 and ISO 14044 provide a structured approach for these assessments, offering clear guidelines and comparison parameters to support the process.

Life-cycle assessment (LCA) builds on the LCIA framework by thoroughly examining a product's inputs, outputs, and environmental impacts across its entire life cycle—from creation to disposal. During the compilation phase, all materials, energy, and resources used in manufacturing, distribution, usage, and disposal are identified, along with any emissions, waste, and byproducts produced at each stage. These inputs and outputs are then evaluated through LCIA, considering environmental indicators such as climate change, pollution, and resource depletion. This process provides a comprehensive view of a product's ecological footprint, helping decision-makers identify areas for improvement in product design, production, and disposal.

LCIA is essential for supporting sustainable development, especially for small-scale industries and MSMEs, who can use this framework to assess their products' environmental impacts. In the future, governments worldwide are expected to mandate LCIA implementation across industries. A lower LCIA score indicates a smaller environmental impact, aligning with sustainability goals.

Piotrowska et al.[3] proposed a structured methodology for conducting environmental evaluation, specifically for LCIA. The process is divided into interconnected components as follows:

1. *Goal and scope:* The goal and scope definition is the starting point of the study, where research objectives and boundaries are established. The goal outlines the desired outcomes, while the scope sets the study's limits, including the target population, time frame, and geographical area. This step is crucial as it directly shapes the subsequent analyses and keeps the assessment focused.
2. *Inventory analysis (LCI):* In this stage, data on inputs (such as raw materials and energy) and outputs (like emissions and waste) is collected, forming an iterative process that may require revisiting the goal and scope. Inventory analysis focuses on identifying and quantifying the environmental loads associated with a specific activity or process, covering materials, energy, emissions, and waste generated. This step provides valuable insights into the environmental impact of the assessed system.
3. *Impact assessment (LCIA):* In this phase, data from the inventory analysis is assessed to identify potential environmental impacts, assigning quantifiable effects to each life-cycle stage. The impact assessment examines emissions, resource use, and waste generation across a product's life cycle, offering a comprehensive evaluation of impacts on indicators like climate change and resource depletion.
4. *Interpretation:* This final step in LCA involves analyzing the impact assessment results to draw meaningful conclusions and make informed decisions. Interpretation includes identifying the main contributors to environmental impacts, acknowledging any uncertainties (such as variations in energy measurements) and limitations (such as lack of specific facilities), and considering adjustments to the goal and scope if necessary. The LCIA groups environmental impacts into categories (e.g., climate change and ozone depletion), classifies inventory items

within these categories, and assesses their effects quantitatively and qualitatively. Improvement analysis plays a crucial role in this phase, examining ways to reduce impacts through alternative materials, production methods, energy sources, and waste management strategies. Additional steps, such as normalization (comparing impacts to reference systems), weighting (assigning importance to each impact category), and aggregation (combining impacts into a single score), help provide a comprehensive evaluation of environmental performance. This structured interpretation process identifies key areas for improvement, supporting efforts to reduce environmental impacts effectively.

Thorough documentation of actions and findings is essential for transparency and advancing future work in LCAs. Clear and detailed documentation ensures that results are reliable and actionable. The ISO has developed specific standards for environmental management, notably ISO 14040 and ISO 14044, which provide comprehensive frameworks and guidelines for conducting LCAs. These standards establish processes and requirements to ensure that assessments are consistent and credible. Following established standards like ISO 14040 and ISO 14044 is vital for maintaining a structured, accurate, and accountable approach in LCAs. Each standard has a distinct focus and scope, working together to offer a complementary framework for guiding LCA practices effectively.

- *ISO 14040: LCA—Principles and framework for environmental management:* ISO 14040 is a foundational standard that outlines the basic principles, concepts, and framework for conducting LCAs. This standard defines key terms and concepts and divides the LCA process into four main phases:
  - *Goal and scope definition:* Establishes the purpose of the study and specifies its boundaries. For example, a company assessing the environmental impact of a new product might define the goal as reducing greenhouse gas emissions throughout the product's life cycle.
  - *Inventory analysis:* Involves gathering data on all relevant inputs (like raw materials and energy) and outputs (such as emissions and waste) within the system boundary. This phase would track resources consumed and byproducts released, like CO_2 emissions or waste from a production line.
  - *Impact assessment:* Analyzes the environmental impacts associated with inventory data, grouping impacts into categories such as climate change, water pollution, and resource depletion. For instance, CO_2 emissions would be classified under climate change.

  - *Interpretation:* This final phase assesses the significance of the findings, identifies the main contributors to environmental impacts, and examines potential areas for improvement. If a certain raw material is found to have a high environmental impact, it might prompt consideration of alternatives.
  - ISO 14040 provides the foundational structure for an LCA study without evaluating the functionality of products or services.
- *ISO 14044: Environmental management—LCA:* Building on ISO 14040 and ISO 14044 provides more detailed requirements and guidance for each phase of an LCA study. It focuses on ensuring the quality and credibility of the study through stringent data collection, data quality requirements, calculation methods, and reporting protocols. This standard serves as an implementation guide for applying ISO 14040 principles in practice. For example, ISO 14044 emphasizes calculating mean values and standard deviations to evaluate data quality, ensuring that collected data is robust and suitable for reliable decision-making. In cases where variability is high (e.g., a mean of 1 with a standard deviation of 0.5), the data quality might be considered low, whereas a high-quality dataset would show lower variability (e.g., a mean of 5 with a standard deviation of 0.1).
  - Following these standards allows for LCAs that are consistent, transparent, and credible. For instance, a manufacturing company using LCA to assess the environmental impacts of different raw materials would benefit from ISO 14044's detailed guidelines on inventory analysis and impact assessment, ensuring that data is accurate and reliable for decision-making.
  - To manage environmental impacts effectively, "data categorization" by environmental impact types—such as effects on human health, air, water, land, or ecosystems—is essential. These categories are aligned with impact indicators, which quantify the extent of an impact. "Characterization" further rates the data on a scale (such as 1 to 1000), helping to quantify impacts like resource depletion or air pollution. For example, energy consumption data could be rated to show its potential contribution to climate change, helping identify the most sustainable production methods.
  - An additional useful approach in LCA is "data normalization," where data is adjusted to fit within specific parameters for easier comparison across different impact indicators. For example, data might be normalized around 13 key indicators relevant to a company's sustainability goals, making it easier to track and compare different environmental impacts. Introducing "weighting factors" allows certain parameters to be prioritized over others

based on their environmental significance. In a car manufacturing study, greenhouse gas emissions might be weighted more heavily than water usage due to their global impact on climate change.

- "Normalization" and "weighting" not only facilitate better comparisons but also clarify which elements of the assessment are essential. They allow decision-makers to focus on the most impactful areas. In some cases, a weighted objective approach might be employed to assign greater importance to indicators with higher environmental significance. For instance, when comparing materials, recycled aluminum might be assigned a higher weight for resource conservation than virgin aluminum due to its lower environmental impact.
- The "interpretation phase" is vital, as it synthesizes all findings to draw meaningful conclusions and guide improvements. Interpretation involves evaluating the main contributors to environmental impacts, considering uncertainties (such as variations in energy use across different scenarios), and identifying limitations (like unavailable data for certain emissions). For example, if energy consumption data shows fluctuations—such as readings of 600, 800, and 300 calories at different times—this uncertainty should be accounted for to improve data accuracy and reliability.
- "Improvement analysis" explores methods to reduce environmental impacts, such as choosing alternative materials, optimizing production methods, adopting renewable energy sources, or enhancing waste management. In practice, if an LCA reveals that a significant amount of CO_2 emissions arises from fossil fuel energy, transitioning to renewable energy sources could reduce the carbon footprint.
- Finally, "documentation and communication" of the LCA findings in a clear, transparent manner is essential for sustainable decision-making. Documenting each step and result, as well as following the established standards (ISO 14040 and ISO 14044), ensures that the LCA study is credible and reproducible. For example, a company might document its LCA findings in a sustainability report to share with stakeholders, demonstrating a commitment to environmental responsibility and providing actionable insights into future improvements.
- In summary, ISO standards, effective categorization, and thorough documentation create a comprehensive framework for LCAs. Adhering to these practices supports informed decision-making, enhances transparency, and promotes sustainability throughout a product's life cycle.

The product's physical life cycle includes stages from raw material extraction to disposal and potential recycling. Figure 3 from Piotrowska et al.[3] outlines the scope of an LCA, detailing the stages in a product's life cycle along with associated inputs and outputs to evaluate environmental impact. This diagram offers a comprehensive view of the LCA framework, showing how resources are used across key life-cycle stages and the resulting environmental outputs. The inputs consist of essential resources needed at each life-cycle stage, such as "Raw Materials—the basic materials extracted from the environment for manufacturing"—and "Energy, which refers to the energy required across stages like production, usage, and recycling." All collected inputs need to be grouped for easier quantification. This categorization helps when comparing products, processes, or methodologies, as it allows for a clear and meaningful analysis. In LCA, evaluating the environmental impact of a product or process across its entire lifetime is essential. However, it is also important to consider the long-term effects beyond a product's life—what might its impact be 20, 50, or even 200 years into the future?

Accounting for these lifetime impacts helps in understanding potential environmental depletion and aids in creating policies to mitigate such effects. Products and processes that are "green"—like those involving renewable energy or sustainable manufacturing—are often incentivized due to their lower emissions, while products with high carbon footprints may face higher taxes as a deterrent. For example, comparing a small, fuel-efficient car that offers 20 km/L with a large luxury car offering only 6 km/L highlights the environmental and economic benefits of choosing efficient, low-emission options.

Such comparisons underscore the need to base product development and market entry on environmentally responsible criteria. From a policy perspective, life-cycle inventory (LCI) considers emissions, resource usage, and waste generation, acknowledging that resources are finite. Effective LCA involves analyzing every stage, from raw material procurement to disposal, and sometimes even beyond the product's life cycle. This comprehensive analysis identifies weak spots, helping to develop more sustainable products and processes.

With continuous innovation, manufacturing processes and materials can be enhanced—for instance, extending a product's lifespan from 5 to 10 years. This improvement reduces raw material and energy consumption, which is highly beneficial. Identifying necessary material modifications allows for targeting weak areas, enabling innovation and better outcomes.

Piotrowska et al.[3] defined a system boundary that includes the primary stages of the life-cycle process, marking the stages analyzed in the LCA. Each stage within this boundary contributes to the product's overall environmental impact:

- *Raw materials acquisition:* In this stage, raw materials are extracted from natural resources. This process includes mining, harvesting, and other extraction methods. This stage involves obtaining raw materials necessary for manufacturing the product, such as mining, drilling, or farming. Raw materials are processed to create components for the product.
- *Production:* Here, raw materials are processed and transformed into the final product through manufacturing activities, often involving energy consumption and emissions. Raw materials are transformed into the final product through manufacturing, fabrication, and assembly operations, ensuring it meets specifications.
- *Use:* This phase covers the operational life of the product, where it is utilized by consumers. It considers any emissions, waste, and energy requirements during the product's active use. The product serves its intended purpose during this stage, meeting user needs. Duration varies based on the product type.
- *End-of-life (recycling):* This stage involves the recycling of the product after its useful life. It aims to recover the materials and energy of the product in an environmentally responsible way. Some products can be recycled, reducing the need for new raw materials and minimizing waste.

In addition to the four main stages, there are three more stages in the product life cycle:

- *Storage:* This stage includes any storage requirements throughout the product's life cycle, especially if storage generates emissions or consumes energy.
- *Distribution:* This phase covers the transportation and distribution of the product from production facilities to points of sale or use, which involves energy use and emissions.
- *Disposal:* The final stage, where the product is discarded if not recycled, considers any lasting environmental impacts from landfill or other disposal methods. At the end of its useful life, the product is disposed of through landfilling, incineration, or other waste management methods, requiring environmentally responsible practices.

In Figure 3 from Piotrowska et al.,[3] the right side lists various outputs associated with each stage of the life cycle. These outputs include both useful and potentially harmful byproducts, with an emphasis on minimizing the harmful ones. When assessing inventory load, it is important to identify and quantify what is released into the environment, such as emissions, waste, or other byproducts. While functional outputs are necessary, additional outputs like emissions, waste needing disposal, or used oil needing periodic replacement should be carefully categorized and accounted for. These can be organized into impact categories, including climate change, water pollution, and resource depletion, for a structured and thorough assessment.

The environmental impacts of these outputs can be categorized as follows:

- *Atmospheric emissions:* Emissions released into the air, such as greenhouse gases (e.g., CO_2) and pollutants, contributing to climate change and air pollution.
- *Waterborne wastes:* Wastes discharged into water bodies, potentially leading to water pollution and harm to aquatic ecosystems.
- *Solid wastes:* Physical waste materials generated at different stages, including production scraps, end-of-life product waste, and residuals from recycling.
- *Other releases:* Additional environmental outputs, including any hazardous byproducts or chemical releases that may impact soil, plants, animals, or human health.

This structured approach to categorizing and managing outputs is essential for a comprehensive environmental assessment.

Based on the framework suggested by Piotrowska et al.,[3] a key question arises: How can one measure the materials, energy, and resources required at each stage—production, distribution, use, and disposal? Much of this data typically originates from manufacturers and involves an iterative process. It is not a one-time task; various documents and studies may need to be referenced to systematically approach these measurements.

Another important consideration is waste: What type of waste will the product generate? What emissions will it produce, and what byproducts? These aspects are essential for evaluating the product's sustainability. Waste, emissions, and byproducts that affect air, water, or land must be accounted for, along with the energy consumed throughout production, usage, and disposal.

9.3.2 Case study on life-cycle assessment

Piotrowska et al.[3] conducted an LCA on car tires to analyze their environmental impact throughout the product's life cycle. Car tires, while essential for vehicle operation, have continuous environmental effects due to material consumption, wear particle emissions, and energy use. This study examined the production, use, and recycling stages of a tire's life cycle, with a specific focus on the tire size P205/55/R16 commonly used in passenger vehicles. The size designation P205/55/R16 means:

- P indicates that it is a passenger vehicle tire.
- 205 refers to the tire's width in millimeters. In this case, the tire is 205 mm wide.

- 55 represents the aspect ratio of the sidewall height to the tire's width. Here, the sidewall height is 55% of the width or 112.75 mm.
- R stands for radial construction, meaning the internal layers of cords are arranged radially from the tire's center.
- 16 indicates the wheel or rim diameter in inches. This tire is made to fit a 16-inch diameter wheel.

In summary, a tire with the size P205/55/R16 is 205 mm wide, has a sidewall height that is 55% of its width, is of radial construction, and fits a 16-inch wheel diameter. The composition of tire material is as follows: synthetic rubber (24.17% of tire mass), natural rubber (18.21), carbon black (19), coated steel wires (11.4), precipitated silica (9.65), oil (6.12), textiles (4.7), zinc oxide (1.58), sulfur (1.28), stearic acid (0.96), and recycled rubber (0.5). The LCA aimed to quantify the environmental impacts across each stage, highlighting areas with the most significant effects.

- *Production stage:* This stage involves sourcing raw materials like synthetic and natural rubber, sulfur compounds, carbon black, and mineral oils, all of which are energy- and resource-intensive. The manufacturing process contributes to environmental impact through emissions and waste generation. For instance, each tire requires about 6 MJ of electrical energy and 45 L of water and generates approximately 0.5 kg of waste during production.
- *Use stage:* During its operational phase, the tire contributes to environmental impact mainly through fuel consumption, as it influences vehicle efficiency. Typical emissions in this stage include CO_2, nitrogen dioxide, and sulfur oxides, which contribute to climate change and air pollution. Over a tire's estimated life span of 50,000 km, it is associated with around 750 L of fuel consumption, making this stage a primary contributor to overall environmental impact.
- *Recycling stage:* At the end of its life, a tire undergoes recycling to reduce waste. Recycling involves either energy recovery through incineration or shredding for material reuse. While recycling is environmentally favorable compared to disposal, it still requires resources; for instance, shredding consumes 7.4 MJ of electricity and 1.5 kg of water per tire. Although the recycling stage has a relatively lower environmental impact than production and use, it plays a crucial role in waste reduction.

The study used three main methods—Eco-Indicator 99, cumulative energy demand (CED), and Intergovernmental Panel on Climate Change (IPCC)—to assess environmental impacts on human health, ecosystem quality, and resource depletion. Eco-Indicator 99 measures impact across multiple environmental categories, CED calculates the total energy demand throughout the tire's life cycle, and IPCC evaluates greenhouse gas emissions. Together, these indicators provide an overview of the tire's sustainability and help identify areas for improvement. The analysis covered various impact categories, such as climate change, respiratory effects, resource extraction, and biodiversity loss. The findings showed that the use phase is the most energy-intensive, while the production phase significantly contributes to resource depletion and toxic emissions.

Key findings:

- *Impact categories:* Major impact categories identified included fossil fuel depletion (i.e., fossil fuel consumption, contributing around 66%), respiratory inorganics (contributing 22%), and climate change (contributing 9%), particularly prominent during the use and production stages. Production activities, driven by materials like rubber and steel, account for over 80% of potential negative impacts due to the intensive processes involved.
- *Environmental improvements:* Extending the lifespan of a tire can significantly mitigate environmental impacts by reducing the need for new production and minimizing waste. This approach could lessen emissions and resource consumption globally.
- *Data quality and validation:* The study emphasized the importance of data accuracy and reliability, employing data quality indicators (DQIs) to assess the dependability of inputs. Standardized software like SimaPro was used for data processing, including simulations to account for variability in environmental impact results.

In summary, this case study of a car tire's life cycle underscores the substantial environmental effects associated with material use, energy consumption, and emissions throughout its life span. The eco index for the manufacturing, use, and recycling of car tires was measured at approximately 153 points, with a standard deviation of 37.2. This high variation suggests a need for improved methods to reduce the environmental impact of tires. Enhancing energy efficiency in recycling systems and lowering energy consumption during manufacturing and use phases could help achieve this. Manufacturing alone contributes around 20% of total energy use. Reducing energy consumption during vehicle operation is also important, making fuel-efficient, smaller vehicles a better choice. These vehicles could benefit from lower taxation compared to larger, less fuel-efficient ones. By analyzing each stage, the LCA provides actionable insights for reducing the tire's ecological footprint, particularly through strategies like extending tire life and improving recycling processes. This research highlights the value of LCA in promoting sustainable practices in product design and management.

9.4 SURFACE ENGINEERING TECHNIQUES

Surface engineering addresses a critical aspect of material performance, given that a significant majority—about 70% to 80%—of material failures initiate at the surface where the material meets its environment. This high incidence of surface-originated damage highlights the need for effective surface modification techniques to reduce degradation. The goal of surface engineering is to adjust surface characteristics, such as texture and roughness, to enhance properties like wear resistance, erosion protection, and corrosion resistance. By refining these surface qualities, materials gain improved durability and functional performance, directly impacting their longevity and reliability in various applications.

9.4.1 Understanding the surface

To fully appreciate surface engineering, it is essential to clarify what we mean by "surface," as definitions vary across scientific sources. Some literature defines the surface as only the top 2 to 10 atomic or molecular layers, approximately 0.5 to 3 nm thick, while other sources consider any layer under 100 nm as the surface, with anything thicker categorized as bulk material. However, if surface roughness itself measures around 1 μm, defining the surface only by the atomic or nanometer-scale layers may not capture all critical factors.

In our approach, we take a broader view: The surface is the layer where the composition and properties differ from those of the underlying bulk material, especially regarding chemical composition and structure. This surface layer could extend up to 50 μm and serves as the first line of defense against environmental factors, playing a key role in durability, wear resistance, and overall performance. For instance, a thin layer can be applied to a substrate to provide protective or functional properties. Common surface engineering techniques, such as physical vapor deposition (PVD), chemical vapor deposition (CVD), electroplating, and thermal spraying, are used to add coatings that enhance surface characteristics. These methods are fundamental in surface engineering and help tailor material surfaces to meet specific performance needs.

9.4.2 Understanding surface engineering

Understanding the nature of a surface sets the stage for engineering it to achieve specific, enhanced properties. Surface engineering is the process of modifying critical qualities of the surface layer—such as chemical composition, morphology, and atomic structure—to reduce friction, wear, corrosion, and erosion. These modifications improve material performance and extend durability, making surface engineering essential across various industries. This field requires extensive research and continuous improvement to tailor surfaces for demanding applications.

The ultimate goal is to create a protective interface between the material and its environment, maintaining functionality while minimizing environmental degradation. Surface engineering techniques can operate on atomic, molecular, or microstructural scales, from 0.5 to 3 nm for atomic layers to 100 nm (0.1 μm) for microstructural modifications. This approach allows for the precise adjustment of properties such as hardness, strength, and corrosion resistance.

To achieve these enhancements, surface engineering employs three primary strategies: coating, patterning, and treatment. Coatings add layers or subsurface structures as thin as 100 nm or a few microns to impart desired qualities. Surface patterning, achieved through techniques like lithography and etching, creates textures or structures that provide unique properties, such as antireflective or superhydrophobic effects. Surface treatments—including polishing, shot peening, anodizing, and plasma treatment—refine topography and roughness, optimizing surfaces for specific applications.

Further modification techniques, such as ion implantation, laser surface treatment, and diffusion, adjust the surface's chemical and physical composition, enhancing hardness, adhesion, and resilience. Together, these strategies in surface engineering make it possible to create tailored, high-performance surfaces that withstand harsh environments and deliver long-term functionality.

9.4.3 Understanding surface roughness

Surface roughness, as defined by ISO 4287, refers to the small-scale irregularities on a surface, often at the microscopic level, which typically contribute to most surface damage. These irregularities, visible as peaks and valleys at the micron or nanoscale, represent the extent to which a surface deviates from an ideally smooth profile. Common causes of roughness include scratches, tool marks, pits, and grooves. Such flaws can lead to stress concentrations (as real contact area due to roughness reduced to 1% to 10% of the apparent area), crack initiation, and accelerated corrosion, ultimately impacting material durability.

Surface roughness can be analyzed using both 2D and 3D profiles. While a 2D profile may aid in conceptual understanding, a 3D profile is typically preferred for a more complete representation, as it captures the 3D nature of surface interactions. Rougher surfaces show greater deviations from the mean line, reflecting higher roughness levels. A 3D view is especially useful for digitization and data-driven analysis, providing a detailed picture of surface characteristics. The choice of stylus and profilometer also affects accuracy and cost. Basic models may be more economical, whereas high-precision, 3D-capable profilometers with advanced features tend to be significantly pricier. The selection ultimately depends on the required accuracy and specific parameters needed for design or analysis.

Profilometers, which come in both contact and noncontact forms, accurately measure these surface irregularities, helping to control surface quality and improve material performance and longevity. Figure 9.6 illustrates this concept with a steel cube displaying visible surface roughness on three faces—a common result of wear or minimal finishing.

A contact stylus profilometer is commonly used to measure surface roughness. This device, featuring a diamond-tipped stylus, moves horizontally across the surface in a single direction, recording vertical distances as it encounters peaks and valleys. In practical applications, surface asperities often undergo plastic deformation, which can impact roughness measurements. Figure 9.6 illustrates a stylus moving from its starting to ending position, capturing roughness along the path. The accuracy of these measurements is influenced by the radius of the stylus tip: a sharp, pinpoint tip provides a more precise reading of the surface's true profile, while a larger-radius tip may miss finer details, reducing precision. As it traverses the surface, the stylus records height variations, building a comprehensive profile of surface roughness. The resulting profile, often represented by a red line, traces the surface's contours, illustrating the peaks and valleys encountered along the way. As the stylus travels across the surface, it measures key parameters, such as:

- R_{max}: The maximum peak-to-valley height within the evaluation length.
- R_t: The total height of roughness, representing the vertical range of irregularities from the highest peak to the lowest valley, within the evaluation length.
- *Evaluation length:* The distance over which the roughness is measured.

In Figure 9.6, these parameters are illustrated, with additional labels such as the *direction of travel* for the stylus, which moves across the surface to capture the profile. The profilometer records the vertical movement of the stylus, providing essential data on surface roughness that helps quantify surface quality and assess durability. ISO 4287 specifies various roughness parameters, commonly used to describe surface roughness quantitatively:

1. *Ra:* Arithmetic mean roughness: The average deviation of the roughness profile from the mean line. Ra is calculated as the arithmetic average of the absolute values of all measured profile heights within a *sampling length*.
2. *Rz:* Ten-point mean roughness (Rz) is calculated as the average of the absolute values of the five highest peaks and the five lowest valleys

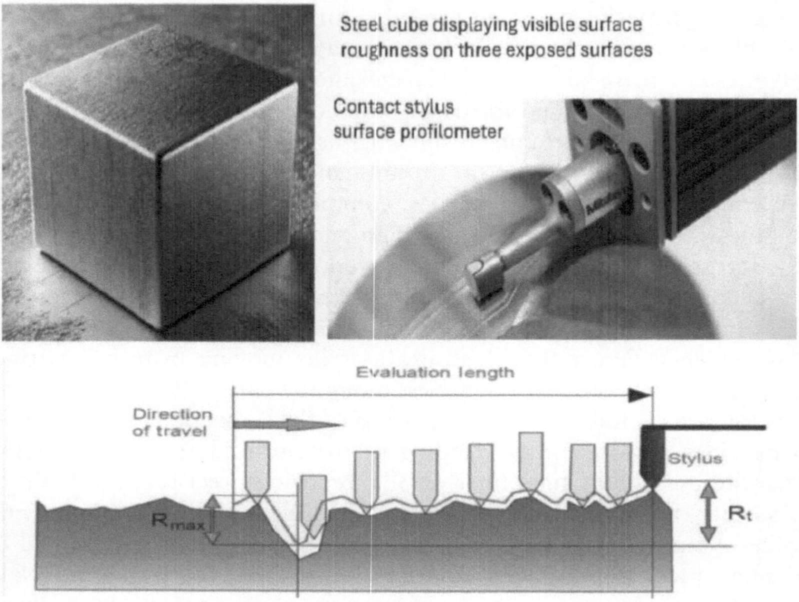

Figure 9.6 Surface roughness measurement using a contact stylus profilometer on a rough metal cube.

within the evaluation length. This metric provides an indication of the height difference between the highest peak and lowest valley within the evaluated section, offering insight into surface texture variations.

3. *Rq:* Root mean square roughness (Rq) is calculated as the root mean square of profile height deviations from the mean line within a sampling length. This measure provides an indication of the overall variation in surface roughness.
4. *Rt:* Total height roughness (Rt) represents the vertical distance between the highest peak and lowest valley within the evaluation length. Rt provides insights into the extreme height differences on the surface.
5. *Rp:* Maximum peak height (Rp) within the evaluation length represents the tallest peak on the surface, providing insight into the surface's highest elevation.

These parameters are measured in micrometers (µm) or nanometers (nm) and offer a detailed quantitative evaluation of surface roughness. It is important to note that, in previous sections, we referred to two distinct

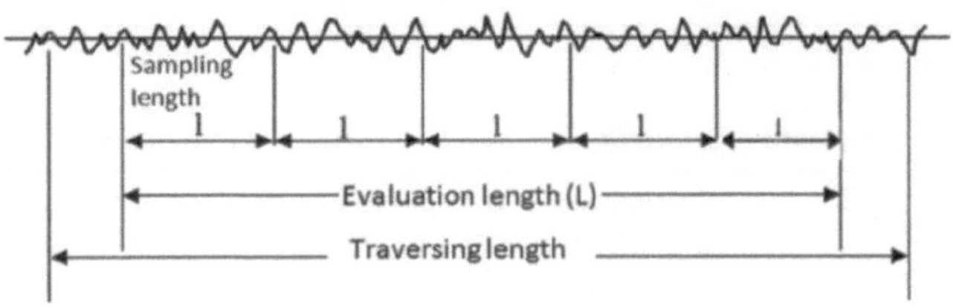

Figure 9.7 Important lengths in the analysis of surface roughness.

measurement lengths: "sampling length" and "evaluation length." The distinction between these two lengths is illustrated in Figure 9.7.

Figure 9.7 represents a surface roughness measurement profile with:

1. *Sampling length:* Often labeled as "*l*" representing individual sampling lengths. These are short segments along the surface over which roughness measurements are taken to gather localized data.
2. *Evaluation length (L):* This is the cumulative length (L = 5l) used for calculating surface roughness. It comprises five sampling lengths combined to provide statistical mean values for Ra and Rq.
3. *Traversing length:* This is the total distance traversed by the profilometer's stylus or measuring device as it moves along the surface to capture the roughness profile. It includes extra lengths before and after the evaluation length. For accurate calculations, the first and last portions of the sampling lengths within the traversing length are discarded to ensure stable and precise measurements across the area of interest.

According to the ISO 4287 standard, accurately measuring surface roughness is not as simple as reading off a single number, like the average roughness (Ra). Instead, it is essential to specify the "sampling length," which is the length of the surface over which the roughness measurement is taken. The sampling length directly affects the roughness values you get because surface irregularities might vary across different areas of a surface. When measuring surface roughness, using a single arbitrary sampling length could lead to inconsistent or misleading results. For instance, if you measure a small area of a surface with a very fine texture, you might get a different roughness value than if you measured a larger area that includes coarser sections. Therefore, the ISO standard specifies an appropriate sampling length based on the average roughness (Ra) of the surface.

Table 9.3 presents surface roughness grades along with the corresponding sampling length, roughness average (Ra), surface finish characteristics, and suitable surface treatments. The process of determining the correct sampling

Table 9.3 Surface roughness grades with corresponding sampling length

Grade	*Sampling length (mm)*	*Ra (µm)*	*Surface finish features*	*Surface treatment*
N12	8	50	Obvious visible tool processing mark	Rough turning, rough milling, rough planning, drilling, and rough grinding wheel processing
N11	8	25	Visible tool processing mark	Rough turning, rough milling, rough planning, drilling, and rough grinding wheel processing
N10	2.5	12.5	Minor tool marks	Rough turning, rough milling, rough planning, drilling, and rough grinding wheel processing
N9	2.5	6.3	Visible processing mark	
N8	2.5	3.2	Minor processing mark	Fine turning, fine milling, and rough grinding
N7	0.8	1.6	Invisible processing mark	
N6	0.8	0.8	Recognizable direction of processing traces	
N5	0.25	0.4	Minor-recognizable direction of processing traces	Grinding and polishing
N4	0.25	0.2	Invisible direction of processing traces	Grinding and polishing
N3	0.25	0.1	Dark glossy surface	Fine grinding, mirror polishing, and super fine grinding
N2	0.25	0.05	Shiny glossy surface	Fine grinding, mirror polishing, and super fine grinding
N1	0.08	0.025	Mirror glossy surface	Fine grinding, mirror polishing, and super fine grinding

length is iterative; after an initial measurement, adjustments to the sampling length may be necessary, either increasing or decreasing it depending on the observed roughness, until the appropriate length that accurately represents the surface is identified.

9.4.4 Importance of surface texture modifications

Surface texturing, such as microdimples organized in arrays on surfaces such as piston rings and thrust bearings, has demonstrated the ability to reduce friction and wear, extending component service life. Textured surfaces serve three essential roles in lubricated applications: debris trapping, lubricant reservoirs, and acting as microbearings. To manage the complexities associated with surface roughness, surfaces are often designed with intentional textures. This deliberate control of surface texture to achieve specific roughness levels is known as "surface roughness design." Achieving the desired texture involves selecting appropriate manufacturing methods, equipment, and abrasives, along with thorough testing to ensure that the surface meets functional requirements. Adjusting surface texture is critical because it directly influences several functional characteristics, such as:

- *Friction reduction:* Textured surfaces can lower friction between interfaces, which is particularly valuable in sliding or rolling contact applications such as bearings, gears, and other tribological components.
- *Wear resistance:* Texturing enhances wear resistance by increasing surface area, enabling more uniform load distribution, and reducing wear in applications subject to mechanical stresses.
- *Lubrication retention:* Surface textures create pockets or channels that retain lubricants, reducing friction and wear by enabling continuous lubrication.
- *Hydrophobicity:* Certain textures create hydrophobic surfaces, useful for applications requiring water repellence, such as marine coatings, aircraft, and structures exposed to water and ice.
- *Optical characteristics:* Texturing can modify a material's reflectivity and diffusion properties, useful in optics and photonics applications.

To achieve these tailored surface properties, several surface texturing and roughening techniques are available, each suitable for different types of materials and desired outcomes. The key techniques include:

- *Mechanical machining:* Techniques like milling, turning, grinding, computer numerical control (CNC), and sandblasting create patterns suitable for larger-scale textures, though precision at the micro- or nanoscale is limited.
- *Chemical etching:* This method uses chemical agents to selectively remove material from the surface. By adjusting the composition and concentration of the etchant, micro- and nanoscale textures can be generated with high precision and repeatability.
- *3D printing (additive manufacturing):* Textures can be integrated directly into 3D-printed components by altering the printing process, offering design flexibility and enabling the creation of complex, customized textures.
- *Laser ablation:* Concentrated laser beams remove material to create textures. By adjusting laser intensity, a range of patterns, shapes, and depths can be achieved, making it ideal for precise micro- and nanoscale texturing across materials such as metals, ceramics, and plastics.

- *Electrochemical etching:* Using electrical current in combination with a chemical solution, this technique precisely removes material to create micro- and nanoscale patterns with consistent depth and shape, offering excellent precision and repeatability.

9.4.5 Discussion on surface texturing techniques

Arslan et al.[4] discussed key features and roles of surface texture in engineering, particularly in improving tribological performance and enhancing cutting tool longevity and effectiveness. They also reviewed the surface texturing manufacturing techniques.

These surface texturing methods allow for customized textures, which improve performance in specific applications and increase durability in challenging environments. The main differences between surface texturing methods come down to scale and precision. Mechanical techniques like milling, turning, grinding, and sandblasting can create surface patterns but are limited in achieving nanoscale or microscale textures. For high precision, chemical etching is often preferred, as it selectively removes material using a mask and etchant, creating detailed patterns and textures while even altering the surface's chemical composition by adjusting etchant concentration. This method is ideal for producing micro- and nanoscale textures with high accuracy, though some surface deviations may still require post-processing.

With chemical etching, we can precisely control how much material is removed, like 25%, 30%, or 50%. This raises the question: Why not use additive manufacturing to build material only where it is needed, reducing waste? Considering LCAs, it is clear that minimizing waste and optimizing material use is important.

Additive manufacturing appears to offer a better alternative, although further development is needed to meet the accuracy and precision required. In the future, 3D printing could create surfaces with a high-quality finish at the nanometer or micron level more quickly. Features as fine as 5 μm (compared to 5 mm) are achievable with additive manufacturing if high-quality CNC machines with strong system stiffness are used and if the deposited material adheres well to prevent delamination. Strong adhesion can overcome many limitations in current additive manufacturing processes, making this technology promising for the future of surface engineering.

Additive manufacturing provides design flexibility, allowing changes in materials, chemical composition, and temperature to create customized textures, even on complex surfaces. While these are significant advantages, a fully reliable additive manufacturing technology is still under development and is likely to emerge soon.

Laser ablation, a widely used texturing method, employs high-energy pulses to melt and vaporize material rapidly. This process includes two mechanisms: pyrolytic and photoelectric. In the pyrolytic process, the material is instantly melted and vaporized, while in the photoelectric (or cold ablation) process, only chemical bonds are broken, minimizing heat-affected zones and preserving material properties. Although laser texturing is precise, it can still cause microbursts, which may require post-processing, such as grinding or polishing, for a smooth finish. Future advancements in laser additive manufacturing could potentially increase precision, reducing the need for additional polishing. Currently, laser texturing can be performed with or without a mask; however, mask-based methods may still require polishing to achieve a refined surface.

Laser micromachining is increasingly applied for surface texturing in tools, bearings, and biomedical implants. Another technique, ultrasonic machining (USM), is effective for brittle materials like ceramics and glass. USM involves a tool oscillating at ultrasonic frequencies to propel abrasive particles against the surface, shaping it by abrasive impact. This process is often used in CNC setups, where the abrasive particles can be collected and reused.

Rotary USM is a cost-effective alternative that combines CNC-controlled USM with diamond grinding. This hybrid approach delivers high precision at a lower cost compared to techniques like electric discharge machine (EDM) or laser machining.

Electrochemical machining (ECM) removes material through anodic dissolution in an electrochemical cell, where the resistance of the electrolyte affects the surface finish. Often mask-based, ECM applies a photoresist layer to selectively dissolve material and create texture. When paired with a scanning tunneling microscope (STM), ECM achieves atomic-level precision, although this approach can be costly. ECM's benefits include low production costs, high efficiency, and minimal tool wear, making it ideal for precise applications.

The future of manufacturing will focus on "functional surfaces" that improve moisture absorption, friction, wear resistance, and optical properties through carefully designed texturing. By optimizing surface design, the lifespan of components can be extended from 5000 hours to 15,000 or even 50,000 hours, illustrating the value of surface engineering. In tribology and lubrication, textured surfaces serve three important functions: trapping debris to reduce wear, acting as lubricant reservoirs, and creating localized pressure pockets that improve load distribution by functioning as microbearings.

9.4.6 Case study: Tribological characteristics of textured surfaces

In the present section, a case study covered by Ali et al.,[5] which investigates the impact of microdimple geometry on friction and wear under various conditions, is presented. This study focuses on the tribological properties of textured surfaces, particularly how microdimple geometries influence performance in lubricated conditions. The research provides insights into optimizing surface textures to improve frictional performance, reduce wear, and enhance overall durability.

The primary goal of the study is to enhance the tribological properties (friction and wear resistance) of cylindrical Al6061-T6 alloy workpieces by applying different microdimple textures. The research examines how specific geometric parameters (shape, size, and depth) of these textures affect friction and wear during lubricated sliding.

9.4.6.1 Methodology

The study uses a "rotating cylinder-on-pin" tribological test setup in "starved lubrication" conditions, where lubrication is limited. This setup simulates high-pressure, near-dry conditions like real-world mechanical applications. Key elements of the methodology include:

- *Surface texturing:* Textures are produced using a dual-frequency resonant vibration method that creates hierarchical microdimples on the surface. These textures are precisely crafted using a polycrystalline diamond (PCD) cutting tool mounted on a CNC lathe.
- *Materials:* Al6061-T6 alloy was selected for its favorable formability and strength-to-weight ratio, despite having poor tribological properties, making it ideal for texturing experiments.
- *Vibration-assisted texturing:* High-frequency (16.3 kHz) 3D vibration combined with low-frequency (230 Hz) 1D vibration creates diverse microdimple surface textures.
- *Tooling:* A PCD cutting tool with a 400 μm nose radius and 7 degrees rake angle was used, resulting in an initial surface roughness of approximately 0.5 μm. The study used various texture depths (e.g., 14 to 17 μm) to identify optimal configurations.
- *Friction testing:* The tribological performance of textured surfaces was assessed using a pin-on-disk setup where a stationary AISI 52100 steel pin contacts the rotating cylinder. Real-time friction coefficient readings were recorded as normal load and sliding speed varied.
- *Stribeck curves:* These curves were generated to analyze transitions between different lubrication regimes and evaluate the effectiveness of textured surfaces under various loads and sliding speeds.

9.4.6.2 Results

The study highlights how different microdimple geometries impact friction and wear characteristics:

- *Effect of load on friction:* At lower loads (10 N), textured surfaces exhibit a higher friction coefficient than untextured surfaces. Conversely, at higher loads (50 N), textured surfaces show reduced friction, attributed to the dimples' ability to retain and release the lubricant under pressure.
- *Impact of microdimple size:* Smaller dimples yield better friction characteristics under high load and high-speed conditions, as larger dimples do not fully engage within the contact zone, limiting hydrodynamic lift and lubricant flow.
- *Hierarchical structures versus single-scale textures:* Hierarchical microdimple structures generally outperform single-scale textures in reducing wear, particularly in high-load, high-speed conditions. These hierarchical textures enhance hydrodynamic lift, maintain lubrication, and reduce wear by effectively trapping worn particles. At lower sliding speeds, single-scale microdimples with small sizes showed lower friction coefficients than hierarchical structures. However, as the sliding speed increased under heavier loads, hierarchical microdimples demonstrated lower friction coefficients compared to single-scale textures. The wear rate analysis further revealed that small microdimples exhibited higher specific wear rates due to frequent stress cycles and lubricant film breakdown in conditions with limited lubrication. In contrast, larger microdimples showed a lower specific wear rate due to reduced surface contact. Hierarchical microdimples also achieved lower specific wear rates than single-scale ones, benefiting from improved lubricant flow into the contact area, which helped minimize wear under increased loads. In summary, hierarchical microstructures in microdimples show significant potential for enhancing frictional performance and reducing wear rates, particularly at certain sliding speeds and load conditions.

9.4.7 Discussion on surface texturing

Surface texturing must be tailored to specific operating conditions, as no universal approach exists for minimizing surface degradation. Each scenario requires a detailed analysis to determine the optimal texture pattern. Adjusting the texture dimensions can enhance oil retention, though results vary based on lubricant properties. For instance, textures that perform well in aluminum may be less effective with steel due to higher contact forces. Thus, thorough research is essential to understand the interactions between textures and materials, and pilot studies are often necessary to identify the ideal pattern for each application. By carefully adjusting microdimple geometry, engineers can reduce friction, increase wear resistance, and improve mechanical efficiency, especially in high-friction environments.

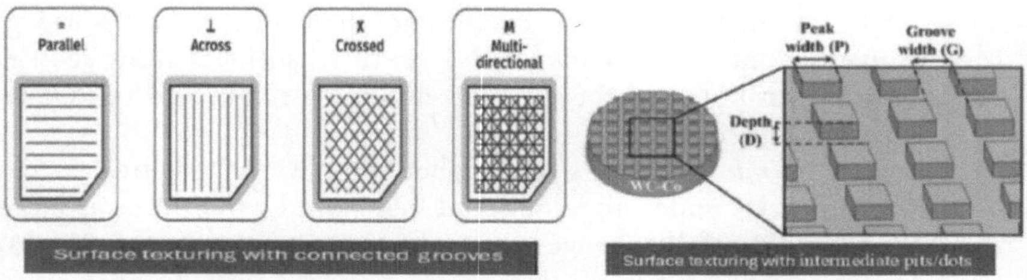

Figure 9.8 Types of surface texturing: connected grooves and intermediate pits/dots.

Surface textures are broadly categorized as grooves or intermediate pits/dots (as illustrated in Figure 9.8). Disconnected dimples, which are separate from each other, often reduce wear more effectively than connected grooves. Grooved textures may allow lubricant to escape more easily, leading to higher wear rates. By embedding solid lubricants within disconnected textures, friction and wear can be significantly reduced as the lubricant adheres to the contact surface, forming a dynamic film. Textures with embedded lubricants show improved retention within the texture compared to smooth surfaces. For example, MoS_2-filled channels enhance wear resistance by trapping lubricant within the texture.

Connected textures like grooves, however, tend to allow the lubricant to flow out of the contact area, reducing lubrication effectiveness and hydrodynamic pressure. Instead, cross-patterned or perpendicular textures can better retain lubricant and lower friction.

Different texture types, such as pits, dots, and banded grooves, each have distinct widths and depths, influencing their performance. For instance, dot-type textures reduce friction and normal force compared to untextured surfaces but are less effective in retaining lubricant than pit-type textures. Adding perpendicular patterns further improves lubricant retention and friction reduction.

9.5 SURFACE COATINGS

Coating is widely used across many applications, with most components requiring some form of it. In some cases, however, the coating serves an aesthetic purpose rather than a functional one. For example, a coating applied purely for color or pattern, such as a pink hue or floral design, is decorative. Additionally, coatings may be used for reconditioning or restoring the appearance of a component, enhancing its surface without altering its functional properties. In such instances, the objective might be simply to achieve a clean surface, bridge pores, or reduce surface roughness.

Surface coating is a process in which a polymer, metal, or ceramic powder is applied in layers to a substrate to create a protective barrier. This barrier, with thicknesses ranging from a few nanometers to several microns, shields the substrate from various types of surface degradation, including wear, corrosion, erosion, tribo-corrosion, and, to some extent, high-temperature damage. The substrate, or the underlying material, serves as the surface when uncoated; however, once coated, the coating itself becomes the new surface. A key factor in coating effectiveness is strong adhesion between the coating and substrate, often achieved with an interlayer to ensure a durable bond. This robust adhesion enhances the coating's longevity and overall durability, providing numerous benefits for protecting and extending the life of materials. Surface coatings offer the following notable benefits:

- *Cost-efficiency:* Coatings can reduce production costs by eliminating the need to fabricate new parts, as they allow damaged surfaces to be restored. A key objective of coating is to keep products within the manufacturing cycle, lowering expenses by enabling materials to be refurbished and reused rather than relying on new raw materials. This approach not only reduces costs but also conserves natural resources. By reconditioning materials, coatings help avoid the cost of producing new parts. Advanced coatings, such as PVD and CVD thin films, also open up opportunities for improved functionality with multilayer applications. Today, coatings just a few microns thick can renew older components and systems, making them ready for reuse. These advancements highlight the importance of applying coatings thoughtfully to maximize their benefits.
- *Durability:* Coatings provide extensive options for long-term protection, enhancing the functionality and lifespan of components even under harsh operating conditions.
- *Enhancing metal properties:* Coatings can increase surface hardness, improve wear and corrosion resistance, and modify wettability and the ability to impart hydrophobic or hydrophilic properties.
- *Shifting toward surface-level solutions:* Another important consideration is that newly developed materials, especially nanomaterials, offer advanced coating options that can significantly extend the lifespan of components. For example, components previously designed to last 5 years can now last 10 or even 15 years with the right coating. Coatings are therefore essential for achieving these extended durability goals. Interestingly, there is a shift from 3D systems back to 2D, as many processes now primarily occur at the surface level. In today's applications, full 3D systems are often unnecessary because surface-level processes meet most functional needs. By focusing on high-quality

surface coatings, the choice of substrate material becomes less critical, as even older or less durable materials can gain enhanced properties like corrosion resistance and extended longevity. This allows materials with otherwise limited lifespans to remain in use for years.

9.5.1 Surface coating techniques

A variety of surface coating techniques are available to protect components from mechanical and chemical degradation, with each method offering unique characteristics suited to specific applications and environments. Key techniques include PVD, CVD, thermal spraying, sol–gel processing, micro-arc oxidation, and electrodeposition. Each method provides specific benefits and may have limitations, necessitating careful selection based on the intended application.

These coatings can be made from a broad range of materials, including metallic alloys, ceramics, bio-glasses, polymers, and engineered plastics, allowing engineers and designers ample flexibility to create durable, protective surfaces. By choosing the appropriate combination of coating materials and methods, the performance, durability, and resilience of components across various industries can be significantly enhanced. One primary advantage of coatings is their potential to reduce manufacturing costs, as they eliminate the need to produce new parts by restoring or protecting existing surfaces. The choice of coating technique ultimately depends on the functional needs of the application, with a range of materials and methods available to deliver tailored, long-lasting protection.

There are some coatings (which will not be part of this book), named "organic compound coatings," such as paints, varnishes, and epoxy resins, which are applied in liquid form, allowing the film to spread across surfaces through intermolecular forces like surface tension. Upon exposure to environmental triggers, such as atmospheric oxygen or ultraviolet light, a polymerization process occurs within the liquid film, transforming it into a solid, durable layer. These coatings, made specifically from organic compounds, are popular for their ease of application, flexibility, and adaptability to various surfaces, making them effective for protecting against environmental damage. Unlike common surface coatings, which can also include inorganic materials like metals, ceramics, and polymers and may use advanced application techniques (e.g., PVD, CVD, thermal spraying), organic coatings are particularly valued for applications requiring a strong yet flexible barrier with simpler application methods.

9.5.2 Customizing coating performance

To meet specific requirements to prevent surface degradation, a coated surface must possess a well-balanced combination of properties, including

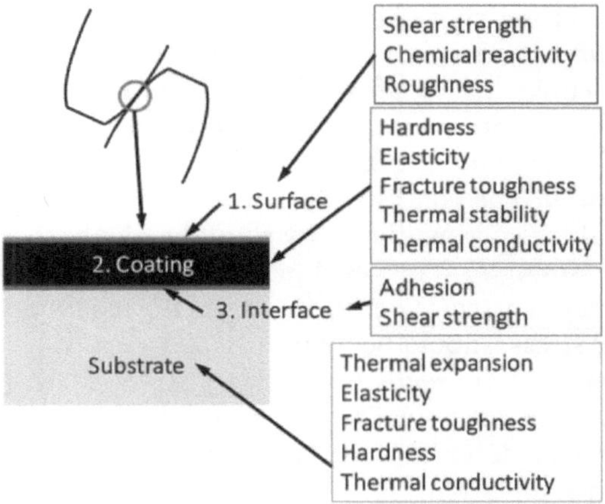

Figure 9.9 Important properties in different zones of coated surface.

hardness, elasticity, shear strength, fracture toughness, thermal expansion, and adhesion. Figure 9.9 illustrates four distinct zones within the coating system, each with unique properties that must be carefully considered. The desired characteristics of both the substrate and coating are influenced by factors such as material composition, microstructure, porosity, and homogeneity. At the interface between the substrate and coating, adhesion and shear strength play critical roles, while at the coating surface, additional factors like chemical reactivity, roughness, and shear strength are significant.

Figure 9.9 outlines a layered structure of a coated material, highlighting the properties specific to each layer within the surface coating system:

- *Surface layer requirements*
  - *Shear strength:* The surface's resistance to shear forces should be minimal.
  - *Chemical reactivity:* The tendency of the surface to interact chemically with the environment or other materials should be minimal.
  - *Roughness:* Surface texture, which influences friction, adhesion, and overall performance, should be carefully designed.
- *Coating layer requirements*
  - *Hardness:* The ability of the coating to resist deformation and scratching should be high.
  - *Elasticity:* The coating's capacity to elastically deform and return to its original shape should be high.
  - *Fracture toughness:* The coating's resistance to crack propagation under stress should be high.

- *Thermal stability:* The ability to maintain properties at high temperatures should be high.
- *Thermal conductivity:* The efficiency of heat conduction through the coating, which affects heat dissipation and performance in high-temperature environments should be high.

- *Interface layer requirements*
  - *Adhesion:* The bond strength between the coating and substrate, crucial for coating durability should be high.
  - *Shear strength:* The resistance to shear forces at the coating-substrate interface, ensuring that the coating stays attached under stress, should be high.
- *Substrate layer requirements*
  - *Thermal expansion:* The degree to which the substrate expands when heated, which must be compatible with the coating to prevent delamination.
  - *Elasticity:* The substrate's ability to elastically deform under load should be reasonable and adjustable to the coating layer.
  - *Fracture toughness:* The resistance of the substrate to cracking under stress should be high.

Each layer in a coating system provides unique properties that enhance the overall durability, strength, and resistance of the material against environmental and mechanical stresses. Achieving optimal performance requires a careful balance of these properties, as certain characteristics may conflict. For instance, strong adhesion at the coating/substrate interface is essential, but it may also increase surface interactions, and while high hardness is often desired, it can reduce both toughness and adhesion. Effective coating design must therefore weigh these technical requirements alongside economic considerations related to the deposition process.

The substrate surface that interfaces with the coating should ideally have some roughness to enhance bonding and ensure lasting durability. In contrast, the outer coating surface, exposed to the environment, should be smooth to optimize performance and longevity. Thermal compatibility is also vital; significant differences in thermal expansion between the substrate and coating can lead to delamination and premature detachment. Matching the thermal expansion of both layers, along with similar elasticity, high fracture toughness, and appropriate hardness, creates a strong, stable bond. If the coating is considerably harder than the substrate, it may cause depressions in the substrate, increasing the risk of coating fractures.

Thermal conductivity is another important factor; if the coating has a conductivity compatible with the substrate, it will better support the component's overall functionality. Compatibility at the interface is also essential, especially if an additional layer is included, as this layer should promote adhesion and possess high shear strength to prevent detachment.

Strong adhesion and high shear strength at the interface are crucial, ideally with the interface having higher shear strength than either the coating or substrate individually. This ensures that the coating remains intact even under stress—it may puncture, but it will not detach from the substrate.

The coating should also possess high hardness, elasticity, fracture toughness, and thermal stability, withstanding elevated temperatures (400°C to 500°C or more) and matching the substrate's thermal conductivity. While high shear strength is needed at the coating-substrate interface for strong adhesion, the coating's outer surface should have low shear strength to minimize friction and reduce wear. Balancing these opposing needs is key to effective coating performance.

Additionally, minimal chemical reactivity is necessary to prevent unwanted reactions with the environment or nearby materials, which can accelerate wear. A high-quality coating should also offer a smooth finish, reducing the need for further finishing operations and providing low roughness, low chemical reactivity, and low shear strength for improved performance and durability.

9.5.3 Classification of coating methods by state

Surface coating based on their manufacturing methods is broadly classified into four main types based on the state of the coating material:

- *Gaseous-state coating:* In gaseous-state coating, the coating material is introduced in a gaseous form and deposited onto the substrate surface as atoms, molecules, or ions. "PVD" is a prominent technique within this category and includes processes such as evaporation and sputtering. During PVD, the material is vaporized from a solid source and condenses onto the substrate to form a thin, controlled film layer. PVD coatings are valued for their precision in thickness and composition and are widely used for thin film deposition. Sometimes these "gaseous-state coatings" are termed as "solid-state coating" as the process involves applying the coating material in a solid or powdered form.
- *Solution-state coating:* Solution-state coating involves dissolving the coating material in a liquid solution, which often includes resins, pigments, solvents, and additives, before applying it to a surface. These coatings are applied through chemical or electrochemical processes, with common techniques like chemical solution deposition and electrochemical deposition (similar to electroplating, where material transfers from the anode to the cathode). Solution-state coatings are versatile and allow for the creation of uniform layers with varied compositions, making them ideal for applications

requiring corrosion protection or decorative finishes. A few common methods in this category are:

- *Spray coating:* The liquid solution is sprayed onto the substrate using air or electrostatic spraying.
- *Dip coating:* The substrate is immersed in the solution and then withdrawn, allowing excess solution to drain.
- *Spin coating:* A rotating platform spreads the coating solution uniformly due to centrifugal force. This technique allows for the easy application of coatings on a variety of surfaces, with control over thickness and uniformity.

- *Molten-state coating:* Molten-state coating methods apply the coating material in a liquid or molten state, commonly used for metal coatings. These techniques are highly effective for thick and durable coatings, often used in industrial applications requiring high wear resistance. Key processes include:
  - *Hot-dip coating:* The substrate is immersed in a bath of molten coating material, which adheres upon cooling.
  - *Flame spraying:* A heat source melts the coating material, which is then propelled onto the substrate by compressed air or gas.
- *Semi-molten-state coating:* In semi-molten-state coating, the material is partially melted and applied to the surface, making it pliable but not fully liquid. The partially melted coating material, often in powder or wire form, is propelled onto the substrate using a heat source termed "thermal spraying." The material solidifies upon impact. This category includes methods like "flame spraying," "arc spraying," and "plasma spraying." Semi-molten coatings are suitable for creating thick, robust layers that provide excellent wear and corrosion resistance.

Each method has its unique applications and advantages depending on the desired properties and characteristics of the coating. Table 9.4 compares various coating processes.[4]

9.5.4 Key coating techniques

- *CVD:* It is a widely utilized coating technique, particularly in the semiconductor industry, for creating high-quality, wear-resistant layers on substrates. This process occurs in a vacuum chamber, where volatile precursors are introduced and undergo a chemical reaction on the substrate's surface, resulting in a dense and adherent coating. CVD can deposit various materials, including carbides, nitrides, and carbon forms (such as fluorocarbons, diamonds, and polymers). Common CVD types include "atmospheric pressure CVD," "low-pressure CVD," and "ultra-high vacuum CVD," with

Table 9.4 Comparative characteristics of coating methods

	Gaseous state		*Solution state*			*Molten and semi-molten*
	PVD	*CVD*	*Chemical solution deposition*	*Electrochemical deposition*	*Sol-gel*	*Thermal spraying*
Deposition rate (kg/h)	0.5	1	0.2 to 0.45	0.1 to 0.5	0.1 to 0.5	0.1 to 10
Component size	Limited to chamber		Limited to solution bath			—
Control of deposit thickness	Good	Fair/good	Fair/good	Fair/good	Fair/good	Manual-variable, automated-good
Uniformity	Good	Very good	Fair/good	Fair/good	Fair/good	Variable
Bonding mechanism	Atomic	Atomic	Surface force			Chemical/mechanical
Distortion of substrate	Low	Can be high	Low	Low	Low	Low

"aerosol-assisted CVD," "plasma-enhanced CVD," and "photo-assisted CVD" offering further specialization. Unlike PVD, CVD can selectively coat specific areas, minimizing material waste, although its high operational temperatures limit its use on heat-sensitive substrates. Known for producing uniform and durable coatings, CVD is critical in industries like microelectronics, optics, and tooling, where precision and endurance are essential. However, the necessity of a controlled vacuum environment and the associated costs restrict CVD's use in high-performance applications where the quality and durability of the coating outweigh the expenses. In CVD, materials are heated to react chemically with other substances, forming a robust and long-lasting coating. This method is particularly valued in the semiconductor industry, where ultra-thin, high-quality coatings are essential, as on wafers for chips, providing high resistance. During the CVD process, byproducts from these reactions are removed via vacuum pumps, maintaining a clean and efficient environment. A notable advantage of CVD is its flexibility, allowing the deposition of a wide range of materials, including carbides, nitrides, oxynitrides, silicon oxides, and diamond-like carbon (DLC) coatings, which have become more cost-effective. Additionally, CVD can produce advanced coatings such as graphene, nanotubes, nanofibers, and various crystallinities, from monocrystalline to polycrystalline and amorphous forms. While the chamber size can limit larger applications, CVD remains ideal for smaller or complex shapes, offering coatings with diverse microstructures tailored for specific applications.

- *PVD:* It is a process that applies a thin, adherent coating to a substrate by physically transferring material within a vacuum environment. This technique creates a durable, wear-resistant finish, making it ideal for metal coatings used in tools, decorative items, and medical devices. Like CVD, PVD requires a vacuum chamber, which can limit the size of items that can be coated and increase costs. However, PVD is highly valued for its corrosion and wear resistance, as well as its ability to produce protective films on various substrates exposed to harsh environments. The process typically involves converting solid or liquid materials into a vapor phase, which then condenses into a dense, solid film. Key PVD methods include "sputtering" and "evaporation," often used to create multilayer coatings due to the thin nature of PVD films. Research indicates that the composition of PVD coatings significantly impacts corrosion resistance; for instance, Ti-based coatings provide superior protection compared to Zr-based coatings, particularly when oxygen content is increased. PVD has broad applications in the aerospace, automotive, biomedical, optics, and firearms industries, though applying it to polymers remains challenging due to the risk of polymer degradation during deposition. As an extension of PVD applications, titanium nitride (TiN) PVD coating is commonly applied to stainless steel equipment to enhance corrosion resistance, particularly in acidic environments where stainless steel alone is insufficient. This additional TiN layer provides robust, long-term protection against corrosion, extending the lifespan of the equipment. PVD coatings, such as TiN, are widely utilized across sectors, including aerospace, automotive, biomedical, and defense, due to their excellent hardness and resistance to corrosion and wear. These properties make PVD coatings invaluable for applications demanding durability in harsh conditions.
- *Chemical solution deposition:* It uses a liquid phase as a medium to transport coating precursors, allowing for lower-temperature applications than solid-state sintering. This technique is cost-effective and ideal for achieving uniform coatings. By adjusting solution parameters such as viscosity and pH, the film's morphology can be controlled, creating coatings that range from amorphous to crystalline. Chemical solution deposition (CSD) is especially useful for large or irregular surfaces, making it a preferred choice for protective and functional coatings in electronics and industrial applications where budget constraints are important. The process begins with a prepared solution, typically in a beaker, where precursors are mixed and then deposited onto the substrate. CSD's ability to deliver the

main material in liquid form, even at low temperatures, simplifies the process and eliminates the need for atomic or molecular conversion, while precise control over parameters like pH and concentration enables flexibility in the coating's final structure.

- *Electrochemical deposition:* It, also known as electrodeposition or electroplating, is a well-established and cost-effective technique for creating nanostructured coatings on conductive substrates. This process enables precise control over coating thickness, uniformity, and composition by adjusting parameters like electrolyte concentration, pH, applied current, and temperature. During electrodeposition, an electric field drives metal ions from the anode toward the substrate, forming a solid layer that can take on varied shapes and morphologies, such as pyramidal and rod-like structures. Widely used in industries like automotive, aerospace, and jewelry, it enhances corrosion resistance, durability, and aesthetic appeal. Notably, this method is time-efficient and economical, with no vacuum required. Advanced simulations in electrochemical deposition also enable accurate predictions of coating thickness based on current and material structure, ensuring consistent results. Careful handling of chemicals and equipment is essential for safety and quality control.
- *Sol–gel coating:* It is widely applied, especially in biomedical fields, to improve corrosion resistance and reduce ion release. The process begins by dissolving calcium–phosphorus precursors in a solution, which transforms into a gel upon heating. This gel can then be applied to substrates through methods like "dip coating," "spray coating," or "spin coating," producing a uniform layer with strong adhesion. Sol–gel coatings are versatile and customizable in chemical composition, making them suitable for complex geometries without requiring high temperatures or vacuum conditions. While cost-effective and adaptable, precise control over factors like dipping speed and drying rates is needed to ensure consistent thickness and prevent defects, especially in multilayer applications. Additionally, sol–gel coatings are valuable for repairing damaged or porous layers, as new layers can be applied directly without removing existing coatings. Precursor solutions, such as calcium phosphate in ethanol or distilled water, undergo gelation with heat, ultimately forming a stable, dry coating. This adaptability makes sol–gel coatings beneficial across diverse industries for both new applications and coating refurbishments.
- *Thermal spraying:* It uses heat sources such as plasma or electric discharges to melt coating materials, which are then sprayed onto surfaces to create protective layers. Renowned for their durability, these coatings provide effective resistance to corrosion and wear. Popular methods include "plasma spraying," "detonation

spraying," "high-velocity oxyfuel (HVOF) spraying," and "wire arc spraying." Thermal spray coatings can range from 20 μm to several millimeters in thickness, making them ideal for large substrates and applications requiring substantial protection. Commonly used in industries like aerospace, automotive, and heavy machinery, thermal spray coating is suitable for large or irregular surfaces, though skilled operation is essential to ensure uniform thickness and strong adhesion. Another aspect of "thermal spray coating" is its ability to convert liquid materials into solid coatings on substrates, with precise control over thickness and properties achieved by adjusting the air mass flow rate. This flexibility allows for the combination of multiple materials, including nanostructured ones, to enhance coating performance. Thermal spray coatings can range significantly in thickness, from as thin as 20 μm to several millimeters, depending on application needs.

As particles bombard the substrate, a layer-by-layer structure forms. A single pass often produces a 50-μm layer, and for thicker coatings, additional passes are added—two passes for 100 μm and three for 150 μm—enabling precise control over the final thickness. Unlike PVD and CVD, which require small vacuum chambers, thermal spray coating does not always necessitate strict vacuum conditions, though some vacuum may be used depending on the specific technique.

- *Dip coating:* It is a popular sol–gel method for creating thin films by immersing a substrate in a colloidal solution and then withdrawing it, allowing a coating to form as it dries through solvent evaporation and drainage. Additional heat treatment can further enhance the coating's properties. Dip coating is valued for its low cost, simplicity, and adaptability, making it suitable for large or irregular substrates. However, maintaining uniform thickness and avoiding edge defects can be challenging. Key parameters like withdrawal speed, sol composition, and solution viscosity are crucial for controlling coating quality and achieving desired structural characteristics. Widely used in mass production, dip coating is applied in industries such as electronics, medical instruments, and wire manufacturing, though it requires skill to ensure a smooth, consistent finish. Dip coating is a cost-effective coating method, where the substrate is immersed in a solution and then withdrawn, leaving a thin liquid layer on the surface. As the solvent evaporates, it deposits a coating layer on the substrate. This process is simpler and more affordable than advanced methods like PVD or CVD, but it may result in a slightly porous or less precise finish. The evaporation rate—and consequently, the coating quality—depends on the solvent's volatility; for example, ethanol evaporates quickly, leaving behind a uniform coating. While dip coating may lack the precision of more advanced techniques, it is an effective solution for applications where high accuracy is not critical.

9.5.5 Considerations for coating selection

Choosing the appropriate coating process and material involves assessing the extensive variety of available options. For specialized requirements, assigning weights to different criteria (as outlined in Chapter 6, Section 6.6.3 of this book) can help prioritize needs and guide the selection process. This structured approach allows for an effective solution by simultaneously considering the process, material, and specific requirements.

Selecting an appropriate coating requires careful evaluation of factors affecting performance, durability, cost, and environmental impacts. Key criteria include hardness for durability, minimized coating thickness to avoid interference with moving parts (particularly in components like piston rings), maximum operating temperature for reliability, cost for economic viability in mass production, and deposition efficiency, which influences both sustainability and production costs by reducing waste. Additional considerations include deposition rate, component size, thickness control, uniformity, bonding strength, and adhesion quality, along with the potential for substrate distortion. PVD typically achieves deposition rates around 0.5 kg/h, though often lower in practice, while CVD can nearly double that rate. Chemical deposition methods are slower but offer controlled application, whereas thermal spray coating reaches up to 10 kg/h with high-velocity guns, with fewer component size restrictions when manipulators or robots are used. Environmental controls, like closed chambers or protective equipment, are crucial to mitigate environmental impact. A comprehensive selection approach, supported by hybrid technologies and structured methodologies, enhances decision-making, resulting in optimized product performance and sustainability.

Distortion is primarily a concern with CVD due to substrate heating, which may cause deformation in temperature-sensitive materials and must be considered in the selection process. For further guidance, please refer to Holmberg and Matthews.[6]

Coating methods, as discussed in Section 9.4.4, differ significantly based on the specific needs of the substrate, desired properties, and operational environment. Each method also has unique requirements for equipment, expertise, and materials, which affect its cost and suitability for different applications. For example:

- *High-performance coatings:* CVD and PVD are excellent for providing wear resistance, adhesion, and corrosion protection, though their vacuum requirements make them more costly. Both processes involve evaporating the coating material and carrying it with an inert gas to deposit onto a nearby substrate, typically within a range of a few millimeters to centimeters. This approach is ideal for creating thin film coatings around 10 to 30 μm, with precise control over thickness and composition due to the vacuum environment. PVD does not require a chemical reaction, relying instead on electron guns to release atoms or molecules from the source material, which are then transported to the substrate by an inert gas. This method is effective for applying a wide range of materials, including titanium-based coatings, which provide strong corrosion resistance, even in highly acidic conditions with pH levels as low as 1 to 3. When a more cost-effective method is required, alternative techniques like dip coating, sol–gel coating, or hydrothermal methods can achieve similar results at a lower price.

A key advantage of CVD is its minimal material waste. This efficiency stems from the localized heating applied only where the chemical reaction is intended, ensuring that material deposition is highly controlled and focused. In contrast, PVD may lead to material overspray, where vaporized material not only deposits on the substrate but may also reach other surfaces, such as the walls of the chamber or even the electron gun. This lack of restriction in PVD can result in higher material waste. CVD's controlled approach, with deposition occurring only where needed, reduces waste significantly and allows for more precise application.

- *Cost-effective solutions:* Solution-based methods, such as chemical solution deposition and sol–gel, provide affordable alternatives, making them ideal for applications with strict budget constraints. While these techniques are less durable than advanced methods like PVD or CVD, they offer significant cost savings. Although sol–gel and dip coating are categorized separately, they are essentially similar processes and can be combined effectively. Despite their affordability, these techniques demand a high level of skill for optimal results. Skilled labor is especially important for high production rates, as the process requires precise timing. Each cycle involves dipping the substrate, allowing it to set, and repeating the steps to build up layers, enabling the liquid to solidify uniformly on the surface. This iterative process benefits greatly from skilled handling and cannot be rushed. However, due to limited automation, sol–gel production rates are generally slower. Additionally, sol–gel coatings may exhibit porosity, often necessitating further repair or recoating to achieve a consistent finish.
- *Durable industrial coatings:* Thermal spraying and electrochemical deposition provide robust, thick layers that are ideal for industrial and high-wear applications, especially in heavy industries where components are subject to extreme wear and require substantial

protection. Thermal spray coating offers significant design flexibility, allowing for extensive coatings over large surface areas and in various environments. However, this process can generate airborne particles, which may contribute to air pollution; therefore, measures should be taken to collect and contain these particles. A key advantage of thermal spray coating is its ability to cover large substrates efficiently, making it suitable for a wide range of applications. Thermal spraying is an umbrella term for methods such as plasma coating and electric coating, which use a thermal source—like chemical combustion or diesel—to melt the coating material. The molten material is then sprayed onto the substrate with a spray gun system that may include manipulators for precision. This process typically involves two main components: a heat source and a spray gun, allowing for varied coverage with techniques like high-velocity oxyfuel (HVOF), flame coating, and wire arc spray coating. Advanced facilities, such as those at IIT Delhi, are equipped with high-velocity thermal spray systems for specialized applications.

- *Environmental and economic impact:* Beyond performance and cost, "environmental and economic impact" must also be considered. Choosing the right coating method requires a balance between environmental efficiency and production costs. For instance:
    - *Energy and material efficiency:* Methods like thermal spraying and molten-state coatings are energy-intensive but reduce material waste.
    - *Waste management:* Electrochemical processes generate hazardous waste that must be carefully managed to minimize environmental damage.
    - *Reusability and lifespan:* High-quality coatings can extend product life, reducing the need for replacements and minimizing overall resource consumption.
- *Hybrid coating technologies:* In many applications, a single deposition process may not suffice. Hybrid technologies that combine two or more traditional methods are increasingly popular, as they leverage the benefits of multiple techniques to enhance performance. For example, combining "electron beam surface hardening" with "PVD hard coatings" increases load-carrying capacity and hardness, which enhances scratch resistance. Similarly, "laser-assisted plasma spraying" merges plasma spraying's versatility with laser remelting precision to minimize porosity and improve mechanical properties, making it suitable for various industrial applications.

Selecting the appropriate coating is challenging due to the complex mechanisms of surface degradation and the influence of factors such as temperature, friction, and substrate compatibility. These variables impact coating performance, making the choice difficult. Traditional selection methods rely on empirical testing, which is often time-consuming and expensive. To streamline this process, researchers can use a data-driven approach based on literature to pre-select coatings that align with application requirements and material properties. This pre-selection narrows the options to 5 to 10 viable coatings, which are then tested in labs and field trials to identify the best choice, thereby optimizing both performance and cost.

9.5.6 Examples of coating applications and innovations

Different coatings offer unique benefits for specific applications:

- *Chromium plating:* Chromium plating is commonly used for piston rings due to its excellent wear resistance, hardness, low friction, high thermal conductivity, and corrosion resistance. Selecting a coating for a piston ring requires careful consideration, as it must withstand high temperatures and significant wear-prone operating conditions. In automotive applications, the coating should be thin, hard, and capable of handling intense operating conditions. High productivity (faster deposition rates) and cost-effectiveness are also important factors. Balancing these criteria helps in making an optimal coating choice, and incorporating hard particles into chromium plating can further enhance performance.
- *Chromium ceramic coating (CKS):* By adding alumina particles to chromium, this coating improves scuff resistance, ideal for high-stress environments.
- *Chromium diamond coating (GDC):* GDC coatings incorporate diamond particles for enhanced scuffing and wear resistance. However, environmental and health concerns due to hexavalent chromium in the plating bath drive research into alternatives.
- *Thermal sprayed molybdenum-based coatings (Mo sprayed):* Known for high thermal conductivity, these coatings are used in piston rings but have limitations in high-temperature environments due to oxidation risk.
- *PVD coatings:* Nitride and carbonitride coatings like ZrN, CrN, and TiN offer excellent erosion and scuffing resistance for heavy-duty applications.
- *Refractory cermet coatings (HVOF):* Known for abrasion and erosion resistance, these coatings, such as WC/Co and CrC/NiCr, serve as alternatives to chromium plating, with applications across industries requiring high wear resistance.

For guidance on selecting a coating using the weighted objective method, refer to Harish Hirani's video lecture, available on Youtube.[7]

9.6 ASSESSING ECONOMIC LOSSES CAUSED BY CORROSION DEGRADATION

This section examines the economic impact of corrosion degradation, a key theme throughout this book. The focus has been on reducing surface degradation to nearly negligible levels. This raises the question: Is it possible to eliminate corrosion entirely and bring degradation down to zero?

9.6.1 Material origins and corrosion tendency

Materials are either synthesized in laboratories or obtained from nature, often through mining. In mining, materials are generally extracted in their oxide forms; for example, iron is commonly found as iron oxide. To increase their usability and purity, these materials undergo various refining processes that require significant energy input. This added energy raises the material's Gibbs energy, making it more reactive and inclined to return to a stable, lower-energy oxide state. This natural tendency is the primary driver behind corrosion.

Corrosion refers to the degradation of material properties due to interactions with the environment. While it is most commonly associated with metals, corrosion also affects polymers, ceramics, and other materials. Despite efforts to prevent it, corrosion is an inevitable process that requires strategic management to minimize its impact. The driving force behind corrosion is the system's tendency to lower its Gibbs energy. Metals, in particular, have a natural inclination to revert to their low-energy oxide forms, which leads to corrosion. Although we cannot completely prevent corrosion, various control methods can slow its progression and manage its effects.

- *Example:* Marine vessels and offshore structures are especially prone to corrosion because of their constant exposure to saltwater. By using protective coatings and cathodic protection techniques, the lifespan of these structures can be extended, reducing the need for frequent maintenance.

9.6.2 Managing corrosion to extend the material lifespan

While preventing corrosion entirely is impossible, various measures can slow it down, extending the lifespan of materials. Corrosion-resistant coatings, environmental control, and corrosion inhibitors all (covered in this book) reduce degradation rates. For example, a metal component that might naturally last only 5 years could last 10 or 15 years with proper corrosion management. This not only improves durability but also reduces repair, replacement, and maintenance costs. Various methods are used to control corrosion effectively. "Protective coatings" act as physical barriers, shielding the material from exposure. "Chemical inhibitors" are added to environments to slow down corrosion reactions. "Alloy modification" enhances corrosion resistance by adjusting the alloy composition, while "surface treatments," such as anodizing or galvanizing, further strengthen corrosion protection.

1. *Example:* Consider bridges exposed to harsh weather. By applying corrosion-resistant coatings and routine monitoring, these structures can last decades longer, resulting in substantial savings and reduced disruptions to public infrastructure.
2. *Example:* Automotive manufacturers use multilayer coating systems on car bodies to protect them from salt and road debris, significantly extending the lifespan of vehicles.

9.6.3 Role of natural oxide layers in corrosion management

An important corrosion management technique involves allowing a thin, natural oxide layer to form on the metal surface. This layer, typically less than 10 nm thick, serves as a protective barrier. Such "sacrificial" layers are particularly beneficial for materials like ferrous metals and aluminum, allowing them to be used in applications without additional corrosion protection. Aluminum, for instance, forms a protective oxide layer immediately after finishing, which significantly increases its resistance to corrosion.

9.6.4 Corrosion mechanisms and predictive strategies

Corrosion (covered in Chapter 3 of this book) is a complex process influenced by various environmental factors. Corrosion is categorized into several types, each with unique characteristics and specific control methods. "Uniform corrosion" is the most common, occurring evenly across a surface and typically controlled through painting or sacrificial protection. "Pitting corrosion" is a localized form that creates small pits and can be prevented with protective coatings and regular inspections. "Crevice corrosion" develops in confined spaces and is managed by sealing and reducing crevices. "Galvanic corrosion" occurs when dissimilar metals are in contact and is avoided by using similar metals or electrical insulation. "Stress corrosion cracking" arises from the combined effects of stress and corrosion and is controlled by reducing stress and selecting corrosion-resistant alloys. Understanding these mechanisms enables engineers to develop effective strategies to slow corrosion. Techniques include applying protective coatings, adjusting alloy composition to improve resistance, and using surface treatments. Complex multilayer coating systems, as seen in applications like automobiles and aircraft, serve as the first line of defense against corrosion.

Corrosion typically involves two main electrochemical reactions: anodic (oxidation) and cathodic (reduction). Inhibitors like chromates, phosphates, nitrates, molybdates, and organic compounds are added to electrolytes to slow these reactions, thereby reducing corrosion rates at anodic and cathodic sites.

3. *Example:* Galvanic corrosion occurs when steel fasteners are used in conjunction with aluminum components, especially in marine environments. Using compatible materials or applying insulating coatings reduces the corrosion rate.
4. *Example:* In oil pipelines, inhibitors are used in the transported fluid to reduce internal corrosion, minimizing the risk of leaks and costly repairs.

9.6.5 Real-time data and condition monitoring in corrosion management

Accurate prediction of corrosion is challenging, making real-time or in situ data capture essential. By continuously collecting and analyzing data, early identification of flaws or faults in protective coatings is possible, allowing timely interventions. This increases material performance and working conditions, often creating opportunities for additional surface treatments to further extend the component's lifespan.

5. *Example:* Condition monitoring systems on aircraft use sensors to detect early signs of corrosion on critical components, enabling preventative maintenance and avoiding costly breakdowns.

9.6.5.1 Data-driven approaches for corrosion control

A data-driven approach (covered in Chapter 7 of this book) is critical for effective corrosion control. Condition monitoring systems (discussed in Section 9.2 of this chapter) use sensors to monitor equipment conditions, followed by data analysis to detect any abnormalities or faults. In addition, adopting a prognostic approach allows for optimized maintenance strategies, further extending the product's service life.

9.6.6 Challenges and economic implications of corrosion

Understanding the wide-ranging impacts of corrosion on various economic sectors is critical for developing effective mitigation strategies and maintaining infrastructure integrity. Proactive measures and targeted interventions can significantly reduce the adverse effects of corrosion, supporting sustainable growth across industries. Corrosion occurs due to the unavoidable presence of water and humidity, both of which are fundamental to life and ever-present in the environment. These factors, coupled with the extensive use of metals in structures, buildings, and artifacts, make corrosion inevitable.

It was once thought that corrosion primarily affected metals, but recent research has shown that ceramics, polymers, and composites are also susceptible to corrosion. This means that nearly all materials degrade over time. Numerous studies have highlighted the significant financial losses caused by the corrosion process. It is essential to recognize that such losses often result from the accelerated material degradation associated with corrosion. While corrosion alone may not directly increase surface damage, corrosion or corrosion-assisted wear significantly speeds up the degradation process. Addressing this issue is vital for extending the lifespan of systems and mitigating the effects of corrosion. Addressing these conditions and their role in corrosion is essential for creating effective management strategies.

9.6.6.1 Understanding the economics of corrosion

Corrosion poses significant challenges across industries, leading to substantial direct costs, such as repair and maintenance, and indirect costs, including productivity losses and safety risks. The economic burden of corrosion extends beyond immediate repair and maintenance costs, affecting public health, environmental sustainability, and national security. By adopting comprehensive and effective corrosion control measures, industries and governments can mitigate these impacts, ensuring long-term economic stability and infrastructure integrity.

In recent years, many countries have conducted studies to estimate the cost of corrosion. While methodologies vary widely, these studies demonstrate a growing awareness of corrosion's economic impact. However, differences in study rigor, sectoral focus, and data availability have led to discrepancies in reported costs.[8] In some reports, these costs are comparable to the economic losses caused by major natural disasters, such as Hurricane Katrina in 2005. However, unlike natural disasters, corrosion is a continuous, recurring issue, making its financial impact more persistent and pressing. A 2011 report by the National Research Council[9] highlights the far-reaching consequences of corrosion, not only on the economy but also on national security. Corrosion reduces the operational readiness of critical military equipment, such as tanks and missile launchers, which could compromise defense capabilities. This finding underscores the urgent need for increased attention to corrosion management and its broader societal impacts. Beyond financial implications, corrosion poses serious risks to public health, environment, and sustainability. For example, the degradation of medical implants due to interactions with physiological fluids can lead to severe health complications, highlighting the importance of corrosion-resistant materials in biomedical applications. Corrosion in infrastructure, such as bridges and water supply systems, can result in environmental contamination, while degradation in energy systems leads to inefficiencies and increased operational costs. These far-reaching impacts stress the need for comprehensive and proactive corrosion management strategies.

Broadly, the global cost of corrosion was estimated at approximately \$2.5 trillion or 3.4% of the global gross domestic product (GDP). Studies indicate that financial losses due to corrosion typically account for about 3% of the GDP in developed countries like the United States and as much as 5.95% (direct costs, which account for 2.3%, and indirect costs, such as delays and failures, which contribute an additional 3.65%) in India. This disparity warrants further analysis. Even in the United States, where corrosion costs are lower as a percentage of GDP, the financial impact remains significant, estimated at \$2 trillion to \$4 trillion per decade or \$0.2 trillion to \$0.4 trillion annually. Effective corrosion control measures could reduce these costs by 15% to 35%, translating into annual savings of \$375 billion to \$875 billion globally. Koch[8] summarized the economic impacts of corrosion across various sectors, as illustrated in Figure 9.10. The figure presents country names along the X-axis and percentages (ranging from 0% to 100%) along the Y-axis, with three bars representing the contribution of corrosion cost related to agriculture, industry, and services to each

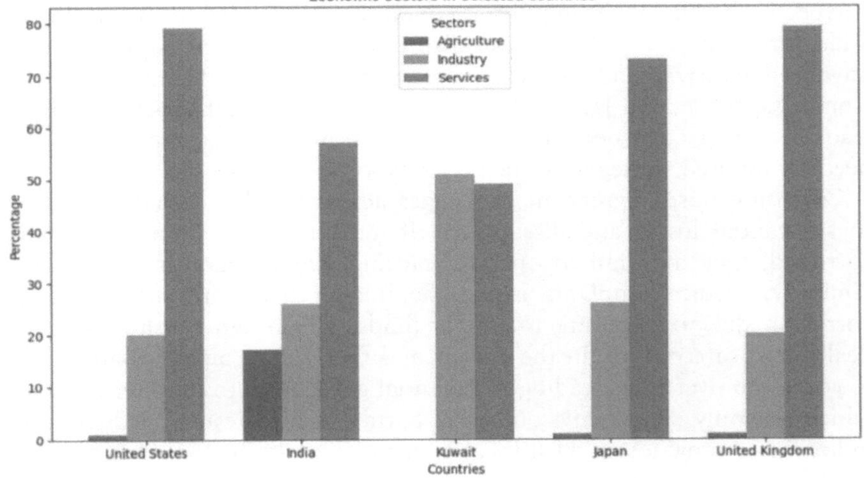

Figure 9.10 Sectoral contributions to the economy in selected countries: a focus on agriculture, industry, and services.

country's economy. Agriculture related to corrosion cost generally accounts for a low percentage, except in India, where it reaches 17%. The industrial sector shows the highest share in Kuwait at 51%, while the services sector dominates in most countries, particularly in the United States and the United Kingdom, where it accounts for 79%. This highlights the critical need for substantial investments in prevention and research to mitigate the economic burden of corrosion.

Koch[3] clearly indicates that corrosion costs across the United States, Japan, and the United Kingdom are dominated by the service sector. For instance, in the United States, the service sector accounts for 79% of these costs, while in Japan, it contributes 73%. Agriculture plays a minor role in these nations, with only 1% of corrosion costs attributed to agriculture in the United States. Conversely, India, with its strong agricultural foundation, sees agriculture contributing around 17% to the total corrosion costs.

In countries with minimal agricultural activity but a significant oil industry presence, such as Kuwait, the industrial sector is the primary contributor, responsible for about 51% of corrosion-related failures. Meanwhile, in developed nations, the industrial sector accounts for roughly 20% of corrosion costs—20% in the United Kingdom, 26% in Japan, and similarly 26% in India. These variations are closely tied to the dominant industries in each country. Therefore, rather than focusing solely on the overall corrosion cost percentages (such as 3%, 5%, or 6%), it is essential to analyze the specific sectors contributing to these costs.

By breaking down the data by sectors, it becomes clear which areas are more critical and require focused efforts to mitigate corrosion impacts. This perspective highlights the importance of targeted/sector-based approaches in addressing corrosion. The analysis also reveals a consistent pattern among advanced industrial and service-driven economies like the United States, the United Kingdom, and Japan. However, India's substantial agricultural base leads to higher corrosion rates in the agricultural sector, emphasizing the need for tailored strategies in such regions.

Corrosion poses significant challenges across various domains, leading to substantial losses and disruptions. If machines or equipment become corroded, they may fail to operate, causing major operational setbacks. Similarly, medical implants inside the human body are susceptible to corrosion due to exposure to bodily fluids, which can result in serious health and safety risks. In the energy and fuel sector, pipelines are prone to corrosion over time, leading to potential gas leaks into the environment, which not only causes environmental harm but also results in direct and indirect economic losses. Historical incidents like the Bhopal tragedy have underscored the devastating consequences of such failures.

Corrosion impacts various sectors of the economy, each facing unique challenges and significant costs.

- In the United States, the oil and gas sector illustrates the high economic burden of corrosion. Expenses related to pipeline repairs, maintenance, and production downtime during shutdowns contribute heavily to costs. Robust corrosion management programs, such as those implemented in oil refineries, can significantly reduce repair frequencies and associated expenses.
- In India, corrosion-related challenges affect the agricultural, industrial, and service sectors distinctly. The agricultural sector accounts for 6.1% of total corrosion costs, while the industrial sector, encompassing mining, manufacturing, and production, incurs 5% to 6% of these costs, with manufacturing and production alone contributing 5.7%. The service sector, vital for transportation, infrastructure, and healthcare, also faces substantial expenses. Transportation-related corrosion costs stand at approximately 3%, and infrastructure-related costs, including railways and bridges, amount to 3.2%. These issues not only pose safety risks but also cause significant economic damage.
- Electricity generation and transmission emerge as the most affected sector, accounting for 10.1% of all corrosion costs across economic activities. This underscores the critical importance of targeted corrosion management efforts to minimize its effects and enhance sustainability across industries and infrastructure.

9.6.6.2 Implementing predictive maintenance with RUL estimation

Predictive maintenance relies on "RUL estimation" to determine how long equipment can operate before needing repair. Accurate RUL estimation is critical in decision-making across industries such as automotive, aerospace, and nuclear. This approach integrates CBM, system health management, and reliability modeling.

6. *Example:* In nuclear plants, accurate RUL estimation for critical components allows for timely maintenance, preventing failures and ensuring plant safety.

The question arises: Can we truly manage CBM and RUL without a data-driven approach? The answer is no, and the reason lies in the complexity of industrial and manufacturing processes, where various forms of degradation occur but are not immediately apparent. This underscores the need for nondestructive methods to detect hidden issues. Traditional life or failure time estimations may not always be accurate, which is why a data-driven approach is essential. Theoretical calculations and models must be continuously refined using iterative methods to improve accuracy. Additionally, understanding the relationship between condition-based monitoring and the degradation process is crucial. By modeling degradation and collecting data, we can estimate the RUL of components. Economics also plays a vital role, as it influences decision-making related to failure behavior. The interconnectedness of condition monitoring, degradation modeling, and economic considerations are essentially required to enable informed and effective decision-making. By integrating these key streams, one can achieve more favorable outcomes.

9.6.6.3 Spreading knowledge to apply corrosion-related processes for beneficial purposes

According to the referenced data,[10] the annual cost of corrosion in India is estimated to be around Rs. 1.67 lakh crores. To put this into perspective, this figure exceeds the entire national budget for education,[11] which is approximately Rs. 1.13 lakh crores. This stark comparison highlights the urgent need to prioritize education and awareness about corrosion management. By increasing knowledge and implementing effective mitigation strategies, significant reductions in corrosion-related expenses can be achieved, benefiting both the economy and infrastructure resilience.

The processes that cause corrosion can also be utilized for beneficial purposes, presenting valuable opportunities when thoroughly studied and understood. Corrosion science offers insights into material removal and deposition processes that have significantly enhanced manufacturing efficiency. This is particularly evident in advanced techniques such as semiconductor manufacturing, which relies on controlled etching and oxidation. These processes, closely linked to corrosion, require precise nanoscale control to achieve desired results.

For example, the semiconductor industry involves repeated cycles of etching and material deposition, resulting in oxide formation on surfaces. Understanding corrosion mechanisms enables effective management of these processes and precise parameter adjustments at the nanoscale. Natural oxide layers on metallic surfaces, often only about 10 nm thick, underscore the importance of precision beyond traditional microscale approaches. Advancing knowledge at this level has transformative potential for modern technologies, reducing reliance on bulky 3D systems.

The principles of corrosion science extend beyond semiconductors to fields like material synthesis, performance analysis, sensor development, biodegradability, battery design, and recycling. A deeper understanding of corrosion mechanisms aids in fault detection and the development of protective barriers, improving material durability and functionality in various environments. By applying these principles, industries can address challenges effectively while promoting sustainability and efficiency.

Research and education in corrosion science are essential for equipping industries with tools to manage risks, innovate solutions, and harness the beneficial aspects of corrosion processes. For instance, corrosion techniques are used to create nanoporous metals for catalysis and fuel cells, where precise material control is crucial for achieving efficiency and optimal performance.

9.6.6.4 Advancing sustainable and innovative technologies

By understanding the mechanisms, impacts, and control methods of corrosion, industries worldwide can better manage material degradation, conserve resources, and protect both the environment and public health. Advanced corrosion research supports innovations in biodegradability, recycling, and renewable energy technologies. Fuel cells, gas separation membranes, and nanoporous catalysts benefit from an in-depth understanding of corrosion processes, driving forward materials science and sustainable engineering practices.

7. *Example:* Corrosion research contributes to the development of biodegradable polymers that degrade predictably in the environment, reducing waste and supporting sustainability goals.

9.6.6.5 Methodologies for economic evaluation

Accurately assessing the economic impact of corrosion is a challenging task due to the lack of standardized methodologies and the complexity of balancing direct and indirect costs. Indirect costs, such as productivity losses and environmental impacts, are often underestimated despite their significant economic implications. Corrosion impacts productivity significantly. For instance, when bridges require maintenance, road closures can last for days or even weeks, forcing commuters to take lengthy detours. This not only wastes time but also reduces overall productivity, highlighting the far-reaching consequences of corrosion in everyday life. Traffic congestion and environmental pollution caused by infrastructure maintenance or repairs must also be factored into the broader impact of corrosion. While some models project these effects, they require further validation and refinement to ensure accuracy before being shared with the public. It is essential that decisions regarding infrastructure benefit society as a whole rather than merely serving the interests of contractors managing the projects. Ultimately, the real burden falls on the public.

Corrosion also poses risks to medical devices, such as needles or implants. A corroded needle could harm the individual receiving the injection, and corrosion in implants may result in severe health complications. Despite the importance of these issues, comprehensive documentation of the indirect effects of corrosion is lacking, even in countries like the United States, where the estimated 3% of GDP attributed to corrosion primarily reflects direct costs. The broader societal and environmental impacts remain underreported. A comprehensive evaluation framework is essential to effectively address these challenges and guide resource allocation.

9.6.6.5.1 Established methodologies for corrosion cost estimation

Establishing standardized evaluation criteria is essential for accurately comparing corrosion costs across countries and regions, ensuring consistency and reducing discrepancies. Such criteria enable a holistic understanding of corrosion's economic impact by encompassing all relevant costs, including environmental and societal consequences. Without a comprehensive framework, critical factors may be overlooked, leading to suboptimal decisions and less effective corrosion management strategies.

Corrosion costs are generally divided into direct and indirect expenses. "Direct costs" include expenditures related to design, manufacturing, repair, and maintenance, such as those for corrosion-resistant materials, protective coatings, and cathodic protection systems. "Indirect costs," often less apparent but equally significant, encompass broader societal impacts like traffic delays from bridge repairs, lost productivity during downtime, and environmental cleanup efforts. For example, corrosion-related delays in infrastructure repairs can disrupt economies by reducing productivity and increasing operational expenses, highlighting the comprehensive economic burden of corrosion.

In the present section, we will learn three methodologies to provide foundational frameworks for estimating corrosion costs:

- *Uhlig method:* Focuses on the expenses related to corrosion prevention techniques, distinguishing between direct costs, such as protective coatings and cathodic protection, and indirect costs, like maintenance of equipment and replacements.
- *Hoar method:* Considers the specific characteristics and requirements of various industries, offering sector-specific cost estimates that reflect the unique nature of corrosion challenges in different sectors.
- *Input/output method:* Uses an input/output (I/O) table to analyze economic interdependencies and estimate corrosion-related expenses across sectors, providing a macroeconomic view of corrosion costs.

We also touched on the concept of over-design, where higher safety factors are applied to account for material degradation. While this approach ensures safety, it often leads to excessive use of materials and inflated costs. To better evaluate these challenges, methods such as the Olig method, how method, I/O method can be utilized. We will explore these three approaches in detail during this and the next lecture. Notably, the Olig method, first introduced in 2001, has demonstrated promising results and remains a valuable tool for such assessments.

The interplay of direct and indirect costs complicates precise decision-making, primarily due to the lack of reliable data. Direct costs, such as those associated with anticorrosion measures like coatings, surface treatments, or the inclusion of nanoadditives, are relatively straightforward to calculate. These tangible expenses allow for more confident and informed decision-making compared to the often unpredictable indirect costs.

Direct costs typically encompass design, manufacturing, fabrication, and overhead expenses like management charges. Indirect costs, however, represent the societal burden, including disruptions, environmental impacts, and penalties. For instance, in long-term contracts like highway projects, maintenance costs often fall partially on the managing company but are ultimately borne by society through taxes or user fees. Understanding these dynamics is vital for equitable and efficient cost distribution.

9.6.6.5.2 Comprehensive corrosion management tools

Corrosion presents significant challenges in the oil and gas and infrastructure sectors, leading to notable economic, operational, and societal impacts. In the oil and gas industry, pipelines and equipment are frequently exposed to harsh environments like seawater and chemical-laden fluids, resulting in costly maintenance and operational disruptions. These challenges can be effectively managed through the use of corrosion inhibitors, which slow down the degradation process, and advanced condition monitoring systems that allow for early detection and targeted intervention. In the infrastructure sector, assets such as bridges and water systems face similar risks, with corrosion driving up repair costs and compromising safety. Life-cycle cost analysis (LCCA) plays a pivotal role in optimizing resource allocation by guiding decisions on protective coatings and selecting durable materials like weathering steel or corrosion-resistant alloys. By adopting comprehensive corrosion management strategies, industries can enhance asset longevity, reduce repair frequencies, and minimize both economic losses and societal burdens, ensuring sustainable and cost-effective solutions.

1. *Financial tools for cost analysis*
    a. *Cost of corrosion analysis:* This tool identifies areas with significant corrosion-related expenses and focuses efforts on reducing these costs through targeted prevention measures.
    b. *LCCA:* LCCA assesses the comprehensive cost of an asset across its entire life cycle, encompassing acquisition, operation, maintenance, and disposal. By considering initial investment, ongoing operational expenses, and eventual disposal costs, it aids in making informed decisions regarding material selection and maintenance strategies. LCCA also incorporates factors

such as inflation and interest rates, employing "net present value (NPV)" to translate future costs and benefits into present-day terms, enabling a clearer understanding of long-term financial implications.

   c. *Return on investment analysis:* Return on investment (ROI) assesses the effectiveness of corrosion prevention measures by comparing their costs to the savings generated through reduced maintenance and extended asset lifespan. It enables industries to prioritize investments based on potential returns, balancing upfront expenses with anticipated long-term benefits and financial savings.
   d. *Cost–benefit analysis:* This method evaluates the economic viability of corrosion control measures by weighing their costs against the benefits, such as enhanced durability, minimized downtime, and improved safety. By adjusting future costs and benefits to present-day values, it provides a clear framework for assessing the practicality of implementing corrosion-resistant strategies. Accurate quantification of these factors is essential for informed and effective decision-making.
   e. *Risk assessment and management:* Risk assessment evaluates the probability and impact of corrosion-related failures, guiding the development of strategies to mitigate these risks effectively.
   f. *Maintenance planning and optimization:* Optimized maintenance practices help reduce corrosion-related costs by adopting proactive and efficient approaches.
   g. *Performance-based contracts:* These contracts align service providers' incentives with corrosion control outcomes by tying rewards to reduced corrosion rates and extended asset lifespans.
2. *Life-cycle cost components*
   a. *Initial costs:* These include the purchase and installation expenses of an asset or system.
   b. *Recurring costs:* Recurring costs encompass ongoing operational and maintenance expenses throughout the asset's life cycle.
   c. *Operating costs:* These are costs associated with regular asset use, such as fuel, labor, and energy consumption.
   d. *Maintenance expenses:* Routine inspections, repairs, and part replacements are essential to ensure functionality and longevity.
   e. *Disposal costs:* These are incurred during asset decommissioning, which may involve removal and environmentally responsible disposal. Disposal costs can be positive (expenses incurred) or negative (returns from selling used assets).
   f. *Residual value:* The estimated monetary value of an asset at the end of its useful life can offset some initial investment costs.

3. *Beyond financial losses: Societal and environmental impacts*
   Corrosion's impact extends beyond financial losses to public health, safety, and environmental sustainability.
   a. *Medical devices:* Corrosion in implants can cause health complications, emphasizing the need for corrosion-resistant materials in biomedical applications.
   b. *Energy systems:* Degradation of electrical grids reduces energy efficiency and reliability.
   c. *Infrastructure:* Corroded water systems increase repair costs and contribute to environmental contamination.
4. *Quantifying corrosion costs:* Proactive corrosion management is vital for reducing financial losses, enhancing asset durability, and minimizing societal and environmental impacts. Tools like ROI, LCCA, and cost–benefit analysis enable industries to make data-driven decisions. With continuous improvement in data collection and modeling, these approaches ensure better resource allocation, sustainable practices, and public safety.
   a. *Key modeling methods:* There are three approaches—Uhlig, Hoar, and I/O, each offers unique insights into understanding and managing corrosion costs, emphasizing prevention, sector-specific analysis, and economic interdependencies, respectively. Combined, they provide a robust framework for developing effective strategies to mitigate the financial and operational impacts of corrosion.
      i. *Uhlig approach:* The Uhlig approach is one of the foundational methodologies for estimating corrosion costs, focusing primarily on the expenses associated with preventive strategies. It was the first systematic attempt to distinguish between direct and indirect corrosion expenses. Direct costs, borne by owners and operators, include expenditures for corrosion prevention techniques such as the application of protective coatings, the use of corrosion-resistant materials, and cathodic protection systems. These measures are proactive investments aimed at minimizing degradation and extending the lifespan of assets. Indirect costs, on the other hand, are typically incurred by end-users and reflect the economic burden of corrosion-related failures, such as replacing water heaters, maintaining internal combustion engines, or replacing automotive silencers. By categorizing these expenses, the Uhlig approach highlights the dual financial dimensions of corrosion and emphasizes the importance of prevention.
      ii. *Hoar approach:* The Hoar approach builds on the Uhlig method by incorporating a more nuanced assessment of the

economic impact of corrosion. It evaluates the costs based on the individual characteristics and specific requirements of various industrial sectors. Rather than focusing solely on broad preventive strategies, this approach considers how corrosion uniquely affects different industries, tailoring the analysis to sector-specific conditions. This allows for a more precise understanding of how corrosion influences operational efficiency, material selection, and maintenance practices in each industry.

iii. *I/O approach:* The I/O approach takes a more data-intensive perspective, estimating corrosion costs using a detailed I/O table that categorizes corrosion-related expenditures across industries and sectors. This method examines the flow of economic resources between different sectors and quantifies how corrosion impacts production, maintenance, and other operational activities. By identifying the interdependencies between various industrial activities and their associated corrosion costs, the I/O approach provides a comprehensive view of the economic implications of corrosion. It requires substantial data to accurately reflect the costs and benefits of corrosion prevention and management, making it a more detailed but resource-intensive method.

b. *Challenges and strategies in corrosion data collection and economic analysis:*

i. *Challenges in data collection:* Despite recognizing the extensive economic losses caused by corrosion, the development of a robust methodology for comprehensive cost assessment remains a significant challenge. The absence of consistent and detailed data hampers accurate analysis. Continuous data collection and systematic analysis are essential to refine existing models, providing more precise insights and enabling informed decision-making.

- *Comprehensive data collection:* Effective data collection requires integrating multiple sources. These include:
- *Incident statistics:* Records of corrosion-related failures and maintenance activities.
- *Published literature:* Research papers, industry reports, and case studies.
- *I/O tables:* Detailed data on inter-industry relationships and corrosion-related expenditures.

It is crucial to validate claims, such as those asserting that coatings or other measures extend system lifespans by five times, through rigorous testing. Accelerated tests, real-world validations, and continuous updates ensure realistic and actionable insights.

ii. *Data processing and analysis:* Collected data must undergo thorough processing to ensure accuracy and relevance. Key steps include:
   - *Data organization:* Cleaning and resolving inconsistencies.
   - *Statistical analysis:* Identifying patterns, trends, and correlations.
   - *Economic modeling:* Employing techniques like I/O analysis and cost–benefit analysis to quantify corrosion's economic impact.

   While quantifying corrosion remains inherently complex, adopting a data-driven approach ensures assessments are updated and models remain reliable.

iii. *Dissemination of data:* Sharing processed and analyzed data with relevant stakeholders is crucial for impactful decision-making. Effective strategies include:
   - *Publications and reports:* Detailed documents outlining findings and recommendations.
   - *Online databases:* Platforms providing accessible, centralized data for industries and policymakers.
   - *Collaborations:* Partnerships with industry associations, research institutions, and government bodies to leverage shared insights.

   Wide accessibility to this information empowers stakeholders to develop effective anticorrosion strategies.

9.6.7 Financial analysis of life-cycle costs and materials degradation

9.6.7.1 Example

John plans to purchase a car worth \$12,000, which he intends to sell after 5 years for a residual value of \$3000. Annual costs include \$1000 for maintenance and repairs and \$3500 for gas. The applicable interest rate is 8%.

- *Initial cost:* \$12,000
- *Recurring costs:* Maintenance and repair (\$1000/year) + gas (\$3500/year) = \$4500/year
- *Total recurring costs (over 5 years):* \$4500 × 5 = \$22,500 (if we do not count discounted rate)
- *Residual value:* \$3000

Present values of these components are calculated as follows:

1. *Present value of recurring costs:*

$$\text{PVrecur} = \sum_{i=1}^{n} \frac{\text{cost}}{(1+r)^n} = \sum_{i=1}^{5} \frac{4500}{(1+0.08)^5} = 17967$$

2. *Present value of residual value:* $\text{PVresid} = \dfrac{3000}{(1+0.08)^5} = 2041.7$

Finally, the life-cycle cost of the car is the sum of the initial cost and the present values of the recurring cost and residual value:

Life-cycle cost (LCC): $\text{LCC} = \text{Initial Cost} + PV_{recur} - PV_{resid}$

$= 12{,}000 + 17967 - 2041.7$

LCC = $27,925

Comments: The recurring cost, if calculated directly over 5 years without considering the present value, would be 4500 × 5 = 22,500, which is significantly higher. However, when discounted to present value at an 8% rate, it reduces to approximately $17,967. Similarly, for the residual value of $3000, its present value comes out to approximately $2042 (rounded).

By summing these components—the present value of the initial cost, recurring costs, and the residual value—the total life-cycle cost over the 5-year period is calculated as approximately $27,925. This approach highlights how discounting future costs and benefits provides a more accurate representation of their current financial impact, enabling better decision-making.

9.6.7.2 Example: Project feasibility using NPV

This comprehensive LCCA evaluates the benefits of the device or equipment in question, helping us determine whether the cost-to-benefit ratio justifies the investment. By quantifying the benefits, we can make an informed decision on whether to proceed with the purchase. This analysis reflects the present value of the investment required for equipment, such as a car or a machine, offering a clearer picture of its financial impact. In this scenario, we use a 6% interest rate, compared to the 8% used in the previous example.

The chief financial officer (CFO) of Briddles Inc. evaluates a project with an upfront cost of $100,000 and expected annual benefits over 3 years:

- Year 1: $50,000
- Year 2: $30,000
- Year 3: $60,000

Discount rate: 6%

Present value of benefits for each year:
To calculate the NPV, we discount each year's benefits back to their present value using the discount rate of 6%. Then, we subtract the initial cost from the sum of the present values of the benefits. Here is how it is done:

Year 1: \$50,000 / (1 + 0.06)^1 = \$47,170
Year 2: \$30,000 / (1 + 0.06)^2 = \$26,700
Year 3: \$60,000 / (1 + 0.06)^3 = \$50,377

Now, summing up the present values of the benefits: PV = \$124247

NPV: ΣPV benefits – Initial cost = 124,247 – 100,000 = \$24,247

As the NPV is positive (\$24,247), the project is considered viable.

Comments: The analysis incorporates the initial cost, enabling a comparison of benefits against the total expenditure. This structured approach underscores the necessity of understanding both costs and benefits when evaluating investment decisions, providing a framework to assess the financial viability of an asset. The resulting benefit-to-cost ratio, approximately 1.24, indicates a favorable ROI and suggests that the project is economically viable. However, discrepancies, such as the reduced benefit of \$30,000 in the second year compared to other years, highlight potential areas for improvement. Analyzing and documenting these variations can help identify opportunities to optimize system performance and enhance overall returns.

The analysis further emphasizes the significance of economic factors, such as minimizing surface degradation, in resource allocation. Making informed decisions requires a comprehensive understanding of variables like degradation and maintenance, as they significantly influence long-term benefits. Accurate documentation and calculations are essential for developing strategies that maximize economic returns and improve system efficiency.

9.6.8 Comprehensive economic analysis of corrosion

Corrosion presents substantial economic challenges across industries, adversely affecting productivity, maintenance costs, and societal well-being. Addressing this issue requires a systematic and structured approach to understand and mitigate corrosion costs effectively. Different countries and sectors exhibit varying vulnerabilities to corrosion, with certain subsectors facing a disproportionately higher impact. To target these high-risk areas, it is vital to conduct detailed analyses. Leveraging existing resources, such as annual reports and IPCC findings, and integrating them into the evaluation process—either manually or using specialized software—are crucial first steps.

Expert insights are equally important. Over time, professionals accumulate tacit knowledge—critical information not formally documented. Extracting this knowledge through interviews with industry experts, government officials, and regulatory authorities can significantly enhance understanding. This involves identifying priority sectors or subsectors and consulting experts who can provide specialized data.

Collected data, whether from literature or expert opinions, must be processed and analyzed rigorously. This enables the development of precise corrosion models, the estimation of direct and indirect costs, and the assessment of societal impacts. Combining documented insights with expert contributions allows for the creation of a robust framework to address corrosion's economic and social ramifications. A systematic economic analysis not only identifies areas for intervention but also informs the design of cost-effective strategies for corrosion management. Below, a detailed examination of the process, supported by illustrative examples and mathematical models, further clarifies this approach.

9.6.8.1 Key steps in economic analysis of corrosion

1. *Identification of sectors and subsectors:* The first step is to categorize the industries affected by corrosion, such as manufacturing, infrastructure, transportation, energy, and defense. For instance, the oil and gas sector faces corrosion in pipelines due to seawater exposure, while infrastructure sectors like bridges and water systems require significant repairs due to material degradation.
2. *Data collection:* Data collection involves surveys, expert inputs, and statistical information from reliable sources:
   - *Surveys:* Questionnaires for industries to gather insights on corrosion types, frequency, and associated costs.
   - *Published data:* Research papers, industry reports, and case studies.
   - *I/O tables:* Comprehensive datasets showing inter-industry interactions and expenses.

   For example, in a factory, data on annual maintenance costs, number of shutdowns due to corrosion, and material replacement rates can be collected.
3. *Data processing and analysis:* After data collection, statistical and economic modeling techniques are applied to analyze the information:
   - *Cost–benefit analysis:* To compare the costs of corrosion control measures with their benefits.
   - *I/O analysis:* To evaluate sector-specific interdependencies and the ripple effects of corrosion-related costs.

Example calculation: Suppose a company spends \$50,000 annually on protective coatings, which reduces downtime by 20%. If the original downtime cost was \$100,000, the *benefit-cost ratio (BCR)* is calculated as:

$$\text{BCR} = \frac{\text{Saving}}{\text{Cost}} = \frac{0.2 * 100{,}000}{50{,}000} = 0.4$$

BCR over 1 denotes a good investment. As it is less than 1 as in the current situation, investment is not a good one.

4. *Estimation of corrosion costs:* Corrosion costs are divided into:
   - *Direct costs:* Repair, replacement, and protective measures.
   - *Indirect costs:* Productivity losses, downtime, and environmental impacts.

   Example: For a bridge repair project, direct costs include material and labor expenses (\$500,000), while indirect costs cover traffic delays costing \$200,000. The total estimated corrosion cost would be:

   $$\begin{aligned}\text{Total cost} &= \text{direct costs} + \text{indirect costs}\\ &= 500{,}000 + 200{,}000 = 700{,}000\end{aligned}$$

5. *Impact on national economy:* By scaling the corrosion costs across all affected sectors, their overall impact on the national GDP can be assessed. For example, if total corrosion costs in a country are \$20 billion and the GDP is \$500 billion, the percentage impact is:

   $$\text{Impact percentage} = \frac{20}{500} * 100 = 4\%$$

6. *Raising awareness:* Public and industrial awareness campaigns educate stakeholders on corrosion prevention strategies. For example, organizing workshops on the use of anticorrosion coatings can reduce future maintenance costs.
7. *Anticorrosion policy formulation:* Based on the analysis, governments and industries can implement policies such as mandatory use of corrosion-resistant materials, tax benefits for adopting preventive measures, and stricter inspection protocols.

Comments: Corrosion significantly impacts the national economy, necessitating a multifaceted approach that includes knowledge diffusion, societal awareness, and the development of anticorrosion policies. The initial step in addressing this issue is to estimate the overall cost of corrosion. With data on corrosion mechanisms and their impact across sectors, we can better assess the economic ramifications. This information can then be used to raise societal awareness, propose effective anti-corrosion policies, and disseminate actionable insights to enable informed decision-making.

For example, environmental factors like water, humidity, and chemicals are key contributors to corrosion. Proper management—such as preventing exposure of reactive materials to these elements—plays a critical role in mitigating damage. From a broader perspective, economic analysis of corrosion requires a comprehensive examination of its effects across multiple economic sectors. Identifying the most severely affected sectors is crucial for prioritizing intervention.

A robust approach hinges on effective data collection. Expert input, particularly tacit knowledge gained through years of experience, is invaluable. This knowledge is often undocumented, presenting a challenge in capturing it systematically. Gathering expert insights is a standard management tool, complemented by thorough literature reviews, to ensure a holistic understanding of the issue. Following data collection, the next steps involve processing and analyzing the data to uncover patterns, correlations, and trends.

Using these insights, data-driven models can be developed and continuously updated. Such models not only refine our understanding but also identify research gaps and inform strategic interventions. Assessing the national economic impact—whether direct or indirect—is a vital component of this process. Widespread dissemination of knowledge and data empowers the public to participate meaningfully in addressing the issue, ultimately fostering more effective societal and industrial responses to corrosion-related challenges.

9.6.8.2 Example: LCCA

LCCA evaluates the total cost of an asset from acquisition to disposal, including maintenance and operational costs. Using *NPV*, future costs are discounted to present value for informed decision-making. Assume a pipeline costs $1 million to install, with annual maintenance costs of $50,000 for (n=) 20 years. Assuming an interest rate of 5%, the NPV of maintenance costs is:

$$\text{NPV} = \sum_{i=1}^{n} \frac{\text{cost}}{(1+r)^{n}} = \sum_{i=1}^{20} \frac{50000}{(1+0.05)^{20}}$$

Using MATLAB coding: NPV = 0.0; for i = 1:n NPV = NPV + 50000/(1 + 0.05)^i; end

Answer: NPV = 6.2311e + 05

Adding the initial cost, the total life-cycle cost is: Total LCCA = 1,000,000 + 623,110 = 1,623,110

Continuous refinement through data: Even with limited initial data, a data-driven approach allows iterative refinement of models, improving decision-making over time. For example, tracking the performance of a new coating can provide valuable data for future applications.

9.6.9 Corrosion analysis and mitigation strategies

A structured evaluation of corrosion-related factors is essential for identifying critical areas that require intervention, enabling informed decision-making, and minimizing losses while enhancing operational efficiency. By integrating corrosion cost analysis and mitigation strategies, societies can lay the foundation for sustainable development. Through the diffusion of knowledge, collaborative research efforts, and the adoption of advanced technologies, corrosion challenges can be effectively addressed, paving the way for long-term progress, economic growth, and environmental sustainability.

9.6.9.1 Framework for corrosion cost analysis

- *LCCA:* As described in the previous paragraphs LCCA includes: (1) procurement costs, (2) operational and maintenance costs, and (3) disposal costs.
- *Economic impact assessment:* After calculating corrosion costs, their effect on the national economy must be assessed. This aids policymakers and stakeholders in understanding implications for: (1) economic growth, (2) employment, and (3) competitiveness.
- *Raising awareness:* Educating industries and the public about the adverse impacts of corrosion encourages proactive measures. Information on economic losses can motivate micro, small, and medium enterprises (MSMEs) to invest in corrosion prevention.
- *Development and enforcement of anticorrosion policies:* A comprehensive anticorrosion policy includes: (1) preventive maintenance practices, (2) use of corrosion-resistant materials, (3) enhancement of inspection and monitoring systems, and (4) encouragement of research and development in corrosion mitigation technologies. These policies must be adopted by authorities and monitored for compliance. Noncompliance should be addressed through education and intervention to align practices with societal benefits.

- *Data diffusion for informed decision-making:* The data acquired during economic analysis should be made accessible to all stakeholders through (1) publications and online databases and (2) collaborations with industry associations and academic institutions. This transparency enables policymakers, researchers, and industry professionals to make informed decisions and develop effective corrosion control strategies.
- *Key areas of research and development:*
  - *Understanding corrosion mechanisms:* Studying electrochemical processes and environmental factors contributing to localized corrosion, galvanic corrosion, and stress corrosion cracking.
  - *Corrosion prevention and protection:* Developing advanced materials, coatings, and inhibitors.
  - *Monitoring and inspection:* Implementing sensors and non-destructive evaluation methods for real-time corrosion detection.
  - *Sustainable mitigation:* Creating eco-friendly inhibitors and sustainable material design.
  - *Challenging environments:* Addressing corrosion in extreme conditions such as high temperatures, pressures, and radiation.
  - *Modeling and prediction:* Utilizing computational modeling and predictive tools for accurate forecasting.
  - *Education and workforce development:* Enhancing training programs and certifications in corrosion science and engineering.
- *Multidisciplinary collaboration:* Addressing these challenges requires partnerships between academia, industry, and government agencies. Collaboration across materials science, chemistry, and engineering disciplines is essential for developing innovative solutions.
- *Integration of data-driven approaches:*
  - *Life-cycle assessments and computational modeling:* The integration of LCA and computational tools like MATLAB, machine learning, and artificial intelligence is critical. These approaches refine predictive models, improve mitigation strategies, and support long-term planning.
  - *Real-world applications:* By applying these tools, stakeholders can:
    - Evaluate mitigation costs versus benefits.
    - Extend asset lifespans.
    - Enhance overall system efficiency.
- *Economic growth, sustainability, and technological progress:* By implementing comprehensive corrosion management strategies, societies can achieve:
  - *Economic growth:* Reducing financial losses caused by corrosion.
  - *Employment opportunities:* Creating jobs in research, manufacturing, and mitigation efforts.
  - *Competitiveness:* Enhancing industrial efficiency and innovation.

9.7 QUANTIFYING ECONOMIC LOSSES ATTRIBUTED TO CORROSION DEGRADATION

The final section of the book "Corrosion, Environmental Degradation, and Surface Engineering: Understanding, Mitigation, and Prevention" summarizes key insights on the economic impacts of corrosion and effective prevention strategies. Corrosion is a pervasive issue, but various technologies are available to combat its effects. These include coatings and surface treatments to prevent or delay degradation, corrosion-resistant materials for long-term protection, and inhibitors effective in harsh environments with chloride ions, acetic acid, or moisture. Emerging solutions like nanoadditives enhance resistance further, while oils and greases create protective barriers, reducing exposure to corrosive elements. For instance, calcium-based grease in rolling bearings prevents water infiltration and enhances reliability. Cathodic protection is another proven method, significantly slowing the corrosion process by protecting the anode. These strategies collectively contribute to mitigating corrosion, extending asset lifespans, and reducing associated economic losses.

Figure 9.11 illustrates key methodologies for assessing and managing the financial implications of corrosion, as discussed in Section 9.5 of this chapter. These tools emphasize a structured approach to evaluation, cost analysis, and proactive strategies aimed at minimizing the economic and operational effects of corrosion.

The "methodological framework for evaluating corrosion costs," depicted in Figure 9.12 and discussed in Section 9.5 of this chapter, is based on the detailed methodology presented by Biezma and San Cristobal[12] in their

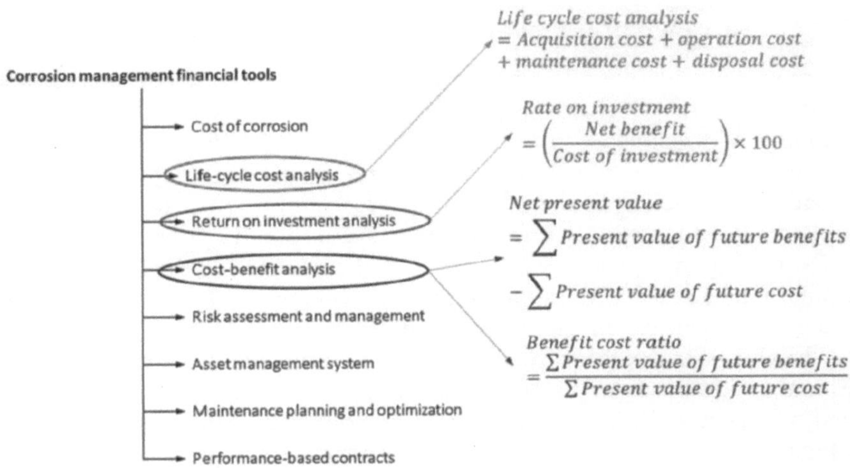

Figure 9.11 Framework for financial analysis and decision-making.

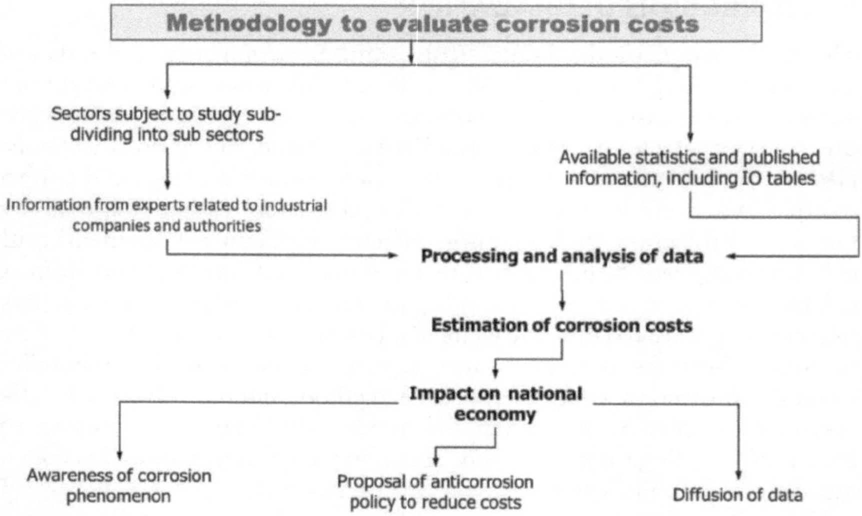

Figure 9.12 Methodological framework for evaluating corrosion costs.

study. This framework provides a systematic approach to understanding and mitigating the economic impacts of corrosion. It focuses on collecting comprehensive data from experts, industry reports, and I/O tables, while also highlighting the importance of assessing the national economic impact in terms of growth, employment, and competitiveness.

9.7.1 Framework for evaluating corrosion costs

Despite the availability of advanced anticorrosion technologies, corrosion remains a persistent challenge with significant economic repercussions. For example, a developed country has reported GDP losses of approximately 3% due to environmental corrosion and degradation. To fully grasp the broader implications of these losses, it is essential to evaluate how various anti-corrosion technologies are implemented across the globe.

Examining the approaches of major economies such as the United States, China, Japan, Germany, and India provides valuable insights into the effectiveness of corrosion management strategies.

The United States leads in GDP per capita, followed by Germany, Japan, China, and India. In terms of overall economic size, the United States ranks first, China second, Japan third, Germany fourth, and India fifth. Notably, China surpasses Germany in overall economic output, securing the second spot, but Germany ranks higher in GDP per capita.

Understanding the economic strengths of these countries provides insights into their global prominence. The United States leads due to its adoption of advanced technologies, extensive research initiatives, and a diverse industrial base. China, ranked second, distinguishes itself with a robust manufacturing sector, export-oriented policies, and substantial investments in infrastructure. Japan, in third place, excels through its technological innovation, efficient operations, and thriving industries like automotive, electronics, and machinery, supported by a highly skilled workforce and consistent focus on research and development. Germany, the fourth-largest economy globally and the leader in Europe, is renowned for precision engineering, high-quality exports, and strong industrial sectors such as automotive manufacturing, benefiting from a well-developed infrastructure and innovation-driven workforce. India, the world's fifth-largest economy and fastest-growing major economy, capitalizes on its expanding middle class, vibrant IT-driven service sector, and substantial domestic market. These economies owe their success to strategic policies, innovation, infrastructure, and skilled labor, collectively driving their global influence.

The key aspect revolves around technology, particularly cutting-edge advancements. A data-driven approach allows for continuous technological improvement. By gathering data, anticipating potential issues, and implementing advanced solutions, progress becomes more efficient and strategic. With comprehensive knowledge, effective execution, and robust policies or management frameworks, countries can achieve significant advancements. For instance, in India, leveraging its substantial domestic consumption by prioritizing local manufacturing could significantly boost economic growth and enhance stability.

Regarding quantification, the various methods of cost evaluation were discussed previously. These include the "Uhlig approach," "Hoar approach," and "I/O method." The I/O method, while highly rigorous, requires extensive data collection and analysis. The "Hoar method," in contrast, employs a creative problem-solving approach, focusing on specific sectors or subsectors to extract detailed insights through a process of vertical thinking—diving deeper into one area to gather comprehensive information. This method often involves classifying costs into direct and indirect categories, acknowledging the approximations inherent in such evaluations.

9.7.1.1 Uhlig

Uhlig classified corrosion costs into two broad categories: "direct costs" and "indirect costs." This classification provides a clear understanding of the financial burden associated with corrosion and its impact on industries. Corrosion's harmful effects on industrial operations manifest in reduced productivity, extended downtime, and potential safety hazards.

Uhlig emphasized that these direct and indirect costs represent a substantial burden on industries, necessitating the development of effective corrosion prevention and control strategies.

- *Direct costs:* Direct costs refer to expenses incurred due to the need for materials and measures to prevent or mitigate corrosion:
    - *Use of alternative materials:* Iron and steel, being highly vulnerable to corrosion, often necessitate the use of non-ferrous metals or alloy steels. The additional expense of these materials represents a direct loss attributed to corrosion. Uhlig noted that, in a corrosion-free scenario, such materials would not be required.
    - *Protective coatings:* A significant proportion of protective measures involves the application of paints and phosphating solutions to shield metals from corrosion. These coatings account for approximately 50% of the annual global paint production, reflecting their critical role in combating corrosion.
- *Indirect costs:* Indirect costs arise from the broader operational and safety implications of corrosion:
    - *Operational downtime and equipment design:* Corrosion necessitates over-designing equipment to compensate for potential degradation and often leads to unexpected shutdowns. These contribute to significant financial and productivity losses.
    - *Loss of efficiency and output:* Corrosion negatively impacts the efficiency of processes and machinery, reducing output and increasing maintenance demands.
    - *Accidents and contamination:* Corrosion can lead to equipment failure, explosions, and contamination of products, posing safety risks and further financial repercussions.

9.7.1.2 Hoar method

The "Hoar method" provides a structured approach to estimate corrosion costs within specific industrial sectors. This method examines not only the direct costs associated with corrosion but also the expenses involved in implementing preventive measures and the financial implications of capital investments for corrosion control.

Key components of the Hoar method:

- *Comprehensive analysis of corrosion processes:* The method involves a detailed evaluation of the entire corrosion process within an industrial sector. This includes:
    - Identifying areas prone to corrosion.

- Assessing the extent of damage caused by corrosion.
- Calculating the resulting financial costs.
- *Direct corrosion costs:* These expenses arise from the replacement or repair of damaged equipment and structures due to corrosion. They include:
  - Costs for procuring materials and labor for maintenance and repairs.
  - Financial losses caused by production halts or shutdowns due to corrosion-related failures.
- *Costs of countermeasures:* The method factors in the implementation of protective and preventive measures aimed at mitigating corrosion risks. Such measures include:
  - Applying protective coatings to shield materials from corrosive environments.
  - Utilizing corrosion-resistant materials in construction and machinery.
  - Installing systems like cathodic protection for enhanced durability.
  - Conducting regular inspections and maintenance activities.
  - Expenses related to materials, installation, and ongoing upkeep are considered part of the total corrosion cost.
- *Capital expenses for corrosion control:* This component accounts for the financial burden of financing corrosion-related projects. It includes:
  - Borrowing costs for funds used in corrosion mitigation.
  - Opportunity costs of allocating company resources to corrosion control rather than alternative investments.

It is worth noting that the "Hoar method" integrates all relevant cost components—direct corrosion costs, preventive measures, and capital expenses—to offer a holistic view of the financial impact of corrosion on industrial sectors. By doing so, it allows industries to identify areas of vulnerability, evaluate the effectiveness of countermeasures, and make informed decisions about resource allocation. This structured approach ensures that both immediate and long-term financial implications of corrosion are comprehensively addressed.

9.7.1.3 I/O model

This model evaluates the impact of corrosion on industries by analyzing domestic commercial interactions. In this framework, the GDP is calculated for three scenarios:

- *Universe I:* The real world, where corrosion exists, and its effects are present.

- *Universe II:* A hypothetical world with no corrosion, assuming corrosion-related effects are entirely eliminated.
- *Universe III:* An ideal world where corrosion is effectively controlled through preventive measures.

Using the GDP values from these scenarios, the costs associated with corrosion are determined as follows:

- *Corrosion cost:* The economic loss caused by corrosion, calculated as the difference between the GDP in a world without corrosion (Universe II) and the GDP in the real world (Universe I):
    - Corrosion cost = GDP (Universe II) – GDP (Universe I).
- *Avoidable corrosion cost:* The economic savings achievable through corrosion control measures, calculated as the difference between the GDP in an ideal corrosion-inhibited world (Universe III) and the GDP in the real world (Universe I):
    - Avoidable corrosion cost = GDP (Universe III) – GDP (Universe I).
- *Research scope:* The gap between an ideal corrosion-inhibited world and a world without corrosion indicates opportunities for further improvements and innovation. It is given by:
    - Research scope = GDP (Universe III) – GDP (Universe II).

It is worthy to note that the "I/O model" provides a structured approach to quantify the economic impact of corrosion on industries. By comparing GDP values across these three scenarios, it highlights the direct costs of corrosion, the savings from preventive measures, and the scope for further research and development in corrosion mitigation.

Table 9.5 provides a clear comparison of the three models, highlighting their unique features, scope, and applications in evaluating and managing corrosion costs.

9.7.2 Case study I

Hou et al.[13] analyzed corrosion costs in China using the "Uhlig method." Hou et al.[13] conducted a comprehensive study focusing on evaluating the costs associated with implementing various anticorrosion measures rather than accounting for the replacement of carbon steels due to corrosion. The study provides insights into the economic burden of corrosion and highlights the financial implications of protective measures across multiple categories. For clarity, "RMB" refers to Renminbi, the official currency of China.

Table 9.5 Comparison among "Uhlig," "Hoar," and "I/O" methods

Aspect	*Uhlig model*	*Hoar method*	*I/O model*
Scope	Classifies costs into direct and indirect categories for a broad understanding of corrosion impact.	Focuses on estimating specific industrial sectors, considering preventive measures and capital investments.	Evaluates corrosion's economic impact through GDP analysis under three scenarios: real world, no corrosion, and controlled corrosion.
Direct costs	Use of alternative materials (e.g., non-ferrous metals or alloy steels) due to iron and steel vulnerability. Protective coatings (e.g., paints and phosphating solutions, accounting for ~50% of annual paint costs). Replacement or repair costs for corroded materials and labor for maintenance. Losses from downtime caused by corrosion.	Downtime, over-design of equipment, efficiency loss, reduced output, and increased maintenance. Accidents, contamination, and safety hazards countermeasures such as protective coatings, materials, cathodic protection, and regular maintenance.	Calculated as the difference between GDP in a non-corrosion world (Universe I) and the real world (Universe II). Corrosion cost = GDP (Universe I) – GDP (Universe II).
Indirect costs	Not explicitly detailed but implied through protective measures.	Accounts for borrowing and opportunity costs for corrosion control projects.	Highlights savings from preventive measures, calculated as the difference between GDP in an ideal corrosion-inhibited world (Universe III) and the real world (Universe II). Avoidable corrosion cost = GDP (Universe III) – GDP (Universe II).
Countermeasure costs	Not explicitly detailed but implied through protective measures.	Explicitly considers countermeasure costs, including materials, installation, and maintenance expenses.	Indirectly reflected in the savings from moving toward an ideal corrosion-inhibited world (Universe III).
Capital expenses	Not addressed explicitly.	Includes borrowing and opportunity costs for corrosion management projects.	Not addressed directly but inferred through GDP.

(*continued*)

Table 9.5 (*Cont.*)

Aspect	*Uhlig model*	*Hoar method*	*I/O model*
Comprehensive analysis	Provides a high-level overview of corrosion's financial and operational impact.	Offers a detailed evaluation of corrosion-prone areas, damage severity, and financial costs within specific sectors.	Quantifies corrosion's impact on the economy by comparing GDP scenarios, offering a macroscopic view.
Long-term focus	Highlights the need for effective prevention and control strategies.	Enables industries to assess effectiveness and plan resource allocation.	Identifies opportunities for innovation and future research. Research scope = GDP (Universe III) – GDP (Universe II).
Best suited for	Industries requiring a broad categorization of corrosion costs.	Sector-specific analysis of control measures.	Evaluating economic impacts of corrosion at a national/macroeconomic level.

9.7.2.1 *Key components of the study*

The analysis categorized costs into six key areas: "coatings," "surface treatments," "corrosion-resistant materials," "corrosion inhibitors," "rust-preventing oils and greases," and "electrochemical protection." The "direct costs" from all the categories combined were estimated at "1063.91 billion RMB" in 2014. Below is a detailed breakdown:

- *Coatings:* In 2014, China produced "16,481,900 metric tons of paints," with anti-corrosion paints accounting for around "50%" of the total production. With this approximation, the estimated cost associated with anti-corrosion paints was "703.78 billion RMB," representing "66.15%" of the total corrosion protection expenses.
  - *Assumptions:* Corrosion cost estimation includes the idea that materials like iron and simple steel, which are highly susceptible to corrosion (e.g., rusting), could have been widely used for various applications. However, due to their vulnerability, alternatives such as non-ferrous metals or alloy steels are often chosen, leading to higher costs directly attributable to corrosion. The cost difference incurred by switching to these alternatives represents a direct economic loss caused by corrosion. Additionally, when using materials like iron and steel, protective measures such as applying coatings, paints, or phosphating are often implemented to mitigate corrosion. It is estimated that approximately 50% of the paint produced in a country is used specifically for corrosion resistance, highlighting the significant expense associated with these preventive measures.
- *Surface treatments:* Surface treatments included "galvanized steels, tinplating, electroplating, and surface treatments for aluminum alloys." Based on consumption rates and price differences from cold-rolled steels, the total cost for surface treatments was calculated to be "140.82 billion RMB," accounting for "13.24%" of the total.
- *Corrosion-resistant materials:* Materials such as "stainless steels, weathering steels, titanium and titanium alloys, and engineering plastics" were evaluated. The total cost for these materials in 2014 was estimated to be "205.81 billion RMB," contributing to "19.34%" of the total.
- *Corrosion inhibitors:* Examples include volatile inhibitors and pickling inhibitors. These measures contribute to reducing corrosion effects, although they represent a smaller portion of the total cost. The detailed cost for inhibitors is integrated into the overall framework. The cost for corrosion inhibitors was "5 billion RMB," representing "0.47%" of the total.

- *Rust-preventing oils and greases:* China consumed "0.2 million tonnes" of rust-preventing oils and greases in 2014, with an associated cost of approximately "2.2 billion RMB," or "0.21%" of the total.
- *Electrochemical protection:* Costs for electrochemical protection were assessed based on sales data from the cathodic protection industry, amounting to "6.3 billion RMB," or "0.59%" of the total.

However, the study did not account for "Indirect corrosion costs," which encompass:

- Compensation for loss of productivity.
- Environmental contamination.
- Casualties and safety hazards.
- Additional damages arising from corrosion-related failures.

When indirect costs were included, the total cost of corrosion in China was estimated to be "2127.8 billion RMB," representing approximately "3.34% of the country's GDP" in 2014.

9.7.2.2 Assumption

If data on the production or synthesis of paints in a specific country is available, it can be estimated that approximately 50% of these costs are directly associated with corrosion prevention. Costs associated with shutdowns, over-designed equipment, loss of output due to machine downtime, efficiency reductions, accidents, or contamination from wear or corrosion are classified as indirect losses. It is often assumed that indirect costs are equivalent to direct costs, meaning the overall corrosion cost can be estimated as 50% direct costs and 50% indirect costs.

The study employing the "Uhlig method" highlights the substantial economic impact of corrosion in China, underscoring the urgent need for effective anti-corrosion technologies and measures. It provides a foundation for policymakers and industry leaders to optimize resources, select corrosion-resistant materials, and implement robust control strategies to minimize financial losses and enhance industrial sustainability. This study adopted 50–50 assumptions for direct and indirect costs. While this approach relies on approximations, it offers a practical framework for understanding and estimating corrosion-related losses, delivering valuable insights despite its inherent limitations.

A more precise and detailed analysis is facilitated by the Hoar method (discussed in Case Studies 3 and 4), which emphasizes sector-specific quantification. Through creative problem-solving techniques, such as brainstorming or synaptic approaches, the method focuses on narrowing investigations to specific industries or subsectors. For instance, within the agricultural sector, analyzing machinery as a subsector allows for an in-depth examination of corrosion challenges. This targeted approach yields more actionable and detailed insights, enabling tailored strategies to mitigate corrosion within distinct industry contexts.

9.7.3 Case study 2

India requires comprehensive studies on the economic costs of corrosion. Few early assessments, such as Rajagopalan's work in 1958[14] provide valuable insights. Using Uhlig's method, Rajagopalan estimated the annual cost of corrosion in India for 1960–1961 at a conservative figure of Rs. 1538 million. His study highlighted key materials used to combat corrosion during this period:

- *Copper and alloys:* The predominant corrosion-resistant materials in India, due to their effectiveness in combating degradation. In this case, the cost contribution from copper alloys, which act as corrosion-resistant materials, is significantly higher at 33%. This highlights the importance of increasing the use of coatings, as these 2D materials are more cost-effective and can significantly reduce overall expenses.
- *Protective coatings/paints:* Significant contributors to corrosion prevention, consuming a substantial portion of resources such as paints, varnishes, and lacquers.
- *Stainless steel:* Ranked third among corrosion-resistant materials, underscoring its critical role in industries.

These findings emphasized the corrosion challenges faced by India as a developing nation and underscored the need for employing corrosion-resistant materials and protective coatings to mitigate industrial losses. Notably, the figures pertain to the 1960–1961 period and do not reflect contemporary conditions. However, adjusting these values to current prices requires applying a multiplication factor for inflation, which is currently estimated at 90.64. This factor, derived from the consumer price index (CPI),[15] equates to approximately Rs. 139,404.32 million, highlighting the growing economic burden of corrosion over time.

Currently, the emphasis is primarily on coatings, as the treatments utilized in the 1960s have undergone significant advancements. With the rapid progress in technology, there is a growing need for updated case studies, particularly in India, to enable data comparison and the implementation of more effective corrosion prevention measures.

In developed countries, the cost of corrosion typically accounts for approximately 3% of GDP, whereas in developing countries, it is closer to 6%. This disparity is largely attributed to limited awareness and insufficient implementation of effective corrosion control measures. Bridging this gap necessitates utilizing existing scientific knowledge, conducting comprehensive data collection, performing unbiased data analysis, and implementing robust corrosion prevention strategies. By increasing awareness and adopting more effective preventive measures, significant reductions in the economic burden of corrosion can be achieved in developing nations, fostering greater efficiency and sustainability.

A significant challenge lies in collecting accurate data, which is further compounded by the lack of expert interpretation. Information is often sought through questionnaires distributed to various industries. However, many industries are reluctant to disclose indirect costs, such as those associated with environmental pollution, casualties, or injuries resulting from atmospheric contamination.

9.7.4 Case study 3

In this case study, Hoar method[14] was utilized to estimate the cost of corrosion across industrial sectors in China to assess the financial impact of corrosion in 2014. This method analyzed data from five key economic sectors: infrastructure, energy, transportation, water, and manufacturing/public services, using representative surveys and questionnaires. The study primarily focused on "direct corrosion costs" due to challenges in quantifying indirect costs like environmental compensation or loss of productivity.

- *Infrastructure:* Covered roads, bridges, ports, and water conservancy projects.
  - The total direct corrosion cost was "74.91 billion RMB," accounting for "6.68%" of the sector's gross output. For example, roads and bridges alone incurred a corrosion cost of "62.37 billion RMB," representing "4.03%" of their total investment.
- *Energy:* Included coal mining, fossil fuel power, oil, and gas industries.
  - Direct corrosion costs totaled "229.33 billion RMB" ("20.45%" of the sector). For coal mining, "84.70 billion RMB" was attributed to corrosion, accounting for "4.67%" of annual production value due to maintenance, asset depreciation, and repair costs.
- *Transportation:* Encompassed automobiles, railways, shipbuilding, and aviation.
  - This sector faced the second-highest cost at "268.72 billion RMB," making up "23.97%" of the sector's gross output. Shipbuilding alone incurred "58.00 billion RMB" ("9.16%" of the industry's revenue) due to the harsh operating environment requiring significant investment in anticorrosion measures, repairs, and asset depreciation.

- *Water Sector:* Focused on water supply, drainage, and reservoirs.
  - The direct corrosion cost was "9.69 billion RMB," representing "0.86%" of the sector's value. Costs stemmed from pipeline replacements and corrosion mitigation in urban water systems.
- *Manufacturing and public services:* The largest contributor, this sector's cost reached "538.57 billion RMB," or "48.03%" of the total. Significant corrosion costs were reported in the chemical industry ("147.10 billion RMB") and electronics ("224.80 billion RMB"), where corrosion impacts efficiency, maintenance, and product durability.

The Hoar method offers a detailed assessment of corrosion-related costs, highlighting direct expenses totaling **1109.02 billion RMB** across surveyed sectors in China. These findings emphasize the urgent need for effective corrosion control strategies, targeted investments, and increased industry awareness to reduce the economic impact of corrosion.

A focused approach within specific industrial subsectors can provide valuable insights into corrosion challenges. For example, identifying whether corrosion is primarily caused by sand, soil, or water allows for a more tailored understanding of the problem. This vertical strategy enables the collection of detailed information about the causes and effects of corrosion, facilitating more effective solutions.

Corrosion costs are typically categorized to enhance clarity. Direct costs include expenses directly attributed to corrosion damage, such as repairs and replacements. Countermeasure costs, including protective coatings and preventative actions, also form a significant portion of the overall expenditure. Additionally, costs associated with research, data collection, and analysis are essential for a comprehensive understanding of corrosion's financial impacts.

This approach involves a systematic examination of the corrosion process, starting from identifying vulnerable areas to assessing potential damage and financial losses. By deeply analyzing the elemental causes and impacts of corrosion, tailored solutions can be developed. This focused methodology enables a thorough understanding of corrosion within specific contexts. It supports informed decision-making and the development of more effective prevention strategies, ultimately reducing the economic burden of corrosion.

9.7.5 Case study 4: Sector-wise corrosion costs in India

The economic burden of corrosion in India has been analyzed using data from the NACE[10] economic impact report and relevant sector-specific studies. Corrosion is a significant issue that impacts various sectors, including agriculture, industry, and services, causing economic losses and impeding growth.

- *Agriculture sector:* Agriculture and allied activities, including forestry, fisheries, and crop production, contribute significantly to India's economy, with a GDP contribution of $203,934 million. Corrosion-related costs in this sector are estimated at $12,496 million, representing 6.1% of the sector's GDP. Key contributors to corrosion costs in agriculture include the deterioration of infrastructure, irrigation systems, storage facilities, and agricultural equipment. These challenges highlight the urgent need for effective anticorrosion measures to enhance productivity and sustainability in the agriculture sector.
- *Industry sector:* The industrial sector in India, which encompasses mining, manufacturing, construction, electricity, and water supply, accounts for substantial corrosion-related expenses. The total corrosion cost for this sector is estimated at $22,805 million or 4.7% of its GDP contribution of $488,110 million. Key highlights include:
    - *Mining and quarrying:* Corrosion costs account for $1619 million, or 6.1% of GDP, primarily due to the deterioration of mining equipment and pipelines.
    - *Manufacturing:* This subsector faces corrosion costs of $10,277 million (5.7% of GDP) from damaged tools, machinery, and infrastructure.
    - *Construction:* With a corrosion cost of $8015 million (3.2% of sector GDP), challenges include the decay of bridges, roads, and buildings.
    - *Electricity:* Power generation and distribution incur $1721 million (~11%) in corrosion costs due to eroding power plants and transmission lines.
    - *Water supply:* Corrosion costs in this subsector amount to $792 million (~6%), driven by the degradation of water treatment plants, pipelines, and distribution systems.
- *Services sector:* The services sector, a major driver of India's economic growth, also incurs corrosion costs, estimated at $13,471 million, which accounts for 3.4% of the sector's GDP of $396,093 million. Specific areas include:
    - *Accommodation and food service activities:* It costs around 23%.
    - *Transportation and storage:* Corrosion-related costs, totaling $2392 million, are driven by vehicle maintenance, infrastructure decay, and terminal repairs.
    - *Accommodation and hospitality:* Infrastructure deterioration adds to operational expenses.
    - *Information technology and telecommunications:* Equipment wear and maintenance contribute to corrosion costs.

This study highlights the profound impact of corrosion on key sectors such as manufacturing, agriculture, and infrastructure, underscoring the urgent need for effective prevention strategies. In agriculture, corrosion severely affects irrigation systems and storage facilities, necessitating investments in anticorrosion technologies. Similarly, the manufacturing sector incurs significant costs due to equipment degradation, with corrosion often rendering machinery inefficient or unusable, leading to substantial productivity losses. Additionally, tool damage compromises product quality, making it difficult to meet stringent standards like zero-defect production.

The water supply sector also faces considerable challenges due to corrosion in pipelines and treatment facilities, which are critical to the national economy. This necessitates comprehensive management policies to minimize corrosion-related losses and enhance system efficiency. Across these sectors, implementing robust anticorrosion measures is essential for reducing economic burdens and ensuring sustainable operations.

The I/O method, a versatile economic model, is applied to analyze such scenarios. This model is not limited to corrosion but can be used to assess other systemic aspects by illustrating how different sectors interact economically—quantifying how much each sector contributes to meet the requirements of others. By disseminating this information, students and researchers can identify hurdles and work toward solutions that benefit industries and the public alike.

9.7.6 Case study 5: Estimation of corrosion costs in India using I/O analysis

- *Overview of I/O method:* The I/O model, developed by Nobel laureate Wassily Leontief, is an economic framework that examines the interactions between various sectors to produce goods and services. By employing national I/O matrices, the model determines the input requirements from different sectors to generate a specific output, such as the resources needed to produce \$1 worth of output in a given sector. For instance, if \$1 million is allocated for corrosion prevention through paint, the I/O model estimates the total cross-sector activity required to produce it, accounting for both direct inputs (within the same sector) and indirect inputs (from other sectors).

This model provides a detailed understanding of sectoral interconnections, illustrating how economic activities across sectors contribute to overall production. It highlights the importance of considering both direct and indirect costs when evaluating sector-specific measures and their broader implications. In the context of corrosion prevention, for example, the I/

O model can trace activities across multiple sectors for a plant worth $1 million, identifying where inputs originate, where outputs are directed, and which stages are vulnerable to corrosion. By pinpointing these critical areas, the model helps identify opportunities for system improvements, promoting more sustainable and efficient operations.

9.7.6.1 Application in India

India adopted the I/O model to analyze the costs of corrosion across 22 key industrial sectors using production data and corrosion coefficients derived from the National Bureau of Standards-Battelle Columbus Laboratories (NBS-BCL) model. The analysis considered:

- *Direct costs of corrosion:* Costs directly incurred due to corrosion, including material replacement, maintenance, and repairs.
- *Avoidable costs of corrosion:* Costs that could be reduced with effective corrosion prevention strategies.

The I/O economic model is a flexible tool that can be adapted to estimate corrosion costs by analyzing gross national product (GNP) across three hypothetical scenarios. "World I" represents the real-world economy where corrosion exists and its effects are evident. "World II" envisions a hypothetical scenario with no corrosion, serving as a baseline to measure the impact. "World III" illustrates an ideal scenario where efficient corrosion prevention strategies are universally applied. The "total cost of corrosion" is calculated as the difference between the GNP of World II (no corrosion) and World I (real-world scenario), while the "avoidable cost of corrosion," reflecting potential savings from effective prevention measures, is determined as the difference between the GNP of World III (optimal prevention) and World I. This structured approach offers a clear framework for quantifying the economic burden of corrosion and the advantages of implementing preventive strategies. One typical example of this approach is listed in the Table 9.6.

Table 9.6 Estimated corrosion costs in India for 1984–1985[13]

Sector	*Value of production (Rs. million)*	*Direct cost of corrosion (Rs. million)*	*Avoidable cost of corrosion (Rs. million)*
Agriculture, livestock, forestry	825,000	18,000	11,000
Petroleum refining and chemical products	150,000	1250	250
Transportation equipment	47,000	940	40
Construction	240,000	3000	1000
Trade and business	300,000	6000	3500
Total	2,240,900	40,760	18,040

- *Total direct cost of corrosion:* Rs. 40,760 million.
    - *Economic impact:* Corrosion imposes a significant financial burden on India's economy, particularly in agriculture, petroleum refining, transportation, and construction.
- *Avoidable cost of corrosion:* Rs. 18,040 million, representing 44% of the total cost.
    - *Savings potential:* The study highlights the opportunity to save nearly half of the corrosion-related costs through effective prevention measures.
    - *Preventive strategies:* Raising awareness and adopting advanced corrosion control technologies, such as protective coatings and material innovations, can mitigate these costs.

The I/O method provides a detailed approach to assessing the economic impact of corrosion. In India, it highlights the need for efficient corrosion prevention strategies to minimize financial losses, enhance productivity, and promote sustainability across major sectors. For corrosion cost estimation, experts play a vital role in providing accurate insights on expenses related to prevention, repairs, or replacements. Given that data can vary between sectors or companies, balancing this information is essential to address inaccuracies. For instance, when evaluating pipeline coating costs, a portion is attributed to corrosion resistance. Similarly, for stainless steel pipes, costs are divided to account for both corrosion resistance and properties like fracture toughness and hardness. This approach ensures a thorough and precise understanding of corrosion's economic effects across industries.

9.7.7 Comprehensive corrosion management system framework

A "comprehensive corrosion management system (CMS)" is vital for effectively addressing corrosion in industries. The CMS plays a critical role in mitigating risks and ensuring safety across high-risk industries. In the "oil and gas sector," corrosion-related leaks can lead to catastrophic accidents and severe environmental damage. Within "infrastructure," corrosion-induced failures, such as bridge collapses, result in significant fatalities and disrupt essential operations. Similarly, "water supply systems" are heavily impacted by corrosion in pipelines and treatment facilities, posing threats to national economies and public health. Implementing CMS effectively in these industries is vital for reducing risks, enhancing operational reliability, and safeguarding communities.

The CMS framework (ISO 31000) provides strategies for planning, implementing, and continuously improving corrosion management practices for both existing assets and future infrastructure. Its key focus includes assessing the likelihood of corrosion events and their consequences, which involve monetary losses due to safety risks, environmental impacts, and asset integrity issues.

Figure 9.13 illustrates CMS framework, which is available in open source.[10] The key components of the CMS framework are:

- *Corrosion-specific processes*
  - *Implementation, inspection, and mitigation strategies:* Incorporates specific plans for corrosion detection, prevention, and reduction.
  - *Life-cycle analysis and ROI:* Evaluates the cost-effectiveness of corrosion control measures.
  - *Regulatory compliance:* Ensures adherence to relevant standards, such as ISO 31000, for systematic risk-based planning.

    ISO 31000 is an international standard for *risk management* that provides principles, a framework, and guidelines for managing risks effectively. It involves "*Scope and Context* (provides universal guidelines for identifying, analyzing, evaluating, and addressing risks)," "*Threat Assessment* (involves identifying potential hazards, estimating their likelihood, and evaluating their impact)," "*Prevention or Mitigation options* (risk treatment: focuses on choosing and implementing appropriate strategies to minimize or eliminate risks)," "*Monitoring and Review* (ensures continuous improvement and adaptation to changing circumstances)," and "*Integration* (embeds risk management into organizational processes and decision-making)."
- *Management-specific processes*
  - *Financial losses and asset integrity:* Addresses direct costs, such as repairs, and indirect costs, including environmental impact and downtime.
  - *Safety risks and environmental concerns:* Focuses on preventing accidents and mitigating hazards associated with corrosion, especially in sectors like oil, gas, and infrastructure.
- *Integration with standard management systems:* The CMS framework integrates seamlessly with standard management systems, particularly aligning with "ISO 31000 standards," which emphasize a risk-based planning approach. This integration involves clearly defining the scope of work, systematically identifying potential threats, and developing effective mitigation plans. By adopting this structured methodology, organizations can proactively address corrosion risks, ensure compliance with international standards, and enhance overall operational resilience.
  - *Risk-based planning:* CMS emphasizes a structured and proactive risk-based planning approach. This involves identifying the assets, processes, and systems susceptible to corrosion and

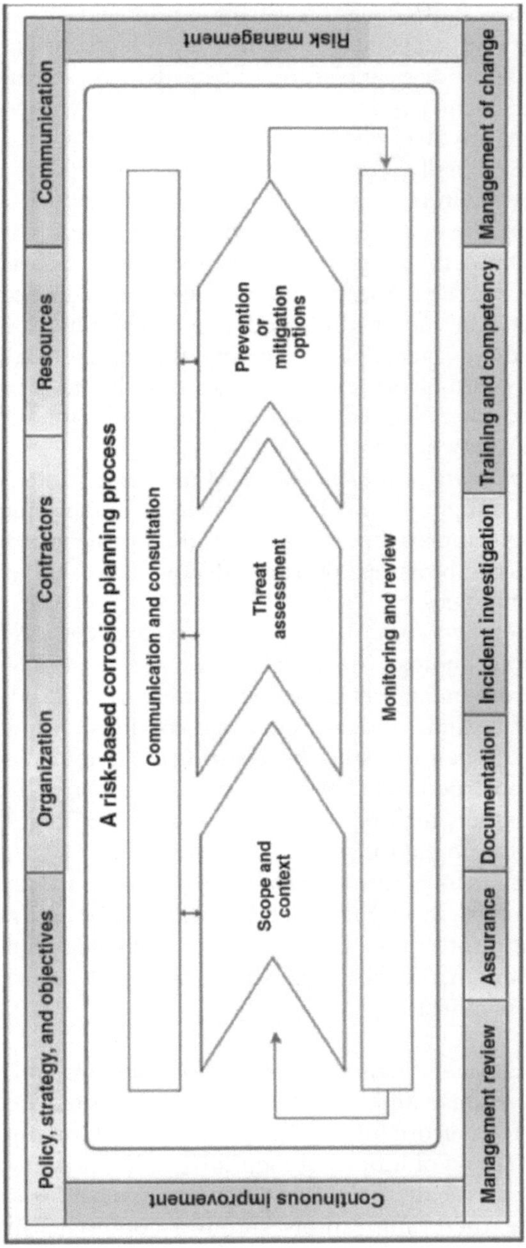

Figure 9.13 CMS framework.[10]

outlining the objectives for mitigation and prevention. For example, in an oil refinery, the scope may include pipelines, storage tanks, and processing equipment that are exposed to corrosive substances. The next step is identifying threats, such as exposure to moisture, chemicals, or extreme temperatures, which can accelerate corrosion. In the refinery context, this might involve assessing risks from acidic crude oil or saline water. Finally, creating mitigation plans entails developing and implementing strategies to minimize these risks, such as applying protective coatings, using corrosion-resistant materials, or introducing cathodic protection systems. For instance, a mitigation plan for the refinery could involve regular inspections, real-time corrosion monitoring, and deploying inhibitors to neutralize corrosive agents. By following this methodical approach, the CMS framework ensures targeted and efficient management of corrosion risks, reducing operational downtime, financial losses, and safety hazards.

- *Monitoring and evaluation:* Monitoring strategies vary by risk level, ranging from occasional checks to continuous (24/7) data-driven approaches. Regular evaluations ensure that corrosion control measures remain effective and align with evolving technologies.
- *Nanotechnology in corrosion management:* Advancements in nanotechnology have revolutionized materials and coatings, offering enhanced resistance to corrosion. Innovations such as "self-healing and self-cleaning coatings" significantly extend the lifespan of assets by resisting wear and tear. Additionally, "nanostructured thin films" provide superior scratch and wear resistance, making them highly effective in harsh environments. Another breakthrough is the development of "smart coatings," which incorporate nanoparticles to enable the controlled release of corrosion inhibitors and biocides, offering targeted and efficient protection. These advancements highlight the potential of nanomaterials to transform corrosion prevention strategies across industries.

Future applications of nanotechnology hold immense promise for industries such as aerospace and automotive, where improved corrosion resistance is critical for performance and longevity. Additionally, advancements in nanosensors are revolutionizing monitoring and inspection technologies, enabling more precise and proactive corrosion management.

By integrating these innovations within a comprehensive CMS framework, industries can effectively mitigate corrosion-related risks and reduce associated costs. Continuous improvement and adherence to international standards ensure resilience against corrosion while promoting sustainability and operational efficiency. These comprehensive strategies are essential for minimizing economic losses, enhancing safety, and driving industrial innovation.

9.8 CONCLUSIONS

The chapter concludes by reinforcing the critical need for proactive maintenance and surface engineering to address surface degradation and its associated costs. It highlights how structured maintenance strategies, such as condition-based monitoring and data-driven approaches like FMEA, enable the early detection of potential failures, reducing downtime and enhancing operational safety. The integration of advanced surface engineering techniques, including coatings and nanotechnology, provides effective solutions for extending material life and improving performance. LCIA emerges as a vital framework for evaluating and minimizing the environmental footprint of products and processes. Together, these strategies form a comprehensive approach to mitigating surface degradation, contributing to enhanced reliability, cost efficiency, and sustainability in industrial operations. Finally, the recommendations to incorporate nanotechnology innovations like smart coatings and nanosensors, which enhance corrosion prevention, reduce costs, improve safety, ensure sustainability, and drive industrial innovation with the help of the CMS framework, have been recommended.

REFERENCES

1. Okoh, P., & Haugen, S. (2014). A study of maintenance-related major accident cases in the 21st century. *Process Safety and Environmental Protection*, *92*(4), 346–356.
2. Filz, M. A., Langner, J. E. B., Herrmann, C., & Thiede, S. (2021). Data-driven failure mode and effect analysis (FMEA) to enhance maintenance planning. *Computers in Industry*, *129*, 103451.
3. Piotrowska, K., Kruszelnicka, W., Bałdowska-Witos, P., Kasner, R., Rudnicki, J., Tomporowski, A., Flizikowski, J., & Opielak, M. (2019). Assessment of the environmental impact of a car tire throughout its lifecycle using the LCA method. *Materials*, *12*(24), 177.
5. Ali, S., Kurniawan, R., Moran, X., Ahmed, F., Danish, M., & Aslantas, K. (2022). Effect of micro-dimple geometry on the tribological characteristics of textured surfaces. *Lubricants*, *10*, 328.
4. Arslan, A., Masjuki, H. H., Kalam, M. A., Varman, M., Mufti, R. A., Mosarof, M. H., Khuong, L. S., & Quazi, M. M. (2016). Surface texture manufacturing techniques and tribological effect of surface texturing on cutting tool performance: a review. *Critical Reviews in Solid State and Materials Sciences*, *41*(6), 447–481.

6. Holmberg, K., & Matthews, A. (2009). *Coatings Tribology: Properties, Mechanisms, Techniques and Applications in Surface Engineering*. Elsevier.
7. NPTEL Lecture by Harish Hirani, www.youtube.com/watch?v=K68ZumhwJVQ&list/=PLp6ek2hDcoNDiczxIkf_F500ChBxDNu5w&index=34
8. Koch, G. (2017). Cost of corrosion. *Trends in Oil and Gas Corrosion Research and Technologies*, 3–30.
9. Council, National & Sciences, Division & Board, National & Engineering, Committee. (2011). *Research Opportunities in Corrosion Science and Engineering*. 10.17226/13032
10. NACE International. *Economic Impact Report: International Measures of Prevention, Application, and Economics of Corrosion Technologies Study*. http://impact.nace.org/economic-impact.aspx
11. Expenditure of Government of India, www.indiabudget.gov.in/doc/Budget_at_Glance/bag6.pdf
12. Biezma, M. V., & San Cristobal, J. R. (2005). Methodology to study cost of corrosion. *Corrosion Engineering, Science and Technology*, *40*(4), 344–352.
13. Hou, B., Li, X., Ma, X., Du, C., Zhang, D., Zheng, M., Xu, W., Lu, D., & Ma, F. (2017). The cost of corrosion in China. *npj Materials Degradation*, *1*(1), 4.
14. Bhaskaran, R., Palaniswamy, N., Rengaswamy, N. S., & Jayachandran, M. (2005). A review of differing approaches used to estimate the cost of corrosion and their relevance in the development of modern corrosion prevention and control strategies. *Anti-Corrosion Methods and Materials*, *52*(1), 29–41.
15. India Inflation Calculator, www.in2013dollars.com/india/inflation#:~:text=The%20rupee%20had%20an%20average,consumer%20price%20index%20for%20India

Index

Abrasive particles, 162, 164–165
Abrasive wear, 1–2
  assisted corrosion, 136
  examples of, 43–44
  hardness on, 43
  lubrication, 42
  roundness factor, 42
Accelerated testing, 212
Accurate evaluations, 208–209
Acetic acid (ethanoic acid), 177
Acoustic emission testing (AE), 373
  advantages, 253
  comprehensive database, 253
  corrosion testing, 254
  definition, 253
  SCC, 254, 255
  sensors, 255
  suppliers, 254
  transition phase, 254
Active corrosion, 112, 113
Active thermography, 250–251
Adatoms, 111
Additive manufacturing, 402
Adhesive wear, 4
  Archard's wear model, 31
  area of contact, 28
  contamination layers, metal surface, 30
  electron transfer adhesion, 29
  localized stress and relative motion, 33
  mechanism, 31
  occurrence in metals, 28
  practical surfaces, micrometer scale, 29
  progressive sequence of events, 153
  relative tangential motion, 30
  stainless steel shaft and carbon graphite seal, 152
  wear volume, 32
Aerosol-assisted CVD, 413
Aerospace industry, 143
AE spectrum interpretation, 256
Agglomerated nanoparticles (NPs), 197–198
Aging hardening process, 138
Airborne particles, 172
Airborne pollutants, 194
Air filtration systems, 197
Akhtar, M., 138
Ali, S., 403
Alloy modification, 422
Alternating current (AC), 158
Aluminum alloys, 111
Aluminum 7075 matrix hybrid composites, 294, 296
American Society for Testing and Materials (ASTM), 55, 208–209
American Society of Mechanical Engineers (ASME), 227
Annual net return, 14
Anodic protection, 110
Anodization process, 201–202
AO-resistant materials, 163
Aqueous corrosion, 110–112
Arafin, M. A., 159
Archard's wear model, 31
Arc spraying, 412
Arithmetic mean roughness, 397
Arslan, A., 402
Artificial intelligence (AI), 322
Artificial neural networks (ANNs), 382
Asperities, 25
Asset management, 288
Atmospheric-assisted surface degradation, aircraft, 193–195
Atmospheric corrosion
  definition, 168
  discoloration and corrosion, 166
  dust and airborne particles, 172
  environmental factors, 165–166
  environmental variability, 169
  erosion and material removal, 166
  factors, 168
  fire environment/high temperatures, 168
  high-temperature environments, 168

humidity, 168
initial exposure conditions, 172
load-bearing capacity over time, 165
mechanical stress, 166
peeling, delamination/cracking, 166
pollutants, 173
precipitation, 172
sheltering, 172–173
temperature, 172
wetness duration, 172
wind velocity, 173
Atmospheric emissions, 392
Atmospheric factors, 165
Atmospheric pressure CVD, 412
Atomic oxygen (AO), 163
Audio and video digitization, 290
Automation, 272–273
Average daily truck traffic (ADTT) over bridges, 192

Back-to-back gear arrangement, 257
Bathtub curve concept, 23, 24, 46, 47, 368, 369
Beach marks, 80, 150
The Bhopal Gas Tragedy, 11, 13, 364
Blikharskyy, Y., 261
Bond breaking, 132
Boundary element method (BEM), 280
Breakdown maintenance (run-to-failure), 371
Broadband nanoindentation creep (BNC), 316

Cabral-Miramontes, J., 201, 202
Cascade classifiers, 299
Cathédrale St-André (UNESCO-listed historic site), 197
Cathodic protection systems, 110, 170, 443
Cause-and-effect, 313
Cavitation wear, 45, 56–59
CBM/predictive maintenance, 367
Charminar Express, derailment of, 11, 13
Charpy impact test, 219
Charpy test, 219
Chemical and heat/radiation synergy, 138–142
Chemical deposition methods, 417
Chemical domain, 4
Chemical etching, 401
Chemical exposure, 164
Chemical inhibitors, 422
Chemical passivation, 107, 113
Chemical solution deposition (CSD), 414–415
Chemical vapor deposition (CVD), 395, 412–414
Chevron lines, 80, 149
Chloride-induced surface degradation
aluminum, 201
canisters, 200–201
coastal and industrial areas, 200
crevice corrosion, 200
microstructural analysis, 201
standard testing procedures, 201
Chlorine (Cl_2), 170
Chromium ceramic coating (CKS), 420
Chromium diamond coating (GDC), 420
Chromium plating, 420
Circular ridges, 80
Classification and prediction module, 307
Classification of NDT methods
AE, 253–258
description, 223
ECT (*see* Eddy current testing (ECT))
evolution of, 225
LPT, 225, 228–229
MT, 225, 229–233
RT (*see* Radiography testing (RT))
TT, 250–253
UT (*see* Ultrasonic testing (UT))
visual inspection, 225–228
Classification techniques
applications and benefits, 299
cascade classifiers, 299
CNNs, 299
data augmentation, 299
predefined groups, 298
wear particle, 298
Clearance factor, 354
Cleavage marks, 150
Clustering technique
corrosion, 300
description, 299
mechanical design, 299
surface engineering, 300
Cobalt–chromium (CoCr) femoral head, 155, 157
Coincidence site lattice (CSL) boundaries, 159
Cold joining, 4
Cold welding, 4
Common welding flaws, 271–272
Comprehensive corrosion management system (CMS), 459–463
beyond financial losses, 433
challenges, 431
corrosion costs, 433
economic analysis, 431–432, 438–440
financial tools, cost analysis, 431–432
LCCA, 440–441
life-cycle cost components, 432
Computational cost, 303
Concrete structures, 4–5
Condition-based maintenance (CBM), 236, 365, 371–372

Condition monitoring, 319
  definition, 373
  high-risk industries, 373
  informed decisions, 372
  of spur gears
    analytics, 374
    comprehensive test setup, 375, 376
    data acquisition, 373
    documentation, 374
    experimental procedure, 378
    fault isolation, 373
    Fe ppm and wear mass, 378, 379
    ferrous particle size distribution, 379–380
    gearbox, 377, 378
    Metallic Wear Debris Sensor, 378
    methods of, 373
    nanoadditives, 381
    oil analysis, 375
    setup, 375, 377
    vibration analysis, 375
    wear debris analysis, 374
Contact stylus profilometer, 397
Continuous cooling transformation (CCT), 91
Convolutional neural networks (CNNs), 299
Copper and alloys, 177, 453
Corrective/breakdown maintenance, 367–368
Corrective maintenance, 363
Corrosion analysis, 2–3, 7, 9, 304–305; *see also* Galvanic corrosion; General (uniform) corrosion; Localized corrosion
  ample warning signs before catastrophic failure, 188
  clustering technique, 300
  coating systems, 108
  and cracks on steel bridges, 190–193
  electrochemical reactions, 104
  formation of products, 105–106
  fundamentals
    Experiment Setup by Evans, 109
    iron samples, 109
    Pourbaix diagram, 110
    types of, 108–109
  inner ring of roller bearing and connecting rod bearing, 107
  mitigation of, 104–105
  nanocoatings, 189–190
  non-metals, 107
  phosphating, 108
  physical barriers, 108
  primary atmospheric contaminants, 173–174
  ramifications of, 104
  rate of substance, 106
  reactive metals, 107
  sacrificial coatings, 108
  safety issues, 104
  specialized coatings, 105
  U.S. Coast Guard records, 188
Corrosion degradation
  advancing sustainable and innovative technologies, 429
  assumption, 452–453
  beneficial purposes, 428–429
  cathodic protection, 443
  costs evaluation, 444
  data collection and economic analysis, 434–435
  data-driven approach, 445
  economic evaluation, 429–430
  economics of corrosion, 424–427
  emerging solutions, 443
  financial analysis and decision-making, 443
  Hoar method, 446–447, 449–450
  I/O model, 447–450
  material lifespan, 421–422
  material origins and corrosion tendency, 421
  natural oxide layers, 422
  and predictive strategies, 422–423
  real-time data and condition monitoring, 423
  Uhlig, 445–446
Corrosion-fatigue (C-F), 192
Corrosion pits, 159
Corrosion rates
  indoor environment, 171
  industrial environment, 170
  marine environment, 170
  rural environment, 169
  urban environment, 169–170
Corrosion testing, 210, 213, 294
Corrosion zones, 110
Corrosive acids, 169
Corrosive and sliding wear mechanisms, 134, 135
Cost-effective solutions, 418
Covariance matrix, 349, 360
Crack formation, 3, 313, 315
Crack initiation, 313, 315
Crack tip opening displacement (CTOD) testing, 146, 147
Cradle-to-cradle approach, 345
Cradle-to-gate approach, 345
Cradle-to-grave approach, 345
Creep fractography, 81
Creep testing, 212
Crest factor, 354
Crevice corrosion, 118, 192, 200, 422
  assessment, 121, 122
  ASTM G78 standard, 121

definition, 118, 120
EDS study, 122
fissure and external environment, 120
inaccessible/concealed locations, 120
perform comprehensive testing, 123
proactive measures, 122
PTFE tape, 121
in welded joint, 121
Crystalline TG failure, 83
Cumulative damage, 132
Cumulative energy demand (CED), 393
Cusick, S. K., 163
Cyber–physical production system (CPPS), 382
Cyber World, 383
Cyclic loading, 132

Damage modes, 158
Damage tolerance approach, 99
Data analysis and plotting, 291
Data augmentation, 299
Data categorization, 388
Data-driven methodologies, 281, 303, 320
Data-driven simulations
advantages of, 302–303
challenges of, 303
description, 301
environmental management and safety, 302
future of, 303
integration of, 301
process optimization, 302
product development, 302
Data extraction, 291
Data normalization, 388
Data quality indicators (DQIs), 394
Degree of difficulty (DOD), 276
Demagnetization, 231
Design FMEA (DFMEA), 324–325
Destructive testing (DT), 216–217
controlled failure, material assessment, 218–219
Detection effectiveness, 328
Detonation spraying, 415–416
Diamond-like carbon (DLC), 413
Digital image correlation (DIC), 221, 373
Digital logic method, 275
Digital radiography, 244, 245
Digital sensor, 283
Digital twin technology, 159, 160, 287, 289, 303
Digitization process
audio and video, 290
definition, 280
document scanning, 290
IDP, 290
image, 290–291
MATLAB® 2020b plotting, 281, 292–296
Origin Software, 281, 290–291
trendline fitting, Excel 365, 291–292
"Trend lines", 280
Dimensionality reduction technique, 300
Dip coating, 412, 416–417
D–I–P–F curve, 366–367
Direct costs, 430, 431, 446
Direct current (DC), 201
Direct grain removal, 36–38
Direct losses, 9, 10
Discoloration and corrosion, 166
Disconnected dimples, 406
Distortion, 417
Document scanning, 290
3D printing (additive manufacturing), 401
Drilling and blasting method, 52
Durable industrial coatings, 418–419
Dust particles, 5, 172
Dynamic configurations, 251
Dynamometer tests, 282

Eco-Indicator 99, 393–394
Eddy current sensors, 373
Eddy current testing (ECT), 220
corrosion, 236–237
definition, 233
influencing parameters, 234–235
process, 235
structural failures, 234
working principle, 234
Effect of rate of returns on profitability, 18
Effect of roughness on wear, 14
Efficiency ratio (ER), 276
Eigendecomposition, covariance matrix, 350
Eigenvalues, 360
Electric discharge machine (EDM), 403
Electrochemical cell generation, 3
Electrochemical deposition, 415
Electrochemical etching, 402
Electrochemical machining (ECM), 403
Electrodeposition, 415
Electron transfer adhesion, 29
Electroplating, 395, 415
Elevated temperatures, 127, 140, 317
Embrittlement, 138–139
hydrogen, 139
radiation-induced, 139
thermal, 139
End-of-life (recycling), 391
Energy and material efficiency, 419
Energy dispersive X-ray spectroscopy (EDS) study, 122, 198
Energy transfer modes, 251

Environmental factors, 2
  of surface degradation
    abrasive particles, 164–165
    acidic conditions, 196
    chemical exposure, 164
    historical perspectives, 166–169
    in surface damage, 195–196
    temperature variations, 164
    UNESCO World Heritage Sites, 196
    UV radiation, 164
    water and humidity, 164
Erosion, 194
  corrosion mechanism, 136
  and material removal, 166
Erosion wear, 4, 44–45, 48
  direct impacts, 49
  ductile and brittle modes, 50
  ductile mode, 48
  illustration of, 49
  water, 49
Erosive resistance, 54–56
EtO-sterilized cups, 157

Failure analysis
  benefits of, 321
  challenges and future directions, 321–322
  degradation mechanisms, 309 (*see also* Failure detection)
Failure detection
  indicators, 311
  KPIs, 312
  mechanical failure, 312
Failure mode and effect analysis (FMEA), 310, 314
  benefits of, 325, 384
  causes and current controls, 323
  challenges in, 325–326
  Cyber World, 383
  data acquisition phase, 382–383
  data collection and preparation, 381
  decision support, 382, 383–384
  DFMEA, 324–325
  effects of, 323
  identification, 323
  integration, 382
  Journal bearing system (*see* Journal bearing system)
  maintenance planning, 381
  model selection and optimization, 381–382
  PCA, 322
  PGMEA, 325
  physical world, 382
  role in, 322
  RPN, 322, 381
  scope of analysis, 323
  severity, occurrence and detection ratings, 323–324
  System FMEA, 325
  technological advancements, 326
Fast Fourier transform (FFT), 256
Fatal oxidation, 6
Fatigue and fracture mechanisms
  plastic ratcheting, 100
  probabilistic methodologies, 100
  reversible stress fluctuations, 98
  sliding contact, 98
  S–N diagram/S–N criteria, 99
  wear phenomena, 97
Fatigue failure, 3
Fatigue pitting, 306
Fatigue testing, 212
Fatigue wear, 4
  crack formation and grain structure evolution, 63
  environmental conditions, 65
  influence of, 62
  localized friction forces, 64
  lubrication, 63
  pit formation rates, 64
  repeated loading and unloading, 59–60
  rolling bearings, 64
  sliding motion, 60
  stacking fault energy, 63
  surface shear strain, grain structure, 61
  temporal variation, contact area, 60
Fault analysis, 365
Fault detection and diagnostics (FDD), 319
Fault isolation, 373
Fault localization, 257
Fault monitoring
  data collection and feature extraction, 352–353
  gear conditions, types of, 352
  industrial machinery, 351
  MATLAB code, 355, 357–358
  PCA, use of, 353–354
  variation trends, originally extracted features, 356
Fault tree analysis (FTA), 310, 315
Feature extractor, 307
Feng, K., 306
Field boundaries, 285
Filtration, 42
Filz, M. A., 381, 382
Finite element method (FEM), 259, 280
Fire corrosion, 168
Fishbone diagram, 215
Flame spraying, 412
Flow-accelerated corrosion (FAC), 136
Flow-assisted corrosion, 6
Fluorinated ethylene propylene (FEP), 181
Foreign object damage (FOD), 194
Formic acid (methanoic acid), 176
Four-ball tester, 46

Fractography
cleavage microstructure, 145
CTOD, 146, 147
definition, 144
dimple microstructure, 144–145
erosion and wear mechanisms, 148–150
fracture, 150–151
of fractured surface, 145
impact fracture, 146
magnified images, 144
magnified regions, 149
micro-voids, 145–146
multimode fractures, 145, 147
patterns of fracture
beach marks, 150
Chevron lines, 149
cleavage marks, 150
multiple cracks, 150
ratchet marks, 150
ridge marks, 150
shear lips, 150
SEM, 144
small cracks, 144
and wear debris, 151–155
Fracture mechanics approach, 99
Fracture surface degradation
beach marks, 80
brittle and ductile fractures, 77, 78
chevrons, 80
combined brittle and ductile fracture, 84, 85
creep deformation, 81
crystalline TG failure, 83
definition, 75–76
ductile materials, 76
fracture failure through three stages, 86
granular/transgranular crack paths, 82
macroscopic fractography, 79
microstructure observation, 82
minimal percentage elongation/area reduction, 76
moisture, 78
Ni– Cr–P–Si active powder, 88
non-crystalline TG failure, 83
nucleation, 86
plastic deformation and necking, 76
prolonged observation, 78–79
ratchet marks, 80
shear lips, 81
slow propagation, 87
surface geometry, 77
UV radiation, 78
Fracture toughness, 212
Fretting corrosion, 137
Fretting wear, 4, 65–70
Friction reduction, 401
Friction testing, 404
F46 second-surface mirrors, 180
Functional failure, 365, 367
Fuselage erosion, 51

Galvanic corrosion, 422
anode and cathode metals, 117
description, 116
diverse techniques, 116
electrochemical potential, 116, 117
electrochemical potential disparity, 116
metal functions, 116
minimal corrosion, 116
rapid and severe corrosion, 117
Gas chromatograph thermal desorption analysis, 143–144
Gaseous-state coatings, 411
Gate-to-gate approach, 345
Gear debris analysis
abrasive wear, 305
definition, 305
high-fidelity digital twin model, 306–307
SEM, 305
statistical measures, 306
transfer learning algorithm, 307
General (uniform) corrosion, 108
active, 112, 113
aluminum alloys, 111
anodizing aluminum, 114
aqueous corrosion, 111, 112
chemical and physical layer, 114
chemical passivation, 113
definition, 110–111
deterioration, 114
electrochemical, aqueous solutions, 115
passive corrosion protection, 112
passive *vs.* active, 113
rusting of chain and jaw-coupling, 111, 112
Generation of free radicals, 132
Greek philosophy, 166, 168
Grosse Cloche (UNESCO-listed historic site), 197
Gupta, K. K., 298

Hailstone, 194
Hardness on abrasive wear, 43
Haugen, S., 364, 373
High-fidelity digital twin model, 306–307
High-performance coatings, 417–418
High-quality fabrication, 191
High-speed flight scenarios, 194–195
High-speed nanoindenters, 316
High-velocity oxyfuel (HVOF) flame coating, 416, 419
Historical steel bridges, 191
Hoar approach, 430, 433–434, 446–447, 449–450, 455

Hot-dip coating, 412
Hou, B., 448
Hu, X., 175, 204
Hybrid coating technologies, 419
Hybrid composite formation, 294
Hydrochloric acid (HCl), 170
Hydrogen embrittlement phenomenon, 139–142, 146–147
  atoms, 203
  crack initiation and propagation, 203
  infiltration, 203
  tensile tests, 203–204
Hydrogen gas evolution, 256
Hydrogen- induced cracking (HIC), 158
Hydrogen-induced material degradation, 140
Hydrogen ingress, 140
Hydrophobicity, 401

Image digitization
  calibration, 291
  data analysis and plotting, 291
  data extraction, 291
  definition, 290
  export and save, 291
  import, 290–291
  tracing, 291
Image tracing, 291
Impact fracture, 146
Improvement analysis, 389
Impulse factor, 354
Indirect costs, 430, 431, 446
Indirect losses, 9, 10
Indium tin oxide (ITO), 181
Indoor environment, 171
Indoor *vs.* outdoor specific approaches, 175
Industrial environment, 170
Industrial fluids, 162
Infant mortality, 24
Informed decision-making, *see* Condition monitoring
Input/output method, 430
In situ non-destructive testing (NDT) technique, 315
Intelligent Document Processing (IDP), 290
Intergovernmental Panel on Climate Change (IPCC), 393
Intergranular corrosion (IGC), 117, 118
International Organization for Standardization (ISO), 385
Internet of Things (IoT) devices, 159, 311
Interpretation phase, 389
I/O approach, 434, 447–448, 449–450, 457–459
Ionizing radiation, 127
Ishikawa diagram, 313

Journal bearing system
  components, 329
  control/ prevention (detect ranking), 328
  correction requirement, 329
  detectability guidelines, 333
  end/iteration, 329
  failure modes with high RPN values, 328, 331, 337
  FMEA process flowchart, 326, 327
  FMEA worksheet
    for bearing, 336
    for journal, 334
    lubrication system component, 335
  occurrence guidelines, 332
  RPN calculation, 329
  setup, 329, 330
  severity guidelines, 332
  simplified overview, 330
  study product and sub-components, 327

Kaewunruen, S., 304
Kapton, 181
Key performance indicators (KPIs), 312, 382
Kurtosis, 354

Labrador Current, 183
Laser ablation, 401, 402–403
Laser machining, 403
Laser micromachining, 403
Laser texturing, 403
LEO satellites
  AO degradation, 178
  design considerations, 179–182
  high-speed environment, 178
  mechanical, chemical and thermal risks, 179
  protective coatings, 179
  reactivity of AO, 178
  services, 178
Life-cycle assessment (LCA), 13
  applications of, 346
  benefits of, 345–346
  CED and IPCC, 393–394
  challenges in, 346–347
  contributory events, 317
  cradle-to-grave approach, 344
  data quality and validation, 394
  detailed mechanisms, 318–319
  disposal, 391
  distribution, 391
  documentation and communication, 389
  end-of-life (recycling), 391
  environmental impacts, 310, 392
  environmental improvements, 394

future directions, 347
impact categories, 394
ISO guidelines, 344–345, 385
mapping creep, 317–318
material characteristics, 317
modeling, 312
multi-faceted approach, 313
production stage, 391, 393
product's inputs, outputs and environmental impacts, 385–386
raw materials acquisition, 391
recycling stage, 393
size designation P205/55/R16, 392–393
small-scale industries and MSMEs, 386
stages of, 311
storage, 391
stress levels, 317
types of, 345
use stage, 391, 393
Life-cycle cost analysis (LCCA), 431
and materials degradation, 435–437
Life-cycle impact assessment (LCIA) approach
definition, 384
emissions, 385
energy consumption, 385
goal and scope, 386
interpretation, 386–387
inventory analysis, 386
ISO 14040 and ISO 14044, 387–388
LCA (*see* Life-cycle assessment (LCA))
material use, 385
sustainable development, 385
waste production, 385
Life-cycle inventory (LCI), 390
Li, L., 230, 231
Liquid penetration testing (LPT), 220
application of developer, 229
application of liquid penetrant, 228
definition, 225, 228
dwell time, 228
effectiveness of, 229
excess penetrant removal, 228–229
inspection and evaluation, 229
limitations, 229
purpose and importance, 228
surface preparation, 228
Liu, T., 180, 181
Li, W., 351, 354
Localized corrosion, 108
crevice, 118, 120–123
IGC, 117, 118
pitting, 118–120
SCC, 123–125
scratches/pre-existing corrosion sites, 117
Lock-in thermography (LT), 252
Low-angle grain boundaries (LAGBs), 159
Low Earth orbit (LEO), 163
Lubrication, 42
retention, 401

Machine learning algorithms, 321
Macro fiber composite (MFC) transducers, 241
Macroscopic fractography, 79
Magnetic particle testing (MT)
aircraft maintenance and safety assurance, 232
automotive industry, 232
definition, 225, 229
demagnetization, 231
ferromagnetic particles, 231
magnetic field, 230
magnetization, 230–231
pipeline inspection, 232
post-processing, 231
preprocessing, 230
railroad inspection, 233
surface cracks, 232
test reliability, 230
trace assessment, 231
welded joints, 232
Magnetic trace assessment, 231
Magnetorheological (MR) material, 148
Maintenance strategies; *see also* Failure mode and effect analysis (FMEA)
along P–F curve
bathtub curve concept, 368, 369
CBM/predictive, 367
corrective/breakdown, 367–368
corrective/breakdown maintenance, 370
D–I–P–F curve, 366–367
functional failure, 365
illustration of, 366
potential failure, 365
precision, 367
preventive, 367
RUL, 368
Bhopal gas tragedy, 364
CBM, 371–372
comprehensive risk analysis, 365
definition, 363
lack of barrier, 364
preventive and corrective, 363
protocols and critical safety equipment, 364
time-dependent *vs.* time-independent, 370–371
Manual data handling and analysis
accuracy and efficiency, 287
asset management, 288
conventional methods, 286
cyber and physical systems, 289
data integration, 289

"Digital Twin" models, 287
predictive maintenance and real-time insights, 289–290
skilled professionals, 287
time-deteriorated system, 289
worn-out surfaces, 288
Manufacturing processes, 14
Marine environment, 170
Marquez, Garcia, 241
Martensite–austenite (MA) constituents, 90
Material degradation, 8, 9
MATLAB® 2020b plotting
excel plot and fitting, 294, 295
functions, 292
plot of sine, 293
Matrix factorization, 252
Maximum peak height, 398
Maximum peak value, 354
MCR 102 Rheometer, 378
Mean value, 353
Mechanical actions, 21, 127
chemical actions
corrosive and sliding wear mechanisms, 134, 135
definition, 134
mechanisms, 136
wear-assisted corrosion mechanisms, 134, 135
and heat/radiation actions
creep, 137
high-speed rubbing/sliding, 137
tensile and impact tests, 138
T6 heat treatment cycle, 138
thermal fatigue, 137–138
heat/radiation and chemical actions, 142–144
Mechanical loading, 131
Mechanical machining, 401
Mechanical rock-cutting techniques, 53
Mechanical wear
defnition, 21
instances of patterns, 23
Melting, 89
wear, 70–75
Metal and alloy degradation, 167
Metal dissolution, 256
Metallic Wear Debris Sensor, 378
Methyl isocyanate (MIC) gas, 364
Microbial-induced corrosion (MIC), 108
Micro-cutting wear phenomenon, 35
Micro-fatigue wear phenomenon, 35
Micro-fracture wear phenomenon, 35
Micro-voids, 145–146
Minimal corrosion, 116
Mitigating synergistic degradation, 143
Mitigation strategies and corrosion analysis, 441
Mohtadi-Bonab, M. A., 158
Molten-state coating, 412
Multiple cracks, 150
Multi-station Hip Simulator tool, 155–158
Multivariate regressions, 382
Munoz, Gomez, 241

Nanoadditives, 381
Nanocoatings, 163, 189–190
Near-surface linear defects, 263–264
Near-surface volumetric defects, 264
Nitrogen oxides (NO_x), 170
Noise reduction techniques, 260
Non-corrosion zones, 110
Non-crystalline TG failure, 83
Non-destructive testing (NDT), 51, 209, 215–217, 319; *see also* Classification of NDT methods
destructive and non-destructive testing methods, 224, 225
integrity without compromising material
applications of, 220–221
DT and, 221–222
economics of, 221
purpose of, 219–220
role of, 220
limitations, 264–266
near-surface volumetric defects, 264
selection criteria
delamination, 262
surface-breaking volumetric cracks, 262–263
types of, 262
specific defect characteristics, 263–264
subsurface linear defects, 264
subsurface volumetric defects, 264
Non-destructive testing (NDT) technique, 281
No observed adverse effect level (NOAEL), 177
Normalization, 389
Nuclear reactors, 142–143
Nucleation, 86, 120

Oil analysis, 375
Oil quality index (OQI), 378
Oil Sensor Suite, 378
Okoh, P., 364, 373
Oliveira, M. L., 197
Operating time (OT), 276
Operational optimization, 301
Optical Character Recognition (OCR), 290
Optical microscopy (OM), 81, 144
Orbital debris (OD), 179
Organic compound coatings, 408

Organic materials, 199
Origin Software, 281, 290–291
Oxidation, 6
Ozonated lignin, 199

Palladium, 181–182
Parametric matrix and MATLAB code, 274
Particulate matter (PM), 170
Passivation, 180
  zones, 110
Passive corrosion protection, 112
Passive thermal control systems, 179
Passive thermography, 250
Passive *vs*. active corrosion, 113
Peeling, delamination/cracking, 166
Peel-off tests, 219
Performance degradation over time, 9, 10
Permanent magnetic bearings (PMBs), 337–344, 383
Phantom YF-4M aircraft, 195
Photo-assisted CVD, 413
Photocatalytic process, 139
Photochemical degradation, 92, 131–132
Photochemical deterioration, 5
Photodegradation, 131–132
Photoionization, 93
Physical domain, 4
Physical vapor deposition (PVD), 395, 414
Pin-on-disc tribometer, 46
Piotrowska, K., 386, 390, 392
"Pitch and catch" technique, 241
Pitting corrosion, 192
  definition, 118, 119
  metal ions, 120
  nucleation, 120
  stages of, 119
Plasma-enhanced CVD, 413
Plasma spraying, 412, 415
Plastic ratcheting, 100–101
Polishing wear
  carbon graphite material, 26
  definition, 25
  numerous factors, 27
  pit formation mechanism, 28
  surface texture, 25, 26
  worn rollers, 27
Pollutants, 173
Polytetrafluoroethylene (PTFE), 157, 158
Portable IR cameras, 252
Potential failure, 365, 367
Pourbaix diagram, 110
Precipitation, 172
Precision maintenance, 367
Precursor solutions, 415
Predictive maintenance, 319
  analytical techniques and benefits of, 301
    classification, 298–299
    clustering, 299–300
    dimensionality reduction, 300
    regression analysis, 298
  condition monitoring, 297
  data-driven simulations, 301–303
  energy sector, 297
  equipment failures, 300
  historical and real-time data, 296
  machine learning models, 297
  with RUL estimation, 427–428
Present value, 14
Preventive maintenance, 363, 367, 371
Principal component analysis (PCA), 252, 300, 304, 322
  anomaly identification and pattern recognition, 348
  benefits of, 350–351
  challenges in, 351
  covariance matrix computation, 349
  data quality and signal-to-noise ratio, 348
  eigendecomposition, 349
  fault monitoring (*see* Fault monitoring)
  future directions, 351
  mathematical foundations, 350
  projection, 349
  selection, 349
  standardization, 348
  statistical method, 347
Probabilistic methods, 99–100
Probability of failure (POF), 327
Process FMEA (PFMEA), 325
Process optimization, 302
Product development, 302
Profilometers, 397, 398
Protective coatings/paints, 170, 422, 453
Pulsed phase thermography (PPT), 252
Pulsed thermal testing, 252
Pulsed thermography (PT), 252
Pulse echo method, 240, 241

Quenching process, 138

The Rabinowicz equation, 41
Radiation, 140
  exposure, 93–94
  induced embrittlement, 139
  intensity, 131–132
Radiography testing (RT)
  advantages, 245
  corrosion, 246
  definition, 244
  digital radiography, 244, 245
  gamma rays, 244
  healing methods and outcomes, 248–249
  internal cracks, failures and surface damage, 246

IQIs, 247
multiple testing methods, 247
radiographic imaging, 244
role of polyurethane, 249
traditional photographic methods, 248
voids and cracks, 249
Ratchet marks, 80, 150
Real-world scenarios, 38
Recalibration, 239
Refractory cermet coatings (HVOF), 420
Regression analysis, 298
Regular maintenance, 192
Reliability engineers, 25
Remaining useful life (RUL), 301, 365–366, 368
Reusability and lifespan, 419
Ridge marks, 150
Rigorous real-world testing, 214
Risk management matrix, 12
Risk mitigation, 301
Risk Priority Number (RPN), 322, 381
Robust filtration systems, 303
Rocket deformation analysis, 260
Rockwell C hardness (HRC), 96, 97
Root cause analysis (RCA), 311, 319
Root mean square (RMS)
roughness, 398
value, 354
Rotary USM, 403
Roundness factor, 42
Running-in period and data capture, 258
Run-to-failure strategy, 370
Rupture of passive film, 256
Rural environment, 169

SABRE aircraft, 195
Sacrificial coatings, 108
Safety factor (FoS) curves, 282
Sandardization, 348
Scanning electron microscopy (SEM), 144, 148, 199, 305, 319
Scanning tunneling microscope (STM), 403
Sector-wise corrosion costs in India, 455–457
Semi-molten-state coating, 412
Semi-solid metal (SSM), 181
Sensors, 159
array, 236
data, 289
placement, 257
Shape factor, 354
Shear lips, 150
Sheltering, 172–173
Signal reception, 240
Signal-to-noise ratio (SNR), 241, 304
Simonov, D., 304
Simulation data, 289
Skewness, 354
Sliding wear-assisted corrosion, 137
Slow crack propagation, 316
Slurry pumps, 162
Solar ultraviolet (SUV) radiation, 181
Sol–gel coating, 415
Solid-state coating, 411–412
Solid wastes, 392
Solution treatment process, 138
Specimen preparation, 294
Spin coating, 412
Spray coating, 412
Square pulse thermography (SPT), 252
Stainless steel, 453
Standard deviation, 353
Static configurations, 251
Steel bridge structural failures, 192
Step heating thermography (SHT), 252
Strain–life (e–N) approach, 99
Stress corrosion cracking (SCC), 109, 142, 158, 254, 255, 422
ASTM G36 standard, 124
in chain, 123
crack densities, 125
fracture, 124
hydrogen embrittlement, 123
initiation, 123
morphologies of, 124–5
propagation, 123
role of inclusions, 159
suitable materials selection, 124
Stress–life (S–N) approach, 99
Stress-relief techniques, 143
Stress–strain DIC testing
advantages of, 260
computational considerations, 260
conventional strain gauges, 259
definition, 258
experimental setup, digital imaging, 261
FEM, 259
high-resolution digital photographs, 258–259
point matching, 259
strain calculations, 259
Striation, 158
Stribeck curves, 404
Subsurface linear defects, 264
Subsurface volumetric defects, 264
Sulfur dioxide (SO_2), 170
Sulfuric acid (H_2SO_4), 170
Superior image quality indicators (IQIs), 247
Suresh, S., 292
Surface-breaking linear defects, 263
Surface-breaking volumetric defects, 263

Surface coatings
  applications
    chromium plating, 420
    CKS, 420
    GDC, 420
    HVOF, 420
    PVD coatings, 420
  classification of methods, 411–412
  coating effectiveness, 407
  cost-efficiency, 407
  CSD, 414–415
  CVD, 412–414
  definition, 406, 407
  dip coating, 416–417
  durability, 407
  electrochemical deposition, 415
  enhancing metal properties, 407
  materials, 408
  mechanical and chemical degradation, 408
  organic compound coatings, 408
  performance, 408–411
  PVD, 414
  sol–gel coating, 415
  surface-level solutions, 407–408
  thermal spraying, 415–416
Surface contact, 132
Surface debris, 158
Surface degradation; *see also* Atmospheric corrosion; Corrosion; LEO satellites; Mitigation strategies
  acetic acid (ethanoic acid), 177
  automobiles, gear systems, oil pipelines, bridge failure and steel structures, 1
  carbonyl compounds, effects of, 176
  commercial perspectives
    material degradation, 8
    3D optical profilometer, 9
  corrosion testing, 210
  due to chloride, 200–202
  due to ozone, 198–200
  environmental factors (*see* Environmental factors of surface degradation)
  failure analysis and life cycle assessment, 1
  failure mechanisms, 4
  fishbone diagram, 1
  formaldehyde (methanal), 176
  formic acid (methanoic acid), 176
  heat and radiation, 89–97
  indoor *vs.* outdoor specific approaches, 175
  mechanical systems, 1
  mechanisms of, 7
  mitigation techniques, 211
  testing methods, 210
  universal *vs.* location-specific approaches, 174–175
Surface deterioration, 128
Surface engineering techniques, 300; *see also* Surface roughness; Surface texture
  definitions, 395
  material performance, 395–396
Surface finish factor, 283, 284
Surface roughness, 132–133
  causes of, 396
  definition, 396
  evaluation length, 399
  grades with, sampling length, 400
  maximum peak height, 399
  parameters, 397
  profilometers, 397, 398
  sampling length, 399
  traversing length, 399
  2D and 3D profiles, 396
Surface texturing
  chemical etching, 401
  connected grooves and intermediate pits/dots, 406
  electrochemical etching, 402
  friction reduction, 401
  hierarchical *vs.* single-scale textures, 405
  hydrophobicity, 401
  laser ablation, 401
  load on friction, 405
  lubrication retention, 401
  mechanical machining, 401
  microdimples, 400
  microdimple size, 405
  optical characteristics, 401
  surface roughness design, 400–401
  tribological characteristics, 403–405
  3D printing (additive manufacturing), 401
  wear resistance, 401
Surface visualization, 3
Synergistically combined degradation mechanisms
  composition, 130
  cumulative damage, 132
  environmental factors, 131
  failure mechanisms, 134
  framework for combined effects, 132, 133
  impact loading, 131
  material properties, 128
  mechanical and chemical actions, 134–137
  mechanical and heat/radiation actions, 137–138
  mechanical loading, 131

microstructure, 130
parameters, 128, 129
radiation intensity, 131–132
service environment, 132
surface coatings/treatments, 130
surface contact, 132
temperature, 130
thermal properties, 130
System boundary, 390
System failures, 1
System FMEA, 325

Temperature, 6, 157, 172
assisted oxidation, 204
control layers, 163
variations, 164
Ten-point mean roughness, 397–398
Testing methods in engineering
component-level and assembly-level testing, 217
DT *vs.* NDT, 217–223
realistic performance evaluations
laboratory *vs.* real-world testing, 213–214
scaled-down laboratory models
accelerated testing, 212
corrosion testing, 213
crack initiation and propagation, 212
creep testing, 212
description, 211
fatigue testing, 212
fracture toughness, 212
friction coefficient, 211
material testing, 211–212
power losses and energy efficiency, 213
wear rate, 211
wind tunnel testing, 213
sustainability, 214–216
Tétreault, J., 175, 177
Thane Creek Bridge, distress of, 11, 13
Thermal compatibility, 410
Thermal conductivity, 410
Thermal embrittlement, 139
Thermal insulation, 182
Thermal inversion, 182, 184, 185
Thermal mechanical fatigue, 204
Thermal spray coating, 416
Thermal sprayed molybdenum-based coatings, 420
Thermal spraying, 395, 412, 415–416
Thermographic analysis, 303–304
Thermographic imaging, 373
Thermographic non-destructive testing (TT)
advantages and limitations, 253
aerospace and automotive sectors, 250
cracks, pores/voids, 251
definition, 250
energy transfer modes, 251
fatigue testing machine, 251–252
IR cameras, 250
material integrity, 253
medical applications, 250
passive and active methods, 250
portable IR cameras, 252
pulsed thermal testing, 252
spur gears, 251
static and dynamic configurations, 251
Three-body abrasion
characteristics, 38
description, 38
1100 Vickers hardness (VHN), 39
Three-dimensional (3D) optical profilometer, 9
Time-critical analysis (TCA), 378
Time-dependent maintenance strategies, 370–371
Time-independent maintenance strategies, 371
Tire wear, 22
Titanic's failure (DiLisi)
actual and false horizon, 184
atmospheric refraction phenomena, 184
evacuation procedures, 185
human error, 187
inferior mirages, 184
Labrador Current, 183
lifeboats, 186
lower-grade steel, 186
material and design flaws, 186
refractive phenomena, 185
safety briefings, 187
superior mirages, 184
thermal inversion, 182, 184, 185
wreckage location, 183
Titanium nitride (TiN) PVD coating, 414
Total height roughness, 398
Toutountzakis, T., 257, 258
Traditional curve representations, engineering
data analysis, 283
model-building process, 283
surface finish factor *versus* UTS, 283, 284
Wear Map (*see* Wear Map)
Traditional photographic methods, 248
Transfer learning algorithm, 307
Transmission electron microscopy (TEM), 81, 144
"Trend lines", 280
Tunnel Boring Machine (TBM), 52
Tunneling process, 52, 54
Turbine blade superalloy, 204
Two-body abrasion, 39–42
direct grain removal, 36–38

illustration of, 44
micro-cutting, 35
micro-fatigue, 35
micro-fracture, 36
relative motion, 34
tribological conditions, 34

Uhlig approach, 430, 433, 445–446
Ultrafine particles (UFPs), 198
Ultra-high molecular weight polyethylene (UHMWPE), 155, 156
Ultra-high vacuum CVD, 412
Ultrasonic machining (USM), 403
Ultrasonic sensors, 373
Ultrasonic testing (UT)
attenuation, 239
definition, 237
frequency scanning, 242
functions, 240–241
modern technology, 237
pulse echo method, 240, 241
recalibration, 239
through transmission method, 238
WTB, 242, 243
Ultrasonic wave generation, 240
Ultraviolet (UV) radiation, 2, 78, 162
Uncoated weathering steel (UWS), 192
Uniform corrosion, 422
Union Carbide India Limited (UCIL) pesticide plant, 364
Universal testing machine (UTM), 218
Universal *vs.* location-specific approaches, 174–175
Unsupervised machine learning method, 299
Urban environment, 169–170
UV radiation, 131, 164
induced degradation, 94

Van Tittelboom, K., 248
Variable ratios, 294
Vehicle tire wear, 22
Vibration analysis, 375
Videoscopes, 227
Visual inspection
definition, 225
videoscopes, 227
VT, 226–227
Visual testing (VT), 226–227
Volatile organic compounds (VOCs), 170

Wang, F., 56
Wang, H., 199
Waste management, 419
Water and humidity, 164
Waterborne wastes, 392
Water erosion, 49
Water-induced erosion on rock, 51
Wear-assisted corrosion mechanisms, 134, 135
Wear bars, 22
Wear debris analysis, 289, 374
abrasive wear, 154
adhesive wear, 152
definition, 151
early warning system, 151–152
fractographers, 155
material's degradation, 152
monitoring and analysis, 151
observable warning signs, 155
types of, 154
Wear Map
description, 284–285
field boundaries, 285
influencing factors, oxide layer thickness, 285
machine learning algorithms, 286
material selection and operational adjustments, 286
mechanisms, 285–286
pressure and relative velocity, 285
wear rate contours, 285
Wear mechanisms, 257–258
dimensionless parameters, 45–47
Wear resistance, 401
Weighted objective (WO) method
case study, 270–271
definition, 266–267
method for NDT, 267
methodology, 273–275
parameters, 273
process of, 267–270
Weighting, 389
Wetness duration, 172
Wind tunnel testing, 213
Wind turbine blades (WTBs), 241
Wind velocity, 173
Wire arc spraying, 416
Worn rollers, 27
Wu, H., 124

Xie, R., 235, 236
X-ray spectroscopy (EDX), 197
Xu, Y., 143

Yeh, C. P., 200

Zero wear, 23
Zhang, Y., 192
Zinc galvanizing process, 108
Zr-based coatings, 414